Symbolic Data Analysis and the SODAS Software

Symbolic Data Analysis and the SODAS Software

Edited by

Edwin Diday
Université de Paris IX - Dauphine, France

Monique Noirhomme-Fraiture
University of Namur, Belgium

John Wiley & Sons, Ltd

Telephone (+44) 1243 779777

Email (for orders and customer service enquiries): cs-books@wiley.co.uk
Visit our Home Page on www.wiley.com

Other Wiley Editorial Offices

John Wiley & Sons Inc., 111 River Street, Hoboken, NJ 07030, USA

Jossey-Bass, 989 Market Street, San Francisco, CA 94103-1741, USA

Wiley-VCH Verlag GmbH, Boschstr. 12, D-69469 Weinheim, Germany

John Wiley & Sons Australia Ltd, 42 McDougall Street, Milton, Queensland 4064, Australia

John Wiley & Sons (Asia) Pte Ltd, 2 Clementi Loop #02-01, Jin Xing Distripark, Singapore 129809

John Wiley & Sons Canada Ltd, 6045 Freemont Blvd, Mississauga, ONT, L5R 4J3

Wiley also publishes its books in a variety of electronic formats. Some content that appears in print may not be available in electronic books.

Library of Congress Cataloging in Publication Data

Symbolic data analysis and the SODAS software/edited by Edwin Diday,
Monique Noirhomme-Fraiture.
p. cm.
Includes bibliographical references and index.
ISBN 978-0-470-01883-5 (cloth)
1. Data mining. I. Diday, E. II. Noirhomme-Fraiture, Monique.
QA76.9.D343S933 2008
005.74—dc22
2007045552

British Library Cataloguing in Publication Data

A catalogue record for this book is available from the British Library

ISBN 978-0-470-01883-5

Typeset in 10/12pt Times by Integra Software Services Pvt. Ltd, Pondicherry, India

This book is printed on acid-free paper responsibly manufactured from sustainable forestry in which at least two trees are planted for each one used for paper production.

Contents

Contributors

Séverine Adans, Facultés Universitaires Notre-Dame de la Paix, Déprt. de Mathematique, Rempart de la Vierge, 8, Namur, Belgium, B-5000

Filipe Afonso, Université Paris IX-Dauphine, LISE-CEREMADE, Place du Maréchal de Lattre de Tassigny, Paris Cedex 16, France, F-75775

Annalisa Appice, Floriana Esposito Universita degli Studi di Bari, Dipartimento di Informatica, v. Orabona, 4, Bari, Italy, I-70125

Anne de Baenst-Vandenbroucke, Facultés Universitaires Notre-Dame de la Paix, Faculté d'Informatique, Rue Grandgagnage, 21, Namur, Belgium, B-5000, adb@info.fundp.ac.be

Ghazi Bel Mufti, ESSEC de Tunis, 4 rue Abou Zakaria El Hafsi, Montfleury 1089, Tunis, Tunisia, belmufti@yahoo.com

Patrice Bertrand, Ceremade, Université Paris IX-Dauphine, Place du Maréchal de Lattre de Tassigny, Paris Cedex 16, France, F-75775, Patrice.Bertrand@ceremade.dauphine.fr

Lynne Billard, University of Georgia, Athens, USA, GA 30602-1952, lynne@stat.uga.edu

Hans-Hermann Bock, Rheinisch-Westfälische Technische Hochschule Aachen, Institut für Statistik und Wirtschaftmathematik, Wüllnerstr. 3, Aachen, Germany, D-52056, bock@stochastik.rwth-aachen.de

Maria Paula de Pinho de Brito, Faculdade de Economia do Porto, LIACC, Rua Dr. Roberto Frias, Porto, Portugal, P-4200-464, mpbrito@fep.up.pt

Brieuc Conan-Guez, LITA EA3097, Université de Metz, Ile de Saulcy, F-57045, Metz, France, Brieuc.Conan-Suez@univ-metz.fr

Marc Csernel, INRIA, Unité de Recherche de Roquencourt, Domaine de Voluceau, BP 105, Le Chesnay Cedex, France, F-78153, Marc.Cesrnel@inria.fr

Francisco de A.T. de Carvalho, Universade Federale de Pernambuco, Centro de Informatica, Av. Prof. Luis Freire s/n - Citade Universitaria, Recife-PE, Brasil, 50740-540, fatc@cin.ufpe.br

Edwin Diday, Université Paris IX-Dauphine, LISE-CEREMADE, Place du Marechal de Lattre de Tassigny, Paris Cedex 16, France F-75775, diday@ceremade.dauphine.fr

Aicha El Golli, INRIA Paris, Unité de Recherche de Roquencourt, Domaine de Voluceau, BP 105, Le Chesnay Cedex, France, F-78153, aicha.elgolli@inria.fr

Floriana Esposito, Universita degli Studi di Bari, Dipartimento di Informatica, v. Orabona, 4, Bari, Italy, I-70125, esposito@di.uniba.it

André Hardy, Facultés Universitaires Notre-Dame de la Paix, Départment de Mathématique, Rempart de la Vièrge, 8, Namur, Belgium, B-5000, andre.hardy@math.fundp.ac.be

Georges Hébrail, Laboratoire LTCI UMR 5141, Ecole Nationale Supériewe des Télécommunications, 46 rue Barrault, 75013 Paris, France, hebrail@enst.fr

Antonio Irpino, Universita Frederico II, Dipartimento di Mathematica e Statistica, Via Cinthia, Monte Sant'Angelo, Napoli, Italy I-80126, irpino@unina.it

Seppo Laaksonen, Statistics Finland, Box 68, University of Helsinki, Finland, FIN 00014, Seppo.Laaksonen@Helsinki.Fi

Pascale Lallemand, Facultés Universitaires Notre-Dame de la Paix, Départment de Mathématique, Rempart de la Vièrge, 8, Namur, Belgium, B-5000

Natale Carlo Lauro, Universita Frederico II, Dipartimento di Mathematica e Statistica, Via Cinthia, Monte Sant'Angelo, Napoli, Italy I-80126, clauro@unina.it

Yves Lechevallier, INRIA, Unité de Recherche de Roquencourt, Domaine de Voluceau, BP 105, Le Chesnay Cedex, France, F-78153, Yves.Lechevallier@inria.fr

Mehdi Limam, Université Paris IX-Dauphine, LISE-CEREMADE, Place du Maréchal de Lattre de Tassigny, Paris Cedex 16, France, F-75775

Donato Malerba, Universita degli Studi di Bari, Dipartimento di Informatica, v. Orabona, 4, Bari, Italy, I-70125, malerba@di.uniba.it

Marta Mas, Asistencia Tecnica y Metodológica, Beato Tomás de Zumarraga, 52, 3°-Izda, Vitoria-Gasteiz, Spain, E-01009, Marta_Mas@terra.es

Soile Mustjärvi, Statistics Finland, Finland, FIN-00022

Adolphe Nahimana, Facultés Universitaires Notre-Dame de la Paix, Faculté d'Informatique, Rue Grandgagnage, 21, Namur, Belgium, B-5000

Monique Noirhomme-Fraiture, Facultés Universitaires Notre-Dame de la Paix, Faculté d'Informatique, Rue Grandgagnage, 21, Namur, Belgium, B-5000, mno@info.fundp.ac.be

Haritz Olaeta, Euskal Estatistika Erakundea, Area de Metodologia, Donostia, 1, Vitoria-Gasteiz, Spain, E-010010, haritz.olaeta@uniqual.es

Haralambos Papageorgiou, University of Athens, Department of Mathematics, Panepistemiopolis, Athens, Greece, EL-15784, hpapageo@cc.uoa.gr

Jean-Yves Pirçon, Facultés Universitaires Notre-Dame de la Paix, Déprt. de Mathematique, Rempart de la Vièrge, 8, Namur, Belgium, B-5000

Jean-Paul Rasson, Facultés Universitaires Notre-Dame de la Paix, Départ. de Mathematique, Rempart de la Vièrge, 8, Namur, Belgium, B-5000, jpr@math.fundp.ac.be

Fabrice Rossi, Projet AxIS, INRIA, Centre de Rechoche Paris-Roquencourt, Domaine de Volucean, BP 105, Le Chesney Cedex, France F-78153, Fabrice.Rossi@inria.fr

Maria Vardaki, University of Athens, Department of Mathematics, Panepistemiopolis, Athens, Greece, EL-15784 mvardaki@cc.uoa.gr

Rosanna Verde, Dipartimento di Studi Europei e Mediterranei, Seconda Universitádegli Studi di Napoli, Via del Setificio, 15 Complesso Monumentale Belvedere di S. Leucio, 81100 Caserta, Italy

Foreword

It is a great pleasure for me to preface this imposing work which establishes, with *Analysis of Symbolic Data* (Bock and Diday, 2000) and *Symbolic Data Analysis* (Billard and Diday, 2006), a true bible as well as a practical handbook.

Since the pioneering work of Diday at the end of the 1980s, symbolic data analysis has spread beyond a restricted circle of researchers to attain a stature attested to by many publications and conferences. Projects have been supported by Eurostat (the statistical office of the European Union), and this recognition of symbolic data analysis as a tool for official statistics was also a crucial step forward.

Symbolic data analysis is part of the great movement of interaction between statistics and data processing. In the 1960s, under the impetus of J. Tukey and of J.P. Benzécri, exploratory data analysis was made possible by progress in computation and by the need to process the data which one stored. During this time, large data sets were tables of a few thousand units described by a few tens of variables. The goal was to have the data speak directly without using overly simplified models. With the development of relational databases and data warehouses, the problem changed dimension, and one might say that it gave birth on the one hand to data mining and on the other hand to symbolic data analysis. However, this convenient opposition or distinction is somewhat artificial.

Data mining and machine learning techniques look for patterns (exploratory or unsupervised) and models (supervised) by processing almost all the available data: the statistical unit remains unchanged and the concept of model takes on a very special meaning. A model is no longer a parsimonious representation of reality resulting from a physical, biological, or economic theory being put in the form of a simple relation between variables, but a forecasting algorithm (often a black box) whose quality is measured by its ability to generalize to new data (which must follow the same distribution). Statistical learning theory provides the theoretical framework for these methods, but the data remain traditional data, represented in the form of a rectangular table of variables and individuals where each cell is a value of a numeric variable or a category of a qualitative variable assumed to be measured without error.

Symbolic data analysis was often presented as a way to process data of another kind, taking variability into account: matrix cells do not necessarily contain a single value but an interval of values, a histogram, a distribution. This vision is exact but reductive, and this book shows quite well that symbolic data analysis corresponds to the results of database operations intended to obtain conceptual information (knowledge extraction). In this respect symbolic data analysis can be considered as the statistical theory associated with relational databases. It is not surprising that symbolic data analysis found an important field of application in official statistics where it is essential to present data at a high level of aggregation rather than to reason on individual data. For that, a rigorous mathematical framework has been developed which is presented in a comprehensive way in the important introductory chapter of this book.

Once this framework was set, it was necessary to adapt the traditional methods to this new type of data, and Parts II and III show how to do this both in exploratory analysis (including very sophisticated methods such as generalized canonical analysis) and in supervised analysis where the problem is the interrelation and prediction of symbolic variables. The chapters dedicated to cluster analysis are of great importance. The methods and developments gathered together in this book are impressive and show well that symbolic data analysis has reached full maturity.

In an earlier paragraph I spoke of the artificial opposition of data mining and symbolic data analysis. One will find in this book symbolic generalizations of methods which are typical of data mining such as association rules, neural networks, Kohonen maps, and classification trees. The border between the two fields is thus fairly fluid.

What is the use of sound statistical methods if users do not have application software at their disposal? It is one of the strengths of the team which contributed to this book that they have also developed the freely available software SODAS2. I strongly advise the reader to use SODAS2 at the same time as he reads this book. One can only congratulate the editors and the authors who have brought together in this work such an accumulation of knowledge in a homogeneous and accessible language. This book will mark a milestone in the history of data analysis.

Gilbert Saporta

Preface

This book is a result of the European ASSO project (IST-2000-25161) http://www.assoproject.be on an Analysis System of Symbolic Official data, within the Fifth Framework Programme. Some 10 research teams, three national institutes for statistics and two private companies have cooperated in this project. The project began in January 2001 and ended in December 2003. It was the follow-up to a first SODAS project, also financed by the European Community.

The aim of the ASSO project was to design methods, methodology and software tools for extracting knowledge from multidimensional complex data (www.assoproject.be). As a result of the project, new symbolic data analysis methods were produced and new software, SODAS2, was designed. In this book, the methods are described, with instructive examples, making the book a good basis for getting to grips with the methods used in the SODAS2 software, complementing the tutorial and help guide. The software and methods highlight the crossover between statistics and computer science, with a particular emphasis on data mining.

The book is aimed at practitioners of symbolic data analysis – statisticians and economists within the public (e.g. national statistics institutes) and private sectors. It will also be of interest to postgraduate students and researchers within data mining.

Acknowledgements

The editors are grateful to ASSO partners and external authors for their careful work and contributions. All authors were asked to review chapters written by colleagues so that we could benefit from internal cross-validation. In this regard, we wish especially to thank Paula Brito, who revised most of the chapters. Her help was invaluable. We also thank Pierre Cazes who offered much substantive advice. Thanks are due to Photis Nanopulos and Daniel Defays of the European Commission who encouraged us in this project and also to our project officer, Knut Utwik, for his patience during this creative period.

Edwin Diday
Monique Noirhomme-Fraiture

ASSO Partners

FUNDP Facultés Universitaires Notre-Dame de la Paix, Institut d'Informatique, Rue Grandgagnage, 21, Namur, Belgium, B-5000 mno@info.fundp.ac.be (coordinator)

DAUPHINE Université Paris IX-Dauphine, LISE-CEREMADE, Place du Maréchal de Lattre de Tassigny, Paris Cedex 16, France, F-75775 rossi@ceremade.dauphine.fr

DIB Universita degli Studi di Bari, Dipartimento di Informatica, v. Orabona, 4, Bari, Italy, I-70125 esposito@di.uniba.it

DMS Dipartimento di Mathematica e Statistica, Via Cinthia, Monte Sant'Angelo, Napoli, Italy I-80126 clauro@unina.it

EUSTAT Euskal Estatistika Erakundea, Area de Metodologia, Donostia, 1, Vitoria-Gasteiz, Spain, E-01010 Marina_Ayestaran@eustat.es

FEP Faculdade de Economia do Porto, LIACC, Rua Dr. Roberto Frias, Porto, Portugal, P-4200-464 mpbrito@fep.up.pt

FUNDPMa Facultés Universitaires Notre-Dame de la Paix, Rempart de la Vierge, 8, Namur, Belgium, B-5000 jpr@math.fundp.ac.be

INRIA, Unité de Recherche de Roquencourt, Domaine de Voluceau, BP 105, Le Chesnay Cedex, France, F-78153 Yves.Lechevallier@inria.fr

INS Instituto Nacional de Estatistica, Direcçao Regional de Lisboa e Valo do Tejo, Av. Antonio José de Almeida, Lisboa, Portugal, P-1000-043 carlos.marcelo@ine.pt

RWTH Rheinisch-westfälische Technische Hochschule Aachen, Institut für Statistik und Wirtschaftmathematik, Wülnerstr, 3, Aachen, Germany, D-52056 bock@stochastik.rwth-aachen.de

SPAD Groupe TestAndGO, Rue des petites Ecuries, Paris, France, F-75010 p.pleuvret@decisia.fr

STATFI Statistics Finland, Management Services/R&D Department, Tyoepajakatu, 13, Statistics Finland, Finland, FIN-00022 Seppo.Laaksonen@Stat.fi

TES Institute ASBL, Rue des Bruyères, 3, Howald, Luxembourg, L-1274

UFPE Universade Federale de Pernambuco, Centro de Informatica-Cin, Citade Universitaria, Recife-PE, Brasil, 50740-540 fatc@cin.ufpe.br

UOA University of Athens, Department of Mathematics, Panepistemiopolis, Athens, Greece, EL-15784 hpapageo@cc.uoa.gr

INTRODUCTION

1

The state of the art in symbolic data analysis: overview and future

Edwin Diday

1.1 Introduction

Databases are now ubiquitous in industrial companies and public administrations, and they often grow to an enormous size. They contain units described by variables that are often categorical or numerical (the latter can of course be also transformed into categories). It is then easy to construct categories and their Cartesian product. In symbolic data analysis these categories are considered to be the new statistical units, and the first step is to get these higher-level units and to describe them by taking care of their internal variation. What do we mean by 'internal variation'? For example, the age of a player in a football team is 32 but the age of the players in the team (considered as a category) varies between 22 and 34; the height of the mushroom that I have in my hand is 9 cm but the height of the species (considered as a category) varies between 8 and 15 cm.

A more general example is a clustering process applied to a huge database in order to summarize it. Each cluster obtained can be considered as a category, and therefore each variable value will vary inside each category. Symbolic data represented by structured variables, lists, intervals, distributions and the like, store the 'internal variation' of categories better than do standard data, which they generalize. 'Complex data' are defined as structured data, mixtures of images, sounds, text, categorical data, numerical data, etc. Therefore, symbolic data can be induced from categories of units described by complex data (see Section 1.4.1) and therefore complex data describing units can be considered as a special case of symbolic data describing higher-level units.

Symbolic Data Analysis and the SODAS Software Edited by E. Diday and M. Noirhomme-Fraiture

The aim of symbolic data analysis is to generalize data mining and statistics to higher-level units described by symbolic data. The SODAS2 software, supported by EUROSTAT, extends the standard tools of statistics and data mining to these higher-level units. More precisely, symbolic data analysis extends exploratory data analysis (Tukey, 1958; Benzécri, 1973; Diday *et al.*, 1984; Lebart *et al.*, 1995; Saporta, 2006), and data mining (rule discovery, clustering, factor analysis, discrimination, decision trees, Kohonen maps, neural networks, . . .) from standard data to symbolic data.

Since the first papers announcing the main principles of symbolic data analysis (Diday, 1987a, 1987b, 1989, 1991), much work has been done, up to and including the publication of the collection edited by Bock and Diday (2000) and the book by Billard and Diday (2006). Several papers offering a synthesis of the subject can be mentioned, among them Diday (2000a, 2002a, 2005), Billard and Diday (2003) and Diday and Esposito (2003). The symbolic data analysis framework extends standard statistics and data mining tools to higher-level units and symbolic data. For example, standard descriptive statistics (mean, variance, correlation, distribution, histograms, . . .) are extended in de Carvalho (1995), Bertrand and Goupil (2000), Billard and Diday (2003, 2006), Billard (2004) and Gioia and Lauro (2006a). Standard tools of multidimensional data analysis such as principal component analysis are extended in Cazes *et al.* (1997), Lauro *et al.* (2000), Irpino *et al.* (2003), Irpino (2006) and Gioia and Lauro (2006b). Extensions of dissimilarities to symbolic data can be found in Bock and Diday (2000, Chapter 8), in a series of papers by Esposito *et al.* (1991, 1992) and also in Malerba *et al.* (2002), Diday and Esposito (2003) and Bock (2005). On clustering, recent work by de Souza and de Carvalho (2004), Bock (2005), Diday and Murty (2005) and de Carvalho *et al.* (2006a, 2006b) can be mentioned. The problem of the determination of the optimum number of clusters in clustering of symbolic data has been analysed by Hardy and Lallemand (2001, 2002, 2004), Hardy *et al.* (2002) and Hardy (2004, 2005). On decision trees, there are papers by Ciampi *et al.* (2000), Bravo and García-Santesmases (2000), Limam *et al.* (2003), Bravo Llatas (2004) and Mballo *et al.* (2004). On conceptual Galois lattices, we have Diday and Emilion (2003) and Brito and Polaillon (2005). On hierarchical and pyramidal clustering, there are Brito (2002) and Diday (2004). On discrimination and classification, there are papers by Duarte Silva and Brito (2006), Appice *et al.* (2006). On symbolic regression, we have Afonso *et al.* (2004) and de Carvalho *et al.* (2004). On mixture decomposition of vectors of distributions, papers include Diday and Vrac (2005) and Cuvelier and Noirhomme-Fraiture (2005). On rule extraction, there is the Afonso and Diday (2005) paper. On visualization of symbolic data, we have Noirhomme-Fraiture (2002) and Irpino *et al.* (2003). Finally, we might mention Prudêncio *et al.* (2004) on time series, Soule *et al.* (2004) on flow classification, Vrac *et al.* (2004) on meteorology, Caruso *et al.* (2005) on network traffic, Bezera and de Carvalho (2004) on information filtering, and Da Silva *et al.* (2006) and Meneses and Rodríguez-Rojas (2006) on web mining.

This chapter is organized as follows. Section 1.2 examines the transition from a standard data table to a symbolic data table. This is illustrated by a simple example showing that single birds can be defined by standard numerical or categorical variables but species of birds need symbolic descriptions in order to retain their internal variation. Section 1.3 gives the definitions and properties of basic units such as individuals, categories, classes and concepts. In order to model a category or the intent of a concept from a set of individuals belonging to its extent, a generalization process which produces a symbolic description is used. Explanations are then given of the input of a symbolic data analysis, the five kinds of variable (numerical, categorical, interval, categorical multi-valued, modal), the conceptual variables which describe concepts and the background knowledge by means of taxonomies and rules. Section 1.4 provides some general applications of the symbolic data analysis paradigm. It is shown that from fuzzy or uncertainty data, symbolic descriptions are needed in order to describe classes, categories or concepts. Another application concerns the possibility of fusing data tables with different entries and different variables by using the same concepts and their symbolic description. Finally, it is shown that much information is lost if a symbolic description is transformed into a standard classical description by transforming, for example, an interval into two variables (the maximum and minimum). In Section 1.5 the main steps and principles of a symbolic data analysis are summarized. Section 1.6 provides more details on the method of modelling concepts by symbolic objects based on four spaces (individuals and concepts of the 'real world' and their associated symbolic descriptions and symbolic objects in the 'modelled world'). The definition, extent and syntax of symbolic objects are given. This section ends with some advantages of the use of symbolic objects and how to improve them by means of a learning process. In Section 1.7 it is shown that a generalized kind of conceptual lattice constitutes the underlying structure of symbolic objects (readers not interested in conceptual lattices can skip this section). The chapter concludes with an overview of the chapters of the book and of the SODAS2 software.

1.2 From standard data tables to symbolic data tables

Extracting knowledge from large databases is now fundamental to decision-making. In practice, the aim is often to extract new knowledge from a database by using a standard data table where the entries are a set of units described by a finite set of categorical or quantitative variables. The aim of this book is to show that much information is lost if the units are straitjacketed by such tables and to give a new framework and new tools (implemented in SODAS2) for the extraction of complementary and useful knowledge from the same database. In contrast to the standard data table model, several levels of more or less general units have to be considered as input in any knowledge extraction process. Suppose we have a standard data table giving the species, flight ability and size of 600 birds observed on an island (Table 1.1). Now, if the new units are the species of birds on the island (which are an example of what are called 'higher-level' units), a different answer to the same questions is obtained since, for example, the number of flying birds can be different from the number of flying species. In order to illustrate this simple example with some data, Table 1.2 describes the three species of bird on the island: there are 100 ostriches, 100 penguins and 400 swallows. The frequencies for the variable 'flying' extracted from this table are the reverse of those extracted from Table 1.1, as shown in Figures 1.1 and 1.2. This means that the 'micro' (the birds) and the 'macro' (the species) points of view can give results which are totally different as the frequency of flying birds in the island is 2/3 but the frequency of the flying species is 1/3.

Notice that in Table 1.2 the values taken by the variable 'size' are no longer numbers but intervals, which are a first kind of 'symbolic data' involving a different kind of variable from the standard data. New variables can be added in order to characterize the second-level units such as the variable 'migration' which expresses the fact that 90% of the swallows migrate, all the penguins migrate and the ostriches do not migrate.

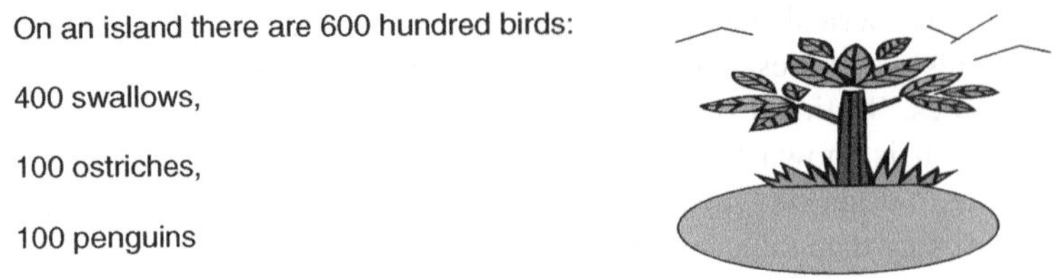

Figure 1.1 Three species of 600 birds together.

Table 1.1 Description of 600 birds by three variables.

Birds	Species	Flying	Size
1	Penguin	No	80
...	...	...	...
599	Swallow	Yes	70
600	Ostrich	No	125

Table 1.2 Description of the three species of birds plus the conceptual variable 'migration'.

Species	Flying	Size	Migration
Swallow	{Yes}	[60, 85]	
Ostrich	{No}	[85, 160]	
Penguin	{No}	[70, 95]	

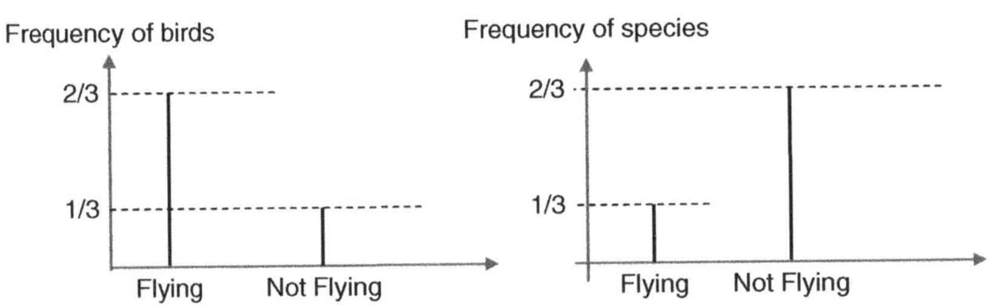

Figure 1.2 Frequencies for birds (individuals) and species (concepts).

As shown in Table 1.3, the need to consider different unit levels can arise in many domains. For example, one might wish to find which variables distinguish species having avian influenza from those that do not. In sport, the first-order units may be the players and the higher-level unit the teams. For example, in football (see Table 1.4), in order to find which variables best explain the number of goals scored, it may be interesting to study first-level units in the form of the players in each team, described by different variables such as their weight, age, height, nationality, place of birth and so on. In this case, the

Table 1.3 Examples of first- and second-order units and requirements on the second-order units (i.e., the concepts).

First-level units (individuals)	Second-level units (concepts)	Requirements on concepts
Birds	Species of birds	Compare species of birds
Players on a team	Teams	Which variables explain the number of goals scored by teams?
Teams	Continent	Which variables explain the winning continent?
Patients	Patients' hospital pathway	Which variables explain risky pathways?
Medical records	Insured patients	Which variables explain patient consumption?
Books	Publishers	Find clusters of publishers
Travellers	Trains	Study trains with the same profit interval
Customers	Shops	Compare shops

Table 1.4 Initial classical data table describing players by three numerical and two categorical variables.

Player	Team	Age	Weight	Height	Nationality_1	...
Fernández	Spain	29	85	1.84	Spanish	
Rodríguez	Spain	23	90	1.92	Brazilian	
Mballo	France	25	82	1.90	Senegalese	
Zidane	France	27	78	1.85	French	

Table 1.5 Symbolic data table obtained from Table 1.4 by describing the concepts 'Spain' and 'France'.

Team	AGE	WEIGHT	HEIGHT	Nationality_3	FUNDS	Number of goals at the World Cup 1998
Spain	[23, 29]	[85, 90]	[1.84, 1.92]	(0.5 S_p, 0.5 Br)	110	18
France	[21, 28]	[85, 90]	[1.84, 1.92]	(0.5 Fr, 0.5 Se)	90	24

task is to explain the number of goals scored by a player. If the task is to find an explanation for the number of goals scored by a team (during the World Cup, for example), then the teams are the units of interest. The teams constitute higher-level units defined by variables whose values will no longer be quantitative (age, weight or height) but intervals (say, the confidence interval or the minimum–maximum interval). For example, Rodríguez is the youngest player and Fernández the oldest in the Spain team in the symbolic data table (Table 1.5) where the new units are teams, and the variable 'age' becomes an interval-valued variable denoted 'AGE' such that AGE(Spain) $= [23, 29]$. The categorical variables (such as the nationality or the place of birth) of Table 1.4 are no longer categorical in the symbolic data table. They are transformed into new variables whose values are several categories with weights defined by the frequencies of the nationalities (before naturalization in some cases), place of birth, etc., in each team. It is also possible to enhance the description of the higher-level units by adding new variables such as the variable 'funds' which concerns the higher-level units (i.e., the teams) and not the lower-level units (i.e., the players). Even if the teams are considered in Table 1.5 as higher-level units described by symbolic data, they can be considered as lower-level units of a higher-level unit which are the continents, in which case the continents become the units of a higher-level symbolic data table.

The concept of a hospital pathway followed by a given patient hospitalized for a disease can be defined by the type of healthcare institution at admission, the type of hospital unit and the chronological order of admission in each unit. When the patients are first-order units described by their medical variables, the pathways of the patients constitute higher-level units as several patients can have the same pathway. Many questions can be asked which concern the pathways and not the patients. For example, to compare the pathways, it may be interesting to compare their symbolic description based on the variables which describe the pathways themselves and on the medical symbolic variables describing the patients on the pathways.

Given the medical records of insured patients for a given period, the first-order units are the records described by their medical consumption (doctor, drugs, ...); the second-order units are the insured patients described by their own characteristics (age, gender, ...) and by the symbolic variables obtained from the first-order units associated with each patient. A natural question which concerns the second-order units and not the first is, for example, what explains the drug consumption of the patients? Three other examples are given in Table 1.3: when the individuals are books, each publisher is associated with a class of books; hence, it is possible to construct the symbolic description of each publisher or of the most successful publishers and compare them; when the individuals are travellers taking a train, it is possible to obtain the symbolic description of classes of travellers taking the same train and study, for example, the symbolic description of trains having the same profit interval. When the individuals are customers of shops, each shop can be associated with the symbolic description of its customers' behaviour and compared with the other shops.

1.3 From individuals to categories, classes and concepts

1.3.1 Individuals, categories, classes, concepts

In the following, the first-order units will be called 'individuals'. The second-level units can be called 'classes', 'categories' or 'concepts'. A second-level unit is a class if it is associated with a set of individuals (like the set of 100 ostriches); it is a category if it is associated with a value of a categorical variable (like the category 'ostrich' for the variable 'species'); it is a concept if it is associated with an 'extent' and an 'intent'. An extent is a set of individuals which satisfy the concept (for the concept 'ostrich' in general, for example, this would be the set of all ostriches which exist, have existed or will exist); an intent is a way to find the extent of a concept (such as the mapping that we have in our mind which allows us to recognize that something is an ostrich). In practice, an approximation of the intent of a concept is modelled mathematically by a generalization process applied to a set of individuals considered to belong to the extent of the concept. The following section aims to explain this process.

1.3.2 The generalization process, symbolic data and symbolic variables

A generalization process is applied to a set of individuals in order to produce a 'symbolic description'. For example, the concept 'swallow' is described (see Table 1.2) by the description vector d = ({yes}, [60, 85], [90% yes, 10% no]). The generalization process must take care of the internal variation of the description of the individuals belonging in the set of individuals that it describes. For example, the 400 swallows on the island vary in size between 60 and 85. The variable 'colour' could also be considered; hence, the colour of the ostriches can be white or black (which expresses a variation of the colour of the ostriches on this island between white and black), the colour of the penguins is always black and white (which does not express any variation but a conjunction valid for all the birds of this species), and the colour of the swallows is black and white or grey. This variation leads to a new kind of variable defined on the set of concepts, as the value of such variables for a concept may be a single numerical or categorical value, but also an interval, a set of ordinal or categorical values that may or may not be weighted, a distribution, etc. Since these values are not purely numerical or categorical, they have been called 'symbolic data'. The associated variables are called 'symbolic variables'.

More generally, in order to find a unique description for a concept, the notion of the 'T-norm of descriptive generalization' can be used (see, for example, Diday, 2005). The T-norm operator (Schweizer and Sklar, 2005) is defined from $[0, 1] \times [0, 1]$ to $[0, 1]$. In order to get a symbolic description d_C of C (i.e., of the concept for which C is an extent), an extension to descriptions of the usual T-norm can be used; this is called a 'T-norm of descriptive generalization'.

Bandemer and Nather (1992) give many examples of T-norms and T-conorms which can be generalized to T-norms and T-conorms of descriptive generalization. For example, it is easy to see that the infimum and supremum (denoted inf and sup) are respectively a T-norm and a T-conorm. They are also a T-norm and T-conorm of descriptive generalization. Let D_C be the set of descriptions of the individuals of C. It follows that the interval $G_y(C) = [\inf(D_C), \sup(D_C)]$ constitutes a good generalization of D_C, as its extent defined by the set of descriptions included in the interval contains C in a good and narrow way.

Let $C = \{w_1, w_2, w_3\}$ and $D_C = \{y(w_1), y(w_2), y(w_3)\} = y(C)$. In each of the following examples, the generalization of C for the variable y is denoted $G_y(C)$.

1. Suppose that y is a standard numerical variable such that $y(w_1) = 2.5, y(w_2) = 3.6, y(w_3) = 7.1$. Let D be the set of values included in the interval [1, 100]. Then $G_y(C) = [2.5, 7.1]$ is the generalization of D_C for the variable y.

2. Suppose that y is a modal variable of ordered (e.g., small, medium, large) or not ordered (e.g., London, Paris, . . .) categories, such that: $y(w_1) = (1(1/3), 2(2/3))$ (where 2(2/3) means that the frequency of the category 2 is 2/3), $y(w_2) = (1(1/2), 2(1/2))$, $y(w_3) = (1(1/4), 2(3/4))$. Then, $G_y(C) = [[1(1/4), 1(1/2)], [2(1/2), 2(3/4)]$ is the generalization of D_C for the variable y.

3. Suppose that y is a variable whose values are intervals such that: $y(w_1) = [1.5, 3.2]$, $y(w_2) = [3.6, 4]$, $y(w_3) = [7.1, 8.4]$. Then, $G_y(C) = [1.5, 8.4]$ is the generalization of D_C for the variable y.

Instead of describing a class by its T-norm and T-conorm, many alternatives are possible by taking account, for instance, of the existence of outliers. A good strategy consists of reducing the size of the boundaries in order to reduce the number of outliers. This is done in DB2SO (see Chapter 2) within the SODAS2 software by using the 'gap test' (see Stéphan, 1998). Another choice in DB2SO is to use the frequencies in the case of categorical variables.

For example, suppose that $C = \{w_1, w_2, w_3\}$, y is a standard unordered categorical variable such that $y(w_1) = 2$, $y(w_2) = 2$, $y(w_3) = 1$, D is the set of probabilities on the values 1, 2. Then $G'(y(C)) = [1(1/3), 2(2/3)]$ is the generalization of $D_C = \{y(w_1), y(w_2), y(w_3)\} = y(C)$ for the variable y.

1.3.3 Inputs of a symbolic data analysis

In this book five main kinds of variables are considered: numerical (e.g., height(Tom) = 1.80); interval (e.g., age(Spain) = [23, 29]); categorical single-valued (e.g., Nationality_1(Mballo) = {Congo}); categorical multi-valued (e.g., Nationality_2(Spain) = {Spanish, Brazilian, French}); and modal (e.g., Nationality_3(Spain) = {(0.8)Spanish, (0.1)Brazilian, (0.1)French}, where there are several categories with weights).

'Conceptual variables' are variables which are added to the variables which describe the second-order units, because they are meaningful for the second-order units but not for the first-order units. For example, the variable 'funds' is a conceptual variable as it is added at the level where the second-order units are the football teams and would have less meaning at the level of the first-order units (the players). In the SODAS2 software, within the DB2SO module, the option ADDSINGLE allows conceptual variables to be added (see Chapter 2).

From lower- to higher-level units missing data diminish in number but may exist. That is why SODAS2 allows them to be taken into account(see Chapter 24). Nonsense data may also occur and can be also introduced in SODAS2 by the so-called 'hierarchically dependent variables' or 'mother–daughter' variables. For example, the variable 'flying' whose answer is 'yes' or 'no', has a hierarchical link with the variable 'speed of flying'. The variable 'flying' can be considered as the mother variable and 'speed of flying' as the daughter. As a result, for a flightless bird, the variable 'speed of flying' is 'not applicable'.

In the SODAS2 software it is also possible to add to the input symbolic data table some background knowledge, such taxonomies and some given or induced rules. For example, the variable 'region' can be given with more or less specificity due, for instance, to confidentiality. This variable describes a set of companies in Table 1.6. The links between its values are given in Figure 1.3 by a taxonomic tree and represented in Table 1.7 by two columns, associating each node of the taxonomic tree with its predecessor.

Logical dependencies can also be introduced as input; for example, 'if age(w) is less than 2 months, then height(w) is less than 80'. As shown in the next section, these kinds of rules can be induced from the initial data table and added as background knowledge to the symbolic data table obtained.

Table 1.6 Region is a taxonomic variable defined by Table 1.7 or by Figure 1.3.

Company	...	Region	...
Comp1		Paris	
Comp2		London	
Comp3		France	
Comp4		England	
Comp5		Glasgow	
Comp6		Europe	

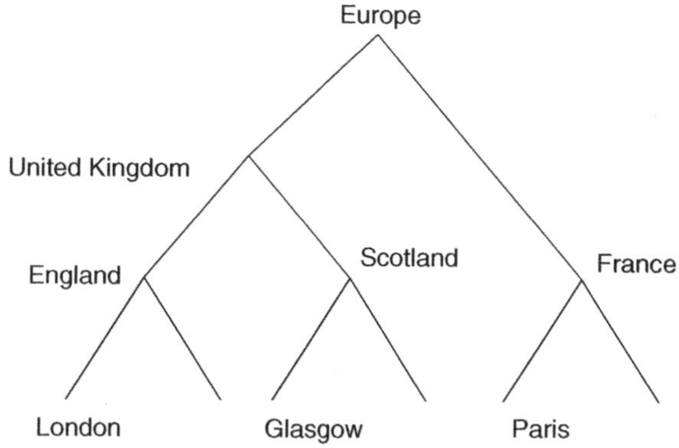

Figure 1.3 Taxonomic tree associated with the variable 'region'.

Table 1.7 Definition of the taxonomic variable 'region'.

Region	Predecessor
Paris	France
London	England
France	Europe
England	Europe
Glasgow	Scotland
Europe	Europe

1.3.4 Retaining background knowledge after the generalization process

At least three kinds of questions are of interest in the generalization process: overgeneralization; the loss of statistical information such as means, mean squares, correlations or contingencies between variables within each generalized class of individuals; and the loss of rules.

Overgeneralization happens, for example, when a class of individuals described by a numerical variable is generalized by an interval containing smaller and greater values. For example, in Table 1.5 the age of the Spain team is described by the interval [23, 29]; this is one of several possible choices. Problems with choosing [min, max] can arise when these extreme values are in fact outliers or when the set of individuals to generalize is in fact composed of subsets of different distributions. These two cases are considered in Chapter 2.

How can correlations lost by generalization be recaptured? It suffices to create new variables associated with pairs of the initial variables. For example, a new variable called $\text{Cor}(y_{i,}y_j)$ can be associated with the variables y_i and y_j. Then the value of such a variable for a class of individuals C_k is the correlation between the variables y_i and y_j on a population reduced to the individuals of this class. For example, in Table 1.8 the players in the World Cup are described by their team, age, weight, height, nationality, etc. and by the categorical variable 'World Cup' which takes the value yes or no depending on the fact that they have played in or have been eliminated from the World Cup. In Table 1.9, the categories defined by the variable 'World Cup' constitute the new unit and the variable Cor(Weight, Height) has been added and calculated. As a result, the correlation between the weight and the height is greater for the players who play in the World Cup than for the others. In the same way, other variables can be added to the set of symbolic variables as variables representing the mean, the mean square, the median and other percentiles associated with each higher-level unit.

In the same way, it can be interesting to retain the contingency between two or more categorical variables describing a concept. This can be done simply by creating new variables which expresses this contingency. For example, if y_1 is a categorical variable with four categories and y_2 is a variable with six categories, a new model variable y_3 with 24 categories which is the Cartesian product of y_1 and y_2 can be created for each concept. In the case of numerical variables, it is also possible to retain the number of units inside an interval or inside a product of intervals describing a concept by adding new variables expressing the number of individuals that they contain. For example, suppose that the set of birds in Table 1.1 is described by two numerical variables, 'size' and 'age', and the species swallow is described by the cross product of two confidence intervals: I_{swallow} (size) and I_{swallow} (age). The extent of the concept 'swallow' described by the cross product I_{swallow} (size) $\times I_{\text{swallow}}$ (age) among the birds can be empty or more or less dense. By keeping the contingencies information among the 600 hundred birds for each concept, a biplot representation of the

Table 1.8 Classical data table describing players by numerical and categorical variables.

Player	Team	Age	Weight	Height	Nationality_1	...	World Cup
Ferández	Spain	29	85	1.84	Spanish		yes
Rodríguez	Spain	23	90	1.92	Brazilian		yes
Mballo	France	25	82	1.90	Senegalese		yes
Zidane	France	27	78	1.85	French		yes
...	...	...	...	...	...		...
Renie	XX	23	91	1. 75	Spanish		no
Olar	XX	29	84	1.84	Brazilian		no
Mirce	XXX	24	83	1.83	Senegalese		no
Rumbum	XXX	30	81	1.81	French		no

Table 1.9 Symbolic data table obtained from Table 1.8 by generalization for the variables age, weight and height and keeping back the correlations between weight and height.

World Cup	Age	Weight	Height	Cov (Weight, Height)
yes	[21, 26]	[78, 90]	[1.85, 1.98]	0.85
no	[23, 30]	[81, 92]	[1.75, 1.85]	0.65

Table 1.10 Initial classical data table where individuals are described by three variables.

Individuals	Concepts	Y_1	Y_2
I_1	C_1	a	2
I_2	C_1	b	1
I_3	C_1	c	2
I_4	C_2	b	1
I_5	C_2	b	3
I_6	C_2	a	2

Table 1.11 Symbolic data table induced from Table 1.10 with background knowledge defined by two rules: $[Y_1 = a] \Rightarrow [Y_2 = 2]$ and $[Y_2 = 1] \Rightarrow [Y_1 = b]$.

	Y_1	Y_2
C_1	$\{a, b, c\}$	$\{1, 2\}$
C_2	$\{a, b\}$	$\{1, 2, 3\}$

species with these two variables will be enhanced by visualizing the density in each obtained rectangle I_{species} (size) $\times I_{\text{species}}$(age) for each species.

Finally, how can rules lost by generalization be recaptured? Rules can be induced from the initial standard data table and then added as background knowledge to the symbolic data table obtained by generalization. For example, from the simple data table in Table 1.10 two rules can be induced: $[Y_1 = a] \Rightarrow [Y_2 = 2]$ and $[Y_2 = 1] \Rightarrow [Y_1 = b]$. These rules can be added as background knowledge to the symbolic data table (Table 1.11) which describes the concepts C_1 and C_2 by categorical multi-valued variables.

1.4 Some general applications of the symbolic data analysis approach

1.4.1 From complex data to symbolic data

Sometimes the term 'complex data' refers to complex objects such as images, video, audio or text documents or any kind of complex object described by standard numerical and/or categorical data. Sometimes it refers to distributed data or structured data – or, more

Table 1.12 Patients described by complex data which can be transformed into classical numerical or categorical variables.

Patient	Category	X-Ray	Radiologist text	Doctor text	Professional category	Age
Patient_1	C_j	X-Ray_1	Rtext_1	Dtext_1	Worker	25
...	...	...	...	...	...	...
Patient_n	C_k	X-Ray_n	Rtext_n	Dtext_n	Self-employed	...

Table 1.13 Describing categories of patients from Table 1.12 using symbolic data.

Categories	X-Ray	Radiologist text	Doctor text	Professional category	Age
C_1	$\{\text{X-Ray}\}_1$	$\{\text{Rtext}\}_1$	$\{\text{Dtext}\}_1$	{worker, unemployed}	[20, 30]
...	...	...	...	...	...
C_k	$\{\text{X-Ray}\}_k$	$\{\text{Rtext}\}_k$	$\{\text{Dtext}\}_k$	{self-employed}	[35, 55]

specifically, spatial-temporal data or heterogeneous data describing, for example, a medical patient using a mixture of images, text documents and socio-demographic information. In practice, complex objects can be modelled by units described by more or less complex data. Hence, when a description of concepts, classes or categories of such complex objects is required, symbolic data can be used. Tables 1.12 and 1.13 show how symbolic data are used to describe categories of patients; the patients are the units of Table 1.12 and the categories of patients are the (higher-level) units of Table 1.13.

In Table 1.12 patients are described by their category of cancer (level of cancer, for example), an X-ray image, a radiologist text file, a doctor text file, their professional category (PC) and age. In Table 1.13, each category of cancer is described by a generalization of the X-ray image (radiologist text file, doctor text file} denoted {X-Ray} ({Rtext}, {Dtext}). When the data in Table 1.12 are transformed into standard numerical or categorical data, the resulting data table, Table 1.13, contains symbolic data obtained by the generalization process. For example, for a given category of patients, the variable PC is transformed into a histogram-valued variable by associating the frequencies of each PC category in this category of patients; the variable age is transformed to an interval variable by associating with each category the confidence interval of the ages of the patients of this category.

1.4.2 From fuzzy, imprecise, or conjunctive data to symbolic data

The use of 'fuzzy data' in data analysis comes from the fact that in many cases users are more interested in meaningful categorical values such as 'small' or 'average' than in actual numerical values. That is to say, they are interested in the membership functions (Zadeh, 1978; Diday, 1995) associated with these categories. Therefore, they transform their data into 'fuzzy data'. Fuzzy data are characterized by fuzzy sets defined by membership functions. For example, the value 1.60 of the numerical variable 'height of men' might be associated with the value 'small' with a weight (i.e. a membership value) of 0.9, 'average' with a weight of 0.1, and 'tall' with a weight of 0.

Table 1.14 Initial data describing mushrooms of different species.

Specimen	Species	Stipe thickness	Stipe length	Cap size	Cap colour
Mushroom_1	*Amanita muscaria*	1.5	21	24 ± 1	red
Mushroom_2	*Amanita muscaria*	2.3	15	18 ± 1	red $\wedge$ white
Mushroom_3	*Amanita phalloides*	1.2	10	7 ± 0.5	olive green
Mushroom_4	*Amanita phalloides*	2.0	19	15 ± 1	olive brown

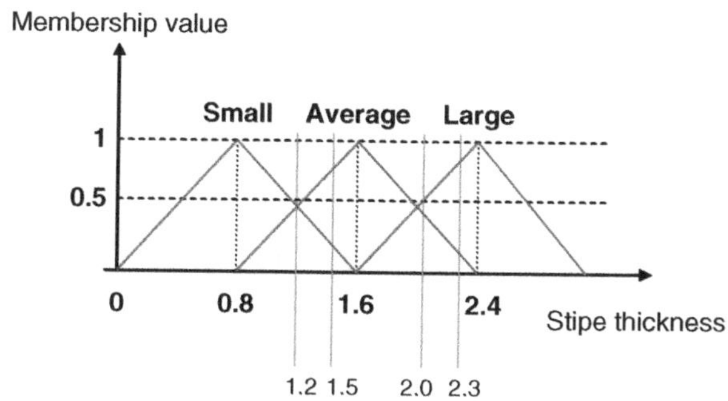

Figure 1.4 From numerical data to fuzzy data: if the stipe thickness is 1.2, then it is (0.5) Small, (0.5) Average, (0) High.

'Imprecise data' are obtained when it is not possible to get an exact measure. For example, it is possible to say that the height of a tree is 10 metres ± 1. This means that its length varies in the interval [9, 11].

'Conjunctive data' designate the case where several categories appear simultaneously. For example, the colour of an apple can be red or green or yellow but it can be also 'red and yellow'.

When individuals are described by fuzzy and/or imprecise and/or conjunctive data, their variation inside a class, category or concept is expressed in term of symbolic data. This is illustrated in Table 1.14, where the individuals are four mushroom specimens; the concepts are their associated species (*Amanita muscaria, Amanita phalloides*). These are described by their stipe thickness, stipe length, cap size and cap colour. The numerical values of the variable 'stipe thickness' are transformed into a fuzzy variable defined by three fuzzy sets denoted 'small', 'average' and 'high'. The membership functions associated with these fuzzy sets are given in Figure 1.4: they take three forms with triangular distribution centred on 0.8, 1.6 and 2.4. From Figure 1.4, it can be observed that the stipe thickness of mushroom_1, whose numerical value is 1.5, has a fuzzy measure of 0.2 that it is small, 0.8 that it is average, and 0 that it is large. The stipe thickness of mushroom_2, whose numerical value is 2.3, has a fuzzy measure of 0 that it is small, 0.1 that it is average. and 0.9 that it is large. In other words, if the stipe thickness is 2.3, then it is (0)Small, (0.1)Average, (0.9) Large. The stipe thicknesses for all four individual mushrooms are expressed as fuzzy data as shown in Table 1.15. For the present purposes, the stipe length is retained as a classical numerical data. The 'cap size' variable values are imprecise and the values of the 'cap colour' variable

Table 1.15 The numerical data given by the variable 'stipe thickness' are transformed into fuzzy data.

Specimen	Species	Stipe thickness			Stipe length	Cap size	Cap colour
		Small	Average	Large			
Mushroom$_1$	*A. muscaria*	0.2	0.8	0	21	24 ± 1	red
Mushroom$_2$	*A. muscaria*	0	0.1	0.9	15	18 ± 1	red $\wedge$ white
Mushroom$_3$	*A. phalloides*	0.5	0.5	0	10	7 ± 0.5	olive green
Mushroom$_4$	*A. phalloides*	0	0.5	0.5	19	15 ± 1	olive brown

Table 1.16 Symbolic data table induced from the fuzzy data of Table 1.15.

Species	Stipe thickness			Stipe length	Cap size	Cap colour
	Small	Average	Large			
A. muscaria	[0, 0.2]	[0.1, 0.8]	[0, 0.9]	[15, 21]	[17, 25]	{red, red $\wedge$ white}
A. phalloides	[0, 0.5]	[0.5, 0.5]	[0, 0.5]	[10, 19]	[6.5, 16]	{olive green, olive brown}

are unique colours or conjunctive colours (i.e., several colours simultaneously, such as 'red $\wedge$white').

Suppose that it is desired to describe the two species of *Amanita* by merging their associated specimens described in Table 1.15. The extent of the first species (*A. muscaria*) is the first two mushrooms (mushroom$_1$, mushroom$_2$). The extent of the second species (*A. phalloides*) is the last two mushrooms (mushroom$_3$, mushroom$_4$). The symbolic data that emerge from a generalization process applied to the fuzzy numerical, imprecise and conjunctive data of Table 1.15 are as shown in Table 1.16.

1.4.3 Uncertainty, randomness and symbolic data

The meaning of symbolic data, such as intervals, is very important in determining how to extend statistics, data analysis or data mining to such data. For example, if we considered the height of a football player without uncertainty we might say that it is 182. But if we were not sure of his height we could say with some uncertainty that it lies in the interval $I_1 = [180, 185]$. That is why, in the following, will say that I_1 is an 'uncertainty height'. If we now consider the random variable X associated with the height of members of the same football team, we can associate with X several kinds of symbolic data such as its distribution, its confidence interval or a sample of heights. If we represent this random variable by its confidence interval $I_2 = [180, 185]$, we can see that $I_1 = I_2$ but their meanings are completely different. This comes from the fact that I_1 expresses the 'uncertainty' given by our own subjective expectation and I_2 expresses the 'variability' of the height of the players in the team. By considering the uncertainty height of all the players of the team we obtain a data table where the individuals are the players and the variable associates an interval (the height with uncertainty) with each player. We can then calculate the 'possibility' (Dubois and Prade, 1988) that a player has a height in a given interval. For example, the possibility that a player has a height in the interval $I = [175, 183]$ is measured by the higher proportion of the intervals of height of players which cut the interval I.

By considering the random variable defined by the height of the players of a team in a competition, we get a random data table where the individuals (of higher level) or concepts are teams and the variable 'height' associates a random variable with each team. We can then create a symbolic data table which contains in each cell associated with the height and a given team, a density distribution (or a histogram or confidence interval) induced by the random variable defined by the height of the players of this team.

Each such density distribution associated with a team expresses the variability of the height of the players of this team. In symbolic data analysis we are interested by the study of the variability of these density distributions. For example, we can calculate the probability that at least one team has a height in the interval $I = [175, 183]$. This probability is called the 'capacity' of the teams in the competition to have a height in the interval $I = [175, 183]$. Probabilities, capacities (or 'belief') and 'possibilities' are compared in Diday (1995). Notice that in practice such probabilities cannot be calculated from the initial random variables but from the symbolic data which represent them. For example, in the national institutes of statistics it is usual to have data where the higher-level units are regions and the symbolic variables are modal variables which give the frequency distribution of the age of the people in a region ([0, 5], [5, 10], ... years old) and, say, the kind of house (farm, house) of each region. In other words, we have the laws of the random variables but not their initial values. In SODAS2, the STAT module calculates capacities and provides tools to enable the study of the variability of distributions. Bertrand and Goupil (2000) and Chapters 3 and 4 of Billard and Diday (2006) provide several tools for the basic statistics of modal variables.

1.4.4 From structured data to symbolic data

There are many kinds of structured data. For example, structured data appear when there are some taxonomic and/or mother–daughter variables or several linked data tables as in a relational database. These cases are considered in Chapter 2 of this book. In this section our aim is to show how several data tables having different individuals and different variables can be merged into a symbolic data table by using common concepts.

This is illustrated in Tables 1.17 and 1.18. In these data tables the units are different and only the variable 'town' is common. Our aim is to show that by using symbolic data it is possible to merge these tables into a unique data table where the units are the towns. In Table 1.17 the individuals are schools, the concepts are towns, and the schools are described by three variables: the number of pupils, the kind of school (public or private) and the coded level of the school. The description of the towns by the school variable, after a generalization process is applied from Table 1.17, is given in Table 1.19. In Table 1.18 the individuals are hospitals, the concepts are towns, and each hospital is described by two variables: its coded number of beds and its coded specialty. The description of the towns by the hospital variable, after a generalization process is applied from Table 1.18, is given in Table 1.20.

Finally, it is easy to merge Tables 1.19 and 1.20 in order to get the symbolic data table shown in Table 1.21, which unifies the initial Tables 1.17 and 1.18 by using the same higher-level unit: the concept of town.

Table 1.17 Classical description of schools.

School	Town	No. of pupils	Type	Level
Jaurès.	Paris	320	Public	1
Condorcet	Paris	450	Public	3
Chevreul	Lyon	200	Public	2
St Hélène	Lyon	380	Private	3
St Sernin	Toulouse	290	Public	1
St Hilaire	Toulouse	210	Private	2

Table 1.18 Classical description of hospitals.

Hospital	Town	Coded no. of beds	Coded specialty
Lariboisière	Paris	750	5
St Louis	Paris	1200	3
Herriot	Lyon	650	3
Besgenettes	Lyon	720	2
Purpan	Toulouse	520	6
Marchant	Toulouse	450	2

Table 1.19 Symbolic description of the towns by the school variable after a generalization process is applied from Table 1.17.

Town	No. of pupils	Type	Level
Paris	[320, 450]	(100%)Public	{1, 3}
Lyon	[200, 380]	(50%)Public, (50%)Private	{2, 3}
Toulouse	[210, 290]	(50%)Public, (50%)Private	{1, 2}

Table 1.20 Symbolic description of the towns by the hospital variable after a generalization process is applied from Table 1.18.

Town	Coded no. of beds	Coded specialty
Paris	[750, 1200]	{3, 5}
Lyon	[650, 720]	{2, 3}
Toulouse	[450, 520]	{2, 6}

Table 1.21 Concatenation of Tables 1.17 and 1.18 by symbolic data versus the concepts of towns.

Town	No. of pupils	Type	Level	Coded no. of beds	Coded specialty
Paris	[320, 450]	(100%)Public	{1, 3}	[750, 1200]	{3, 5}
Lyon	[200, 380]	(50%)Public, (50%)Private	{2, 3}	[650, 720]	{2, 3}
Toulouse	[210, 290]	(50%)Public, (50%)Private	{1, 2}	[450, 520]	{2, 6}

Table 1.22 The four cases of statistical or data mining analysis.

	Classical analysis	Symbolic analysis
Classical data	Case 1	Case 2
Symbolic data	Case 3	Case 4

1.4.5 The four kinds of statistics and data mining

Four kinds of statistics and data mining (see Table 1.22) can be considered: the classical analysis of classical data where variables are standard numerical or categorical (case 1); the symbolic analysis of classical data (case 2); the classical analysis of symbolic data (case 3); and the symbolic analysis of symbolic data (case 4). Case 1 is standard. Case 2 consists of extracting a symbolic description from a standard data table; for example, symbolic descriptions are obtained from a decision tree by describing the leaves by the conjunction of the properties of their associated branches.

In case 3 symbolic data are transformed into standard data in order to apply standard methods. An example of such a transformation is given in Table 1.23, which is a transformation of Table 1.19. It is easily obtained by transforming an interval-valued variable into variables for the minimum and maximum values; the modal variables are transformed into several variables, each one attached to one category. The value of such variables for each unit is the weight of this category. For example, in Table 1.19, the value of the variable 'public' is 50 for Lyon as it is the weight of the category 'public' of the variable 'type' in Table 1.19. The categorical multi-valued variables are transformed into binary variables associated with each category. For example, the variable 'level' of Table 1.19 is transformed in Table 1.23 into three variables: Level 1, Level 2, Level 3. Hence, the multi-categorical

Table 1.23 From the symbolic data table given in Table 1.19 to a classical data table.

Town	Min. no. of pupils	Max. no. of pupils	Public	Private	Level 1	Level 2	Level 3
Paris	320	450	100	0	1	0	1
Lyon	200	380	50	50	0	1	1
Toulouse	210	290	50	50	1	1	0

value of Lyon in Table 1.19 is {2, 3} and is transformed in Table 1.23 to three values: Level 1 (Lyon) = 0, Level 2 (Lyon) = 1, Level 3 (Lyon) = 1. The advantage of obtaining such a standard data table is that the standard methods of statistics or data mining can be applied. The disadvantage is that much information is lost.

Why it is more useful to work on symbolic data than on their standard data table decompositions? For example, a standard dissimilarity between intervals transformed in a standard data table, will only take into account the dissimilarity between the two minimum values and the two maximum values, but not between the maximum and the minimum values, as done for example by the Hausdorff dissimilarity between intervals (see Chapter 11); a principal component analysis on the standard data table will produce just points associated with each concept, but a symbolic principal component analysis (see Chapter 15) will produce a rectangular (or convex) surface whose size will express the internal variation of the individuals of the extent of the concept. For example, if the higher-level units are 'very bad', 'bad', 'average', 'good', 'very good' players, in using the standard data table, each concept will be represented by a point, but in using the symbolic principal component analysis the good players will be represented by a small rectangle, showing that they constitute a homogeneous group. Also a decision tree applied to the standard data table will only select minimum or maximum variables instead of selecting the symbolic variables themselves (see Ciampi *et al.*, 2000). Several cases of these examples are illustrated in Billard and Diday (2006).

Case 4 (symbolic analysis of symbolic data) is developed in several chapters in this book. For example, in Chapter 11 where clusters are obtained from a symbolic data table by the SCLUST module, they are associated with prototypes described by symbolic data.

1.5 The main steps and principles of a symbolic data analysis

1.5.1 Philosophical aspects: the reification process

The aim of a reification process is to transform any subject into an object to be studied. But as Aristotle (1994) said in the *Organon*, any object cannot be defined by 'what it contains or anything that can be asserted on it'. Therefore, we can only give an approximation of its definition. In the same book, Aristotle said that there are two kinds of objects: first-level objects called 'individuals' (e.g., a horse or the chair on which I am sitting) and second-level objects which we have called 'concepts' (e.g., a species of horse or a type of chair). Following Arnault and Nicole (1965), concepts have intent and extent composed of a set of individuals which satisfy the intent. Hence, considering that individuals and concepts are 'objects' has four consequences: they are unique; their description is not unique; their description restricts what they are (what we say on an object is not the object, which is a consequence of Aristotle); and two individuals (or concepts) having the same description can remain different. These important properties allow us to extend standard statistics on standard statistical units to symbolic statistics on concepts.

In statistics, units to be studied (such as households or companies) are reified as 'statistical units' described by the variables of a given standard data table. In symbolic data analysis, these units are reified as first-order objects called 'individuals'. In SODAS2 these individuals are approximately defined first by a categorical variable and then by a set of variables. The categories of the categorical variable are then reified as concepts described by symbolic variables induced from the one which describes the individuals. In such a case the intent of the concept is the category and its extent is the extent of the category (i.e., the set of individuals which satisfy the category). For more details on this reification process see Section 2.3 in Diday (2005).

1.5.2 Symbolic data analysis in eight steps

This book is based on the eight following steps described in Figure 1.5 (which will be detailed in Chapter 24). First, a relational database is assumed to be given, composed of several more or less linked data tables (also called relations).

Second, a set of categories based on the categorical variables values of the database are chosen by an expert. For example, from the Cartesian product age $\times$ gender, with three levels of age (young if less then 35, age between 35 and 70, old if more than 70), six categories are obtained.

Third, from a query to the given relational database, a data table is obtained whose first column represents the individuals (no more than one row for each individual), and whose second column represents a categorical variable and associates with each individual its unique category. Each category is associated with its 'extent' in the database. This extent

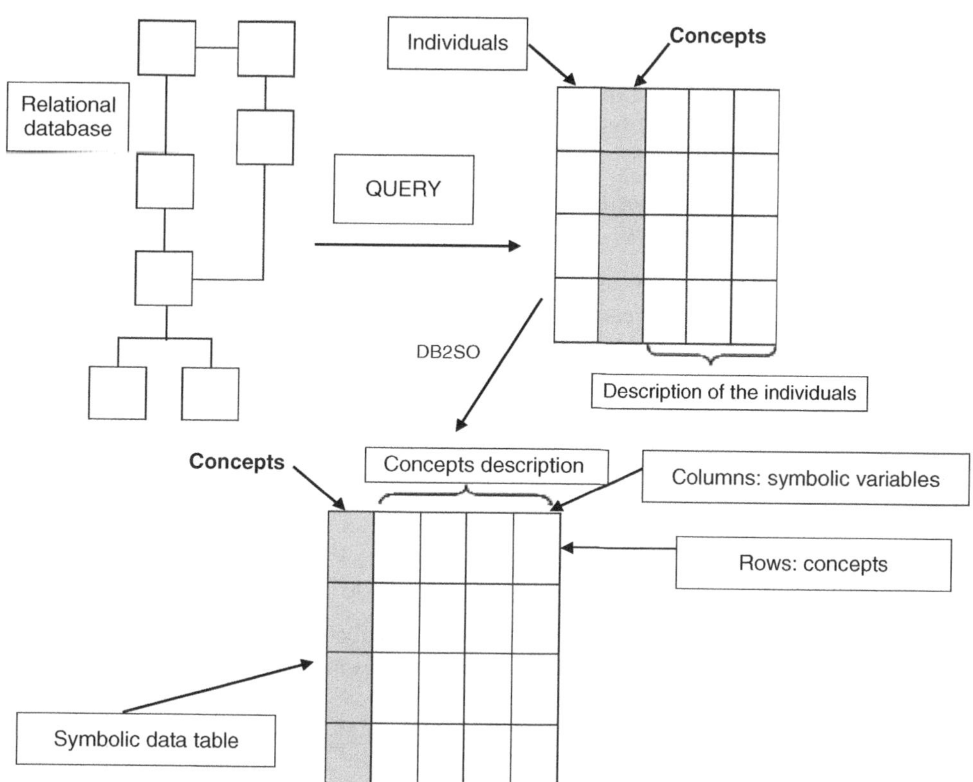

Figure 1.5 From a relational database to a symbolic data table.

is defined by the set of individuals which satisfy the category. For example, a male aged 20 is an individual who satisfies the category (male, young).

Fourth, the class of individuals which defines the extent of a category is considered to be the extent of the concept which reifies this category.

Fifth, a generalization process is defined, in order to associate a description with any subset of individuals.

Sixth, the generalization process is applied to the subset of individuals belonging to the extent of each concept. The description produced by this generalization process is considered to be a description of each concept. This step is provided in SODAS2 by the DB2SO module (see Chapter 2). As already noted, the generalization process must take into account the internal variation of the set of individuals that it describes and leads to symbolic variables and symbolic data.

Seventh, a symbolic data table, where the units are the concepts and the variables are the symbolic variables which describe them, is defined. Conceptual variables can be added at the concept level, as can any required background knowledge (rules and taxonomies).

Eighth, symbolic data analysis tools such as those explained in this book are applied on this symbolic data table by taking the background knowledge into account.

1.5.3 From data mining to knowledge mining by symbolic data analysis: main principles

In order to apply the tools of standard data mining (clustering, decision trees, factor analysis, rule extraction, . . .) to concepts, these tools must be extended to symbolic data. The main principles of a symbolic data analysis can be summarized by the following steps:

1. A symbolic data analysis needs two levels of units. The first level is that of individuals, the second that of concepts.
2. A concept is described by using the symbolic description of a class of individuals defined by the extent of a category.
3. The description of a concept must take account of the variation of the individuals of its extent.
4. A concept can be modelled by a symbolic object which can be improved in a learning process (based on the schema of Figure 1.6 explained in the next section), taking into account the arrival of new individuals.
5. Symbolic data analysis extends standard exploratory data analysis and data mining to the case where the units are concepts described by symbolic data.
6. The output of some methods of symbolic data analysis (clustering, symbolic Galois lattice, decision trees, Kohonen maps, etc.) provides new symbolic objects associated with new categories (i.e., categories of concepts).
7. The new categories can be used in order to define new concepts such as in step 2, and so on.

1.5.4 Is there a general theory of symbolic data analysis?

As it is not possible to say that there is a general theory of standard statistics, it follows that we also cannot say that there is a general theory of symbolic data analysis. The aim, in both cases, is to solve problems arising from the data by using the most appropriate mathematical theory. However, we can say that in symbolic data analysis, like the units, the theory increases in generality. For example, if we consider the marks achieved by Tom, Paul and Paula at each physics and mathematics examination taken during a given year, we are able to define a standard numerical data table with n rows (the number of exams) and two columns defined by two numerical random variables X_P and X_M which associate with each exam taken by Tom (or Paul or Paula), the marks achieved in physics and mathematics. Thus $X_P(\text{exam}_i, \text{Tom}) = 15$ means that Tom scored 15 in mathematics at the ith examination. Now, if we consider that Tom, Paul and Paula are three concepts, then we obtain a new data table with three rows associated with each pupil and two columns associated with two new variables Y_P and Y_M for which the values are the preceding random variables where the name of the pupil is fixed. For example, $Y_P(\text{Tom}) = X_P(\cdot, \text{Tom})$. If we replace, in each cell of this data table, the random variable that it contains with the confidence interval of its associated distribution, we obtain an example of symbolic data table where each variable is interval-valued.

1.6 Modelling concepts by symbolic objects

1.6.1 Modelling description by the joint or by the margins

A natural way to model the intent of a concept (in order to be able to obtain its extent) is to give a description of a class of individuals of its extent (for example, the penguins of an island in order to model the concept of 'penguin'). From a statistical point of view there are two extreme ways to describe such a class.

The first is to describe it by the joint distribution of all the variables (if the variables are a mixture of numerical and categorical, it is always possible to transform them so that they all become categorical). In this case, the problem is that the space tends to become empty as the number of variables becomes large, which is one of the characteristics of data mining.

The second is to describe this class of individuals by the margins associated with all available descriptive variables. In this case, some joint information is lost but easy interpretable results are obtained, as the variables are transparent instead of being hidden in a joint distribution.

In practice, between these two extremes, for some variables the joint distribution can be used, while for others only the marginals can be used. Notice that the use of copulas (Schweizer and Sklar, 2005) can reduce calculation by getting the joint directly from the marginal. It is also possible to add new variables, retaining, for example, correlations as explained in Section 1.3.4. In all these cases, only an approximate description of a concept is obtained. The following section describes the modelling of concepts based on their intent description and defined by the so-called 'symbolic object'.

1.6.2 The four basic spaces for modelling concepts by symbolic objects

In Figure 1.6 the 'set of individuals' and the 'set of concepts' constitute the 'real world' set; we might say that the 'world of objects in the real world: the individuals and concepts' are considered as lower and higher-level objects. The 'modelled world' is composed of the 'set of descriptions' which models individuals (or classes of individuals) and the 'set of symbolic objects' which models concepts. In Figure 1.6(a) the starting point is a concept C whose extent, denoted $\text{Ext}(C/\Omega')$, is known in a sample Ω' of individuals. In Figure 1.6(b) the starting point is a given class of individuals associated with a concept C.

If the concept C is the swallow species, and 30 individual swallows have been captured among a sample Ω' of the swallows of an island, $\text{Ext}(C/\Omega')$ is known to be composed of these 30 swallows (Figure 1.6(a)). Each individual w in the extent of C in Ω' is described in the 'description space' by using the mapping y such that $y(w) = d_w$ describes the individual w.

Given a class of 30 similar birds obtained, for example, from a clustering process, a concept called 'swallow' is associated with this class (Figure 1.6(b)). Each individual w in this class is described in the 'description space' by using the mapping y such that $y(w) = d_w$ describes the individual w.

The concept C is modelled in the set of symbolic objects by the following steps described in Figure 1.6.

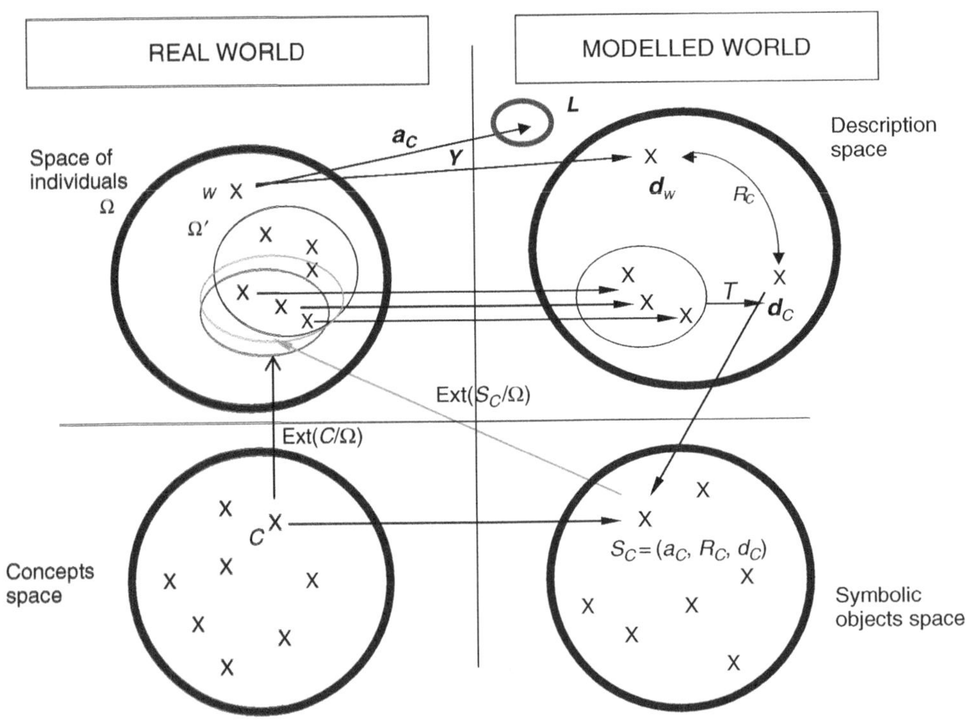

(a)

Figure 1.6 (a) Modellization by a symbolic object of a concept known by its extent.

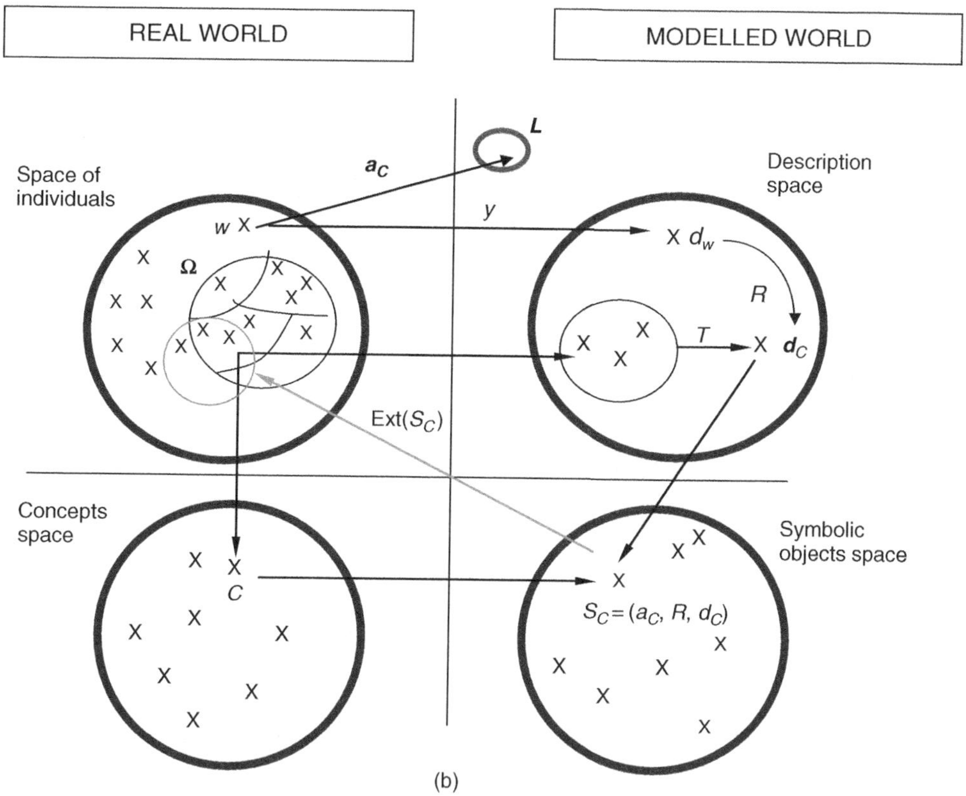

Figure 1.6 (b) Modellization by a concept known by a class of individuals.

First, the set of descriptions of the individuals of Ext(C/Ω') in Figure 1.6(a), or of the class of individuals associated with the concept C in Figure 1.6(b), is generalized with an operator T in order to produce the description d_C (which may be a set of Cartesian products of intervals and/or distributions or just a unique joint distribution, etc.).

Second, the matching relation R can be chosen in relation to the choice of T. For instance, if $T = \cup$, then $R_C =$ '$\subseteq$'; if $T = \cap$, then $R =$ '$\supseteq$' (see Section 1.7). The membership function is then defined by the mapping a_C: $\Omega \rightarrow L$ where L is [0, 1] or {true, false} such that $a_C(w) = [y(w)\ R_C\ d_C]$ which measures the fit or match between the description $y(w) = d_w$ of the unit w and the description d_C of the extent of the concept C in the database. The symbolic object modelling the concept C by the triple $s = (a_C, R, d_C)$ can then be defined. In the next section, this definition is considered again and illustrated by several examples. Sometimes the following question arises: 'why not reduce the modelling of the concept C to just the membership function a_C as it suffices to know a_C in order to obtain the extent?'. The reason is that in this case we would not know which kind of description d_C is associated with the concept nor which kind of matching relation R is used as several d_C or R can produce the same extent. More intuitively, if I wish to know if the object, w, that I have in my hand is a pencil, I will get two descriptions – one from my hand, d_{hand}, and another from my eyes, d_{eyes} – and compare them with the concept of pencil $d_C = (d_{C,\text{hand}}, d_{C,\text{eyes}})$ that I have in my head, by two matching relations R_{hand} and R_{eyes}. Then the membership function a will aggregate both matching results and allow me to say if w is a pencil or not. This can be summarized by the following kind of formula (detailed in the next section): $a(w) = [d_{hand}\ R_{hand}\ d_{C,hand}] \wedge [d_{eyes}\ R_{eyes}\ d_{C,eyes}] \in \{\text{true, false}\}$.

Knowing the extent E_1 of a concept in Ω' and the extent E_2 in Ω' of a symbolic object which models this concept, T, R and a have to be chosen such that E_1 and E_2 become as close as possible. Numerous learning processes and their convergence can been defined depending on the kind of organization or modelling of the concepts. Diday and Emilion (1997, 2003) showed that as the number of individuals increases, the concepts and their Galois lattice structure converge. Diday (2002b) and Diday and Vrac (2005) showed that a partition of concepts modelled by vectors whose entries are distributions converges by improving a given criteria based on copulas.

1.6.3 Extent of concepts and symbolic objects

By using the description of the extent of a concept, symbolic objects can be used to find the individuals of this extent.

For example, if the age of two individuals w_1, w_2 is given by $\text{age}(w_1) = 30$, $\text{age}(w_2) = 35$, the description of the class $C = \{w_1, w_2\}$ obtained by a generalization process can be [30, 35]. The extent of this description contains at least w_1 and w_2 but may contain other individuals. In this simple case, the symbolic object s is defined by a triple $s_C = (a_C, R_C, d_C)$, where $d_C = [30, 35]$, $R_C =$ '$\in$' and a_C is the mapping: $\Omega \to \{\text{true, false}\}$ such that $a_C(w)$ is the true value of '$\text{age}(w) R_C d_C$', written $a_C(w) = [\text{age}(w) \in d_C]$. An individual w is in the extent of s if and only if $a_C(w) = \text{true}$.

More formally, let Ω be a set of individuals, D a set containing descriptions of individuals or of classes of individuals, and y a mapping defined from Ω into D which associates with each $w \in \Omega$ a description $d \in D$ from a given symbolic data table. Let R be a relation defined on D by a subset E of $D \times D$. If $(x, y) \in E$, then x and y are said to be 'connected' by R, written as xRy. More generally, xRy takes its value in a set L, where $L = \{\text{true, false}\}$ or $L = [0, 1]$. When $L = \{\text{true, false}\}$, $[d'\ R\ d] =$ true means that there is a connection between d and d'. When $L = [0, 1]$, $[d'\ R\ d] = x$ means that d is more or less connected to d'. In this case, $[d'\ R\ d]$ can be interpreted as the 'true value' of xRy or the degree to which d' is in relation R with d (see Section 1.5.2 in Bandemer and Nather, 1992, on fuzzy relations).

For instance, $R \in \{=, \equiv, \leqslant, \subseteq\}$ or is an implication, a kind of matching taking account of missing values, etc. R can also use a logical combination of such operators.

A 'symbolic object' is defined by a description d, a relation R for comparing d to the description of an individual, and a mapping a called the 'membership function'. More formally we have the following definition: a symbolic object is a triple $s = (a, R, d)$ where R is a relation between descriptions, d is a description, and a is a mapping defined from Ω in L depending on R and d.

Symbolic data analysis is usually concerned with classes of symbolic objects where R is fixed, d varies among a finite set of comparable descriptions, and a is such that $a(w) = [y(w)\ R\ d]$, which is by definition the result of the comparison of the description of the individual w to d.

In the case where $y(w)$ and d are graphs, R is a relation between graphs.

More generally, many other cases can be considered. For instance, the mapping a may take the form $a(w) = [h_e(y(w))\ h_j(R)h_i(d)]$ where the mappings h_e, h_j and h_i are 'filters' (for more details, see Diday, 2000b).

There are two kinds of symbolic objects: if $[y(w)\ R\ d] \in L =$ {true, false} we have a 'Boolean symbolic object', while if $[y(w)\ R\ d] \in L = [0, 1]$ we have a 'modal symbolic object', also referred to as a 'probabilistic symbolic object'. In both cases, $y(w) = (y_1, \ldots, y_p)$, and the y_i are as defined in Section 1.3.3.

As an example of a Boolean symbolic object, let $a(w) = [y(w)\ R\ d]$ with R: $[d'\ R\ d] = \vee_{i=1,2}[d'_i\ R_i\ d_i]$, where $\vee$ has the standard logical meaning and $R_i = \subseteq$. If $y(w) =$ (colour(w), height(w)), $d = (d_1, d_2) =$ ({red, blue, yellow}, [10,15]) and the individual u is such that colour(u) = {red, yellow}, height(u) = {21}, then $a(u) =$ [colour(u) $\subseteq$ {red, blue, yellow}] $\vee$ [height(u) $\subseteq$ [10,15]]= true $\vee$ false = true.

As an example of a modal symbolic object, let $a(u) = [y(u)Rd]$ where, for instance, R: $[d'\ R\ d] = \max_{i=1,2}[d'_i\ R_i\ d_i]$. The choice of the maximum is one of many possible choices related to probabilist, possibilist, belief, and copula theory (see Diday, 1995, 1998, 2000a). For example, in the case of marginal probabilities the likelihood would use the product. The relation R between two probability distributions can be defined, for example, for two discrete probability distributions $d'_i = r$ and $d_i = q$ of k values by $r\ R_i\ q = \sum_{j=1,k} r_j q_j e^{(r_j - \min(r_j, q_j))}$. Chapter 8 discusses the MATCH module of SODAS2 which allows an asymmetric comparison between symbolic descriptions.

In the Boolean case, the extent of a symbolic object s, denoted Ext(s), is defined by the extent of a, which is Extent(a) = $\{w \in \Omega \mid a(w) = \text{true}\}$. In the modal case, given a threshold α, it is defined by $\text{Ext}_\alpha(s) = \text{Extent}_\alpha(a) = \{w \in \Omega \mid a(w) \geqslant \alpha\}$.

1.6.4 Syntax of symbolic objects in the case of assertions and hoards

If the initial data table contains p variables, we denote $y(w) = (y_1(w), \ldots, y_p(w))$, $D = (D_1, \ldots, D_p)$, $d \in D: d = (d_1, \ldots, d_p)$ and $R' = (R_1, \ldots, R_p)$ where R_i is a relation defined on D_i. An 'assertion' is a special case of a symbolic object defined by $s = (a, R, d)$ where R is defined by $[d'R\ d] = \wedge_{i=1,p}[d'_i R_i d_i]$ in which $\wedge$ has the standard logical meaning in the boolean case and a is defined by $a(w) = [y(w)R\ d]$. Notice that, considering the expression $a(w) = \wedge_{i=1,p}[y_i(w)R_i\ d_i]$, we are able to define the symbolic object $s = (a, R, d)$. This explanatory expression defines a symbolic object called an 'assertion'.

As an example of an assertion, suppose that $a(w)$ = [age(w) $\subseteq$ {12, 20, 28}]$\wedge$ [SPC(w) $\subseteq$ {employee, worker}]. If the individual u is described in the original symbolic data table by age(u) = {12, 20} and SPC(u) = {employee}, then $a(u)$ = [{12, 20} $\subseteq$ {12, 20, 28}]$\wedge$ [{employee} $\subseteq$ {employee, worker}] = true.

In the modal case, the variables are multi-valued and weighted. An example is given by $a(u) = [y(u)\ R\ d]$ with $[d'\ R\ d] = f(\{[y_i(w)R_i\ d_i]\}_{i=1,p})$ where, for instance, $f(\{[y_i(w)R_i\ d_i]\}_{i=1,p}) = \prod_{i=1,2}[d'_i\ R_i\ d_i]$.

Other kinds of symbolic objects have been introduced such as 'hoards' (Diday, 1987a, 1987b), which are used when the variables are not defined on the same units. For example, in the case of national statistical institutes using regions described by the distribution of age and type of dwelling, the concept of people aged between 20 and 30 and living on a farm or smallholding can be written with $w = (u,v)$: $a(w)$ = [age(u) $\in$ [20, 30]] $\wedge$ [dwelling(v) $\in$ {farm, smallholding}] where u is a person and v is a dwelling. Gowda and Diday (1992) develop dissimilarity measures between such hoard objects.

1.6.5 Some advantages in the use of symbolic objects and their learning process

The following are some of the advantages that can be observed in the use of symbolic objects.

- They model the intent of a concept and provide a way of finding its extent.
- They give a summary of the original symbolic data table in an explanatory way (i.e., close to the initial language of the user) by expressing concepts in terms of descriptions based on properties concerning the initial variables or meaningful variables (such as indicators obtained by regression or factor axes). This is developed in Chapter 2.
- They can be easily transformed in terms of a database query and so they can be used in order to propagate concepts between databases (for instance, from one time to another or from one country to another). This is studied in Chapter 3.
- By being independent of the initial data table, they are able to identify the matching with any new individual. This is dealt with in Chapter 8.
- In the use of their descriptive part, they are able to give a new higher-level symbolic data table on which a second-level symbolic data analysis can be applied and so on.
- It is possible to measure their quality, robustness and reliability as explained in the next section.

1.6.6 Quality, robustness and reliability of symbolic objects

Returning to the scheme of Figure 1.6, it is always possible in the boolean case to choose an operator T in such a way that the extent of the symbolic object s_C contains the extent of the concept C, but it can also contain individuals which are not in the extent of C. In the modal case, some individuals belonging to the extent of C may be outside the extent of s_C; this depends on the choice of the threshold α: $\mathrm{Ext}_\alpha(s_C) = \{w|a(w)\} \geqslant \alpha\}$. Therefore, two kinds of errors can occur:

(i) individuals who satisfy the concept and are not in the extent of s_C,

(ii) individuals who do not satisfy the concept but are in the extent of s_C.

The 'quality' and 'robustness' of the symbolic objects can then be defined in several ways. The error percentage in the first and second kind of error are denoted $e_1(\alpha)$ and $e_2(\alpha)$. In order to find the best α, by varying α between 0 and 1, the α which minimizes the product $e_1(\alpha)e_2(\alpha)$ can be retained.

In order to validate the symbolic object $s_{C,}$ the following method can be used. Repeat the following steps n times:

1. Obtain a sample Ω' with replacement from the whole set of given individuals Ω.
2. Calculate the symbolic object s_C by following the scheme given in Figure 1.6.

3. Calculate the extent of s_C in Ω'.
4. Calculate the errors of type (i) and (ii).

The quality and robustness of the symbolic object s_C is highest when the mean and mean square of the two histograms of the frequency of errors of type (i) and (ii) are lowest. In other words, let X_1 be the random variable which associates with each sample the frequency of error of type (i), and X_2 be the random variable which associates with each sample the frequency of error of type (ii). Then, the lower are the mean and mean square of these two random variables, the higher are the quality and robustness of the symbolic object s_C.

The 'reliability' of the membership of an individual w of the extent of s_C can be measured by the mean $m(w)$ of the membership value $a_C(w)$ when Ω' varies . More precisely, if the ith sample gives the value $a_i(w)$, then $m(w) = \sum_{i=1,n} a_i(w)/n$ and the reliability of s_C can be defined by $W(s_C) = \sum_{w \in \mathrm{Ext}(C)} m(w)/|\mathrm{Ext}(C)|$. The higher (i.e. the closer to 1) $W(s_C)$ is, the better is the reliability of s_C. The 'sensitivity' of s_C can be measured by $W'(s_C) = \sum_{w \in \mathrm{Ext}(C)} \sigma(w)/|\mathrm{Ext}(C)|$, where $\sigma(w)^2 = \sum_{i=1,n} (a_i(w) - m(w))^2/n$. These measures seem natural but their precise study remains to be done in comparison with other measures of such kind.

1.7 Underlying structures of symbolic objects: a generalized conceptual lattice

Under some assumptions on the choice of R and T (for instance $T \equiv \max$ if $R \equiv \leqslant$ and $T \equiv \min$ if $R \equiv \geq$), it can be shown that the underlying structure of a set of symbolic objects is a Galois lattice (Diday, 1989, 1991; Brito, 1991, 1994; Diday and Emilion, 1996, 1997, 2003; Polaillon, 1998; Bock and Diday, 2000) where the vertices are closed sets defined below by 'complete symbolic objects'. More precisely, the associated Galois correspondence is defined by two mappings F and G. F is a mapping from $P(\Omega)$ (the power set of Ω) into S (the set of symbolic objects) such that $F(C) = s$ where $s = (a, R, d)$ is defined by $d = T_{c \in C} y(c)$ and so $a(w) = [y(w) R\ T_{c \in C} y(c)]$, for a given R. For example, if $T_{c \in C}\ y(c) = \cup_{c \in C} y(c)$, $R \equiv \subseteq$, $y(u)$ = {pink, blue}, $C = \{c, c'\}$, $y(c)$ = {pink, red} and $y(c')$ = {blue, red}, then $a(u) = [y(w)\ R\ T_{c \in C} y(c)]$ = [{pink, blue} $\subseteq$ {pink, red} $\cup$ {blue, red}})={pink, red, blue}] = true and $u \in \mathrm{Ext}(s)$. G is a mapping from S into $P(\Omega)$ such that: $G(s) = \mathrm{Ext}(s)$.

A complete symbolic object s is such that $F(G(s)) = s$ (Diday, 1989; Brito, 2002). Such objects can be selected from the Galois lattice as well as from a partitioning, from a hierarchical or a pyramidal clustering, from the most influential individuals on a factor axis, from a decision tree, etc. In the following example we show how all the complete symbolic objects can be obtained from a Galois lattice.

Figure 1.7 shows a symbolic data table along with the Galois lattice obtained from it (the theory and algorithm for building such a lattice are given, for instance, in Diday 2000a). Here, the operator of generalization $T \equiv \cup$ and $R \equiv \subseteq$ are chosen. The complete symbolic objects and their extent, which are the nodes of this lattice, are as follows (the missing variables come from the fact that in order to simplify, the events $[y_i(w) \subseteq O_i]$ have been

	y_1	y_2	y_3
w_1	$\{a, b\}$	$\varnothing$	$\{g\}$
w_2	$\varnothing$	$\varnothing$	$\{g, h\}$
w_3	$\{c\}$	$\{e, f\}$	$\{g, h, i\}$
w_4	$\{a, b, c\}$	$\{e\}$	$\{h\}$

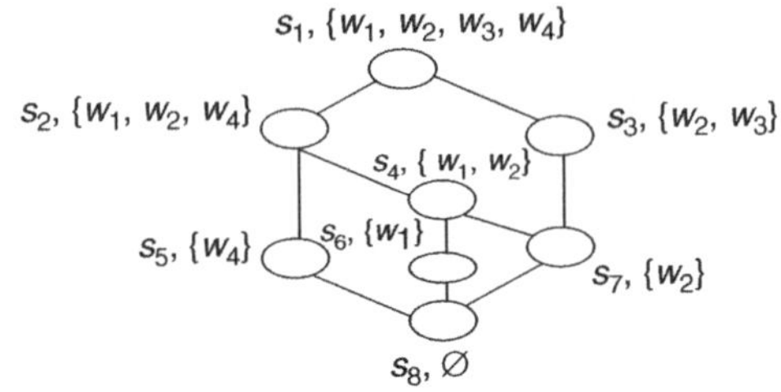

Figure 1.7 The conceptual lattice obtained from the symbolic data table.

deleted (except in s_1), as when this event is deleted the extent of s_i, $i=2, \ldots, 8$, does not change):

$$s_1 : a_1(w) = [y_1(w) \subseteq O_1] \wedge [y_2(w) \subseteq O_2] \wedge [y_3(w) \subseteq O_3], \mathrm{Ext}(s_1) = \{w_1, w_2, w_3, w_4\},$$
$$s_2 : a_2(w) = [y_2(w) \subseteq \{e\}] \wedge [y_3(w) \subseteq \{g, h\}], \mathrm{Ext}(s_2) = \{w_1, w_2, w_4\},$$
$$s_3 : a_3(w) = [y_1(w) \subseteq \{c\}], \mathrm{Ext}(s_3) = \{w_2, w_3\},$$
$$s_4 : a_4(w) = [y_1(w) \subseteq \{a, b\}] \wedge [y_2(w) = \varnothing] \wedge [y_3(w) \subseteq \{g, h\}], \mathrm{Ext}(s_4) = \{w_1, w_2\},$$
$$s_5 : a_5(w) = [y_2(w) \subseteq \{e\}] \wedge [y_3(w) \subseteq \{h\}], \mathrm{Ext}(s_5) = \{w_4\},$$
$$s_6 : a_6(w) = [y_1(w) \subseteq \{a, b\}] \wedge [y_2(w) = \varnothing] \wedge [y_3(w) \subseteq \{g\}], \mathrm{Ext}(s_6) = \{w_1\},$$
$$s_7 : a_7(w) = [y_1(w) = \{\varnothing\}] \wedge [y_2(w) = \varnothing] \wedge [y_3(w) \subseteq \{g, h\}], \mathrm{Ext}(s_7) = \{w_2\},$$
$$s_8 : a_8(w) = [y_1(w) = \varnothing] \wedge [y_2(w) = \varnothing] \wedge [y_3(w) = \varnothing], \mathrm{Ext}(s_8) = \{\varnothing\}.$$

Observe that $s_6 T s_7 = s_6 \cup s_7 = s_4$, by the union of the symbolic descriptions associated with each value. The extent of s_4 is the set of elements $w \in \Omega = \{w_1, w_2, w_3, w_4\}$ such that $a_4(w)$ is true. Therefore, $\mathrm{Ext}(s_4) = \{ w_1, w_2\}$.

1.8 Book overview

The book contains 24 chapters divided into four parts. Part I, consisting of Chapters 2–7, is concerned with the extraction of symbolic objects from a database and the exportation of symbolic objects to a database. It also includes metadata, editing and visualization of symbolic objects. Part II consists of Chapters 8–16 and covers dissimilarities between symbolic descriptions, clustering and factor analysis methods. Part III, which is made up of Chapters 15–20, is devoted to supervised methods and covers decision trees, factor discrimination, regression and neural networks. Part IV, containing Chapters 21–24, discusses several applications and the SODAS2 software.

Chapter 2, by Lechevallier, El Golli and Hebrail, is associated with the DB2SO module of SODAS2. This chapter is devoted to the important question of modelling concepts by symbolic description obtained from a generalization process. In order to avoid overgeneralization and to make the symbolic descriptions simpler, a refinement process by removing some atypical individuals is used.

Symbolic descriptions, constructed by the basic generalization process, may be enriched by metadata, picking up further information from the database: conceptual variables, taxonomies in variable domains, mother–daughter variables and rules. The process of reification of categories in concepts is illustrated by the possibility of adding 'conceptual variables'. For example, starting from a database where the units are players, we can define the teams as concepts described by the initial variables (age, height, nationality,. . .) transformed into symbolic variables; then we can add specific descriptive variables for the teams, such as the variable 'funds'.

By adding a membership or matching function to the symbolic description of a concept, it is possible to export a symbolic object and calculate its extent from one database to another. This is the aim of Chapter 3, by Malerba, Esposito and Appice. It is associated with the SO2DB module of SODAS2. SO2DB includes a graphical user interface which provides the user with facilities for controlling all parameters of the export process and uses the MATCH module to estimate the matching between each symbolic object and each individual stored in a relational database table. MATCH implements both canonical and flexible matching, which are detailed in Chapter 8. Canonical matching compares two structures (e.g. a symbolic object and a unit of a relational table) just for equality and returns either 0 (failure) or 1 (success). Flexible matching compares two structures in order to identify their 'degree of matching' rather than their equality. Finally, retrieved individuals are stored in a table of the input relational database. An application of SO2DB is also discussed.

Chapter 4, by Papageorgiou and Vardaki, concerns 'metadata', or data about data. The aim is to define a general metadata model for symbolic data. They develop step by step a statistical metadata model designed specifically for symbolic data in order to capture the metadata information needed for the symbolic objects, the creation of symbolic data tables and the successful implementation of the methods of symbolic data analysis. The metadata model under consideration should hold meta-information for the classical (original) data (survey variables, statistical units, frame population, etc.) and the symbolic data. More specifically, it should store meta-information both for the main stages of the processes of the classical data analysis, and for the symbolic data analysis procedures. It is also desirable for the model to store the processing history, from the creation of the original variables to the creation of symbolic data tables and the application of certain statistical methods and/or the visualization of the final results. The applicability of the model is illustrated using a data table showing the simultaneous manipulation of both data and metadata, as well as the improvement of the visualization of symbolic objects.

Chapter 5, by Noirhomme-Fraiture, Brito, de Baenst and Nahimana, concerns the editing – that is, creation and transformation – of a symbolic data table. The commonest way to construct a symbolic description is to extract individuals satisfying some criteria on variables (i.e. a query) from a database, as done by DB2SO as described in Chapter 2. Here two other ways are described: the importation from a 'native data' file and interactive creation. A native data file means that aggregation of individual data has been done with another statistical package (SAS, SPSS, . . .). In this case, the file is transformed into an XML one with SODAS2 software. In the second case, the user has already obtained symbolic objects from treatments and wishes to record them in a SODAS file, defining variables and objects. The possible transformations that the SOEDIT module of SODAS2 can carry out are as follows: modification of a particular cell; selection (or projection) of variables; addition of variables; modification of the labels of variables or of categories; selection of variables corresponding to certain rules; selection (or projection) of symbolic objects (considered here as a unit of a SODAS file); addition of symbolic objects; suppression of symbolic objects;

modification of the labels of symbolic objects; and sorting of symbolic objects. It is also possible to merge two SODAS files having the same objects or the same variables. Finally, Section 5.4 describes in detail the process of generalization of symbolic descriptions, that is, the construction of a symbolic description which summarizes (or is more general than) several others. Different cases of variable types, with different kinds of generalization, are considered (by union, by intersection, by maximum, by minimum).

Chapter 6, by Csernel and de Carvalho, discusses the 'normal symbolic form'. Symbolic descriptions can be constrained by domain knowledge expressed in the form of rules. However, taking these rules into account in order to analyse symbolic data usually results in exponential computation time. This chapter shows how their approach, based on a Maxwell–Boltzmann statistic, leads to a polynomial computation time.

The final chapter in Part I is Chapter 7, by Noirhomme-Fraiture and Nahimana. This chapter explains the process of visual data mining with the help of cognitive psychology knowledge. Then it focuses on the case of symbolic data in the symbolic data analysis framework. The basic options of star representation for symbolic objects are recalled. In the zoom star representation, each axis corresponds to a variable in a radial graph. In order to represent a symbolic description it allows variables with intervals, multiple values, weighted values, logical dependencies and taxonomies to be represented. The superimposition of stars and their breakdown are described. This is not only a way to visualize a symbolic description, included in another, but is also a way to construct it interactively. For the purpose of the method, the meaning of inclusion of symbolic objects and of hierarchies of symbolic objects is given. Finally, a visual drill-down function is explained, with an example.

Part II is devoted to unsupervised methods. It opens with Chapter 8, by Esposito, Malerba and Appice, which aims to provide tools for comparing descriptions of individuals or concepts. Different types of comparisons are required for diverse data analysis tasks. In this chapter two types of comparison functions are distinguished: resemblance and matching. The former is defined as a real-valued function that satisfies the symmetric property and can be further characterized into a similarity or a dissimilarity measure. The latter is a directional (or asymmetric) comparison where the two symbolic objects play two different roles: the referent (i.e., intent), which is the symbolic description of a class of individuals, and the subject, which is an individual. The use of the DISS module of SODAS2 for the computation of dissimilarity measures for both boolean and modal symbolic objects is illustrated, together with the VDISS module for the visualization of the dissimilarities by means of bidimensional scatterplots and line graphs. An explanation of the outputs and the results of the MATCH module for the computation of the matching functions completes the chapter.

The next chapters of Part II are devoted to clustering methods applied to symbolic descriptions. Chapter 9, by Rasson, Pirçon, Lallemand and Adans, is devoted to unsupervised divisive classification and associated with the SCLASS module of SODAS2. By definition of a divisive clustering method, the algorithm starts with all individuals described by interval data, in one larger cluster, and successively splits each cluster into (two) smaller ones until a suitable stopping rule prevents further divisions. The original contribution of this method lies in the way nodes are split. Indeed, the cut is based only on the assumption that the distributions of points can be modelled by a non-homogeneous Poisson process, where the intensity will be estimated by the kernel method. The cut will then be made so as to maximize the likelihood function. After the tree-growing algorithm and the pruning procedure, the final clustering tree is obtained. The nodes of the tree represent the binary questions selected by the algorithm and the k leaves of the tree define the k-partition. Each cluster is characterized by a rule, that is, the path in the tree which provided it. The clusters therefore become new symbolic objects defined according to the binary questions leading from the root to the corresponding leaves. The chapter ends with an application which illustrates these methods.

Chapter 10, by Brito and de Carvalho, is devoted to hierarchical and pyramidal clustering of symbolic descriptions. It starts by recalling hierarchical and pyramidal models. It then describes how these models are adapted to a symbolic clustering method, which allows for the formation of self-interpreted clusters by a symbolic description. The corresponding algorithms have an aggregation step where generalization is performed variable-wise and the procedure, including the computation of generality measures, is detailed for different types of variables. The problem of symbolic clustering according to these models in the presence of hierarchical rules, for both categorical and modal data, is also addressed. The HIPYR module of SODAS2 is presented and its use explained. The chapter ends with an application which illustrates these methods.

Chapter 11, by de Carvalho, Lechevallier and Verde, is concerned with clustering methods in symbolic data analysis. It deals with methods for clustering large sets of individuals described by symbolic data into a predefined or optimized (as shown in Chapter 13) number of homogeneous classes. The proposed partitioning algorithms are based on a generalization of the classical 'dynamical clusters method' (*nuées dynamiques*). The general optimized criterion is a measure of the best fit between the partition and the representation of the classes. The clusters obtained are suitably interpreted and represented by generalized prototypes. A prototype is a symbolic description model of representation of a class, and it is an element of the same description space of the individuals to be clustered. The allocation function for the assignment of an individual to a class depends on the nature of the variables which defines the symbolic description. This chapter contains two partitioning methods – symbolic clustering (SCLUST) and clustering on dissimilarity data tables (DCLUST) – and finally a clustering interpretative aid.

Chapter 12, by Bock, considers a (typically very large) collection of individuals, each described by a multi-dimensional interval-type data vector. It presents a method for visualizing these data in the form of a landscape (map) where the similarity of individuals with regard to the recorded data is revealed by their spatial neighbourhood. The mapping obtained is such that individuals with a similar data-specific behaviour are located in the same region of the map and their properties can be visualized, for example, by region-specific zoom stars, or labels. In this way, the user can easily identify groups of similar objects and identify the internal structure of the data set. The method is a generalization of the classical construction process for Kohonen maps: the individuals, and thus also the interval data vectors, are arranged into a number of homogeneous, relatively small-sized clusters which are represented by cluster prototypes. These cluster prototypes are then suitably assigned to the vertices of a square grid on the screen ('neurons') such that similar clusters are assigned to neighbouring vertices. By choosing the number of rows and columns of the grid, the user can control the number of vertices, that is, the number of clusters and thus the degree of resolution in the picture. Finally, the data-specific properties of the clusters and their prototypes are visualized by clicking on the corresponding vertex on the screen. This allows the user to see the properties of the objects and clusters in a selected region of the map in the form of zoom stars, bar diagrams, etc., and to compare the cluster-specific behaviour of different clusters and in different regions of the map (e.g., by simultaneously visualizing all zoom stars). Finally, the resulting grouping of objects and the file of cluster prototypes can be used as input to many other SODAS2 modules for further analysis.

Chapter 13, by Hardy, is devoted to the determination of the number of clusters. Methods for the determination of the number of clusters are applied to hierarchies produced by four symbolic hierarchical clustering methods, and to sets of partitions given by the symbolic clustering procedure SCLUST of Chapter 11. Artificial and real data sets are analysed. Two hypothesis tests for the number of clusters are considered. These are based on the hypervolumes clustering criterion: the hypervolumes test and the gap test. These statistical methods are based on the assumption that the observed points are generated by a homogeneous Poisson process (Karr, 1991) in k disjoint convex sets. The five best stopping rules for the number of clusters analysed by Milligan and Cooper (1985) are considered. Hardy shows how these methods can be extended in order to be applied to units described by interval, multi-valued and modal variables. Three criteria are used in SODAS2 for the SCLUST module and five criteria are associated with the four symbolic hierarchical clustering methods.

Chapter 14, by Bertrand and Bel Mufti, is the last of the chapters on clusterings. It is concerned with stability measures for assessing a partition and its clusters (Bertrand and Bel Mufti, 2006). These stability measures assess either a cluster or a partition, with respect to one or both of the following criteria: cluster isolation and cluster cohesion. Informally speaking, the aim of each of these stability measures is to estimate the reproducibility of clustering results after removal of a few individuals from the class of individuals to be partitioned. This motivation is justified because cluster stability is generally intended to hold if and only if membership of the clusters is not affected by small perturbations of the data set. In the case of symbolic data, it is often appropriate to identify the individuals that are intermediate between two or more clusters: a procedure is proposed to this end. Several recommendations are then given on how to interpret the numerical values taken by the stability measures. Finally, this approach is illustrated on two symbolic data sets. The stability criterion is used in the SCLUST module.

In the final two chapters of Part II attention shifts to factor analysis methods extended to symbolic descriptions. Chapter 15, by Lauro, Verde and Irpino, is concerned with symbolic principal component analysis, which extends classical principal component analysis to individuals described by interval variables. Principal component analysis aims to visualize, synthesize and compare units on factor spaces with a minimum loss of information. Symbolic principal component analysis aims to look for the best representation of a set of individuals described by interval variables on a factor plane. On the other hand, symbolic objects described by interval-valued variables, represented as a hypercube in a multidimensional space, need to be visualized, synthesized and compared on factor spaces, bearing in mind not only their location but also their size and shape.

Chapter 16, again by Lauro, Verde and Irpino, is devoted to generalized canonical analysis extended to symbolic data. It looks for the most suitable factor axes in order to study the relationships between individuals and symbolic variables on a Cartesian plane. The analysis is performed on a coding matrix of the symbolic descriptions where each row identifies the coordinates of the vertices of the hypercube associated with each symbolic description in the original representation space. As for symbolic principal component analysis, in this analysis a cohesion constraint is considered in order to preserve the uniqueness of the symbolic descriptions in the analysis. The optimized criterion is an additive criterion and each component expresses the discriminatory power of each symbolic descriptor. The generalized canonical analysis of symbolic objects can be considered as a general factor analysis procedure. For instance, it has been used in the intermediate, quantification step of factor discriminant analysis (see Chapter 18), as well as in a particular case of multiple correspondence analysis when all the individuals are described by categorical multi-valued variables.

Part III moves on to supervised methods. Chapters 17, by Rasson, Lallemand and Adans, is devoted to the Bayesian decision tree. It is based on a tree-growing algorithm which explicitly treats interval symbolic variables, and its original contribution lies in the way nodes are split. Briefly, the Bayesian decision tree, based on density estimation, aims to classify new objects into one class of a prior partition. Each split is carried out selecting the best discriminant variable, that is, the one which leads to the purest nodes, and the classification step is performed in accordance with the Bayesian decision rule using kernel density estimation. This description requires a class-specific density estimate and an optimal choice of the corresponding window bandwidths. Suitable prior probabilities are also computed. The proposed method gives rules in order to classify the new objects. It is associated with the SBTREE module of SODAS2.

Chapter 18, a third contribution by Lauro, Verde and Irpino, is devoted to factor discriminant analysis. On individuals described by symbolic data, this allows suitable geometrical classification rules to be defined in order to classify individuals into a priori classes. The SFDA (symbolic factor discriminant analysis) module of SODAS2 looks for the best discriminant subspace as a linear combination of the predictors. As a first step, SFDA performs a selection of the predictors according their discrimination power. Then, because the predictors of SFDA can be both quantitative and categorical variables, a quantification step is performed in order to homogenize such variables. Generalized canonical analysis on symbolic data (SGCA) has been proposed as a quantification procedure for the descriptors coding in the SFDA. Furthermore, the addictiveness of the optimized criterion in SGCA makes it possible to get a suitable selection of the best discriminant predictors. Finally, a classical factor discriminant analysis is carried out on the new factors, obtained by the quantification process. The SGCA and SFDA modules in SODAS2 accept as input any kind of variable (single-valued quantitative, interval, single-valued categorical, multi-valued categorical and modal).

The purpose of Chapter 19, by Afonso, Billard, Diday and Limam, is to propose ways to adapt classical ideas in regression analysis to enable regression analysis on symbolic data. It is associated with the SREG module in SODAS2. The present work considers how to implement regression analysis in the presence of interval, modal, taxonomic and hierarchically dependent variables. The method is described and tested on a data set simulated from real statistics.

Chapter 20, by Rossi and Conan-Guez, is concerned with multi-layer perceptrons and symbolic data. In some real-world situations, linear models are not sufficient to accurately represent complex relations between input variables and output variables of a system of interest. Multi-layer perceptrons are one of the most successful non-linear regression tools but they are unfortunately restricted to inputs and outputs that belong to a normalized vector space. In this chapter, a general recoding method is presented that allows use of symbolic data both as inputs and outputs to multi-layer perceptrons. The recoding is quite simple to implement and yet provides a flexible framework that allows the analyst to deal with almost all practical cases.

Part IV presents some applications and closes with an overview of SODAS2. In Chapter 21, by Mustjärvi and Laaksonen, the data are derived from the European Social Survey (ESS), a harmonized, academically driven survey carried out in late 2002 and early 2003 in 22 countries. For this application only the Portuguese, Spanish and Finnish data are used. The aim is to compare a selection of political opinions of Finnish, Spanish and Portuguese citizens. The symbolic data files are created in different ways in order to compare the impact of these. The categories of the first symbolic data table studied are gender and residential area (urban versus rural). The maximum number of symbolic objects for the three countries is $2 \times 2 \times 3 = 12$. The second table also contains three different background variables – gender, employment status and age group ($2 \times 3 \times 3 = 18$) – giving $3 \times 18 = 54$ symbolic objects. Other alternatives for analysing the data are also discussed. The core of the chapter includes an application of two symbolic data analysis methods – divisive classification, and hierarchical and pyramidal clustering. Overall the results show clearly that Finns differ substantially from the Spanish and Portuguese. It is shown that symbolic data analysis has several advantages in this kind of analysis.

Chapter 22, by Laaksonen, again uses the ESS data, but now the aim is to study people's life values and trust components in 20 of the 22 survey countries. The initial micro-data were mainly collected by face-to-face questionnaires. Two sets of variables were thus selected. The first set consists of variables that measure people's trust in their country's government and legal system, and the second concerns Schwartz's life value factors. The 20 countries are the 'concepts', described by symbolic data after an explained generalization process. Then several symbolic data analysis methods are conducted on these data. For example, Poland, Israel, France and Sweden are compared on four symbolic variables, producing interesting, opposite results. Other interesting results such as a symbolic hierarchical clustering of the 20 countries give a nice overview of the thinking in these countries.

Chapter 23, by Mas and Olaeta, studies a time use survey. This survey provides very valuable information on what people do with their time, what proportion of time is spent on economically productive activities, leisure pursuits, personal care or household-related activities in the Basque country. The study of diverse time use patterns between males and females, older and younger generations and other different socio-economic groups is important not only for policy development and programmed planning but also for product and service producers. This chapter focuses on differences in time spent on housework activities and applies symbolic data analysis, and more particularly symbolic visualization and pyramidal clustering, in order to process, filter and identify subpopulations with significant differences in housework behaviour.

Chapter 24, by de Baenst and Lechevallier, is an overview of SODAS2. It shows the facilities of the software: its general human–computer interface, the input and output, with special attention to the different types of (input and output) data, the types of treatments, the proposed visualizations for the interpretation of results, how data can be exported, and the integration of metadata. It then briefly describes the methodological organization of SODAS2 based on a modular approach offering easy access to software components. The components, called modules, are of three types: data management, statistical treatment and visualization. The treatment modules are assembled according to their scientific assessment in generic methods.

1.9 Conclusion: past, present and future of symbolic data analysis

Symbolic data analysis is based on three main ideas. First, Aristotle (1994 Vrin edition), four centuries before the birth of Christ, clearly defined two levels of objects, a first-order level ('this man' or 'this horse') and a second-order level ('man' in general or the 'horse' in general). Second, Arnault and Nicole (1662) considered these second-order objects as 'concepts' defined by an 'intent' and an 'extent'. The third idea originates from Schweizer (1984), who says that 'distributions are the number of the future'.

In this book, from the first idea, our aim is to extend statistics, exploratory data analysis and data mining from first-order objects to second-order objects. From the second idea, the second-order objects are considered as concepts whose intent and whose extent calculation method are modelled by 'symbolic objects'. From the third idea, new variables whose value for each concept is a random variable defined on the lower-level units can be defined. These random variables take account of the internal variation among the units of the extent of each concept. In practice, it is not possible to store the vast amount of information contained in the random variables but it is feasible to store their distribution, as is usefully done by the national statistical institutes. Therefore, in a symbolic data table, a cell can contain a distribution, or interval, or several values, sometimes weighted, and sometimes linked by taxonomies, logical rules, and copula models in order to reduce the lost information. The need to extend standard data analysis methods (exploratory, clustering, factor analysis, discrimination, . . .) to symbolic data tables is increasing in order to get more accurate information and to summarize extensive data sets contained in databases.

In practice, any standard data analysis or standard statistical study can be enhanced by a complementary symbolic data analysis as it is always possible to find classes, categories or concepts related to these data (even if all the variables are numerical, it is always possible to transform them into categories). These classes, categories or concepts considered as 'atoms' of knowledge to be studied as higher-level units need new kinds of descriptions based on 'symbolic data' for which new tools have to be created. Symbolic data analysis not only renews the methodology of statistics, multidimensional data analysis, data mining and their associated computer tools, but also fundamentally modifies our way of thinking about these domains. In these directions much has already been done, but there remains much to do. For example, the proceedings of recent conferences of the International Federation of Classification Societies (Kiers *et al.*, 2000; Jajuga *et al.*, 2002; Banks *et al.*, 2004; Batagelj *et al.*, 2006), published by Springer-Verlag in the 'Studies in Classification, Data Analysis, and Knowledge Organization' series, contain sections with numerous papers on SDA tools and theory. Recent papers can also be found in the electronic *Journal of Symbolic Data Analysis*, edited by R. Verde and Y. Lechevallier, at www.jsda.unina2.it/newjsda/volumes/index.htm.

In the future, from the theoretical point of view, many mathematical results on reducing the information lost by the generalization process used from the first-level to the second-level units (by copula models, rules, taxonomies), the learning of the structure of symbolic objects (by lattices, non-linear pyramids or hierarchies) and their robustness, quality, validity, reliability, etc. have to be obtained. Much remains also to be done by extending statistics, multidimensional data analysis and data mining to symbolic data, for example, in time series, multidimensional scaling, textual data, sequential and stochastic classification, grid distributed data mining, spatial classification, rule extraction, etc., extended to symbolic data.

This book describes the methods and software that have been added to the SODAS2 package since the book prepared by Bock and Diday (2000). A pedagogical volume by Billard and Diday (2006) describes and illustrates in detail the statistical methodology of the symbolic data analysis framework. In that sense all three books complement each other.

References

Afonso, F. and Diday, E. (2005) Extension de l'algorithme Apriori et des règles d'association aux cas des données symboliques diagrammes et intervalles. In *Revue RNTI, Extraction et Gestion des Connaissances* (EGC 2005), Vol. 1, pp 205–210, Toulouse: Cépaduès.

Afonso, F., Billard, L. and Diday, E. (2004) Symbolic linear regression with taxonomies. In D. Banks, L. House, F.R. McMorris, P. Arabie and W. Gaul (eds), *Classification, Clustering, and Data Mining Applications*, pp. 429–437. Berlin: Springer-Verlag

Appice, A., D'Amato, C., Esposito, F. and Malerba, D. (2006) Classification of symbolic objects: a lazy learning approach. *Intelligent Data Analysis*, 10(4): 301–324.

Aristotle (1994) *Organon*, Vol. I: *Catégories*, Vol. II: *De l'interprétation*. Paris: Vrin.

Arnault, A. and Nicole P. (1965) *La Logique ou l'art de penser*. Reprinted by Froman, Stuttgart, 1965.

Bandemer, H. and Nather, W. (1992) *Fuzzy Data Analysis*. Dordrecht: Kluwer.

Banks, D., House, L., McMorris, F.R., Arabie, P. and Gaul, W. (eds) (2004) *Classification, Clustering, and Data Mining Applications*. Berlin: Springer-Verlag.

Batagelj, V., Bock, H.-H., Ferligoj, A. and Ziberna, A. (eds) (2006) *Data Science and Classification*. Berlin: Springer-Verlag.

Bertrand, P. and Bel Mufti, G. (2006) Loevinger's measures of rule quality for assessing cluster stability. *Computational Statistics and Data Analysis*, 50(4): 992–1015.

Bertrand, P. and Goupil, F. (2000) Descriptive statistics for symbolic data. In H.-H. Bock and E. Diday (eds), *Analysis of Symbolic Data: Exploratory Methods for Extracting Statistical Information from Complex Data*. Berlin: Springer-Verlag.

Benzécri, J.P. (ed.) (1973) *L'Analyse de données*, Vol. 1: *La Taxinomie*, Vol. 2: *L'Analyse des correspondances*. Paris: Dunod.

Bezerra, B.L.D. and de Carvalho, F.A.T. (2004) A symbolic approach for content-based information filtering. *Information Processing Letters*, 92(1): 45–52.

Billard, L. (2004) Dependencies in bivariate interval-valued symbolic data. In D. Banks, L. House, F.R. McMorris, P. Arabie and W. Gaul (eds), *Classification, Clustering, and Data Mining Applications*, pp. 319–354. Berlin: Springer-Verlag.

Billard, L. and Diday, E. (2003) From the statistics of data to the statistics of knowledge: symbolic data analysis. *Journal of the American Statistical Association*, 98: 470–487.

Billard, L. and Diday, E. (2006) *Symbolic Data Analysis: Conceptual Statistics and Data Mining*. Chichester: Wiley.

Bock, H.-H. (2005) Optimization in symbolic data analysis: dissimilarities, class centers, and clustering. In D. Baier, R. Decker and L. Schmidt-Thieme (eds), *Data Analysis and Decision Support*, pp. 3–10. Berlin: Springer-Verlag.

Bock, H.-H. and Diday, E. (2000) *Analysis of Symbolic Data: Exploratory Methods for Extracting Statistical Information from Complex Data*. Berlin: Springer-Verlag.

Bravo, M.C. and García-Santesmases, J.M. (2000) Symbolic object description of strata by segmentation trees. *Computational Statistics*, 15: 13–24.

Bravo Llatas M.C. (2004) *Análisis de segmentación en el análisis de datos simbólicos*. Madrid: Universidad Complutense de Madrid. Servicio de Publicaciones. http://www.ucm.es/BUCM/tesis/mat/ucm-t25329.pdf (accessed June 2007).

Brito, P. (1991) Analyse des données symboliques. Pyramides d'héritage. Doctoral thesis, Université Paris IX Dauphine, France.

Brito, P. (1994) Order structure of symbolic assertion objects. *IEEE Transactions on Knowledge and Data Engineering*, 6(5), 830–835.

Brito, P. (2002) Hierarchical and pyramidal clustering for symbolic data. *Journal of the Japanese Society of Computational Statistics*, 15(2): 231–244.

Brito, P. and Polaillon, G. (2005) Structuring probabilistic data by Galois lattices. *Mathématiques et Sciences Humaines*, 169(1): 77–104.

Caruso, C., Malerba, D. and Papagni, D. (2005) Learning the daily model of network traffic. In M.S. Hacid, N.V. Murray, Z.W. Ras and S. Tsumoto (eds), *Foundations of Intelligent Systems*, Lecture Notes in Artificial Intelligence 3488, pp. 131–141. Berlin: Springer-Verlag.

Cazes, P., Chouakria, A., Diday, E. and Schektman, Y. (1997) Extension de l'analyse en composantes principales à des données de type intervalle, *Revue de Statistique Appliquée*, XIV(3): 5–24.

Ciampi, A., Diday, E., Lebbe, J., Perinel, E. and Vignes, R. (2000) Growing a tree classifier with imprecise data. *Pattern Recognition Letters*, 21: 787–803.

Cuvelier, E. and Noirhomme-Fraiture, M. (2005) Clayton copula and mixture decomposition. In J. Janssen and P. Lenca (eds), *Applied Stochastic Models and Data Analysis* (ASMDA 2005), Brest, France, 17–20 May, pp. 699–708.

da Silva, A., de Carvalho, F., Lechevallier, Y. and Trousse, B. (2006) Mining web usage data for discovering navigation clusters. Paper presented at the XIth IEEE Symposium on Computers and Communications (ISCC 2006), Pula-Cagliari, Italy.

de Carvalho, F.A.T. (1995) Histograms in symbolic data analysis. *Annals of Operations Research*, 55(2): 229–322.

de Carvalho, F.A.T., Lima Neto, E. de A. and Tenerio, C.P. (2004) A new method to fit a linear regression model for interval-valued data. In S. Biundo, T. Frühwirth and G. Palm (eds), *KI 2004: Advances in Artificial Intelligence*, pp. 295–306. Berlin: Springer-Verlag.

de Carvalho, F.A.T., de Souza, R., Chavent, M. and Lechevallier, Y. (2006a) Adaptive Hausdorff distances and dynamic clustering of symbolic interval data. *Pattern Recognition Letters*, 27(3): 167–179.

de Carvalho, F.A.T., Brito, P. and Bock H.-H. (2006b) Dynamic clustering for interval data based on L_2 distance. *Computational Statistics*, 21(2): 231–250.

de Souza, R.M.C.R. and de Carvalho, F.A.T. (2004) Clustering of interval data based on city-block distances. *Pattern Recognition Letters*, 25(3): 353–365.

Diday, E. (1987a) The symbolic approach in clustering and related methods of Data Analysis. In H.-H. Bock (ed.), *Classification and Related Methods of Data Analysis*. Amsterdam: North-Holland.

Diday, E. (1987b) Introduction à l'approche symbolique en analyse des données. Première Journées Symbolique-Numérique. Université Paris IX Dauphine, December.

Diday, E. (1989) Introduction à l'analyse des données symboliques. INRIA Research Report 1074, August.

Diday, E. (1991) Des objets de l'analyse des données à ceux de l'analyse des connaissances. In Y. Kodratoff and E. Diday (eds), *Induction symbolique et numérique à partir de données*. Toulouse: Cépaduès.

Diday, E. (1995) Probabilist, possibilist and belief objects for knowledge analysis. *Annals of Operations Research*, 55: 227–276.

Diday, E. (2000a) L'Analyse des données symboliques: un cadre théorique et des outils pour le data mining. In E. Diday, Y. Kodratoff, P. Brito and M. Moulet, *Induction symbolique numérique à partir de données*. Toulouse: Cépaduès.

Diday, E. (2000b) Symbolic data analysis and the SODAS project: purpose, history, perspective. In H.-H. Bock and E. Diday (eds), *Analysis of Symbolic Data: Exploratory Methods for Extracting Statistical Information from Complex Data*. Berlin: Springer-Verlag.

Diday, E. (2002a) An introduction to symbolic data analysis and the SODAS software. *Journal of Symbolic Data Analysis*, 0(0).

Diday, E. (2002b) Mixture decomposition of distributions by copulas. In K. Jajuga, A. Sokolowski and H.-H. Bock (eds), *Classification, Clustering, and Data Analysis: Recent Advances and Applications*. Berlin: Springer-Verlag.

Diday, E. (2004) Spatial pyramidal clustering based on a tessellation. In D. Banks, L. House, F.R. McMorris, P. Arabie, and W. Gaul (eds), *Classification, Clustering, and Data Mining Applications*. Berlin: Springer-Verlag.

Diday, E. (2005) Categorization in symbolic data analysis. In H. Cohen and C. Lefebvre (eds), *Handbook of Categorization in Cognitive Science*. Amsterdam: Elsevier.

Diday E. and Emilion R. (1996) Lattices and capacities in analysis of probabilist objects. In E. Diday, Y. Lechevallier and O. Opitz (eds), *Ordinal and Symbolic Data Analysis*. Berlin: Springer-Verlag.
Diday, E. and Emilion, R. (1997) Treillis de Galois maximaux et capacités de Choquet. *Comptes Rendus de l'Académie des Sciences, Série I*, 325: 261–266.
Diday, E. and Emilion, R. (2003) Maximal and stochastic Galois lattices. *Journal of Discrete Applied Mathematics*, 127: 271–284.
Diday, E. and Esposito, F. (2003) An introduction to symbolic data analysis and the SODAS software. *Intelligent Data Analysis*, 7(6): 583–601.
Diday, E. and Murty, N. (2005) Symbolic data clustering. In J. Wang (ed.), *Encyclopedia of Data Warehousing and Mining*. Hershey, PA: Idea Group.
Diday, E. and Vrac, M. (2005) Mixture decomposition of distributions by copulas in the symbolic data analysis framework. *Discrete Applied Mathematics*, 147(1): 27–41.
Diday, E., Lemaire, J., Pouget, J. and Testu, G. (1984) *Eléments d'analyse des données*. Paris: Dunod.
Duarte Silva, A.P. and Brito, P. (2006) Linear discriminant analysis for interval data. *Computational Statistics*, 21(2): 289–308.
Dubois, D. and Prade, H. (1988) *Possibility Theory*. New York: Plenum.
Esposito, F., Malerba, D. and Semeraro, G. (1991) Classification of incomplete structural descriptions using a probabilist distance measure. In E. Diday and Y. Lechevallier (eds), *Symbolic-Numeric Data Analysis and Learning*, pp. 469–482. New York: Nova Science.
Esposito, F., Malerba, D. and Semeraro, G. (1992) Classification in noisy environments using a distance measure between structural symbolic descriptions. *IEEE Transactions on Pattern Analysis and Machine Intelligence*, 14(3): 390–402.
Gioia, F. and Lauro, N.C. (2006a) Dependence and interdependence analysis for interval-valued variables. In V. Batagelj, H.-H. Bock, A. Ferligoj and A. Ziberna (eds) *Data Science and Classification*. Berlin: Springer-Verlag.
Gioia, F. and Lauro, N.C. (2006b) Principal component analysis on interval data. *Computational Statistics*, 21(2): 343–363.
Gowda, K.C. and Diday, E. (1992) Symbolic clustering using a new similarity measure. *IEEE Transactions on Systems, Man and Cybernetics*, 22(2): 368–378.
Hardy, A. (2004) Les méthodes de classification et de determination du nombre de classes: du classique au symbolique, In M. Chavent, O. Dordan, C. Lacomblez, M. Langlais and B. Patouille (eds), *Comptes Rendus des Onzièmes Rencontres de la Société Francophone de Classification*, pp. 48–55.
Hardy, A. (2005) Validation in unsupervised symbolic classification. In J. Janssen and P. Lenca (eds), *Applied Stochastic Models and Data Analysis* (ASMDA 2005), Brest, France, 17–20 May, pp. 379–386.
Hardy, A. and Lallemand, P. (2001) Application du test des hypervolumes à des données symboliques de type intervalle. In *Proceedings of EGC 2001 (Extraction et Gestion des Connaissances)*, Vol. 1, No. 4, pp. 287–292. Paris: Hermès.
Hardy, A. and Lallemand, P. (2002) Determination of the number of clusters for symbolic objects described by interval variables. In K. Jajuga, A. Sokolowski and H.-H. Bock (eds), *Classification, Clustering, and Data Analysis: Recent Advances and Applications*, pp. 311–318. Berlin: Springer-Verlag.
Hardy, A. and Lallemand, P. (2004) Clustering of symbolic objects described by multi-valued and modal variables. In D. Banks, L. House, F.R. McMorris, P. Arabie and W. Gaul (eds), *Classification, Clustering, and Data Mining Applications*, pp. 325–332. Berlin: Springer-Verlag.
Hardy, A., Lallemand, P. and Lechevallier Y. (2002) La détermination du nombre de classes pour la méthode de classification symbolique SCLUST. In *Actes des Huitièmes Rencontres de la Société Francophone de Classification*, pp. 27–31.
Irpino, A. (2006) Spaghetti PCA analysis: An extension of principal components analysis to time dependent interval data. *Pattern Recognition Letters*, 27(5): 504–513.

Irpino, A., Verde, R. and Lauro, N.C. (2003) Visualizing symbolic data by closed shapes. In M. Shader, W. Gaul and M. Vichi (eds), *Between Data Science and Applied Data Analysis*, pp. 244–251. Berlin: Springer-Verlag.

Jajuga, K., Sokolowski, A. and Bock, H.-H. (eds) (2002) *Classification, Clustering, and Data Analysis: Recent Advances and Applications*. Berlin: Springer-Verlag.

Karr, A.F. (1991) *Point Processes and Their Statistical Inference*. New York: Marcel Dekker.

Kiers, H.A.L., Rasson, J.-P., Groenen, P.J.F. and Schader, M. (eds) (2000) *Data Analysis, Classification, and Related Methods*. Berlin: Springer-Verlag.

Lauro, N.C., Verde, R. and Palumbo F. (2000) Factorial data analysis on symbolic objects under cohesion constrains. In H.A.L. Kiers, J.-P. Rasson, P.J.F. Groenen and M. Schader (eds), *Data Analysis, Classification, and Related Methods*. Berlin: Springer-Verlag.

Lebart, L., Morineau, A. and Piron M. (1995) *Statistique exploratoire multidimensionnelle*. Paris: Dunod.

Limam, M., Diday, E. and Winsberg, S. (2003) Symbolic class description with interval data. *Journal of Symbolic Data Analysis*, 1(1).

Malerba, D., Esposito, F. and Monopoli M. (2002) Comparing dissimilarity measures for probabilistic symbolic objects. In A. Zanasi, C.A. Brebbia, N.F.F. Ebecken and P. Melli (eds), *Data Mining III, Series Management Information Systems*, Vol. 6, pp. 31–40. Southampton: WIT Press.

Mballo, C., Asseraf, M. and Diday E. (2004) Binary tree for interval and taxonomic variables. *Students*, 5(1): 13–28.

Meneses, E. and Rodríguez-Rojas O. (2006) Using symbolic objects to cluster web documents. In *Proceedings of the 15th International Conference on the World Wide Web*, pp. 967–968. New York: ACM Press,

Milligan, G.W. and Cooper, M.C. (1985) An examination of procedures for determining the number of clusters in a data set. *Psychometrika*, 50: 159–179.

Noirhomme-Fraiture, M. (2002) Visualization of large data sets: the zoom star solution. *Journal of Symbolic Data Analysis*, 0(0).

Pollaillon, G. (1998) Organisation et interprétation par les treillis de Galois de données de type multivalué, intervalle ou histogramme. Doctoral thesis, Université Paris IX Dauphine, France.

Prudêncio, R.B.C., Ludermir T., de Carvalho F.A.T. (2004) A modal symbolic classifier for selecting time series models. *Pattern Recognition Letters*, 25(8): 911–921.

Saporta, G. (2006) *Probabilités, analyse des données et statistique*. Paris: Editions Technip.

Schweizer, B. (1984) Distributions are the numbers of the future. In A. di Nola and A. Ventre (eds), *Proceedings of the Mathematics of Fuzzy Systems Meeting*, pp. 137–149. Naples: University of Naples.

Schweizer, B. and Sklar, A. (2005) *Probabilistic Metric Spaces*. Mineola, NY: Dover.

Soule, A., Salamatian, K., Taft, N., Emilion, R. and Papagiannaki, K. (2004) Flow classfication by histograms: or how to go on safari in the internet. *ACM SIGMETRICS Performance Evaluation Review*, 32(1): 49–60.

Stéphan V. (1998) Construction d'objets symboliques par synthèse des résultats de requêtes. Doctoral thesis, Paris IX Dauphine University.

Tukey, J.W. (1958) *Exploratory Data Analysis*. Reading, MA: Addison Wesley.

Vrac, M., Diday, E. and Chédin, A. (2004) Décomposition de mélange de distributions et application à des données climatiques. *Revue de Statistique Appliquée*, LII (1): 67–96.

Zadeh, L.A. (1978) Fuzzy sets. *Information and Control*, 8: 338–353.

Part I

DATABASES VERSUS SYMBOLIC OBJECTS

2

Improved generation of symbolic objects from relational databases

Yves Lechevallier, Aicha El Golli and George Hébrail

2.1 Introduction

The goal of statistical data analysis is to provide synthesis of large amounts of data or to use them to make decisions. The classical way of doing so is to process directly the detailed raw data which are almost always available. Another approach is to proceed in two steps: a first step transforms the detailed data into summarized descriptions (describing detailed data at a higher level), and a second step is to analyse the summarized descriptions. This approach is very interesting because the analyst can concentrate his/her activity on the core of the analysis, dealing with high-level objects, without being swamped by the large amounts of detailed data. This approach requires the ability (1) to construct these high-level summarized objects and (2) to process such high-level objects. The symbolic data analysis framework (see Bock and Diday, 2000) addresses point (2) by providing a formalism to describe high-level objects (called 'assertions') having the capability to describe groups of individuals along with their variations within each group. Moreover, standard methods of statistical data analysis have been extended to treat assertions so that it is possible to use almost every method of data analysis on high-level objects. As for point (1), Stéphan *et al.* (2000) have developed an approach (and a related computer program called DB2SO) which can extract raw detailed data from relational databases and build high-level summarized objects using a supervised generalization process.

It is known that generalization is a difficult problem both in the case of supervised generalization and in the case of unsupervised generalization. The approach developed in Stéphan *et al.* (2000) was a supervised generalization approach. The supervised approach is interesting because at the beginning of the process the end user defines the semantics associated with the high-level objects. The drawback of this approach is that the available

Symbolic Data Analysis and the SODAS Software Edited by E. Diday and M. Noirhomme-Fraiture

detailed data may not be well separated according to the high-level objects. This problem has been observed on real data sets and a first answer has been given by the application of a refinement process after the generalization phase. The basic idea of this refinement process was to determine an optimized small set of atypical observations which were removed from the input data. Both the generalization and the refinement processes are briefly recalled in Section 2.2.

In this chapter, we propose another improvement for this generalization process. This improvement is based on the addition of an unsupervised phase to the supervised generalization process. In the preceding approach (see Stéphan *et al.*, 2000), only one assertion was generated for each group of individuals defined by the user. In the improved approach, several assertions may be obtained for each group of individuals by a decomposition process based on a clustering algorithm locally applied to each group of individuals. This approach consequently makes it easier to handle heterogeneity within groups of individuals. The chosen clustering algorithm, a divisive one, is presented in Section 2.3.

Section 2.4 shows how the divisive clustering algorithm can be integrated into the generalization process, in addition to the refinement process. Applications are given in Section 2.5 to illustrate the improvement in the generalization process. It is also shown how the new version of DB2SO software handles this decomposition process.

The concluding Section 2.6 gives some new directions to investigate on the use of symbolic objects stemming from such a decomposition.

2.2 Construction of symbolic objects by generalization

In this section, we briefly recall the process of constructing symbolic objects (SOs) from the contents of a relational database. Details can be found in Stéphan *et al.* (2000). It is assumed that the relational database stores:

- a set of individuals described by variables, which corresponds to the available detailed raw data (the variables are assumed to be single-valued but can be numerical or nominal);
- a set of groups where each group defines one SO;
- the relationship between individuals and groups (an individual belongs to only one group so that the groups form a partition of the indiduals).

Each generated description SO is the description of one group of individuals by some variables. The SOs generated are *boolean* and *modal* symbolic objects in the framework of Diday. The variables measuring the variations among the group of individuals are described by:

- the interval of observed values on individuals in the group for numerical variables;
- the list of observed values on individuals in the group for nominal variables;
- the probability distribution of observed values on individuals in the groups for nominal variables, when the user asks for a modal SO.

2.2.1 Basic generalization process

The input to the basic process is a database table (or view) containing one row (tuple) for each individual. Table 2.1 is an example of such a data input. The user has to write an SQL query which returns such a table with the following expected structure: the first column gives the individual ID, the second the group ID for the individual, and other columns represent variables (characteristics) describing individuals.

In our example we use the abalone database (Nash *et al.*, 1994).[1] Individuals are abalone. Their characteristics are their sex, their length/diameter/height, their whole weight, shucked weight, viscera weight and shell weight, and number of rings (Table 2.1). The first and last of these characteristics are used to generate groups. For example, the first row of Figure 2.1 shows group F_4-6, female abalone with 4–6 rings.

Figure 2.2 shows the SOs generated from the previous example, of which there are 24. Symbolic objects are generated using a *generalization* function to aggregate the individuals in each group:

- Numerical variables describing individuals lead to interval variables describing groups: the generalization function is here to generalize a set of values to the interval of values.
- Nominal variables describing individuals lead to either boolean or modal multi-valued variables describing groups. If the user chooses to generate a boolean multi-valued variable, the aggregation function simply constructs the list of observed values within the group. If the user chooses to generate a modal multi-valued variable, the aggregation function constructs the probability distribution of the nominal variable among individuals of the group.

Table 2.1 Characteristics of abalone and statistics for numeric domains.

Name	Data type	Mesure	Description	Min	Max	Mean	SD
Sex	nominal		M, F and I (infant)				
Length	continuous	mm	Longest shell measurement	0.075	0.815	0.524	0.120
Diameter	continuous	mm	Perpendicular to length	0.055	0.650	0.408	0.099
Height	continuous	mm	with meat in shell	0.000	1.130	0.140	0.042
Whole weight	continuous	grams	whole abalone	0.002	2.826	0.829	0.490
Shucked weight	continuous	grams	weight of meat	0.001	1.488	0.359	0.222
Viscera weight	continuous	grams	gut weight (after bleeding)	0.001	0.760	0.181	0.110
Shell weight	continuous	grams	after being dried	0.002	1.005	0.239	0.139
Ring	discrete		Number of rings	1	29	9.934	3.224

[1] See the UCI Machine Learning website, http://www.ics.uci.edu/~mlearn/MLSummary.html

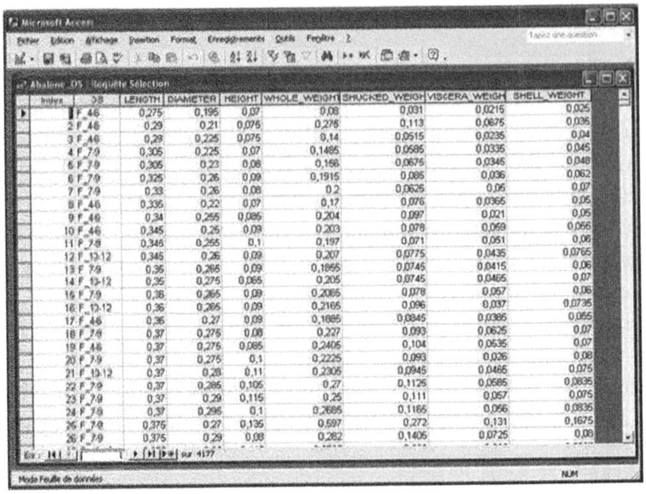

Figure 2.1 Descriptions of groups.

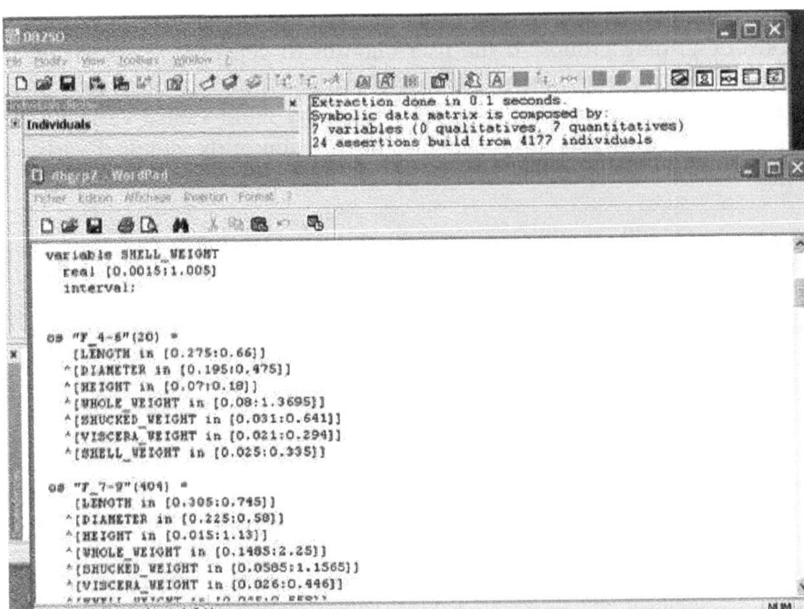

Figure 2.2 Symbolic objects generated by DB2SO.

2.2.2 Symbolic object refinement

The basic generalization functions described in the previous section may lead to group descriptions of poor quality. Since the generalization process is applied separately to each variable, one may obtain very similar descriptions for different groups, even if groups have very different contents. In particular, if groups contain atypical individuals, intervals

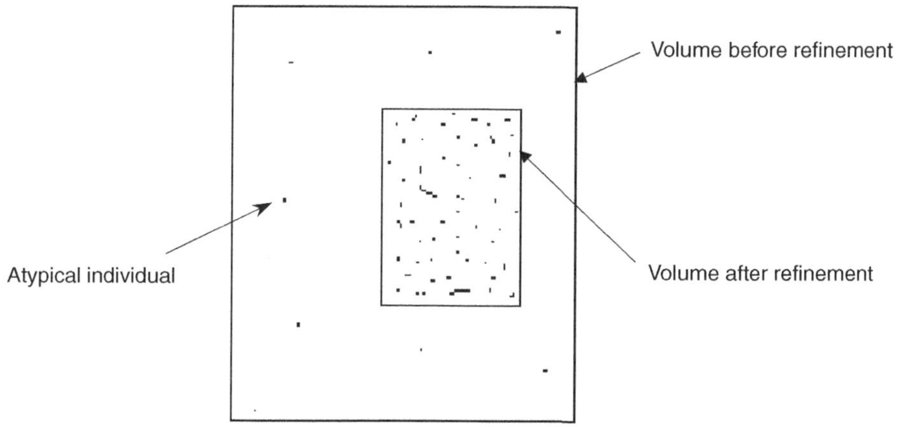

Figure 2.3 Effect of the refinement process in the case of two interval variables.

describing numerical variables may become large as well as list of values associated with nominal variables.

The idea of the refinement process is to remove atypical individuals from each group. Each SO, i.e. each group of individuals, is processed separately. With each SO can be associated a function which says whether an individual is recognized by an SO or not. In the case of interval or boolean multi-valued variables, this function says 'yes' if individual values for every variable are in the interval or list of values of the corresponding symbolic variables. Atypical individuals are removed and simpler SOs are constructed again from the remaining individuals. Figure 2.3 illustrates this refinement process in the case of two interval variables.

The choice of atypical individuals to remove is guided by an optimization criterion. The following optimization constraints are used:

- A minimum threshold is defined on the number of individuals of the group still recognized by the refined SO (this threshold is specified by the user, typically 80%).
- A volume measure is associated with each SO, representing the amount of subspace its variable values occupy in the Cartesian product of variable domains. The choice of atypical individuals to be removed is made to maximize the decrease in volume associated with the refinement process.

Details of this refinement process and a description of the algorithm can be found in Stephan (1998). In particular, the refinement process can be applied to individuals described by several variables which can be either numerical, nominal, or of mixed type.

Once the refinement process has been run, the user can visualize its effect through a graphical representation like that shown in Figure 2.4. The result is presented in the form of a square matrix, in which each cell corresponds to a pair of SOs. Each cell shows:

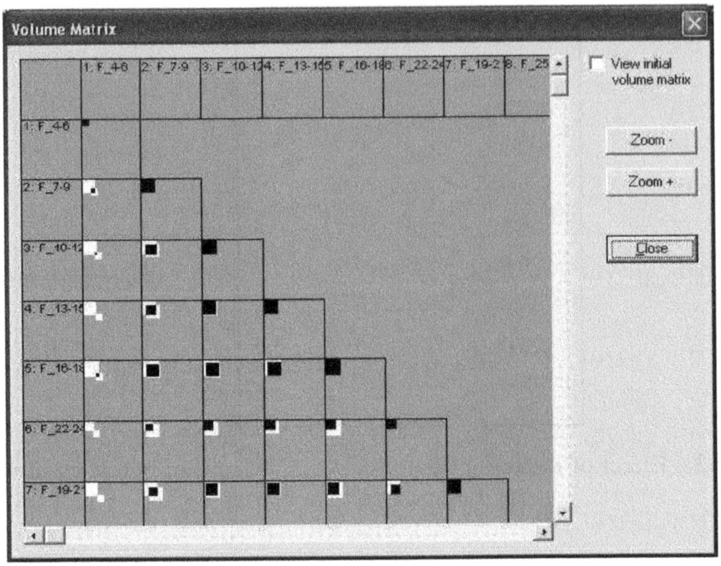

Figure 2.4 Visualization of the effect of the refinement process.

- a white square with area proportional to the volume of the corresponding row description SO;
- another white square with area proportional to the volume of the corresponding column description SO;
- a black square representing the volume of the intersection of the row and column description SOs (note that the two white squares are laid out so that their intersection is the black square).

Two dark grey lined squares show the state of this volume matrix before running the refinement process. This visualization gives an idea of the effect of the process: each description SO volume is expected to decrease as well the overlap between pairs of SOs.

2.3 Criterion-based divisive clustering algorithm

As mentioned in Section 2.1, the description generated may be of poor quality even after the refinement process. If the individuals from a group are heterogeneous, the description may cover much space where there are no individuals observed. Figure 2.5 illustrates this phenomenon in the case of two interval variables. In this example, the group defined by the user covers in reality several natural groups of individuals. The solid rectangle shows the description of the basic generalization process. The solution we propose here to solve this problem is to perform a clustering of individuals of the group and to describe the group by several descriptions of SOs, shown in the figure by dotted rectangles. The rest of this section presents the clustering algorithm we use in this process.

Divisive hierarchical clustering starts with all the objects in one large cluster, successively splitting each cluster into two smaller ones until a suitable stopping rule prevents

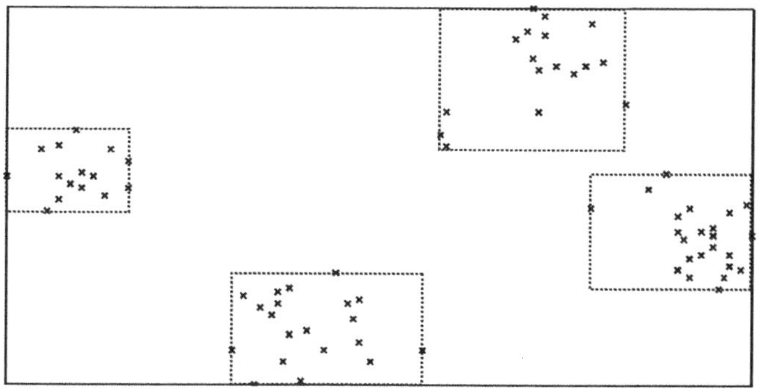

Figure 2.5 Illustration of group heterogeneity.

further divisions. On the other hand, agglomerative clustering reverses the previous process by starting with n singleton clusters, each consisting of one element of Ω, the set to be clustered, and successively merging two clusters on the basis of a similarity measure. Furthermore, the complexity of the two algorithms differs. At the first stage of the agglomerative method, we have to evaluate all the possible aggregations of 2 individuals among the n, so there are $n(n-1)/2$ possibilities. The divisive method based on the complete enumeration evaluates all the divisions of n individuals into two not empty subsets, so there are $2^{n-1}-1$ possibilities. A lot of strategies have been proposed for the divisive method in order to reduce this complexity. We propose a divisive clustering method for single-valued data, quantitative and qualitative, that reduces this complexity to $n-1$ and can be integrated into the process of generating SOs from relational databases.

In this section, we propose a divisive clustering method for single-valued data. A divisive clustering method for symbolic data (interval, multi-valued, modal) was proposed in Chavent (2000). The criterion used by Chavent to evaluate the quality of a partition is an extension of the within-cluster sum-of-squares criterion to the case of a distance matrix. In our case we work with classical data (numerical, nominal) since the algorithm is applied to individuals belonging to the groups. We propose to work with the standard within-cluster criterion based on the centre of gravity.

Let n be the number of individuals in Ω, the set to be clustered. Each individual is described by p quantitative or qualitative variables. At each stage, the division of a cluster is carried out according to the within-cluster inertia criterion. This criterion is minimized among bipartitions induced by a set of binary questions. In fact, we wish to find a bipartition (C_1, C_2) of C such that the within-cluster inertia is minimized. In the classical approach, one chooses the optimal bipartition (C_1, C_2) among the $2^{n-1}-1$ possible bipartitions. It is clear that the number of computations needed when n is large will be prohibitive. In the Chavent (2000) approach, to reduce the complexity, C is divided according to a binary question (Breiman *et al.*, 1984) of the form:

- $Y_j \leq c$ when $Y_j : \Omega \rightarrow R$ is a real variable and $c \in R$ is called the cut point;
- $Y_j \in \{m_i, \ldots, m_k\}$ when Y_j is a qualitative variable, $\{m_i, \ldots, m_k\}$ is a subset of categories of Y_j called the cut subset.

The complexity of our algorithm depends on the total number of possible bipartitions and thus on the type of variable:

- If the variable Y_j is quantitative, we evaluate at most $n-1$ different bipartitions. Indeed, whatever the cut point c between two consecutive observations $Y_j(x_i)$ may be, the bipartition induced is the same. In order to ask only $n-1$ questions to generate all these bipartitions, we decide to use the middle point of two consecutive observations $Y_j(x_i) \in R$. If there are p variables, we choose among the $p(n-1)$ corresponding bipartitions (C_1, C_2), the bipartition having the smallest within-cluster inertia.
- If the variable is ordinal qualitative with m categories, we evaluate at most $m-1$ different bipartitions.
- If the variable is nominal qualitative with m categories, we face the problem of complexity and the number of dichotomies of the domain is equal to $2^{m-1}-1$.

Each individual is described by a vector x_i, and weighted by a real value p_i $(i = 1, \dots, n)$. Usually, the weights are equal to 1 or $1/n$. The inertia I of a cluster C_k is a homogeneity measure equal to

$$I(C_k) = \sum_{x_i \in C_k} p_i d_M^2(x_i, g_k), \tag{2.1}$$

where

$$g_k = \frac{1}{\mu(C_k)} \sum_{x_i \in C_k} p_i x_i$$

in which

$$\mu(C_k) = \sum_{x_i \in C_k} p_i$$

and p_i is the weight of x_i; d_M is

- the Euclidean distance in the case of real variables or ordinal qualitative variables, or
- the ϕ^2 distance in the case of nominal qualitative variables;

M is a symmetric positive definite matrix; and g_k is the centre of gravity of the cluster C_k.

In the quantitative/ordinal qualitative case,

$$\forall x_i \in R^p, \qquad d_M^2(x_i, g_k) = (x_i - g_k)' M (x_i - g_k). \tag{2.2}$$

g_k is the centre of gravity of the cluster C_k:

$$g_k = \frac{1}{\mu(C_k)} \sum_{x_i \in C_k} p_i x_i.$$

In the case where $M = I$,

$$d^2(x_i, g_k) = \sum_{j=1}^{p}(x_i^j - g_k^j)^2.$$

In the nominal qualitative case, the individuals are described by a set of nominal qualitative variables. A first stage allows recoding of the observed values into a binary table formed by values in $B^m = \{0, 1\}^m$ (i.e. for a variable Y_i, $Y_i(x_i^j) = 1$ when category j is observed):

$$\forall x_i \in B^m, \qquad \phi^2(x_i, g_k) = \sum_{j=1}^{m} \frac{1}{f_{.j}} \left(\frac{f_{ij}}{f_{i.}} - g_k^j \right)^2, \tag{2.3}$$

where $f_{ij} = x_i^j / n_{..}$, $f_{i.} = n_{i.}/n_{..}$, $f_{.j} = n_{.j}/n_{..}$, $n_{..} = np$ in which n is the total number of individuals, p the total number of variables, m the total number of categories and x_i^j is the value for individual i for category j in the $B = \{0, 1\}$ space. In addition,

$$n_{.j} = \sum_{i=1}^{n} x_i^j,$$

$$n_{i.} = \sum_{j=1}^{n} x_i^j = p,$$

$$g_k^j = \frac{\sum_{x_i \in C_k} f_{i.}(f_{ij}/f_{i.})}{\sum_{x_i \in C_k} f_{i.}} = \frac{x_i^j(k)}{p * n(k)},$$

where $x_i^j(k) = \sum_{x_i \in C_k} x_i^j$ and $n(k)$ is the total number of individuals in cluster C_k. The ϕ^2 distance is given by

$$\begin{aligned} \phi^2(x_i, g_k) &= \sum_{j=1}^{m} \frac{np}{n_{.j}} \left(\frac{x_i^j}{n_{i.}} - g_k^j \right)^2 \\ &= (n * p) \sum_{j=1}^{m} \frac{1}{n_{.j}} \left(\frac{x_i^j}{p} - g_k^j \right)^2. \end{aligned} \tag{2.4}$$

So the within-cluster criterion of a cluster C_k in the nominal qualitative case is

$$I(C_k) = \sum_{x_i \in C_k} f_{i.} \phi^2(x_i, g_k).$$

Given an Euclidean metric space, according to the Huygens theorem, minimizing the within-cluster inertia associated with a partition into two clusters (bipartition (C_1, C_2)) of a cluster C is equivalent to maximizing the between-cluster inertia B equal to

$$B(C_1, C_2) = \mu(C_1) d^2(g_C, g_{C_1}) + \mu(C_2) d^2(g_C, g_{C_2}).$$

In our case we use the between-cluster criterion to evaluate the bipartition. The use of this criterion allowed a considerable reduction in the complexity of calculation and thus an efficient implementation of the algorithm. The between-cluster criterion corresponds to the Ward criterion (Ward, 1963):

$$B(C_1, C_2) = \frac{\mu(C_1) * \mu(C_2)}{\mu(C_1) + \mu(C_2)} d^2(g_{C_1} \cdot g_{C_2}).$$

The evaluation criterion of a bipartition in our case corresponds to the simple distance between the centre of gravity of the two clusters C_1 and C_2 of C.

The proposed divisive clustering algorithm is as follows:

Initialization:
$P_1 = \Omega$;
$k \leftarrow 1$;
While $k < K - 1$ **do**:

1. For each cluster $C \in P_k$
   For each variable Y_j
   For each cut c, calculate the between criterion $B(C_1, C_2)$ of the bipartition (C_1, C_2) of C,

$$B(C_1, C_2) = \frac{\mu(C_1) * \mu(C_2)}{\mu(C_1) + \mu(C_2)} d^2(g_{C_1}, g_{C_2})$$

   Among all the bipartitions induced, keep the one which maximizes B
2. Choose the cluster $C \in P_k$ that maximizes

$$\Delta(C) = W(P_k) - W(P_{k+1}) = I(C) - I(C_1) - I(C_2)$$

3. $P_{k+1} = P_k \cup \{C_1, C_2\} - \{C\}$
4. $k \leftarrow k + 1$;

End While

The divisions stop after K iterations, where K is the number of clusters specified by the user.

So, first of all, among all the bipartitions we choose the one that induces the highest between-cluster inertia. Secondly, we choose to divide the cluster that induces the smallest within-cluster inertia of the partition into three clusters. These two steps are repeated until the number K of clusters fixed by the user is reached.

The method can handle missing data, which are very common in real-world applications. Many approaches have been proposed to solve this problem. Some methods propose to eliminate the individuals with missing data. Here we use the technique proposed in Jain (1988). The criterion used to choose which cluster to divide is the inertia. This criterion requires the calculation at each stage of the centre of gravity of all the induced bipartitions. The calculation of the centre of gravity is performed on all the non-missing observed values. Then, each individual having a missing value will be assigned to the cluster of the induced bipartition with the closest centre of gravity. The technique chosen to calculate the distance between a vector x_i and the centre of gravity g_k having missing values, is the following (Jain, 1988):

The distance d_j between two vectors, for a variable j, is defined by

$$d_j = \begin{cases} 0, & \text{if } x_i^j \text{ or } g_k^j \text{ is missing,} \\ x_i^j - g_k^j, & \text{otherwise.} \end{cases}$$

Then the distance between x_i and g_k is

$$d^2(x_i, g_k) = \frac{p}{p - d_0} \sum_{j=1}^{p} d_j^2,$$

with d_0 the number of missing values in x_i or g_k or both. When there are no missing values, d is the Euclidean distance.

2.4 Improving the generalization process by decomposition

The generalization process described in Section 2.2 is supervised and consequently sensitive to atypical individuals within each group and to groups containing heterogeneous individuals. As mentioned in Section 2.2, the problem of atypical individuals can be solved by the refinement process described in this same section. Heterogeneity of the groups leads to overgeneralization: the descriptions generated may cover much space which does not correspond to any observed individual. We therefore propose to integrate a decomposition step, based on the divisive clustering algorithm (presented in the preceding section), that improves the generalization process while preserving the symbolic formalism. For each group extracted the decomposition is carried out in order to obtain homogeneous clusters and to reduce the overgeneralization effect. In fact, the decomposition allows the description of each group to be reduced in an unsupervised way and to obtain homogeneous subclusters.

To evaluate the contribution of the decomposition in the process of generating symbolic data, we use the density criterion defined in Stéphan *et al.* (2000) where details can be found. In fact, we measure the quality of the description set d_i of a generalization within a group G_i, by means of a good trade-off between the homogeneity of the distribution of individuals within the description set and the covering set of d_i on G_i. To measure the parsimony of a generalization, we may use a volume criterion which yields the generality index of a description $d_i = (\delta_{i1}, \delta_{i2}, \ldots, \delta_{ip})$ of G_i:

$$\text{vol}(d_i) = \prod_{j=1}^{p} \mu(\delta_{ij}),$$

where

$$\mu(\delta_{ij}) = \begin{cases} \text{card}(\delta_{ij}), & \text{if } \; Y_j \text{ is nominal,} \\ \max(\delta_{ij}) - \min(\delta_{ij}), & \text{if } \; Y_j \text{ is quantitative or ordinal.} \end{cases}$$

We define the density of an assertion s_i as follows:

$$\text{Dens}(s_i) = \frac{|\text{ext}(s_i|G_i)|}{\text{vol}(d_i)},$$

where $|\text{ext}(s_i|G_i)|$ is the number of elements of G_i belonging to the assertion s_i.

The integration of the decomposition step to the generalization process allows the descriptions of the groups obtained by generalization to be improved and structured. We illustrate this improvement by a simplified example.

2.5 Applications

2.5.1 Ruspini data

We consider a group formed by 75 individuals described by two quantitative variables Y_1 and Y_2 (Ruspini, 1970). The generalization process produces the assertion

$$a = [Y_1 \in [2, 131]] \wedge [Y_2 \in [17, 115]]$$

and a problem of overgeneralization (Figure 2.6).

The density of this assertion is approximately 0.006. The integration of the decomposition step based on the proposed divisive method, allows a partition into two, three and four clusters to be found.

At the first stage, the method induces $2(75-1) = 148$ bipartitions (C_1, C_2). We choose the bipartition with the largest between-cluster inertia; this is induced by the binary question '$Y_1 \leq 75.5$'. Notice that the number of bipartitions has been reduced from $2^{75} - 1 = 3.78 \times 10^{22}$ to 148.

At the second stage, we have to choose whether we divide C_1 or C_2. Here, we choose C_1 and its bipartition (C_1^1, C_1^2) because $\Delta(C_1) > \Delta(C_2)$. The binary question is '$Y_2 \leq 54$'.

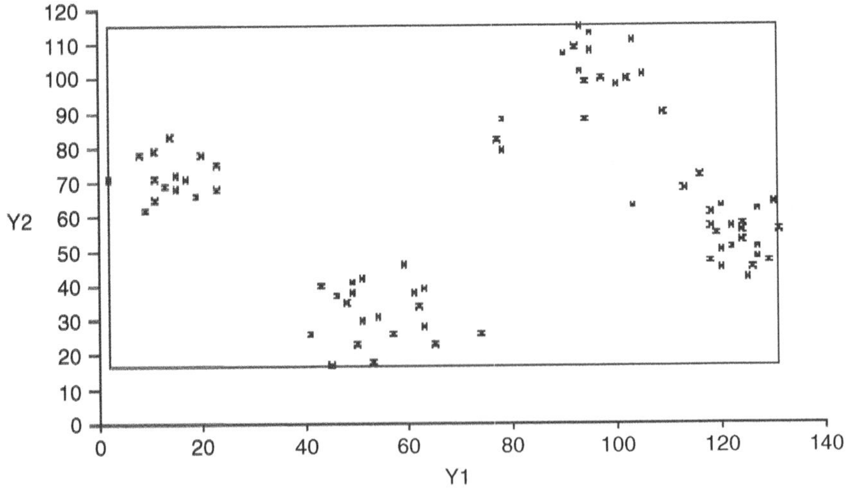

Figure 2.6 The hypercube obtained after generalization: problem of overgeneralization.

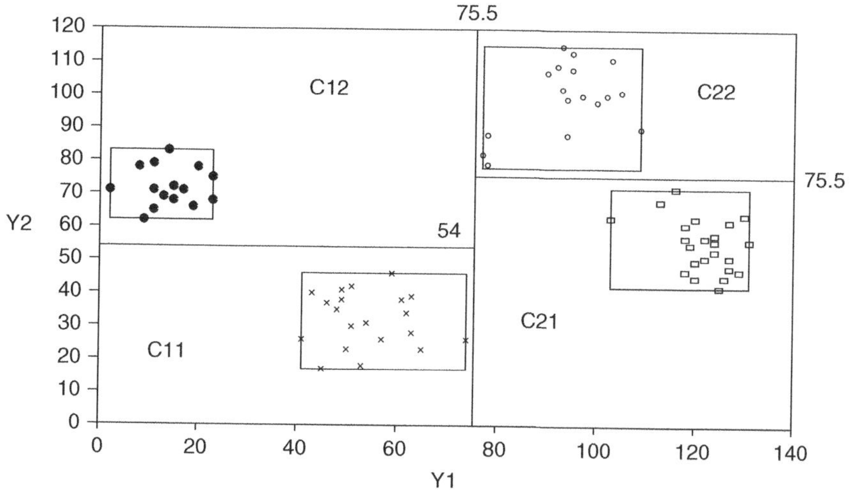

Figure 2.7 The four-cluster partition.

At the third stage, we choose the cluster C_2 and its bipartition (C_2^1, C_2^2). The binary question is '$Y_2 \leq 75.5$'. Finally, the divisive algorithm gives the four subgroups in Figure 2.7.

The generalization process produces the following disjunction of assertions:

$$[Y_1 \in [41, 74] \wedge Y_2 \in [17, 46]] \vee [Y_1 \in [2, 23] \wedge Y_2 \in [62, 83]]$$
$$\vee [Y_1 \in [103, 131] \wedge Y_2 \in [42, 72]] \vee [Y_1 \in [77, 109] \wedge Y_2 \in [79, 115]]$$

The density of this description is approximately 0.1.

2.5.2 Abalone data

We return to the example of the abalone database (see Section 2.2.1). There are 4177 cases, described by nine attributes (one nominal) and there are no missing attribute values. Consider the groups generated by sex and number of rings. We have 24 SOs generated by the generalization process described in Section 2.2.1. The description (assertion) of the second SO, for example, concerning the female abalone with 7–9 rings is the following:

os "F_7-9"(404) = [LENGTH ∈ [0.305, 0.745]
∧DIAMETER ∈ [0.225, 0.58]
∧HEIGHT ∈ [0.015, 1.13]
∧WHOLE_WEIGHT ∈ [0.1485, 2.25]
∧SHUCKED_WEIGHT ∈ [0.0585, 1.1565]
∧VISCERA_WEIGHT ∈ [0.026, 0.446]
∧SHELL_WEIGHT ∈ [0.045,0.558]]

After the decomposition step, the generalization process produces the following disjunction of 2 assertions:

os "F_7-9"(404) = [(Weight = 182)
[LENGTH ∈ [0.305,0.6]
∧[DIAMETER ∈ [0.225,0.485]
∧[HEIGHT ∈ [0.07,1.13]
∧[WHOLE_WEIGHT ∈ [0.1485, 0.8495]
∧[SHUCKED_WEIGHT ∈ [0.0585, 0.4355]
∧[VISCERA_WEIGHT ∈ [0026, 0232]
∧[SHELL_WEIGHT ∈ [0.045, 0.275]]
∨ (Weight = 222)
[LENGTH ∈ [0.485, 0.745]
∧DIAMETER ∈ [0.355, 0.58]
∧HEIGHT ∈ [0.015:0.215]
∧WHOLE_WEIGHT ∈ [0.858, 2.25]
∧SHUCKED_WEIGHT ∈ [0.333, 1.1565]
∧VISCERA_WEIGHT ∈ [0.1255, 0.446]
∧SHELL_WEIGHT ∈ [0.1825, 0.558]]]

The decomposition step performs a division of assertions, to homogenize the description of each assertion. The decomposition process decomposes each assertion automatically into a disjunction of assertions. The user has to specify the number of clusters for each assertion. In the current version, all assertions are decomposed into the same number of clusters.

2.6 Conclusion

In this chapter we have presented a method to improve the construction of symbolic objects from detailed raw data stored in relational databases. The basic approach is a supervised one. Two different improvements have been developed:

- a refinement process which removes atypical individuals in order to reduce the volume of the symbolic objects generated. In this case each group is still described by only one assertion.
- a decomposition process which applies a clustering algorithm within each group and generates one assertion per cluster. In this case each group is described by several assertions.

It is important to note that the decomposition process adds an unsupervised contribution to the process of generating symbolic objects. This is very useful when detailed raw data are not compatible with the supervised approach. On the other hand, the decomposition process produces several assertions for each group defined in a supervised way by the user. Current symbolic data analysis software cannot handle such symbolic objects structured as disjunctions of assertions. Several directions may be followed to use them anyway:

- When the number of groups is not too large, the user can interpret the result of the decomposition process and consider constructing an interpretation for clusters and then use existing symbolic data analysis software on the clusters.

- Distances between disjunctions of assertions can be defined (see De Carvalho 1994) so that standard clustering techniques can be applied. K-nearest-neighbour techniques can also be applied for classification applications.
- Symbolic data analysis methods can be extended to accept as input objects which are described by a disjunction of assertions.

This last point can be investigated either by using the *hoard structure* already introduced by E. Diday, or by considering assertions of a disjunction describing a group as prototypes of the group which are richer than standard prototype individuals.

References

Bock, H.-H. and Diday, E. (eds) (2000) *Analysis of Symbolic Data: Exploratory Methods for Extracting Statistical Information from Complex Data.* Berlin: Springer-Verlag.

Breiman, L., Friedman, J.H., Olshen, R.A. and Stone, C.J. (1984) *Classification and Regression Trees.* Belmont, CA: Wadsworth.

Chavent, M. (2000) Criterion-based divisive clustering for symbolic objects. In H.-H. Bock and E. Diday (eds), *Analysis of Symbolic Data, Exploratory Methods for Extracting Statistical Information from Complex Data*, pp. 299–311. Berlin: Springer-Verlag.

De Carvalho, F.A.T. (1994) Proximity coefficients between boolean symbolic objects. In E. Diday, Y. Lechevallier, M. Schader, P. Bertrand and B. Burtschy (eds), *New Approaches in Classification and Data Analysis*, pp. 387–394. Berlin: Springer-Verlag.

Jain, A.K. and Dubes, R.C. (1988) *Algorithms for Clustering Data.* Englewood Cliffs, NJ: Prentice Hall.

Nash, W.J., Sellers, T.L., Talbot, S.R., Cawthorn A.J. and Ford, W.B. (1994) The population biology of abalone (*Haliotis* species) in Tasmania. I. Blacklip abalone (*H. rubra*) from the North Coast and Islands of Bass Strait. Technical Report No. 48, Sea Fisheries Division, Marine Research Laboratories, Taroona. Department of Primary Industry and Environment.

Ruspini, E.M. (1970) Numerical methods for fuzzy clustering. *Information Science*, 2, 319–350.

Stéphan, V. (1998) Construction d'objets symboliques par synthèse des résultats de requêtes SQL. Doctoral thesis, Université Paris IX Dauphine.

Stéphan, V., Hebrail, G. and Lechevallier, Y. (2000) Generation of symbolic objects from relational databases. In H.-H. Bock and E. Diday (eds), *Analysis of Symbolic Data, Exploratory Methods for Extracting Statistical Information from Complex Data*, pp. 78–105. Berlin: Springer-Verlag.

Ward, J.-H. (1963) Hierarchical grouping to optimize an objective function. *Journal of the American Statistical Association*, 58, 236–244.

3

Exporting symbolic objects to databases

Donato Malerba, Floriana Esposito and Annalisa Appice

3.1 The method

SO2DB is a SODAS module that exports a set of symbolic objects (SOs) to a relational database and determines the individuals in a population Ω that are part of the extent of these SOs. Detailed data (micro-data) on individuals in Ω are stored in a table of a relational database. Therefore, SO2DB is the counterpart of DB2SO which imports the descriptions of SOs by generalizing micro-data stored in a relational database.

Recall from Bock and Diday (2000) and Diday and Esposito (2003) that an SO is defined by a description d, a relation R which compares d to the description d_w of an individual, and a mapping a called the 'membership function'. Hence, the extent of an SO s, denoted by Ext(s) in the Boolean case (i.e., $[y(w)\ R\ d] \in \{\text{true, false}\}$), is defined as the set of all individuals w from a population Ω with $a(w) = \text{true}$. It is identical to the extension of a, denoted by Extension(a). Hence, we have $\text{Ext}(s) = \text{Extension}(a) = \{w \in \Omega | a(w) = \text{true}\}$. Conversely, in the probabilistic case (i.e., $[y(w)\ R\ d] \in [0,1]$), given a threshold α, the extent of an SO s is defined by $\text{Ext}_\alpha(s) = \text{Extension}_\alpha(a) = \{w \in \Omega | a(w) \geq \alpha\}$.

The extent of an SO that is computed on a population Ω can be used either to manage the information that is lost during the import process from Ω or to study the evolution of the retained concept on the same population Ω at a different time. Alternatively, comparing the extents of the same SO computed on several populations (e.g., populations associated with different countries) can be used to investigate the behaviour of some phenomenon in different regions. For instance, let us suppose that the following description associated with an SO s,

$$[gender = \text{F}] \wedge [field = \text{factory}] \wedge [salary = [1.5, 2.1]] \\ \wedge [weekly\_working\_hours = [40, 42]],$$

Symbolic Data Analysis and the SODAS Software Edited by E. Diday and M. Noirhomme-Fraiture

is obtained by generalizing data collected by the Finnish National Statistics Institute on a sample of the Finnish female population working in a factory. In this case, the data analyst may determine the extent of s computed on the populations of other European countries and perform a comparative analysis of these countries on the basis of working conditions of women in factories. This operation, supported by SO2DB and known as *propagation on a database*, is generally useful for discovering new individuals stored in databases of individuals different from those used to generate an SO but with similar characteristics.

The SO2DB module is run by choosing Export... from under Sodas file in the SODAS2 menu bar. This module inputs a set of SOs stored in an SODAS file and matches each SO against the individuals in a population Ω (micro-data) stored in a database table. The correct association between the names of symbolic variables and the attributes of the database table is user-defined. The method returns a new database table describing the matching individuals in Ω. Attributes of the new relational table are denoted with the names of the input symbolic variables.

Input, matching procedure and output are detailed in the next sections.

3.2 Input data of SO2DB

The input SODAS file contains a symbolic data table, whose columns correspond to symbolic variables, while rows represent symbolic descriptions d_1. An SO $s=(d, R, a)$ corresponding to a description d_1 models the underlying concept and it is assumed to be in the form of assertion. Individuals of a population Ω are stored in a single table of a relational database. The current release of SO2DB has been tested on an MS Access database which is accessed through an ODBC driver.

The association between symbolic variables reported in the input SODAS file and columns of the database table is interactively established by the user by means of a graphical user interface (see Figure 3.1). The association may actually involve only a subset of symbolic variables. The user is allowed to perform a selection in order to link each symbolic variable to one table column. In any case, both subsets must have the same cardinality and the user-defined association must be a bijective function. The association of a symbolic variable with a table column is admissible in the following cases:

Type of symbolic variable	*Type of table column*
categorical single-valued	string
categorical multi-valued	string
quantitative single-valued	number (integer or real)
interval	number (integer or real)
modal	string

If all selected symbolic variables are either categorical single/multi-valued or quantitative single-valued or interval then input SOs are handled as Boolean symbolic objects (BSOs) (see Chapter 8). If all selected symbolic variables are modal then input SOs are handled as probabilistic symbolic objects (PSOs). As explained in the next section, matching functions

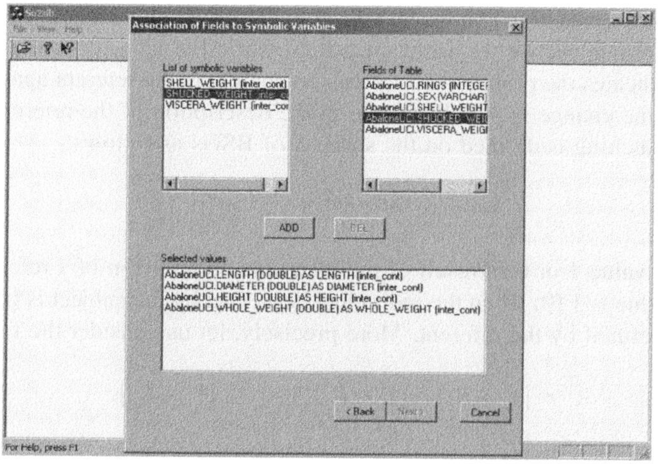

Figure 3.1 An example of an user-interactive association between the symbolic variables reported in an SODAS file and columns of a database table in SO2DB.

implemented in SO2DB are defined for either BSOs or PSOs. In the general case of SOs described by both set-valued and modal variables, a combination of a matching function for BSOs with a matching function for PSOs is used to compute the extent of input SOs.

Finally, SO2DB allows users to select a subset of rows of the symbolic data table to match against individuals of the population Ω. In this case, the extent will be computed only for a subset of input SOs.

3.3 Retrieving the individuals

The exportation of an SO s to a population Ω aims to retrieve the individuals (micro-data) of Ω which are 'instances' of the class description underlying s (i.e., the extent of s computed on Ω). Let us consider:

- an SO s whose description is of the form

$$s : [Y_1 \in v_1] \wedge [Y_2 \in v_2] \wedge \ldots \wedge [Y_m \in v_m],$$

  where each $Y_i (i = 1, \ldots, m)$ is a symbolic variable,
- a population Ω of individuals described by m single-valued (continuous and discrete) attributes Y_i' such that there is a correct association between the name of the symbolic variable Y_i and the attribute Y_i'.

The extent of s is computed by transforming each individual $I \in \Omega$ in an SO s_I that is underlying the description of I and resorting to a matching judgement to establish whether the individual I described by s_I can be considered as an instance of s.

The matching between SOs is defined in Esposito *et al.* (2000) as a directional comparison involving a referent and a subject. The referent is an SO representing a class description, while the subject is an SO that typically corresponds to the description of an individual.

In SODAS, two kinds of matching are available, namely, *canonical matching* and *flexible matching*. The former checks for an exact match, while the latter computes the degree of matching that indicates the probability of precisely matching the referent against the subject, provided that some change is possibly made in the description of the referent.

Canonical matching is defined on the space S of BSOs as follows:

$$CanonicalMatch: S \times S \rightarrow \{0, 1\}.$$

This assigns the value 1 or 0 as result of the matching comparison of a referent r against a subject s. The value is 1 (0) when the individual described by the subject is (not) an instance of the concept defined by the referent. More precisely, let us consider the following pair:

$$r: [Y_1 \in R_1] \wedge [Y_2 \in R_2] \wedge \ldots \wedge [Y_p \in R_p],$$

$$s: [Y_1 \in S_1] \wedge [Y_2 \in S_2] \wedge \ldots \wedge [Y_p \in S_p].$$

Then

$$CanonicalMatch(r, s) = \begin{cases} 1, & \text{if } S_j \subseteq R_j \ \forall j = 1, \ldots, p, \\ 0, & \text{otherwise.} \end{cases}$$

Similarly, flexible matching is defined on S by

$$FlexMatch: S \times S \rightarrow [0, 1],$$

such that

$$FlexMatch(r, s) = \max_{s' \in S(r)} P(s|s'),$$

where $S(r) = \{s' \in S | CanonicalMatch\ (r, s') = 1\}$ and P represents the probability (likelihood) that the observed subject is s when the true subject is s'. The reader may refer to Esposito *et al.* (2000) for more details on both canonical matching and flexible matching.

Finally, the definition of flexible matching can be extended to the space of PSOs as described in Chapter 8.

Notice that, in the case of canonical matching comparison, exporting s to Ω retrieves the individuals $I \in \Omega$ for which $CanonicalMatch(s, s_I) = 1$, while in the case of flexible matching comparison, given a user-defined threshold $fm\text{-}Threshold \in [0, 1]$, exporting s to Ω retrieves the individuals $I \in \Omega$ such that $FlexMatch(s, s_I) \geq fm\text{-}Threshold$.

3.4 Output of SO2DB

SO2DB outputs a new database table that describes the matching individuals in Ω. This table includes both attributes denoted with the names of the input symbolic variables and an additional attribute denoted with 'SO'. Rows of this table describe the individuals in Ω, which have at least one SO matching them.

The matching comparison is performed using either canonical or flexible matching as specified by the user. In particular, for each individual in Ω, the result of the matching comparison is stored in the 'SO' attribute as either one record for each single matching

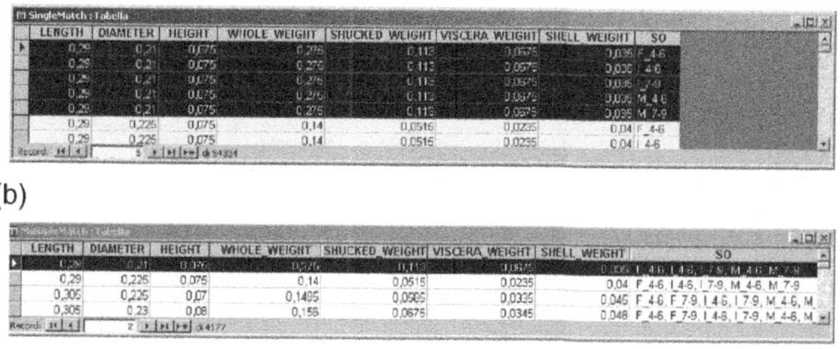

(a)

SingleMatch : Tabella

LENGTH	DIAMETER	HEIGHT	WHOLE WEIGHT	SHUCKED WEIGHT	VISCERA WEIGHT	SHELL WEIGHT	SO
0,29	0,21	0,075	0,275	0,113	0,0675	0,036	F_4-6
0,29	0,21	0,075	0,275	0,113	0,0675	0,036	I_4-6
0,29	0,21	0,075	0,275	0,113	0,0675	0,036	I_7-9
0,29	0,21	0,075	0,275	0,113	0,0675	0,036	M_4-6
0,29	0,21	0,075	0,275	0,113	0,0675	0,036	M_7-9
0,29	0,225	0,075	0,14	0,0515	0,0235	0,04	F_4-6
0,29	0,225	0,075	0,14	0,0515	0,0235	0,04	I_4-6

(b)

LENGTH	DIAMETER	HEIGHT	WHOLE WEIGHT	SHUCKED WEIGHT	VISCERA WEIGHT	SHELL WEIGHT	SO
0,29	0,21	0,075	0,275	0,113	0,0675	0,036	I_4-6, I_4-6, I_7-9, M_4-6, M_7-9
0,29	0,225	0,075	0,14	0,0515	0,0235	0,04	F_4-6, I_4-6, I_7-9, M_4-6, M_7-9
0,305	0,225	0,07	0,1485	0,0585	0,0335	0,045	F_4-6, F_7-9, I_4-6, I_7-9, M_4-6, M
0,305	0,23	0,08	0,156	0,0675	0,0345	0,048	F_4-6, F_7-9, I_4-6, I_7-9, M_4-6, M

Figure 3.2 The result of the SOs exportation to database when the matching comparison is stored in the 'SO' attribute as either (a) one record for each single matching or (b) one record for multiple matching.

or one record for multiple matching (see Figure 3.2). In the former case, for each SO s matching an individual $I \in \Omega$, a row describing I is inserted in the output database table and the 'SO' attribute contains the identifier (name or label) of s as value. This means that when several SOs match the same individual $I \in \Omega$, the output database table contains one row for each SO matching I. In the latter case, a single row of the output database table describes the non-empty list of SOs matching I. In this case, the value assigned to the 'SO' attribute is the list of the identifiers of the SOs matching I.

3.5 An application of SO2DB

In this section, we show how SO2DB is used to export 24 SOs stored in the SODAS file abalone.xml to the abalone database.

The symbolic data involved in this study were generated with DB2SO by generalizing the micro-data collected in the abalone database.[1] In particular, the abalone database contains 4177 cases of marine crustaceans described in terms of nine attributes, namely, sex (discrete), length (continuous), diameter (continuous), height (continuous), whole weight (continuous), shucked weight (continuous), viscera weight (continuous), shell weight (continuous), and number of rings (integer). Symbolic data are then generated by the Cartesian product of sex (F=female, M=male, I=infant) and the range of values for the number of rings ([4, 6], [7, 9], [10, 12], [13, 15], [16, 18], [19, 21], [22, 24], [25, 29]) and resulting SOs are described according to seven interval variables derived from generalizing the continuous attributes length, diameter, height, whole weight, shucked weight, viscera weight, and shell weight.

The file abalone.xml is opened by selecting File/Open from the menu bar of the SO2DB graphical user interface. A wizard allows users to select a subset of the symbolic variables reported in the SODAS file and identify one or more SOs from the SODAS file to be exported to database.

The database is accessed using the Microsoft ODBC facility. In particular, the wizard allows users to choose or create an ODBC data source to access the database, select a table from the list of tables collected in the database and interactively establish a correct

[1] The abalone database is available at the UCI Machine Learning Repository (http://www.ics.uci.edu/~mlearn/MLRepository.html).

	LENGTH	DIAMETER	HEIGHT	WHOLE_WEIGHT	SHUCKED_WEIGHT	VISCERA_WEIGHT	SHELL_WEIGHT
F_4-6	[0.28 : 0.66]	[0.19 : 0.47]	[0.07 : 0.18]	[0.08 : 1.37]	[0.03 : 0.64]	[0.02 : 0.29]	[0.03 : 0.34]
F_7-9	[0.31 : 0.75]	[0.22 : 0.58]	[0.01 : 1.13]	[0.15 : 2.25]	[0.06 : 1.16]	[0.03 : 0.45]	[0.05 : 0.56]
F_10-12	[0.34 : 0.78]	[0.26 : 0.63]	[0.06 : 0.23]	[0.20 : 2.66]	[0.07 : 1.49]	[0.04 : 0.53]	[0.07 : 0.73]
F_13-15	[0.39 : 0.81]	[0.30 : 0.65]	[0.10 : 0.25]	[0.28 : 2.51]	[0.11 : 1.23]	[0.05 : 0.52]	[0.09 : 0.80]
F_16-18	[0.40 : 0.75]	[0.31 : 0.60]	[0.10 : 0.24]	[0.35 : 2.20]	[0.12 : 0.84]	[0.09 : 0.48]	[0.12 : 1.00]
F_22-24	[0.45 : 0.80]	[0.38 : 0.63]	[0.14 : 0.22]	[0.64 : 2.53]	[0.16 : 0.83]	[0.11 : 0.59]	[0.24 : 0.71]
F_19-21	[0.49 : 0.73]	[0.37 : 0.58]	[0.13 : 0.21]	[0.68 : 2.12]	[0.17 : 0.81]	[0.13 : 0.45]	[0.20 : 0.85]
F_25-29	[0.55 : 0.70]	[0.47 : 0.58]	[0.18 : 0.22]	[1.21 : 1.81]	[0.32 : 0.71]	[0.20 : 0.32]	[0.47 : 0.52]
I_1-3	[0.08 : 0.24]	[0.05 : 0.17]	[0.01 : 0.06]	[0.00 : 0.07]	[0.00 : 0.03]	[0.00 : 0.01]	[0.00 : 0.02]
I_4-6	[0.13 : 0.58]	[0.09 : 0.45]	[0.00 : 0.15]	[0.01 : 0.89]	[0.00 : 0.50]	[0.00 : 0.19]	[0.00 : 0.35]
I_7-9	[0.26 : 0.67]	[0.19 : 0.50]	[0.00 : 0.19]	[0.08 : 1.30]	[0.03 : 0.60]	[0.01 : 0.32]	[0.03 : 0.39]
I_13-15	[0.32 : 0.66]	[0.25 : 0.52]	[0.08 : 0.19]	[0.16 : 1.69]	[0.06 : 0.71]	[0.03 : 0.40]	[0.05 : 0.42]
I_10-12	[0.34 : 0.73]	[0.26 : 0.55]	[0.09 : 0.22]	[0.17 : 2.05]	[0.07 : 0.77]	[0.02 : 0.44]	[0.06 : 0.65]
I_16-18	[0.44 : 0.65]	[0.33 : 0.52]	[0.13 : 0.20]	[0.44 : 1.63]	[0.16 : 0.63]	[0.07 : 0.34]	[0.13 : 0.53]
I_19-21	[0.45 : 0.58]	[0.35 : 0.44]	[0.12 : 0.19]	[0.41 : 1.18]	[0.11 : 0.39]	[0.07 : 0.22]	[0.16 : 0.31]
M_1-3	[0.16 : 0.21]	[0.11 : 0.15]	[0.04 : 0.05]	[0.02 : 0.04]	[0.01 : 0.02]	[0.00 : 0.01]	[0.00 : 0.01]

Figure 3.3 Symbolic descriptions of the SOs stored in abalone.xml.

association between the symbolic variables reported in the input SODAS file and the columns of this database table.

Finally, users specify the type of matching (canonical or flexible), the name of the output database table describing the matching individuals and the format of the matching result (i.e., either one record for each single matching or one record for multiple matching).

In this study, we decided to export the 24 SOs underlying the symbolic descriptions stored in abalone.xml to the same micro-data processed to generate the symbolic data. All the symbolic variables associated to the columns of the symbolic data table are selected and the canonical matching is chosen to determine the abalones that exactly match each SO.

Finally, the extents resulting from exporting abalone SOs to the abalone database are compared to identify SOs covered by the same individuals. For instance, we may compare the extents computed on the abalone database of the SO 'F_10-12', whose description is generated by generalizing the abalones with 'sex = F' and 'number of rings ∈ [10, 12]', and the SO 'F_13-15', whose description is generated by generalizing the abalones with 'sex = F' and 'number of rings ∈ [13, 15]' (see Figure 3.3). The export of 'F_10-12' to the abalone database retrieves 3650 cases, while the export of 'F_13-15' to the abalone database retrieves 3391 cases. When we compare the extents of 'F_10-12' and 'F_13-15', we discover that they share exactly 3379 cases. This may suggest a higher level of aggregation (i.e., 'F_10-15') in generating the SOs.

References

Bock, H.-H. and Diday, E. (2000) Symbolic objects. In H.-H. Bock and E. Diday (eds), *Analysis of Symbolic Data: Exploratory Methods for Extracting Statistical Information from Complex Data*, pp. 54–77. Berlin: Springer-Verlag.

Diday, E. and Esposito, F. (2003) An introduction to symbolic data analysis and the SODAS software. *Intelligent Data Analysis* 7(6), 583–602.

Esposito, F., Malerba, D. and Lisi, F.A. (2000) Matching symbolic objects. In H.-H. Bock and E. Diday (eds), *Analysis of Symbolic Data: Exploratory Methods for Extracting Statistical Information from Complex Data*, pp. 186–197. Berlin: Springer-Verlag.

4

A statistical metadata model for symbolic objects

Haralambos Papageorgiou and Maria Vardaki

4.1 Introduction

Symbolic data not only serve to summarize large data sets, but also lead to more complex data tables, thus enabling the manipulation of huge data sets (Billard and Diday, 2003). In order to handle the underlying concepts, a statistical metadata model is required.

In this section we develop step by step a statistical metadata model designed especially for the symbolic data environment in order to capture the metainformation needed for the creation of symbolic objects (SOs) and symbolic data tables (SDTs) and the successful implementation of symbolic data analysis methods. The metadata model under consideration should hold metainformation for the classical (original) data (survey variables, statistical units, frame population, etc.) and the symbolic data. More specifically, it should store metainformation both for the main stages of the processes of the classical data analysis, and for the symbolic data analysis procedures. It is also desirable for the model to store the processing history, from the creation of the original variables to the creation of the SDT and the application of certain statistical methods and/or the visualization of the final results.

Finally, the applicability of the model is verified by providing an example based on a data table recorded in Bock and Diday (2000, p. 58), illustrating the capability of the model in the simultaneous manipulation of both data and metadata, as well as for the improvement of the visualization of SOs.

Symbolic Data Analysis and the SODAS Software Edited by E. Diday and M. Noirhomme-Fraiture

4.2 Metadata, templates and statistical metadata models

The general term 'metadata' or 'metainformation' (used interchangeably in this section) is used in many different sciences and contexts. It is generally defined as 'data about data'. For example, data users may have encountered this term while using a modern database management system (DBMS) such as Oracle or Microsoft Access. In this case, the term 'metadata' simply refers to a set of specific tables also known as the database catalogues (i.e. a kind of contents table). However, in the case of statistics, the term 'metadata' includes much more semantic information (Papageorgiou *et al.*, 1999, 2000a; Sundgren, 1991). The term 'statistical metadata' is defined as the body of information required by a data user to properly understand and use statistical data. They mainly refer to explanations, definitions and procedures that were followed from the design phase until the publication of a survey's results and are generally considered as essential for statistical data processing concerning primary or secondary data analysis. Examples of metadata are the various statistical populations, sampling techniques, definitions of nomenclatures, classifications, and monetary units.

4.2.1 Statistical templates

As is well known, statisticians in national statistical institutes (NSIs) have always captured and produced statistical metadata. This metainformation was initially stored as series of free-text sentences (footnotes to tables). However, free-text notes can be ambiguous or incomplete. Ambiguity in metadata severely jeopardizes data quality, since it creates errors in the interpretation of final results. Furthermore, incomplete footnotes create the problem of missing metainformation. This problem forces statisticians to make assumptions that may later be proved incorrect (Papageorgiou *et al.*, 2000b). Therefore, the use of footnotes is gradually being abandoned.

To solve the problem of ambiguity and missing metainformation, Sundgren (1991, 1996) proposed the use of statistical templates for capturing metadata. Templates are actually empty forms (printed on paper or presented on a computer screen) that a statistician must fill in with metadata. By providing carefully designed documentation for each field of the template, the dangers of ambiguity and missing metainformation are greatly reduced. Furthermore, every piece of metadata captured in a template can easily be stored in a relational DBMS, thus simplifying the task of long-term storage of metadata. Metainformation captured using a template-based approach provides the end-user with the necessary documentation to understand the meaning of his/her data. In addition, this metainformation can subsequently be used in full-text search engines (such as Excite and Infoseek) and in thesaurus-based engines to speed up the task of locating statistical information.

However, templates do have a major drawback. They are mainly used for *entering* metainformation and not for *modelling* metainformation. Thus, although they specify correctly what metainformation is needed and how to store it, they do not specify how to use it, thus reducing its usefulness. Furthermore, machines fail to understand the meaning of the captured metadata and treat them as mere strings of characters. That is, as far as machines are concerned, 'Euros' and 'US Dollars' are text labels, bearing no other meaning or relation. Thus, computers cannot provide intelligent assistance to the end-user for the purpose of statistical data processing. Therefore, in order to enable information systems to use metadata more actively, we need to make machines understand the meaning of the data. A way of achieving this goal is by using a statistical (meta)data model.

4.2.2 Statistical metadata models

The amount of information processed by NSIs is constantly growing, as the demand for timely, accurate and high-quality data is increasing. Institutions are trying to extend the automation of their procedures in order to respond to this challenge. This attempt is influenced by the use of the Internet, as well as by the construction of new metadata-enabled statistical information systems. One of the most important features of these systems was the extended use of the metadata concept, as pointed out, for example, in Eurostat (1993), Froeschl (1997), Grossmann (1999), Grossmann and Papageorgiou (1997) and Sundgren (1996). However, the huge sets of data and their underlying concepts stored in large databases by the NSIs cannot be managed easily.

An attempt to tackle this problem has been made by symbolic data analysis in several fields of activity. Bock and Diday (2000, p. 12) posed the question: 'How can we obtain classes and their descriptions?' A major step was to describe the concepts under consideration by more complex types of data, which are called 'symbolic data' or 'symbolic objects', as they are structured and contain internal variations. This led to the extension of classical data analysis into symbolic data analysis, and the mathematical design of concepts led to the introduction of symbolic objects. Symbolic data not only serve to summarize large data sets (Billard and Diday, 2003), but also lead to more complex data tables, called 'symbolic data tables', where each row is an SO and each column a 'symbolic variable'. Hence, each cell contains not just a single value, as in classical data analysis, but several values, which can be weighted and linked by logical rules and taxonomies (Bock and Diday, 2000, p. 1).

When SDTs are used by advanced systems, their construction and handling will be automatically accompanied by the appropriate documentation (metadata), which in turn greatly improves the quality of the tables produced by reducing the dangers of data and metadata mismatches (Papageorgiou *et al.*, 1999).

In the case of SOs, a prerequisite for enabling new knowledge extraction and the manipulation of the corresponding SDTs and their underlying concepts is the development of an appropriate metadata model that can hold additional information for the construction of an SO and an SDT and their relation to classical data.

4.3 General requirements for the development of the metadata model

4.3.1 What metadata should be included in the statistical model?

According to the needs of SO and SDT creation, certain categories of metadata should be considered which not only describe the process of SO and SDT formation, but also store the history of this process and also allow for further SO manipulations. Two main categories of metadata are deemed necessary.

4.3.1.1 Metadata for the original data

In this category metadata describing the original data used for the creation of an SO and an SDT are considered.

(i) *Survey metadata*, consisting of metadata describing the way a survey is carried out. In particular, the statistical population, describing the individuals participating in the survey, and the sampling method applied, are two examples of survey metadata. Generally, these metadata refer to all the data collected after the survey has finished and not to a specific individual or variable. This means that if we store the survey data in a relational table where each column represents a variable and each row represents an individual, then survey metadata refer to this table as a whole.

(ii) *Metadata: for variables*, referring to a specific variable measured in a survey. For example, if the occupation of individuals is one of the questions required in a survey, we cannot compare them with relevant surveys in the past or in different countries unless we know the exact definition of occupation or the classification used to measure the answers. Hence, these metadata refer to a specific variable of a survey or to a column in a relational table.

A related, more extended metadata model for classical data has been developed in Papageorgiou *et al.* (2001a, 2001b).

4.3.1.2 Symbolic metadata

(i) *Metadata for an SO*. Metadata should describe the process of SO creation by denoting the class membership variables, the operator applied to those variables (average, sum, etc.), the type of condition, the concept with which the SO is associated and the corresponding values (upper and lower limits, thresholds, etc.). One notable difference from the classical metadata setting is that metainformation about individuals (now classes of individuals) is a very important piece of metadata, while in the classical setting one is mainly interested only in the universe of individuals and the variables, and not in the individuals themselves. By composing sets (groups) of individuals, the need to describe them and the process of synthesis provide useful metainformation both for the interpretation of the results and for the handling of the output for further processing. Very important pieces of metadata on the SO are the number of individuals from the sample that form the SO, as well as the equivalent number of individuals that correspond to the whole population using the sampling weights.

Therefore, metadata considered in the metadata model for the SO are the following: name, identifier, code, number of individuals participating, condition of aggregation (type, threshold and restrictions), description, the mapping association, the relation, the size, the SDT row where this SO is described and the related operations (transformations) that can be used in order to obtain the SO. The related classes should also be related to the symbolic data methods applied on an SO.

(ii) *Metadata for the symbolic variables*. The symbolic variables are produced from the operation of object creation. Each object (class of original individuals) is associated with a set of values, rather than a single one. An operation on this set forms the symbolic variable.

Furthermore, metadata should describe how these variables were created from the original variables. This possibility is deemed necessary when we want to create a new classification based on the one previously used (see also Vardaki, 2004).

In addition, the symbolic variables should be named, defined and their domain described by indicating the type of variable among the five possible (Diday,

2000) – (a) single quantitative; (b) single categorical; (c) categorical multi-valued; (d) interval; (e) modal – and the description of the associated domains, which may be a set of sets (for cases (c)–(e)).

More specifically, for each type of symbolic variable we should define: the range; the classification or nomenclature the variable is measured with (i.e for the variable 'Country' a list of countries); the superset of the set of values or categories; the superset of the interval; the number of values and the domain of the associated weights; the SDT column representing the variable.

Therefore, metadata should describe the symbolic variables, their nature, components and domain.

(iii) *Metadata for the symbolic data table*. An SDT looks like a (micro)data table in the sense that it contains rows that correspond to individuals and columns that correspond to (symbolic) variables. However, in the SDT, the original objects (individuals) can be changed (aggregated or disaggregated), using the values of one or more of the original variables such as class membership of a collection of classes of individuals (second level) and of classes of classes of individuals. Statistical metadata that describe the universe, the frame and the sample refer to the original objects. These metadata should be transformed using the metadata that correspond to the class membership variables into ones describing the new set of objects. This will include documenting the relations and denoting the operators and the descriptions.

4.3.2 Properties for the metadata items selected to model

The main properties of the metadata items selected are the following:

- *Completeness*. Data and metadata items should be as complete as possible, meaning that as many relevant metadata as possible must be held, but avoiding duplication.
- *Availability*. Data and metadata content should be primarily selected from the information pools for which data providers can guarantee user friendliness.
- *Integrability*. Data and metadata that will be stored and modelled must comply with the internationally required information for indicators and standards development and have a target of comparison with other statistical metadata models developed from NSIs or other statistical organizations, or through statistical research projects.

4.4 Selection of modelling technique

Having identified the metadata items that will be included in the model, the next step is to define which modelling technique is most appropriate to be used. The entity–relationship modelling technique has proved inadequate in the past since it failed to capture the dynamics of metainformation. Technically, this is due to the fact that there is no way of altering the definition of an already existing entity–relationship model. That is, after constructing the model, no additional information can be added, thus the model lacks flexibility.

To solve the above problem, it is recommended that the SO model under consideration is constructed according to the object-oriented paradigm, because this allows the model to be adapted in the event that consumers require additional metadata. Therefore, it seems that, currently, the best solution for constructing a metadata model is to use a designing tool which follows the object-oriented paradigm, for example the UML (Universal Modelling Language).

4.5 Step-by-step development of the model

As described in the previous section, the metadata model necessary for the Symbolic data creation and manipulation consists of two main parts: the metadata submodel for the original data (survey, original variables); the metadata submodel for the symbolic data (SO, SDT, symbolic variables). The transformations that can be applied to a SO and a SDT should also be considered in order to store the process history of any new SO and SDT creation.

These two submodels provide information about the different stages of the data analysis and the SO creation and manipulation. However, they are strongly interrelated. The main link metadata item between these two submodels is the class *group* which holds information about the set of individuals – drawn from the *sample* – that satisfies a specific condition related to the *original variables*. Note that in describing the model we use italics to denote the classes.

A simplified view of the submodels and their link metadata items is given in Figure 4.1.

4.5.1 The metadata submodel for the original data

Figure 4.2 illustrates the parts and the relationships of the submodel under examination in this section. A *survey* measures one or more *original variables* representing certain characteristics of *individuals* in the *statistical population*. An original variable can be either quantitative or qualitative, depending on the method of measurement used. Its values are measured in terms of a *measure unit*, which is probably related with another measure unit through the *equals* class. This means that if there is a specified relationship between two

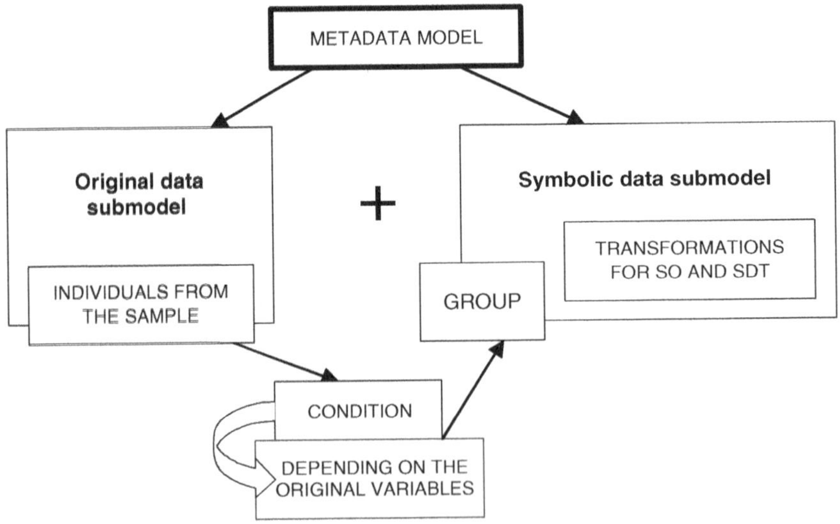

Figure 4.1 Simplified view of the model.

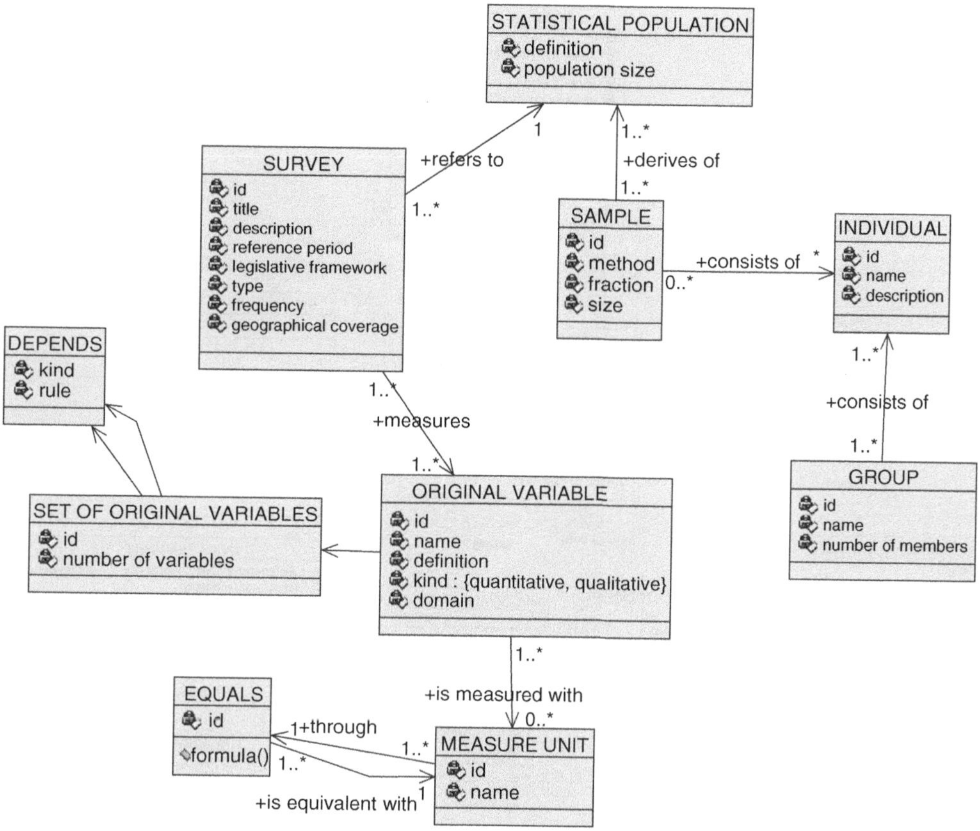

Figure 4.2 The metadata submodel for the original data.

measure units, the one can be transformed into the other and vice versa. For example, yards are related to metres according to a specific formula (1 yard = 0.91 metres), thus values measured in metres can be transformed into new ones measured in yards and vice versa.

The definition of the set of all *individuals* involved in a *survey*, and therefore described by a given set of variables, is the *statistical population* of the *survey*. Since it is not always feasible to examine the entire statistical population, we derive a *sample* through a *sampling method*; therefore, the *individuals* now considered are the ones included in the *sample*.

We can group the individuals according to one or more characteristics (i.e. 'persons employed in manufacturing businesses') and form a *group* of individuals. The 'groups of individuals' are our key relation of the submodel discussed with the submodel for the symbolic analysis data that is presented in the following subsection.

4.5.2 The metadata submodel for the symbolic data

Figure 4.3 illustrates the main parts and relationships of a structured metadata schema for symbolic data (SO, SDT, symbolic variables). We can see how the metadata of individuals are related with the metadata of symbolic objects referring to groups of individuals.

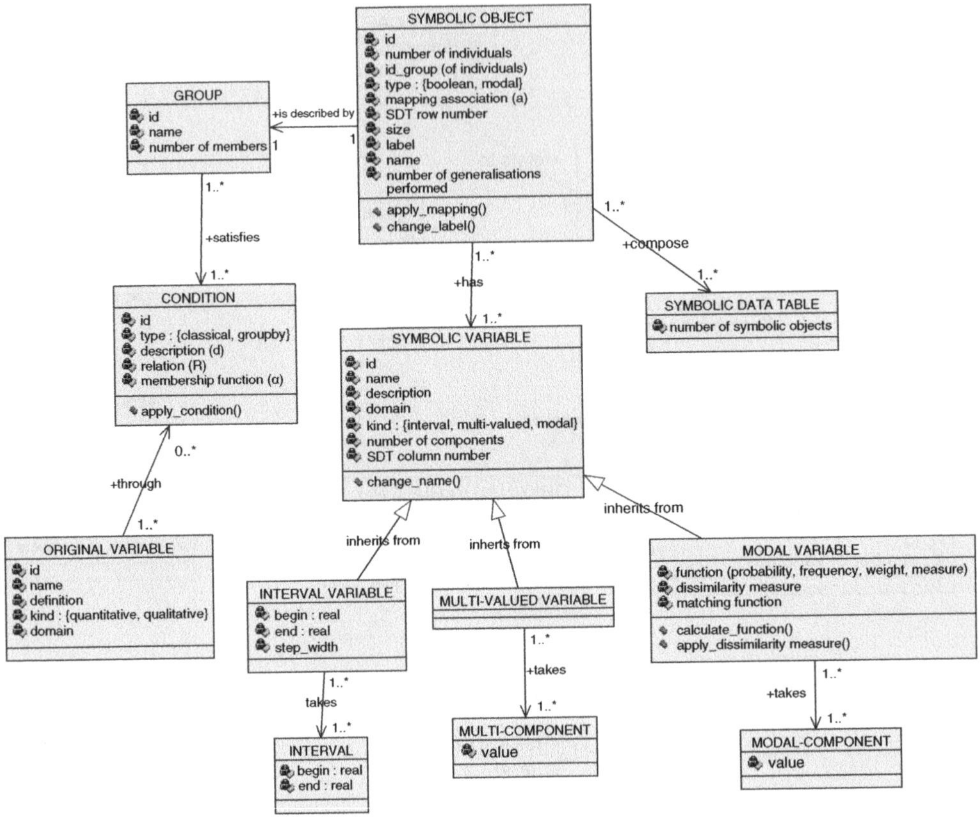

Figure 4.3 The submodel of the symbolic data.

In particular, a symbolic object (SO) is defined by a symbolic description (SD), which is denoted generally by d in our model, an SD relation (denoted generally by R in the model) plus a membership function (denoted by a). This symbolic description is defined by a set of *conditions* satisfied by a *group* of individuals.

Therefore, for each symbolic object the following attributes are stored: the number of individuals considered for its creation, its type (according to the type of the related variable), its symbolic description, the SD relation the mapping association (a), its size, and the SDT row number where this SO is described.

The *conditions* on the *original variables* examined can be either plain value comparisons or a 'group-by' condition meaning that the SOs created are first grouped according to the common value of these variables. For example, assume that from the set of 'types of cars' we wish to create an SO referring to the *group* 'limousines produced in Germany'. In this case, a group-by condition is applied to the *original variable* 'Type of car' (selecting the value=limousine) and an additional group-by condition on the original variable 'Country of production' selecting 'Germany'. It should be noted that the two *conditions* are applied conjunctively, meaning that the attribute operator of the *SO* equals the logical AND (^). As a result, a number of symbolic objects is created having two new *symbolic variables*: the type of car, and the regions of production. The rest of the original variables that may be under consideration in the initial original data table, e.g. year of production, price, brand, become symbolic variables without conditions (or with 'null' conditions). Consequently, a *symbolic object* is related to the *group* of *individuals* it stems from and to its variables. Another important point is that the attribute SDT row number of the *symbolic object* captures its position (row) in the symbolic object table, which is useful for further SDT manipulations.

Additionally, for a *symbolic variable*, its *kind* is stored as an attribute: it may be either an *interval variable* (whose values are intervals of numbers or ordered categorical values), a *multi-valued variable* (whose values are sets of values) or a *modal variable*. The latter is more complex than the others: its values are sets of pairs, each pair consisting of a value observed in the specific SO (group of individuals) and its relative function. The relative function can be calculated using a frequency, probability or weight distribution.

Furthermore, certain operations (transformations) need to be added in case new SOs and SDTs have to be created by a user, altering the already existing ones. These transformations can be applied either to a *symbolic object* or to a *symbolic data table*.

As illustrated in Figure 4.4, the transformations that a user can apply on an SO are as follows:

- *Addition* or *removal* of one symbolic variable (new SO is created).
- *Selection* or *projection* of a set of two or more symbolic variables at the same time (new SO is created).
- *Symbolic analysis methods* applied to a SDT.

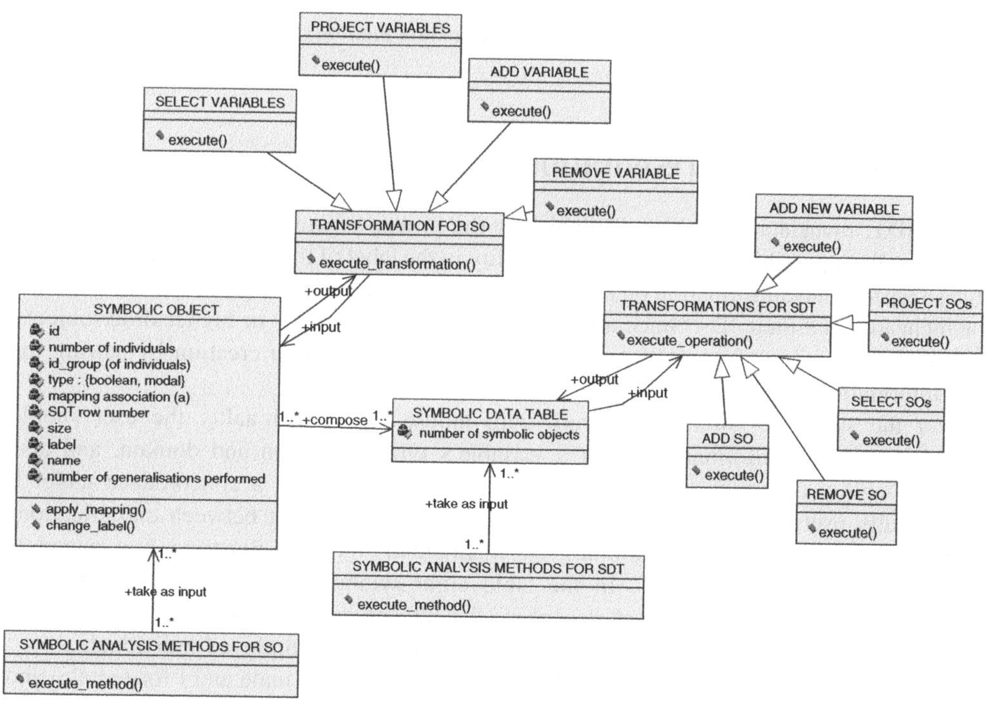

Figure 4.4 Transformations on SO and SDT.

On an SDT the user can apply the following:

- *Addition* or *removal* of one SO (new SDT produced).
- *Selection* or *projection* of a set of more than one SO at the same time (new SDT is produced).
- *Sorting* of symbolic objects contained in a specific symbolic data table (new SDT is produced).
- *Symbolic analysis methods* applied on an SDT.

Such information for the possible transformation should be added in a metadata model developed for symbolic data, also in order to ensure that the history of the SO and SDT is stored. All the operations shown in Figure 4.4 are represented by separate classes inheriting from the class *transformations for SO* and *transformations for SDT* accordingly and they maintain the property of closure, that is, if they are applied on an SDT then the result is a new SDT. For example, if a user removes a *symbolic variable* from a *symbolic object*, then the new *symbolic object* will be related with the previous one through an instance of the class *remove variable*. Thus, the history of processes is also maintained, as one can trace back to find out how the final *symbolic data table* was produced from the previous one. The maintenance of processing history is regarded essential for symbolic data users as it assists them in estimating the quality of SOs already defined and supports the execution of certain transformations in case it is necessary to go back and use the initial microdata tables.

Furthermore, there are classes for *symbolic analysis methods* taking as input a *symbolic object* or a *symbolic data table*.

4.6 Metadata representation

In the SDT, metadata should be available for the groups of individuals and the symbolic variables. With possible mouse-over functionality in the label of the group, the user should be able to retrieve metainformation on how many individuals formed the group, the condition(s) for inclusion, and their IDs. Since a group may consist of second- or higher-order objects, it should be possible to go further and trace the history of group creation by storing the metainformation for each level.

For the symbolic variables as well, with mouse-over functionality the user should be able to retrieve information about a variable's type, definition and domain, and also have the ability (if applicable) to change its classification using a concordance table. It is especially useful, for official statistics data, to be able to shift between classifications especially in the case of modal variables. The example in Figure 4.5 illustrates how metadata could be presented on an SDT. In the table eight symbolic objects (rows) with three variables (columns) are represented, describing the educational level according to the ISCED classification, the occupation and the status of employment of employed individuals.

All SOs have been grouped by sex (taking two values: m for male and f for female) and age (with values between 15 and 65 aggregated into five categories (intervals)).

In the case of graphical representation of the above SDT in a zoom star, for example (Noirhomme-Fraiture and Rouard, 1997), the presentation of metadata shown in Figure 4.6 could be considered.

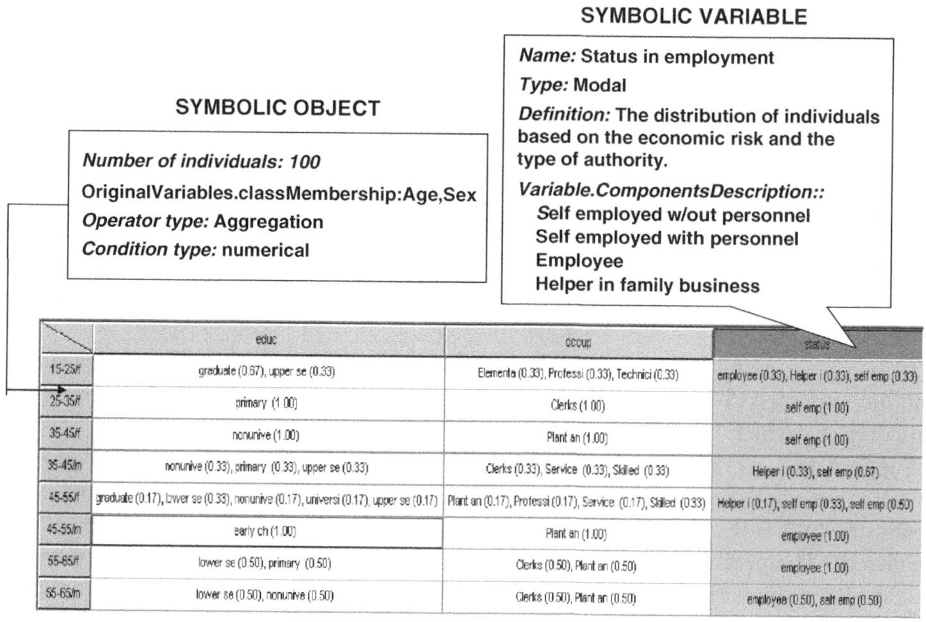

	educ	occup	status
15-25/f	graduate (0.67), upper se (0.33)	Elementa (0.33), Professi (0.33), Technici (0.33)	employee (0.33), Helper i (0.33), self emp (0.33)
25-35/f	primary (1.00)	Clerks (1.00)	self emp (1.00)
35-45/f	nonunive (1.00)	Plant an (1.00)	self emp (1.00)
35-45/m	nonunive (0.33), primary (0.33), upper se (0.33)	Clerks (0.33), Service (0.33), Skilled (0.33)	Helper i (0.33), self emp (0.67)
45-55/f	graduate (0.17), lower se (0.33), nonunive (0.17), universi (0.17), upper se (0.17)	Plant an (0.17), Professi (0.17), Service (0.17), Skilled (0.33)	Helper i (0.17), self emp (0.33), self emp (0.50)
45-55/m	early ch (1.00)	Plant an (1.00)	employee (1.00)
55-65/f	lower se (0.50), primary (0.50)	Clerks (0.50), Plant an (0.50)	employee (1.00)
55-65/m	lower se (0.50), nonunive (0.50)	Clerks (0.50), Plant an (0.50)	employee (0.50), self emp (0.50)

Figure 4.5 Metadata representation on a SDT.

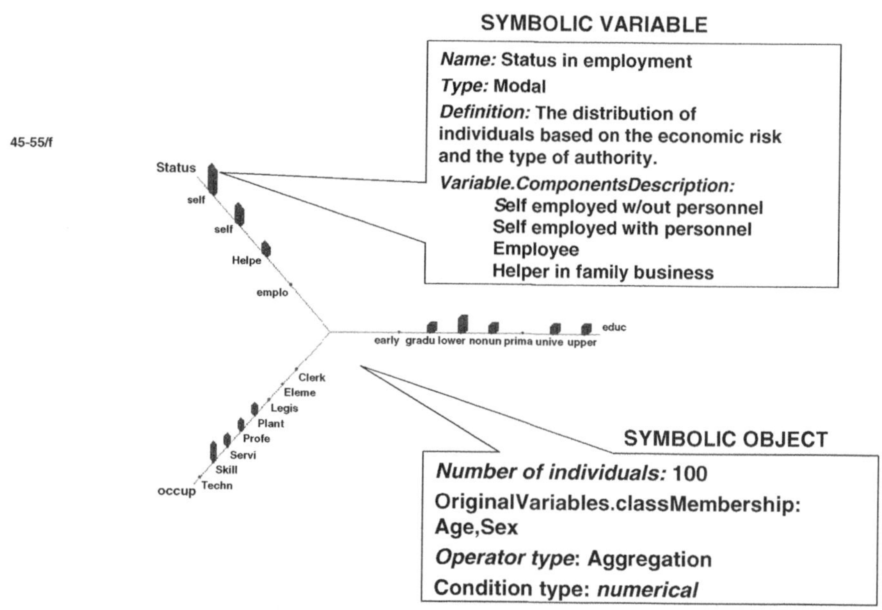

Labour Force Survey, Greece, 1997 Survey Metadata

Relevant Unit:	Unit of Labour Force survey
Surveys frequency:	quarterly
Geographical coverage:	The whole country (Greece)
International comparability:	NACE REV.1, STEP-92 (ISCO-88), ISCED-97

Figure 4.6 Metadata representation on a zoom star.

4.7 Case study

This example is based on Bock and Diday (2000, p. 58, Table 4.3). We consider a classical table which gives, for six countries (Am, Be, Ci, Do, Fu, Ga), the number of inhabitants, the main industrial activities (types of industries are indicated as A, Co, C, E, I, En), the gross national product (GNP in billions of euros) and the political party that won the elections in 1995. In this table, we have: number of individuals = 308 (in millions), classical variables = 5 (y1 = country, y2 = number of inhabitants, y3 = main industries, y4 = political party in the 1995 elections and y5 = GNP), as illustrated in Figure 4.7, where all related information is stored by the relevant attributes of the classes *survey* and *original variables*.

If we aggregate by the variable y1, a new table (an SDT) is derived with the following symbolic variables: Y1 = number of individuals, Y2 = GNP, Y3 = main industries, Y4 = political party, which is aggregated (with a group-by condition) by country.

Suppose that we request the following information: for all countries that are large (Y1>30), which have a considerable GNP (200<Y2<800 in billions), without a type of industry (Y3 = {C, Co, E, En, I}), and no consideration of political parties (Y4 not considered). The procedure, as illustrated in Figure 4.7 (for the sake of simplicity we have not considered the variable y4), is represented by the model in Figure 4.8. The individuals are separated into six groups according to the aggregation variable. This grouping has created the SO and symbolic variables presented in Figure 4.8. We have included only the

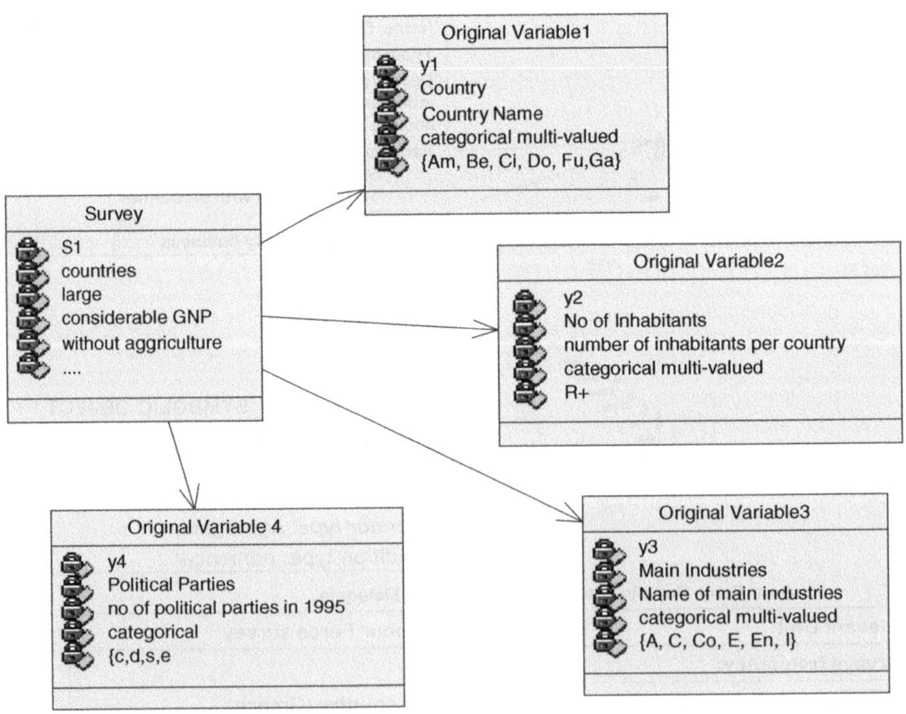

Figure 4.7 Case study – Original variables' representation.

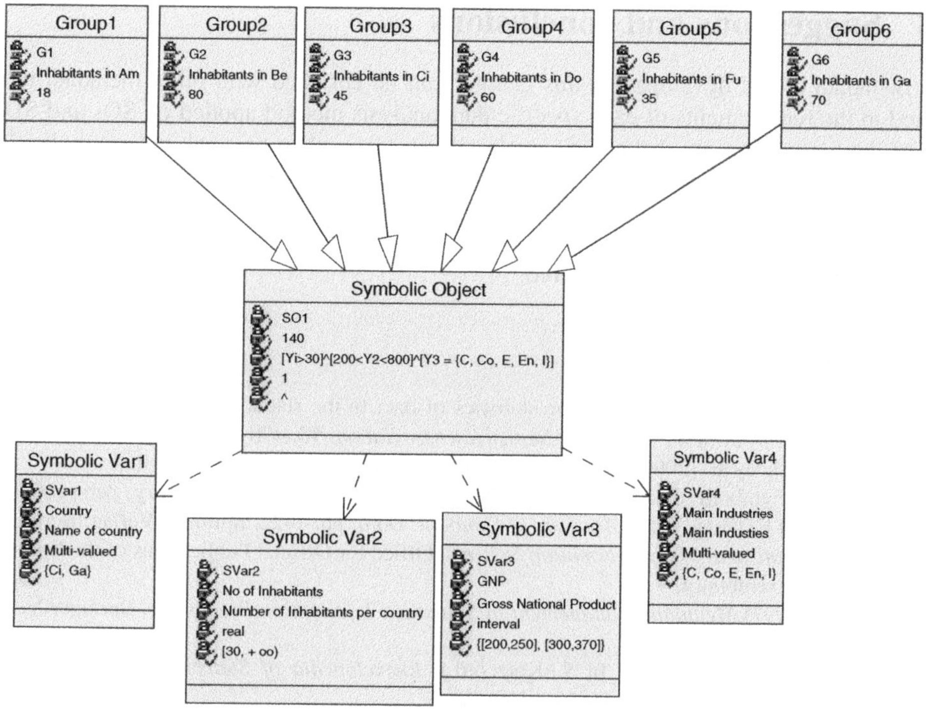

Figure 4.8 Case study – Symbolic variables' representation.

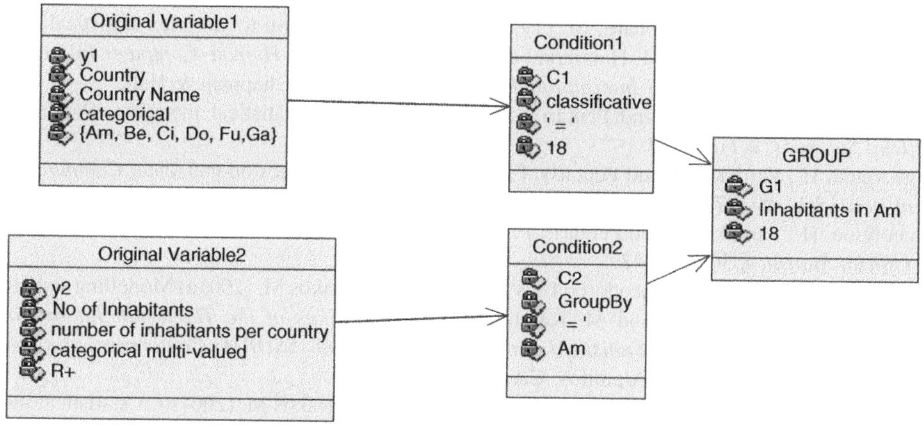

Figure 4.9 Case study – Relation of original variables with group.

SO and the corresponding symbolic variables that are meaningful according to the given constraints. Note that we have omitted the attributes description and mapping association from *symbolic object* class.

We could also illustrate how the *original variables* are related to each *group*. Figure 4.9 illustrates this relationship. It should be noted that only Group 1 of Figure 4.8 is provided, since all others can be developed similarly.

4.8 Suggestions and conclusions

The metadata model developed in this chapter can be enriched with other metadata items related to the requirements of each specific data analysis method applied on SOs and SDTs and to the harmonization of the various classifications used in original and symbolic data tables. In addition, the original data submodel described has incorporated only the minimum metadata items that are mandatory for SO and SDT creation. Metadata items concerned with the 'quality' of the entire process, as well as 'operators' applied for the original data table handling, could be also considered.

References

Billard L. and Diday E. (2003), From the statistics of data to the statistics of knowledge: symbolic data analysis, *Journal of the American Statistical Association*, **98**, 470–487.

Bock, H.-H. and Diday E. (2000) *Analysis of Symbolic Data*. Springer-Verlag, Berlin.

Diday E. (2000) Symbolic data analysis and the SODAS project: Purpose, history, perspective. In H.-H. Bock and E. Diday (eds), *Analysis of Symbolic Data*, pp. 1–22. Springer-Verlag, Berlin.

Eurostat (1993) *Statistical Meta Information Systems*. Office for Official Publications of the European Community, Luxembourg.

Froeschl, K.A. (1997) *Metadata Management in Statistical Information Processing*. Springer-Verlag, Vienna.

Grossmann, W. (1999) Metadata. In S. Kotz (ed.), *Encyclopedia of Statistical Sciences*, Vol. 3, pp. 811–815. John Wiley and Sons, Inc., New York.

Grossmann, W. and Papageorgiou, H. (1997) Data and metadata representation of highly aggregated economic time-series. In *Proceedings of the 51st Session of the International Statistical Institute, Contributed Papers*, Vol. 2, pp. 485–486. ISI, Voorburg.

Noirhomme-Fraiture, M. and Rouard, M. (1997) Zoom Star, a solution to complex statistical object representation. In S. Howard, J. Hammond and G. Lindgaard (eds), *Human–Computer Interaction: Proceedings of the IFIP TC13 International Conference*. London: Chapman & Hall.

Papageorgiou, H. Vardaki, M. and Pentaris, F. (1999) Quality of statistical metadata, *Research in Official Statistics*, **2**(1), 45–57.

Papageorgiou, H., Vardaki, M. and Pentaris, F. (2000a) Recent advances on metadata, *Computational Statistics*, **15**(1), 89–97.

Papageorgiou, H., Vardaki, M. and Pentaris, F. (2000b) Data and Metadata Transformations, *Research in Official Statistics*, **3**(2), 27–43.

Papageorgiou, H., Pentaris, F., Theodorou E., Vardaki M. and Petrakos M. (2001a) Modelling statistical metadata. In L. Kerschberg and M. Kafatos (eds), *Proceedings of the Thirteenth International Conference on Scientific and Statistical Database Management (SSDBM) Conference*, pp. 25–35. IEEE Computer Society, Los Alamitos, CA.

Papageorgiou, H., Pentaris, F., Theodorou E., Vardaki M. and Petrakos M. (2001b) A statistical metadata model for simultaneous manipulation of data and metadata. *Journal of Intelligent Information Systems*, **17**(2/3), 169–192.

Sundgren, B. (1991) What metainformation should accompany statistical macrodata? Statistics Sweden R&D Report.

Sundgren B. (1996) Making statistical data more available. *International Statistical Review*, **64**, 23–38.

Vardaki M. (2004) Metadata for symbolic objects. *Journal of Symbilic Data Analysis*, **2**(1), 1–8.

5

Editing symbolic data

Monique Noirhomme-Fraiture, Paula Brito, Anne de Baenst-Vandenbroucke and Adolphe Nahimana

5.1 Introduction

As summarized in Figure 5.1, a symbolic data table can be created by importation from a database (see Chapter 2) or from native data (data files which have been prepared by aggregating individual data using standard software). A symbolic data table can also be the result of an editing process. This chapter is concerned with this editing process, as well as with the creation of symbolic descriptions.

5.2 Creation of symbolic data

The most common way to construct a symbolic description is to extract statistical units satisfying some criteria on variables from a file or from a database. For example, we may want to extract, by querying the database, all the women aged between 25 and 34 years. To describe this set of units, we transform the initial variables into new ones of more complex form (interval, categorical multi-valued or modal). These variables are called symbolic variables. Usually, we construct more than one symbolic description so that all the symbolic descriptions are presented in a symbolic data table. The SOEDIT module in SODAS2 allows the visualization and editing of such data tables.

In the software, we use the term *symbolic object* to refer to an entity described by a row of the data table and to the description itself. In Section 5.3, to be consistent with the first chapter, we use the term *individual* for each entity and *symbolic description* for the corresponding description. *Stricto sensu*, a symbolic object is a more complex mathematical entity which is presented in Chapter 1.

Symbolic Data Analysis and the SODAS Software Edited by E. Diday and M. Noirhomme-Fraiture

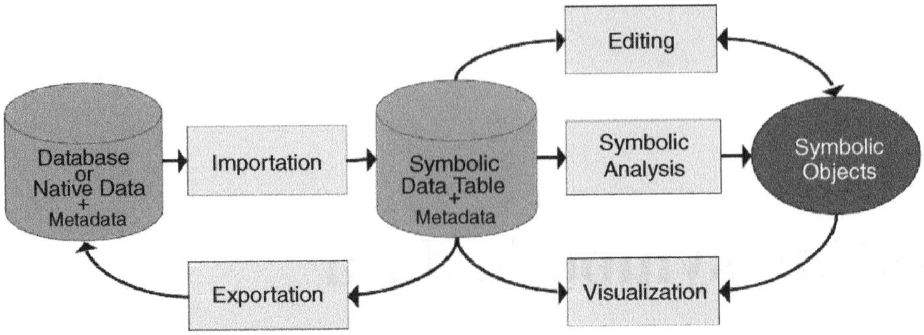

Figure 5.1 The cycle of symbolic data analysis in the ASSO project.

In this chapter, we will consider variables of the following types: quantitative single-valued; interval; categorical single-valued; categorical multi-valued; and modal (see Chapter 1).

5.2.1 Importation from native data file

Most of the standard statistical packages offer a facility for aggregating individual data. It is possible to select individuals satisfying some conditions on the original variables and aggregate them into groups or sets. For example, from an environmental survey, it is possible to group the individuals according to the variables sex and age. For each sex category crossed with an age category (0–15, 15–20, 20–25, etc.), a group of individuals is built. For each group, standard software allows the user to compute the minimum and maximum of quantitative variables and the proportion for categories of qualitative original variables. The aggregated data are usually stored in an ASCII file and presented in a table.

Figure 5.2 shows such a table for the case of a survey on small industries. In this example, columns 4, 5 and 6 give the proportion of individuals in the group for categories pinvest0, pinvest1 and pinvest2 of the variable pinvest, and columns 9 and 10 give the minimum and maximum for the variable conf.

Obs	SIC92	SIZEBAND	pinvest0	pinvest1	pinvest2	mprod	n	minconf	maxconf
1	52111	1	0.71756	0.27481	0.007634	.	.	.	.
2	52111	2	0.92857	0.07143	0.000000	70.616	4	3.134	138.097
3	52111	3	0.50000	0.50000	0.000000	41.412	2	28.543	54.280
4	52111	4	0.37500	0.62500	0.000000	56.126	4	36.157	76.095
5	52111	5	0.00000	1.00000	0.000000	253.273	2	21.600	.
6	52111	6	0.00000	1.00000	0.000000	54.784	9	35.109	74.459
7	52111	7	1.00000	0.00000	0.000000	.	.	.	.
8	52119	1	0.70037	0.29213	0.007491	.	.	.	.
9	52119	2	0.60000	0.40000	0.000000	48.232	35	38.989	57.475
10	52119	3	0.46000	0.54000	0.000000	57.292	20	36.676	77.907
11	52119	4	0.25000	0.75000	0.000000	62.618	8	41.793	83.443
12	52119	5	0.14286	0.85714	0.000000	68.911	6	56.597	81.225
13	52119	6	0.02667	0.97333	0.000000	90.059	62	73.483	106.634
14	52119	7	.	.	.	48.735	1	.	.
15	52119	8	0.00000	1.00000	0.000000	.	.	.	.

Figure 5.2 An ASCII file with native data.

The SODAS module ND2SO (from Native Data to Symbolic Objects) is a tool for transforming an ASCII standard aggregated file into an XML file (which in SODAS2 is called a SODAS file).

The values of the variables in the input file may be:

- quantitative (for quantitative variables),
- minima or maxima (which will be transformed into an interval variable),
- names or numbers (for categorical variables),
- proportions (which will be transformed into a modal variable).

Only columns with number values 0 or 1 are accepted for creating symbolic multi-valued categorical variables. The missing values must be represented by the characters'.'

The software displays the original data in the form of a table. The user selects interactively the appropriate columns in this table and specifies the type and label of the new variable.

A new table is created and displayed with the symbolic variables as columns and the same individual labels. The individuals are now modelled by symbolic descriptions and the new table is a so-called symbolic data table (see Figures 5.3 and 5.4).

5.2.2 Interactive creation of a symbolic data table

In this case, the user has already obtained symbolic descriptions from other treatments and wishes to record them in a SODAS file. This option has been requested by our user partners in order to deal with small examples for beginners in symbolic data analysis. But professionals can also use it when they have obtained symbolic data from other analyses which are not in a SODAS file. It is offered in the module SOEDIT.

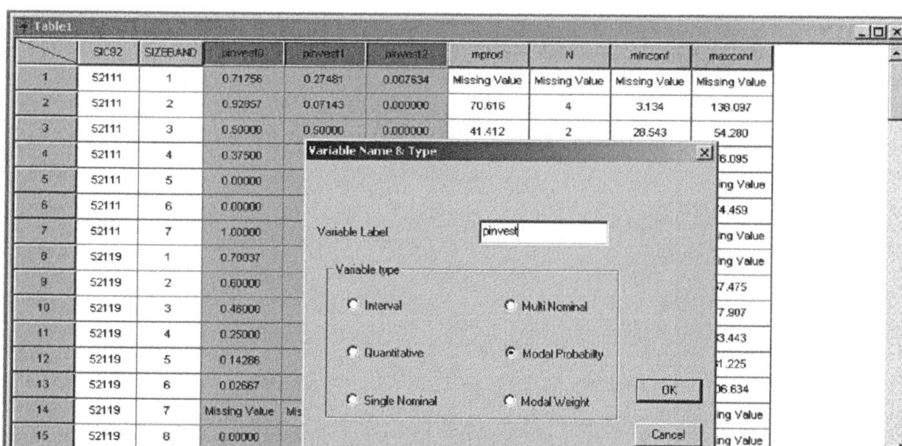

Figure 5.3 Selection of columns and choice of the variable type.

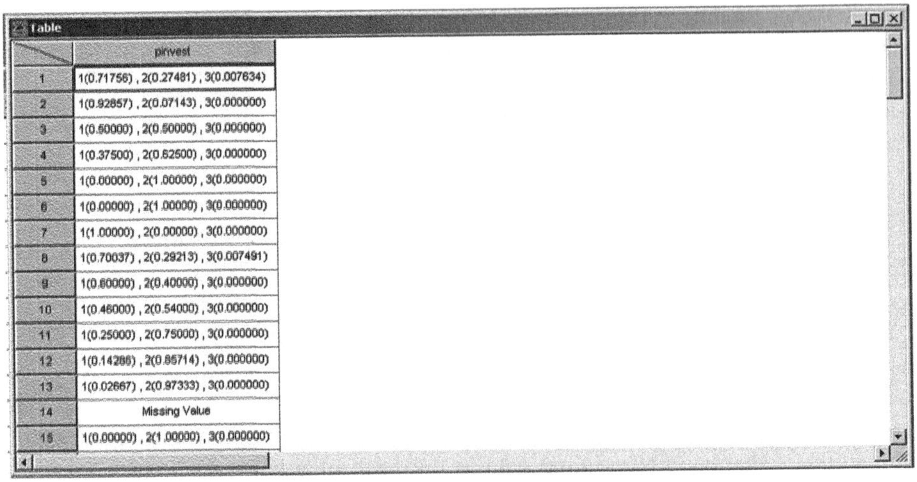

	pinvest
1	1(0.71756) , 2(0.27481) , 3(0.007634)
2	1(0.92857) , 2(0.07143) , 3(0.000000)
3	1(0.50000) , 2(0.50000) , 3(0.000000)
4	1(0.37500) , 2(0.62500) , 3(0.000000)
5	1(0.00000) , 2(1.00000) , 3(0.000000)
6	1(0.00000) , 2(1.00000) , 3(0.000000)
7	1(1.00000) , 2(0.00000) , 3(0.000000)
8	1(0.70037) , 2(0.29213) , 3(0.007491)
9	1(0.60000) , 2(0.40000) , 3(0.000000)
10	1(0.46000) , 2(0.54000) , 3(0.000000)
11	1(0.25000) , 2(0.75000) , 3(0.000000)
12	1(0.14286) , 2(0.85714) , 3(0.000000)
13	1(0.02667) , 2(0.97333) , 3(0.000000)
14	Missing Value
15	1(0.00000) , 2(1.00000) , 3(0.000000)

Figure 5.4 Resulting symbolic data table.

This procedure is of course of interest only if one works with small quantities of objects and variables. Otherwise, it is recommended to present the data in an ASCII file as in Figure 5.2 and to use the ND2SO module. The user gives the number of rows, then the codes and the labels for all symbolic descriptions (see Figure 5.5). This will allow the creation of the new symbolic data table. For each new variable, the user gives its code, label, type, number of categories (if any), codes and labels of categories (or modalities). Then, he fills

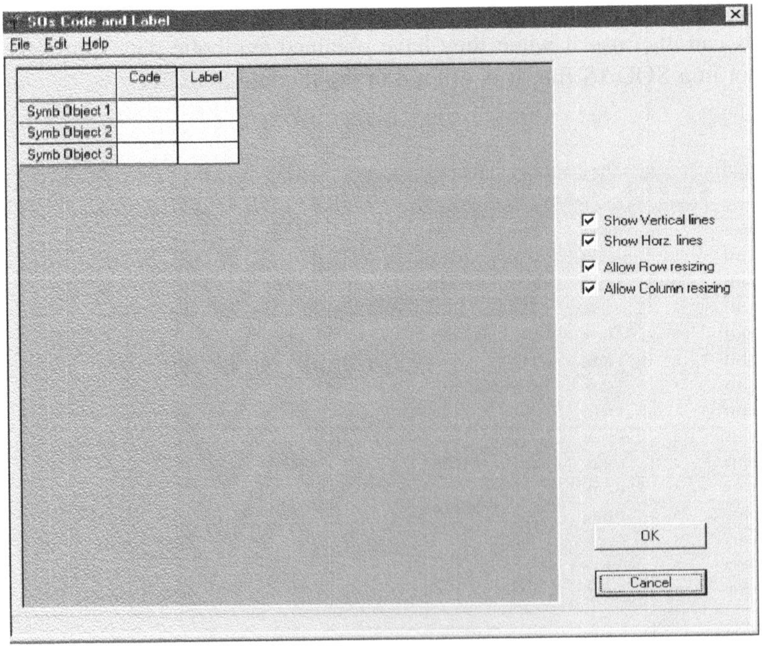

Figure 5.5 Dialogue box for coding and labelling symbolic objects.

in a table with the required values (min, max for interval variables; weights or probabilities for modal variables). An on-line 'Help' file gives all the details of the process, with the corresponding screens.

5.2.3 Breakdown or drilldown process

This process constructs a new symbolic description, using a database search, through interaction with a visual representation. As this option involves the star visualization, it is explained in Chapter 7.

5.3 Transformation of symbolic data

5.3.1 Introduction

Starting with a given symbolic data table, the user may want to modify the content of a cell, add or remove a row or column, or modify the labels. The result of such a transformation will be a new symbolic data table. In this way, it will be possible to keep track of all the modifications operated on the data. Of course, all the attributes relating to the original symbolic data table are maintained or updated.

The SOEDIT module facilitates such a process. The possible transformations included in SOEDIT are the following:

- modification of a particular cell;
- selection of variables;
- addition of new variables;
- relabelling of variables or categories;
- selection of variables obeying specified rules;
- selection of individuals;
- addition of individuals and their symbolic descriptions;
- modification of the labels of individuals;
- sorting of the individuals.

Moreover, it is possible to combine several symbolic descriptions into a new one. This transformation is called 'generalization' and is explained in Section 5.4. It is also possible to merge two given symbolic data tables. This process is explained in Section 5.5.

5.3.2 Module transformation

The SOEDIT module has implemented all these transformations in order to allow the user as much interaction with the table as possible. To modify a cell, the user clicks on the cell and replaces the content as required, the same way as in Excel (Noirhomme-Fraiture and Rouard, 2000). If a variable is categorical or modal, the categories are listed in a separate window. To modify the label of an individual (respectively variable), the user double-clicks on the head of the row (or column) and can then proceed with the required modification. To select rows (i.e. individuals) or columns (i.e. variables), the user clicks on the head of the row or column. He can also choose Selection from the menu bar and then Symbolic Objects. The Selection dialogue box appears: choosing the Symbolic Objects tab gives two lists containing available and selected elements. Objects can then be selected and moved between boxes as desired.

The selection is saved in a new symbolic file by choosing File from the menu bar and clicking on Save As. It is possible to select individuals satisfying a given rule on the values of a particular quantitative variable (single or interval). This rule can be of the form min $\geq$ or $\leq$ and/or max $\geq$ or $\leq$ 'value'. It is also possible to select individuals for which a categorical variable (single or multiple) has a particular categorical value.

To add a new individual with its symbolic description, the user chooses the Add Symbolic Object option under File in the menu bar and replaces the value of the newly created individual by the correct one. This gives a model for the values that he wishes to use.

To add a new symbolic variable, after selecting the respective item and the type of variable, the user fills in a predefined window with the values describing the symbolic variable, depending on its type.

To modify the order of symbolic descriptions, individuals can be sorted alphabetically. They can also be rearranged by the values of a quantitative single variable or of the min value of an interval variable. The order can also, of course, be modified manually.

5.4 Generalization of symbolic descriptions

5.4.1 Introduction

It is sometimes required to build a symbolic description which is more general than each symbolic description of a given set. This is done by considering the corresponding symbolic objects. One symbolic object is more general than another if its extension contains the extension of this other symbolic object (Diday, 1989).

There are not one but several ways to consider generalization. It depends on the knowledge we have on the initial symbolic objects and on the type of the variables which describe them.

The basic idea of generalization is to merge the information given by several symbolic objects into one. For example, given symbolic objects describing geographical statistical units, we may want to merge these descriptions into one. We will speak of a *union* of symbolic descriptions. This can be considered as a description of the set of all the individuals who are included in the different geographical statistical units. More conceptually, it can also be considered as the description of a concept which generalizes the other ones.

When several concepts are described, we may also be interested in a concept which is included in each one of the given concepts. We will speak in this case about generalization by intersection.

It is not possible to compute the generalization by a union or by an intersection for modal variables when the individual descriptions are lost. Nevertheless, we will still speak of generalization, by the maximum or by the minimum, in this case, as explained later.

5.4.2 Description of the method

Let s_i, $i=1, \ldots, n$, be the symbolic objects to be generalized. Generalization of a set of symbolic objects is performed variablewise. Therefore we must define how a generalization is performed for different kinds of variables.

5.4.2.1 Interval variables

In the presence of an interval variable y, the assertions associated with the s_i contain events of the form $e_i = [y \in [l_i, u_i]]$, $i=1, \ldots, n$. Two possibilities arise.

In the case of *generalization by union*, in the generalized description, the interval associated with variable y is $[\min\{l_i\}, \max\{u_i\}]$, since this is the smallest interval that contains all the intervals $[l_i, u_i]$, $i=1, \ldots, n$.

For example, consider a set of assertions, defined on variables age and salary:

$$a_1 = [\text{age} \in [20, 45]] \wedge [\text{salary} \in [1000, 3000]],$$

$$a_2 = [\text{age} \in [35, 40]] \wedge [\text{salary} \in [1200, 3500]],$$

$$a_3 = [\text{age} \in [25, 45]] \wedge [\text{salary} \in [2000, 4000]],$$

$$a_4 = [\text{age} \in [30, 50]] \wedge [\text{salary} \in [2000, 3200]].$$

These correspond to the following group descriptions:

	age	salary
Group1	[20, 45]	[1000, 3000]
Group2	[35, 40]	[1200, 3500]
Group3	[25, 45]	[2000, 4000]
Group4	[30, 50]	[2000, 3200]

The generalized assertion is

$$a = [\text{age} \in [20, 50]] \wedge [\text{salary} \in [1000, 4000]],$$

corresponding to the description

$$([20, 50]], [1000, 4000]).$$

In the case of *generalization by intersection*, in the generalized description, the interval associated with variable y is $[\max\{l_i\}, \min\{u_i\}]$, if $\max\{l_i\} \leq \min\{u_i\}$, otherwise it is the empty set, $\emptyset$, since $[\max\{l_i\}, \min\{u_i\}]$ is the largest interval contained in all the intervals $[l_i, u_i]$, $i=1, \ldots, n$.

Consider again the data of the previous example. The generalized description in this case is:

$$([35, 40], [2000, 3000]).$$

5.4.2.2 Categorical single and multi-valued variables

For categorical single or multi-valued variables, the events in the assertions have the form: $e_i = [y = v_i]$ (where y is a categorical single variable) or $e_i = [y \subseteq V_i]$ (y a categorical multi-valued variable). For generalization purposes, $e_i = [y = v_i]$ is equivalent to $e_i = [y \subseteq \{v_i\}]$, so both cases can be considered together. Again, two situations may arise.

In the case of *generalization by union*, generalization is done by performing set union on the sets associated with the corresponding events; in the generalized description, the set corresponding to the variable in question will be $V = \bigcup_{i=1}^{n} V_i$, since V is the smallest set that contains all the V_i, $i = 1, \ldots, n$.

For example, consider a set of assertions, each corresponding to a group of people, defined on variables sex and nationality:

$$a_1 = [\text{sex} \subseteq \{\text{M}\}] \wedge [\text{nationality} \subseteq \{\text{French, English}\}],$$

$$a_2 = [\text{sex} \subseteq \{\text{M, F}\}] \wedge [\text{nationality} \subseteq \{\text{French, German, English}\}],$$

$$a_3 = [\text{sex} \subseteq \{\text{M, F}\}] \wedge [\text{nationality} \subseteq \{\text{French, Belgian}\}],$$

$$a_4 = [\text{sex} \subseteq \{\text{M, F}\}] \wedge [\text{nationality} \subseteq \{\text{French, Portuguese}\}].$$

The generalized description is:

$$(\{\text{M, F}\}, \{\text{French, Belgian, German, English, Portuguese}\}).$$

In the case of *generalization by intersection*, generalization is done by performing set intersection on the sets associated to the corresponding events; in the generalized description, the set corresponding to the variable in question will be $V = \bigcap_{i=1}^{n} V_i$, since V is the largest set contained in all the V_i, $i = 1, \ldots, n$.

Consider again the data of the previous example. The generalized description is:

$$(\{M\}, \{\text{French}\}).$$

5.4.2.3 Modal variables

For modal variables, generalization can be done in three ways, each with a particular semantics. Let y be a modal variable with underlying domain $O = \{m_1, \ldots, m_k\}$ (i.e., $m_1, \ldots, m_k$ are the possible categories of y).

It is easy to carry out *generalization when size is known*, that is, if we know $N_1, \ldots, N_n$, the number of individuals fitting the description associated with $s_1, \ldots, s_n$.

Let the distribution of variable y for description associated with s_i be given by $\{m_1(p_1^i), \ldots, m_k(p_k^i)\}$ where $p_1^i, \ldots, p_k^i$ are the probabilities (or frequencies) of the corresponding categories for the description associated with s_i. We have

$$p_1^i + \cdots + p_k^i = 1, \qquad \forall\, i.$$

The generalized description will be defined, as concerns variable y, by

$$m_1(P_1), \quad \ldots, \quad m_k(P_k),$$

where

$$P_j = \frac{p_j^1 N_1 + p_j^2 N_2 + \cdots + p_j^n N_n}{N_1 + N_2 + \cdots + N_n}.$$

For *example*, let x indicate, in each case, the size of the corresponding description and

$$d_1 = (100, \{\text{administration}(30\%), \text{teaching}(70\%)\}),$$
$$d_2 = (200, \{\text{administration}(60\%), \text{teaching}(20\%), \text{secretariat}(20\%)\}).$$

Then the generalized description D will be defined by

$$D = ([x = 300] \wedge [y = \{\text{administration}(150/300), \text{teaching}(110/300), \text{secretariat}(40/300)\}]).$$

In the case of *generalization by the maximum*, generalization of events is performed by taking, for each category m_j, the maximum of its probabilities. The generalization of

$$e_1 = [y \leq \{m_1(p_1^1), \ldots, m_k(p_k^1)\}], \ldots, e_n = [y \leq \{m_1(p_1^n), \ldots, m_k(p_k^n)\}]$$

is

$$e = [y \leq \{m_1(p_1), \ldots, m_k(p_k)\}]$$

where $p_\ell = \max\{p_\ell^i, i = 1, \ldots, n\}$, $\ell = 1, \ldots, k$. Notice that the generalized event is typically no longer probabilistic, in the sense that the distribution obtained is no longer a probability distribution, since $p_1 + \cdots + p_k > 1$ is possible.

The extension of such an elementary event e, as concerns variable y, is the set of modal objects defined by

$$\text{Ext}\, e = \{a \mid (p_j^a \leqslant p_j, j = 1, \ldots, k\}$$

and consists of modal objects such that for each category of variable y, the probability/frequency of its presence is at most p_j.

For *example*, the generalization of

$$[y = \{\text{administration}(30\%), \text{teaching}(70\%)\}] \text{ and}$$
$$[y = \{\text{administration}(60\%), \text{teaching}(20\%), \text{secretariat}(20\%)\}]$$

is

$$[y \leq \{\text{administration}(60\%), \text{teaching}(70\%), \text{secretariat}(20\%)\}].$$

In the case of *generalization by the minimum*, generalization of events is performed by taking, for each category, the minimum of its probabilities. The generalization of

$$e_1 = [y \geq \{m_1(p_1^1), \ldots, m_k(p_k^1)\}], \ldots, e_q = [y \geq \{m_1(p_1^n), \ldots, m_k(p_k^n)\}]$$

is

$$e = [y \geq \{m_1(p_1), \ldots, m_k(p_k)\}] \text{ where } p_\ell = \min\{p^i_\ell, i = 1, \ldots, n\}, \ell = 1, \ldots, k.$$

The generalized event is typically no longer probabilistic since in this case $p_1 + \cdots + p_k \leq 1$ is possible.

The extension of such an elementary event e, as concerns variable y, is the set of modal objects defined by $\text{Ext}\, e = \{a \mid p_j \leqslant p^a_j, j = 1, \ldots, k\}$ such that for each category of variable y, the probability (frequency) of its presence is at least p_j.

For example, the generalization of

$$[y = \{\text{administration}(30\%), \text{teaching}(70\%)\}] \text{ and}$$
$$[y = \{\text{administration}(60\%), \text{teaching}(20\%), \text{secretariat}(20\%)\}]$$

is

$$[y \geq \{\text{administration}(30\%), \text{teaching}(20\%)\}].$$

5.4.2.4 Taxonomic variables

Let $y_i : \Omega \to O_i$, $i = 1, \ldots, p$. y_i is a taxonomic variable if its underlying domain O_i is ordered into a tree structure. An example is shown in Figure 5.6. Two different cases may arise: either the leaves are unordered, or the leaves constitute an ordered set.

Taxonomies may be taken into account in generalizing: first, we proceed as in the case of multi-valued variables, then a set of values of O_i is replaced by a value placed higher in the taxonomy. We choose to go up to level h when at least two successors of h are present:

$$[\text{Form} = \{\text{triangle, rectangle}\}] \to [\text{Form} = \{\text{polygon}\}].$$

5.4.2.5 Hierarchically dependent variables

Hierarchical rules reduce the dimension of the description space given by the variables, and a variable y_j becomes not-applicable if and only if another one y_i takes values within a given set:

$$y_i \text{ takes values in } S_i \iff y_j \text{ is not applicable.}$$

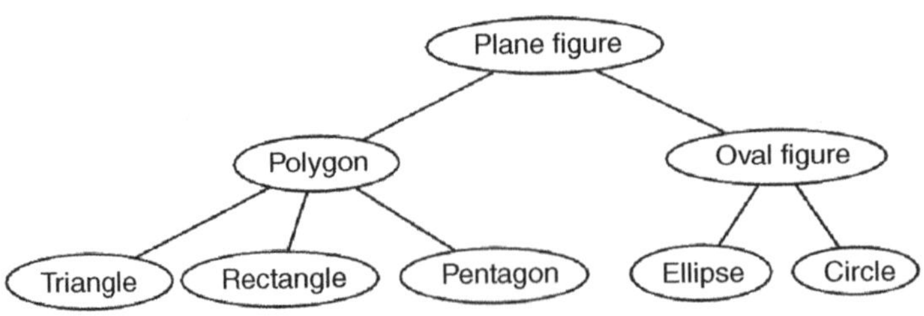

Figure 5.6 Taxonomy on variable 'Form'

In the case of single dependence, the hierarchical dependence is expressed by the rules

$$r_1 : [y_i \in S_i \subseteq O_i] \quad \Longrightarrow [y_j = \text{NA}]$$

and

$$r_2 : [y_j = \text{NA}] \quad \Longrightarrow [y_i \in S_i \subseteq O_i]$$

To perform a generalization in the presence of hierarchical rules, NA is treated as any other category. However, if in the obtained description, we have for the variable triggering of the rule, both values for which the rule must be applied and values for which it is not, then in the generalized description the rule must be included.

Consider, for example, two symbolic objects defined on variables 'First job' and 'Type of previous job' associated with the assertions:

$$[\text{FirstJob} = \{\text{no}\}] \wedge [TPJ = \{\text{administration, teaching}\}]$$

and

$$[\text{FirstJob} = \{\text{yes}\}] \wedge [TPJ = \text{NA}].$$

The assertion corresponding to the generalized symbolic object is:

$$[\text{FirstJob} = \{\text{yes, no}\}] \wedge [TPJ = \{\text{administration, teaching, NA}\}] \wedge [\text{FirstJob} \in \{\text{yes}\}] \Longleftrightarrow [TPJ = \text{NA}].$$

5.4.3 Implementation

The generalization process is implemented in SOEDIT. The user selects individuals in the table, then chooses the Generalization option under Edit in the menu bar. Finally, he chooses the type of generalization he wishes to perform.

5.5 Merging of symbolic data tables

It often happens that users want to merge information from several sources. This information may concern different objects described by the same variables or the same objects described by different variables. Both types of merging can be done by SOEDIT. We will present these two cases separately.

5.5.1 Merging for same symbolic variables

Let us describe this functionality with an example. Suppose that symbolic descriptions have been designed in different European countries using the same variables. It is, for example, the case for the Labour Force Survey, which has been made uniform by Eurostat. Each country has the same kind of statistics and can design symbolic descriptions by aggregating survey data by region crossed with age and gender. The European administration may be interested in considering all this information together and comparing the different regions, using analysis methods such as clustering. It is thus very useful to merge the files coming from the different statistical institutes. This option is of course relevant only if the symbolic description are described by the same symbolic variables.

5.5.2 Merging for same individuals

In this case, the same individuals are described by different variables. For example, a first survey can describe individuals by socio-economic data. When aggregating these data, by region crossed with gender and age, we obtain individuals described, for example, by level of education, marital status, and level of income. Or, we can get the results of surveys on time use, or on ecological attitude, or on mobility. In standard statistics, it is very difficult to merge these different data if the individuals are not identical in both surveys.

In symbolic data analysis, working at object level, we can merge different sources of information because the information is given at the level of concepts. Each object (region × gender × age) will be described by two (or more) kinds of symbolic variables: socio-economic and answers to surveys. The descriptive variables will be given on the form of distributions or of intervals (often confidence intervals). After merging the different files, it is possible to analyse the information globally.

References

Diday, E. (1989) Introduction à l'approche symbolique en analyse des données. *RAIRO*, 23 (2).

Noirhomme-Fraiture, M. and Rouard, M. (2000) Visualizing and editing symbolic objects. In H.-H. Bock and E. Diday (eds), *Analysis of Symbolic Data*. Berlin: Spinger-Verlag.

6

The normal symbolic form

Marc Csernel and Francisco de A.T. de Carvalho

6.1 Introduction

Symbolic descriptions can be constrained by domain knowledge expressed in the form of rules. Taking these rules into account in order to analyse the symbolic data usually requires exponential computation time.

We present an approach called the normal symbolic form (NSF) for managing symbolic data in the presence of rules within polynomial computation time. The aim of this approach is to implement a decomposition of a symbolic description in such a way that only coherent parts of the description (i.e. which do not contradict the rules) are represented. Once this decomposition is achieved, the computation can be done in polynomial time as described in Csernel (1998), although in some rare cases it can lead to an increased memory requirement. Nevertheless, we obtain mostly a reduction rather than an increase, as can easily be demonstrated.

The idea of the normal symbolic form is naturally related to the normal forms as used in the field of data bases (Codd, 1972), especially the third normal form; see Section 6.2.1 below.

In the following, we first define the kind of rules we are able to deal with and we explain why the presence of such rules can induce an exponential growth in the computation time. Then we explain why we transform the data in what we call the normal symbolic form, we explain what the NSF is, and, after considering the time complexity, the advantages generated by the NSF. In order to be more concrete in our explanation, we will take an example of computation which is simple but rather useful, especially, but not only, in the field of distance computation: the computation of the description potential (i.e. a measure defined on the coherent part of a symbolic description).

Symbolic Data Analysis and the SODAS Software Edited by E. Diday and M. Noirhomme-Fraiture

We then consider the possible memory growth both locally and globally. We explain which combination of rules and variables can generate memory growth, and which can generate a memory reduction.

Before concluding we will look at a real example and see that there is a space reduction ranging from 55% to 80% instead of a growth.

In this chapter, we mostly focus on categorical multi-valued variables, but the case of quantitative and interval data is also considered. The extension of the NSF to also manage modal symbolic data is still a research subject.

6.1.1 Constraints on symbolic descriptions

Symbolic descriptions can be constrained by dependencies between pairs of variables, expressed by rules. Such rules can be considered as constraints present in the description space; they produce some 'holes' in it because they forbid some points to be considered as a part of the virtual extension of a symbolic description. We take into account two kinds of dependencies: hierarchical and logical. A variable associated with the premise (or conclusion) of each rule will be called a *premise* (*conclusion*) *variable*.

Let y_1 and y_2 be two categorical multi-valued variables whose domains are respectively $\mathcal{D}_1$ and $\mathcal{D}_2$. A hierarchical dependency between y_1 and y_2 is expressed by the following kind of rule:

$$y_1 \subseteq D_1 \Longrightarrow y_2 = \text{NA},$$

where $D_1 \subset \mathcal{D}_1$ and NA stands for 'not applicable', hence the variable does not exist. With this kind of dependency, we sometimes talk about mother–daughter variables. The following rule is an example of such a rule:

$$\text{Wings} \subseteq \{\text{absent}\} \Longrightarrow \text{Wings\_colour} = \text{NA} \qquad (r_1)$$

A logical dependency between the variables y_1 and y_2 is expressed by the following kind of rule:

$$y_1 \subseteq D_1 \Longrightarrow y_2 \subseteq D_2.$$

An example of a logical rule is:

$$\text{Wings\_colour} \subseteq \{\text{red}\} \Longrightarrow \text{Thorax\_colour} \subseteq \{\text{blue}\} \qquad (r_2)$$

In this chapter, we will deal mostly with hierarchical rules; further treatment of logical rules can be found in Csernel and de Carvalho (2002).

Both rules reduce the number of individual descriptions belonging to the extension of a symbolic description, but the first kind of rule reduces the number of dimensions (i.e. variables), whereas the second does not. It has been demonstrated in De Carvalho (1998) that computation using rules leads to exponential computation time depending on the number of rules. In order to avoid this exponential computation time we introduce the normal symbolic form.

Remark. Few works have considered the semantics associated with the NA term, but we can mention the works of N. Lerat (see Lerat and Lipski, 1986) which are mostly in the field of databases.

6.2 Coherence within symbolic description

The exponential growth in computation time needed to manage symbolic data constrained by rules is closely linked to the notion of coherence. Given a symbolic description and domain knowledge expressed by a set of rules, we say that:

- a symbolic description is *fully coherent* when it is either not affected by the rules or if the rules apply to the whole description;
- a symbolic description is *coherent* when it has a part which is not affected by the rules.

In any other case a symbolic description is said to be *not coherent.*

For example, if we have the following rule:

$$\text{if Wings} \subseteq \{\text{absent}\} \text{ then Wings\_colour} = \text{NA} \qquad (r_3)$$

and the symbolic data table shown in Table 6.1, then:

- d_1 is not coherent because when Wings = absent then Wings_colour should be NA, according to rule r_1, and this is not the case in this description.
- d_2 is only coherent because Wings = {absent, present} and Wings_colour has a value. According to the rule, this value is meaningful only when Wings = present, but not when Wings = absent.
- d_3 is fully coherent because since Wings = {present} and Wings_colour has a value, the rule does not apply.
- d_4 is fully coherent because since Wings = {absent} and Wings_colour has no value (NA), the rule applies.

Because computations need to be done only on the coherent part of a symbolic description, the computation time has exponential complexity:

- The different algorithms need to compute precisely which part of a symbolic description is coherent in order to carry out computations only using the coherent part of the description.
- Hence, the idea underlying the NSF is to represent only the fully coherent part of a symbolic description, in order to avoid the previously mentioned calculus.

Table 6.1 Symbolic data table.

Description	Wings	Wings_colour
d_1	{absent}	{blue, red, yellow}
d_2	{absent, present}	{blue, red, yellow}
d_3	{present}	{blue, red, yellow}
d_4	{absent}	{NA}

If we can achieve such a goal at a reasonable cost, then all the computations will achieve a linear complexity (as if no rule existed).

6.2.1 Comparison of symbolic description with database relations and normal forms

Symbolic descriptions in their mathematical presentation are close to relations as they are used in databases.

The relational model was created by Codd (1972), and at the very beginning it took into account normalization problems in order to simplify the updating of databases and reduce redundancy. Of the three normal forms established by Codd, we will focus mostly on the first, which forbids multiple values as well as composite values, and the third, which forbids functional dependencies between one non-key field and another. The first is illustrated by the example shown in Table 6.2. The original relation describes a person, but the relation does not follow the first normal form: first, because the address (one field) can be decomposed into four different fields (number, street, city, and postcode); and second, because the forename is multi-valued.

To follow the first normal form the person relation needs to be decomposed into two relations. The first one is the person relation, but with a *normalized address*, the second one is the *forename relation*.

An example of Codd's third normal form can be found in Table 6.3. We see in the car relation that the owner name and owner address are functionally dependent on the owner ID (a number): if the owner (ID) changes the owner name and the owner address will change as

Table 6.2 Example of Codd's first normal form.

p_number	surname	forename	Adress
222	Smith	Peter, Paul, John	33, Wentworth Street, London E1
223	Watson	John, Hamish	221b, Baker Street, London NW1

person: *original relation*

p_number	surname	s_num	street	city	post-code
222	Smith	33	Wentworth Street	London	E1
223	Watson	221b	Baker Street	London	NW1

person: *normalized address*

p_number	surname
222	Peter
222	Paul
222	John
223	John
223	Hamish

person: *normalized forename*

Table 6.3 Example of Codd's third normal form.

car_number	brand	model	owner_id	owner_name	owner_city
222	Austin	Healey	220	Brown	London
223	Aston Martin	DB4	220	Brown	London

car: *original relation*

car_number	brand	model	owner_id
222	Austin	Healey	220
223	Aston Martin	DB4	220

car: *normalized relation*

owner_id	owner_name	owner_city
220	Brown	London

owner: *new relation*

well. But if the owner moves, the owner address of the relation car will have to be changed in the two tuples. To follow the third normal form, the relation must be decomposed into two: the car *normalized relation* and a new relation, *owner*. The new set of two relations follows Codd's third normal form.

The major difference between relations, as they are used in databases, and symbolic descriptions is the following. All the tuples (rows) of a relational table *must* be different, and in most cases all tuples must be identified using a single value (the *key*) which can be made up of one or several columns (variables). More precisely:

- All the different symbolic descriptions have a name; this name is independent of their content. The *key* notion is absent.
- A set of symbolic descriptions (or *knowledge base*) can contain as many identical descriptions as needed. In a relational database all rows must differ, at least in their key values.
- A knowledge base can contain in each cell a set of values. In a relational database, this is forbidden by the first normal form.
- Each symbolic description represents an intention, while a relational database contains only the elements of an extension.

Remark. To facilitate the use of symbolic descriptions, particularly to prevent a user giving the same name to different concepts, some tools such as DB2SO automatically generate a different name for each symbolic description. This name has no meaning, it is not at all a part of the data, and therefore it cannot be considered as a key, even if, because of its uniqueness, it looks like a key.

6.3 Computation of the description potential

The description potential (de Carvalho, 1998) is a measure frequently used for computation dealing with symbolic description, especially in the case of distance computation. It is the measure of the coherent part of the hypervolume generated by a symbolic description, that is, where all the points respect the rules. Its formal definition is as follows. Let $s = (a, R, d)$

be a symbolic object, where $a = \wedge_{i=1}^{p}[y_i \subseteq D_i]$ is the assertion associated with s and $d = D_1, \ldots, D_p$ is the symbolic description also associated with s, D_i being a subset of the domain of definition $\mathcal{D}_i$ of the variable y_i. We denote by $\pi(a)$ the description potential of a, which is the measure of the coherent part of the Cartesian product $R = D_1 \times \cdots \times D_p$. In this section we will describe how to compute the description potential.

In the absence of rules, the computation of the description potential is exactly equivalent to the computation of a volume. If a is the assertion associated to s, then

$$\pi(a) = \prod_{i=1}^{p} \mu(D_i),$$

where

$$\mu(D_i) = \begin{cases} \text{cardinal}(D_i), & \text{if } y_i \text{ is discrete,} \\ \text{Range}(D_i), & \text{if } y_i \text{ is a continuous variable,} \end{cases}$$

and Range(D_i) is the value of the difference between the upper bound and the lower bound if D_i is an interval of $\mathbb{R}$. The complexity of the computation of the description potential is a linear function of the number of variables.

In the presence of rules, as the 'holes' generated in the description space can intersect, the complexity of the computation of the description potential quickly becomes exponential in the number of holes which can intersect, i.e. the number of rules. The 'holes' induced by the different rules must be subtracted from the volume, but their pairwise intersections must be added back, which leads to the formula described below.

Let $a = \wedge_{i=1}^{p}[y_i \subseteq D_i]$ be the assertion associated with a symbolic object s constrained by a set of rules $\{r_1, \ldots, r_t\}$, each inducing a dependency among the variables. Then if $\pi(a|\{r_1 \ldots r_t\})$ expresses the value of the description potential of a in the presence of the rules $r_1, \ldots, r_t$,

$$\pi(a|\{r_1 \cdots r_t\}) = \prod_{i=1}^{p} \mu(D_i) - \sum_{j=1}^{t} \pi(a \wedge \neg r_j) + \sum_{j<k} \pi((a \wedge \neg r_j) \wedge \neg r_k) + \cdots$$
$$+ (-1)^{t+1} \pi((a \wedge \neg r_1) \wedge \neg r_2) \wedge \cdots \wedge \neg r_t).$$

In this case, the complexity is exponential in the number of rules, and linear in the number of variables. This formula is quite similar to Poincaré's formula. It can be observed that the formula applies under any circumstances, even if it is not always the best way to proceed. The solution we will present below needs some special conditions to be valid, but if they are fulfilled we have a polynomial complexity in time for the computation of the description potential.

Example 6.1. Let $d = (\{a_1, a_2\}, \{b_1, b_2\}, \{c_1, c_2\}, \{d_1, d_2\})$ be a symbolic description. In the absence of rules the description potential computation is quite straightforward:

$$\pi(d) = 2 \times 2 \times 2 \times 2 = 16.$$

If we have the following two rules:

$$\text{if } y_1 \subseteq \{a_1\} \text{ then } y_2 \subseteq \{b_1\}, \qquad (r_4)$$

$$\text{if } y_3 \subseteq \{c_1\} \text{ then } y_4 \subseteq \{d_1\}, \qquad (r_5)$$

Table 6.4 Coherence table.

No.	Description (individual)	Coherent	No.	Description (individual)	Coherent
1	$a_1b_1c_1d_1$	Y	9	$a_2b_1c_1d_1$	Y
2	$a_1b_1c_1d_2$	N(r_5)	10	$a_2b_1c_1d_2$	N(r_5)
3	$a_1b_1c_2d_1$	Y	11	$a_2b_1c_2d_1$	Y
4	$a_1b_1c_2d_2$	Y	12	$a_2b_1c_2d_2$	Y
5	$a_1b_2c_1d_1$	N (r_4)	13	$a_2b_2c_1d_1$	Y
6	$\underline{a_1b_2c_1d_2}$	$\underline{\text{N}(r_4, r_5)}$	14	$a_2b_2c_1d_2$	N(r_5)
7	$a_1b_2c_2d_2$	N(r_4)	15	$a_2b_2c_2d_1$	Y
8	$a_1b_2c_2d_2$	N(r_4)	16	$a_2b_2c_2d_2$	Y

then we must consider one by one the individuals belonging to the virtual extension of d to verify whether they fit the rules.

In Table 6.4, each row belonging to one half of the array represents an individual which is numbered. The 'coherent' column contains a Y or a N depending on wether the individual's description is coherent or not. If the individual is not coherent, the rule by which the incoherence occurs is indicated in brackets. All rows with an N correspond to the first terms of the formula $-\sum_{j=1}^{t} \pi(a \wedge \neg r_j)$. Row 6, which has N($r_4$, r_5) in the coherence column, also corresponds to the second term of the formula $+\sum_{j<k} \pi((a \wedge \neg r_j) \wedge \neg r_k)$.

Since, in this example, we consider only two rules, the other terms of the formula do not appear. As there is no additivity among the different elements, we must reduce the volume according to each rule, and then add back the volume corresponding to their intersection.

The value of the description potential is equal to the number of coherent individuals described in the previous array, i.e. the number of Y, which is 9.

6.4 Normal symbolic form

In order to provide a better explanation of what the normal symbolic form is, we will consider the symbolic data table shown in Table 6.5. Here, there are two symbolic descriptions d_1, d_2, and three categorical multi-valued variables. The values are constrained by two rules: a hierarchical one (r_6) and a logical one (r_7):

$$\text{Wings} \subseteq \{\text{absent}\} \Longrightarrow \text{Wings\_colour} = \text{NA}, \qquad (r_6)$$

$$\text{Wings\_colour} \subseteq \{\text{red}\} \Longrightarrow \text{Thorax\_colour} \subseteq \{\text{blue}\}. \qquad (r_7)$$

Table 6.5 Original table.

	Wings	Wings_colour	Thorax_colour	Thorax_size
d_1	{absent, present}	{red, blue}	{blue, yellow}	{big, small}
d_2	{absent, present}	{red, green}	{blue, red}	{small}

Table 6.6 Decomposition tables.

	wings_T	Thorax_size
d_1	**{ 1, 3}**	**{big,small}**
d_2	{2,4}	{small}

main table

wings_T	Wings	colour_T
1	**absent**	**4**
2	absent	5
3	**present**	**{ 1, 2 }**
4	present	{ 1, 3 }

wings_T table

colour_T	Wings_colour	Thorax_colour
1	**{ red }**	**{blue }**
2	**{ blue }**	**{ blue, yellow }**
3	{ green }	{blue, red }
4	**NA**	**{ blue, yellow }**
5	NA	{ blue, red }

colour_T table

The result of the NSF transformation is presented in Table 6.6. In these tables the upper left-hand corner contains the table name. A new kind of column appears where the values are integers referring to a row in another table with the same name as the column.

The first table is called the main table, and refers to the initial table. The other tables are called secondary tables, and each one of them has a name corresponding to the premise variable from the rule that induces it.

In each secondary table, a double rule separates the rows where the first variable verifies the premise from the rows where it does not. The rows in bold face correspond to the description of d_1.

We observe a growth in the space needed for these two symbolic descriptions. In the color_T table we need five rows to describe two items. In the original table only two rows where needed. We will analyse this growth in Section 6.7 and show that it is bounded and how.

We now have three tables instead of a single one, but only the valid parts of the objects are represented: the tables now include the rules.

6.5 Computation of description potential with the NSF

The computation of the description potential of a symbolic description under the NSF is straightforward, because under the NSF only the valid parts of the symbolic descriptions are represented. So once under the NSF the complexity (in time) of the description potential is polynomial. We proceed in a recursive way. Each secondary table row describes a coherent hypervolume, and all the rows contributing to the representation of the same object describe hypervolumes which do not intersect (by construction). Therefore, one has to sum together the potentials described by each row of a secondary table.

Table 6.7 Potential computation.

	Wings ...	Thorax_size	pot
d_1	{1,3}	{big,small}	10
d_2	{2,4}	{small}	5

main table

wings_T	Wings	colour	pot
1	absent	4	2
2	absent	5	2
3	present	{1,2}	3
4	present	{1,3}	3

wings_T table

colour_T	Wings_colour	Thorax_colour	pot
1	red	{blue }	1
2	blue	{ blue, yellow }	2
3	green	{blue, red }	2
4	NA	{ blue, yellow }	2
5	NA	{ blue, red }	2

colour_T table

In the example in Table 6.7, the description potential of each row of the colour_T table has to be computed first, then the description potential of the rows of the wings_T table, and, finally, the description potential of each object described in the main table.

For example, row 3 of the wings_T table refers to the rows 1 and 2 of the colour_T table. The description potential is the sum of the description potentials described by these two rows: $1 + 2 = 3$. In the same way, the description potential of d_1 is obtained by multiplying the sum of the description potentials of rows 1 and 3 of the wings_T table $(2 + 3)$ by the description potential due to the variable Thorax_size (2), giving the result 10. The computation of the description potential is done recursively, following the table tree.

6.6 Formal definition

In this section we shall give a more formal definition of the NSF. A knowledge base is under the NSF if it fulfils the following conditions:

- *First NSF condition.* No dependency exists between variables belonging to the same table, but between the first variable and the others.
- *Second NSF condition.* For one symbolic description, all the values of a premise variable must lead to the same conclusion.

Most of the time, in order to fulfil the NSF conditions, a symbolic data table needs to be decomposed, as a relation needs to be decomposed to fulfil Codd's normal forms.

Concerning the first NSF condition, one can remark that:

- The reference to the first variable is only for convenience, any other place will do, as long as it is constant.
- The first condition implies that the NSF can be used with maximum efficiency only when the rules form a tree or a set of trees.
- We have to decompose the original symbolic data table into different tables.
- Because of the table decomposition due to the NSF conditions, we have to introduce new variables called *reference variables*.

The second NSF condition has one main consequence – that we have to decompose each individual within a table into two parts: one part where the premise is verified, in which case all the conclusions appearing in the rules do apply; and one part where the premise is not verified, in which case the values corresponding to the conclusion variables stay unchanged. For case of notation we will denote this consequence CQ2.

The different tables will form a unique tree according to the dependencies. Each dependency tree between the variables forms one branch of the table tree. The root of the table tree is called the main table. To refer from one table to another one, we need to introduce some reference variables, which carry a small space overhead.

All the tables, except the main table, are composed in the following way:

- The first variable is a premise variable, all other variables are conclusion variables.
- In each row the premise variable can lead to a unique conclusion for all conclusion variables.
- If we want to represent different conclusions within a table, we need, for each object, as many rows as there are conclusions.

The main advantage of the NSF is that after this transformation we do not have to worry about rules and there is no longer exponential growth of the computation time necessary to manage symbolic data. Instead, we have polynomial growth as if there were no rules to consider.

For example, if we have the following rule:

$$\text{if Wings\_colour} \subseteq \{\text{red}\} \text{ then Thorax\_colour} \subseteq \{\text{blue}\},$$

the row

No.	Wings_colour	Thorax_colour
1	{red,blue}	{blue,yellow}

becomes

No.	Wings_colour	Thorax_colour
1	{red}	{blue}
2	{blue}	{blue, yellow}

Notice that the union of these two rows implies the initial row.

6.6.1 The reconstruction problem

Having decomposed a given set data in order, for example, to improve computation time, it is hoped that the result will be the same as if no decomposition had been done. At least we hope that it will be easy to retrieve the initial data: this process is called reconstruction.

We have not considered this problem until now, but, bearing in mind the initial condition required for the NSF transformation (dependency between variable forming a tree or a set of trees), the solution is obvious:

- The new tables contain only the initial variables and the reference variables. If they are suppressed, only the initial variables remain.
- The first NSF condition implies that the decomposition process follows the dependency tree between the variables. For the same knowledge base the decomposition, and thus the tables obtained, will always be identical.
- We have seen that the row decomposition generated by the second NSF condition produces a set of rows, and that the union of these rows produces the initial rows.

When the dependencies between variables induced by the rules produce a tree or a set of trees (we only considered that case) the NSF decomposition is unique and the reconstruction obvious.

Things should be different if we ever consider that a variable can be the conclusion variable of more that one premise variable. In that case the NSF decomposition could provide different set of tables according to the order in which the different rules appear, and the decomposition problem should be tackled with a greater care.

6.7 Possible memory growth

We showed in the previous section that the second NSF condition can induce memory growth (consequence CQ2). To measure this possible growth, let S be the size of the biggest secondary table induced by the NSF transformation. For convenience, we will assume here that all the premise variables have only one set of premise values, and consequently, that the size of a table can at most double between two nodes of the table tree.

If D is the depth of the dependency tree and N is the number of initial symbolic descriptions, then $S = N \times 2^D$ at most. If the tree is well balanced then $D = \log_2(T)$, T being the number of premise variables. In this case we have at most $S = N \times 2^{\log_2(T)} = N \times T$. If the tree is not well balanced, in the worst case $D = T$ and S is at most $N \times 2^T$. In the first case we can have polynomial memory growth, in the second case exponential. In the following, we will study this growth in more detail according to the kind of rules used.

6.7.1 Hierarchical dependency

In this subsection, we will consider the possible memory growth, using only categorical values and hierarchical dependencies. We will use two indices S_n and S_d: S_n indicates the maximum possible size of the table due to CQ2, while S_d indicates the maximum possible size of the table according to the domains of the different variables used. And we always have $S = \min(S_n, S_d)$.

In order to describe the growth more accurately, we will generally distinguish two cases: the local case, where the comparison is done by comparing a mother table to one of its daughters; and the global case, where the comparison is done by comparing the original table with a leaf table.

6.7.1.1 When all the conclusions are determined by the same set of premise values

Locally In this case we have one premise variable y and m conclusion variables $x_1, \ldots, x_m$. Each conclusion variable x_j is defined on the domain X_j.

Let the domain of the variable y be divided into two parts: A and $\overline{A}$. A is the set of values which makes the conclusion variables inapplicable. Let N_m be the number of rows in the mother table.

According to CQ2, the size N_d of the daughter table is at most $N_d = 2 \times N_m$. According to the domain of the variables, the size N_d is at most:

$$N_d \leq S_d = (2^{|A|} - 1) + (2^{|\overline{A}|} - 1) \prod_{j=1}^{m} (2^{|X_j|} - 1), \qquad (6.1)$$

where $|A|$ is the cardinality of the set A.

Equation (6.1) has two terms, each of which represents the maximum number of possible rows within the table, the first one when the premise is true, and the second one when the premise is false.

The first term represents the number of all possible combinations of values of the premise variable when the premise is true. This term is not a product because all the conclusions variables have a unique value, NA.

The second term of the equation is a product; the last factor of this product represents the number of all possible combinations of values of the different conclusion variables. The first factor represents all the possible combinations of values of the premise variable when the premise is false.

We define the *local growth factor* F_l as the ratio between the number of rows N_d of a daughter table and the number of rows of its mother table N_m: $F_l = N_d / N_m$. Its upper bound is given by

$$F_l = \frac{N_d}{N_m} = \frac{\min(S_n, S_d)}{N_m} \leq 2. \qquad (6.2)$$

Globally If the set of rules forms a tree, the question arises whether there is an increase in size depending on the depth of the tree. This idea is natural because, if for a single level the number of rows doubles, one may suppose that for two levels the number of rows will quadruple.

In fact, this is not the case. If the number of rows doubles, half of these rows (in fact all the newly introduced rows) are filled with NA values. These rows, because of the semantics of the NA term, cannot refer to any row of another table, so, only some of the N_m initial rows will lead to values in the second table.

We define the *global growth factor* F_g as the ratio between the number of rows of a leaf table and the number of rows of the initial table. Its upper bound is given by

$$F_g = \frac{N_d}{N} = \frac{\min(S_n, S_d)}{N} \leq 2. \tag{6.3}$$

6.7.1.2 When all the conclusions are determined by different sets of values

Locally First we will consider the case where we have y, the premise variable, and two sets of conclusion variables: $\{x_1, \ldots, x_m\}$ and $\{z_1, \ldots, z_t\}$. The domain of x_j and z_k is respectively X_j and Z_k.

Let the domain of y be partitioned into three sets A_1, A_2 and A_3. A_1 is the set of values that make x_j inapplicable, A_2 the set of values that make z_k inapplicable, and $A_3 = \overline{A_1 \cup A_2}$.

According to CQ2 the size N_d of the daughter table is at most $N_n = 3 \times N_m$. According to the domain of the variables the size N_d is at most

$$N_d \leq S_d = (2^{|A_1|} - 1) \prod_{j=1}^{t} (2^{|Z_j|} - 1) + (2^{|A_2|} - 1) \prod_{k=1}^{m} (2^{|X_k|} - 1)$$
$$+ (2^{|A_3|} - 1) \prod_{k=1}^{m} (2^{|X_k|} - 1) \prod_{j=1}^{t} (2^{|Z_j|} - 1) \tag{6.4}$$

So, N_d is at most $\min(S_n, S_d)$ and the upper bound of F_l is

$$F_l = \frac{N_d}{N_m} = \frac{\min(S_n, S_d)}{N_m} \leq 3. \tag{6.5}$$

More generally, if the domain of the variable y is partitioned into n parts, then $F_l \leq n$.

Globally When the premise value is partitioned into three sets, we observe the same kind of phenomenon. Each table is divided into three parts, in the first part all the x_j have a value and all z_k are NA. In the third part all x_j have the same value as in the first part. So, if x_j is a mother variable, it will reference the same row of the daughter table in the first and in the third part, so we have $F_g \leq 3$. More generally, if the domain of the variable y is partitioned into n parts, then $F_g \leq n$.

6.8 Application

The biological knowledge base describes the male phlebotomine sand flies from French Guiana (see Vignes, 1991). There are 73 species (each corresponding to a Boolean symbolic description) described by 53 categorical variables. There are five hierarchical dependencies corresponding to five rules. These rules are represented by three different connected graphs involving eight variables. These variables are categorical multi-valued.

The three secondary tables had 32, 18 and 16 rows. In this particular example the local and global growth factors are identical for each secondary table, because there is only one mother table, the main table. The values of the different growth factors are 32/73, 18/73 and 16/73. As expected, there is a reduction in memory requirement.

The computation time taken to obtain a dissimilarity matrix for the 73 symbolic descriptions, which was around 15 minutes without the use of the NSF, is now down to less than 3 seconds.

6.9 Conclusion

Hitherto we have just indicated how the growth of memory requirement is bounded, using categorical variables and hierarchical rules. But in real applications, we have observed a reduction rather than a growth. This reduction concerns the secondary tables and is due to the fact that with a limited number of variables (as in the secondary tables) some different symbolic descriptions can have the same description which will appear only once in the table. The greater the number of symbolic descriptions, the greater the chance of obtaining a factorization.

In the case of logical dependency a major change occurs. If the local growth factor is exactly the same, the global one is different, because the limitation due to the semantics of the NA term will not appear. The size of each secondary table can grow according to its position within the table tree. However, if the tree is deep and the number of variables within a table is small, then S_d is smaller than S_n and the growth will be bounded.

In the case of continuous variables, the number of rows in a secondary table cannot be bounded by S_d because the domain of each continuous variable is infinite. As a consequence, factorization is less likely to occur. Nevertheless, there is little chance of having continuous variables as premise when using hierarchical rules, but they can just induce a local growth.

To conclude, it appears that we can say that in every case the NSF reduces the computation time necessary to manage symbolic data from exponential to polynomial. If the variables are categorical and there are only hierarchical rules, the possible memory growth is bounded by a small constant factor, and in most cases a memory reduction is obtained. If the rules are logical, the memory growth is usually linear depending on the number of rules, but in the worst cases the memory growth can be exponential but still bounded by S_d. In both cases a factorization may occur and reduce the memory growth. If continuous variables are used there are no longer any bounds to the possible growth, and very little chance of obtaining a factorization. However, they may still be used without introducing an exponential growth in the memory requirement.

Nowadays no method in SODAS2 software uses the NSF, and research concerning the integration of rules in most of the method is not yet fully completed. This integration will need some modifications, for example all methods using the description potential (mostly methods related to distance computation) will need to call a different subprogram. The integration of the NSF (and the related program) in the ASSO library is nearly completed. But to taking constraints into account via the NSF is a task which is as yet beyond most of the methods. However, in the case of distance computation, the reasearch phase seems to be complete, but not the implementation.

6.10 References

Codd, E.F. (1972) Further normalization of the data relational model. In R. Rustin (ed.), *Data Base Systems*, pp. 33–64. Prentice Hall, Englewood Cliffs, NJ.

Csernel, M. (1998) On the complexity of computation with symbolic objects using domain knowledge. In A. Rizzi, M. Vichi and H.-H. Bock (eds), *New Advances in Data Science and Classification*, pp. 403–408. Springer-Verlag, Berlin.

Csernel, M. and de Carvalho, F.A.T. (2002) On memory requirement with normal symbolic form. In M. Schwaiger and O. Opitz (eds), *Exploratory Data Analysis in Empirical Research*, pp. 22–30. Springer-Verlag, Berlin.

De Carvalho, F.A.T. (1998) Extension based proximities between constrained Boolean symbolic objects. In C. Hayashi, K. Yajima, H.-H. Bock, N. Ohsumi, Y. Tanaka and Y. Baba (eds), *Data Science, Classification, and Related Methods*, pp. 370–378. Springer-Verlag, Tokyo.

Lerat, N. and Lipski, W. (1986) Nonapplicable nulls. *Theoretical Computer Science*, 46, 67–82.

Vignes, R. (1991) Caractérisation automatique de groupes biologiques. Doctoral thesis, Université Paris VI, Paris.

7

Visualization

Monique Noirhomme-Fraiture and Adolphe Nahimana

7.1 Visualization: from cognitive psychology to data mining

Visual perception is, for humans, a highly developed tool for recognizing complex information. Humans can distinguish huge amounts of detail, interpret complex shapes, and mentally reconstruct hidden details and fuzzy images. This ability is vital for data exploration and is used in a method now called 'visual data mining'.

It is generally recognized that, in the first weeks after birth, a child begins to notice contours and discontinuities. These soon begin to be organized into objects and become meaningful (Harris, 1999). An adult sees objects directly, because his long training and experience help him to recognize them, but the primitive starting points are still contours, discontinuities and contrasts. He can easily compare complex shapes, or recognize a form, as soon as he has memorized it.

In fact, the perception process is only the first step in visualization. A human perceptor model (Norman and Draper, 1986; Eysenck, 1993) presents the process as a series of steps, relying on memory. In response to a visual stimulus, a perceptive subsystem stores the image in an entry buffer. This is followed by consciousness, comprehension, analysis, decision-making and action.

These successive stages are supported by different kinds of memory: perceptive memory, short-term memory and long-term memory (Eysenck, 1993). Let us examine the importance of short-term memory in data mining.

Data analysis and especially data mining are focused on a knowledge acquisition process; they are concerned with the building up of knowledge from numerous and complex data. In this interpretative process, visualization can play an important part. Relying on the power of visual treatment, data exploration and the results of analysis can be exploited to improve understanding.

Symbolic Data Analysis and the SODAS Software Edited by E. Diday and M. Noirhomme-Fraiture

Indeed, reading and understanding large amounts of data is a long and tedious activity. It demands close attention. Humans are only able to cope with a few numbers at a time. It has indeed been proved that our short-term memory can only deal simultaneously with 7 ± 2 elements (Miller, 1956). Working with a large quantity of numbers involves considerable mental strain. However, if they are presented graphically, we can easily distinguish and interpret trends, detect abnormal values, and recognize interesting patterns.

All these tasks are greatly facilitated if discrete points are joined, building continuous lines and contours, and giving birth to forms, more easily recognizable. Psychologists have discussed form recognition at considerable length. Models of recognition by template, by schema, and by decomposition into segments have been proposed (Eysenck, 1993). In any case, we are able to easily recognize rather complex forms.

But, when the given information is too important, it induces excessive mental load which can divert from essential meaning. For this reason, it is better not to present on the screen too much information but to show it in stages, as required. The powerful interactivity now offered by computers is very interesting in this respect.

As visual perception goes from general shape to particular details (Lemaire, 1999), it is a good idea to begin with a global view and only then to delve into specifics. Many authors have emphasized the attraction of such an approach (Schneiderman, 2002). To reduce mental load, when the number of variables is very large, it is also possible to select parts of variables and create a link between groups of variables.

Nevertheless, in an exploratory phase, reducing the choice of variables and objects is dangerous. It can generate error in understanding and interpretation, because it does not allow a sufficiently large view of the problem.

7.2 Visualization in the process of analysis

As already stated, visualization is first an exploratory tool, which allows us to better acquaint ourselves with data, to derive preliminary results, to take decisions on what to do next. But the visualization process is also used when analysing statistical results. These can also be complex and difficult to interpret, even if they come from a synthesis of rough information.

Visualization techniques are then very useful, not only for analysing data resulting from statistical treatment, but also for transmitting knowledge to other users. Synthetic

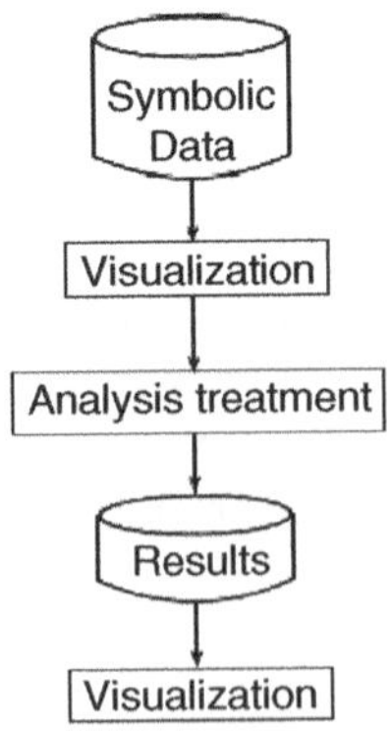

Figure 7.1 Sequence of treatments in symbolic data analysis.

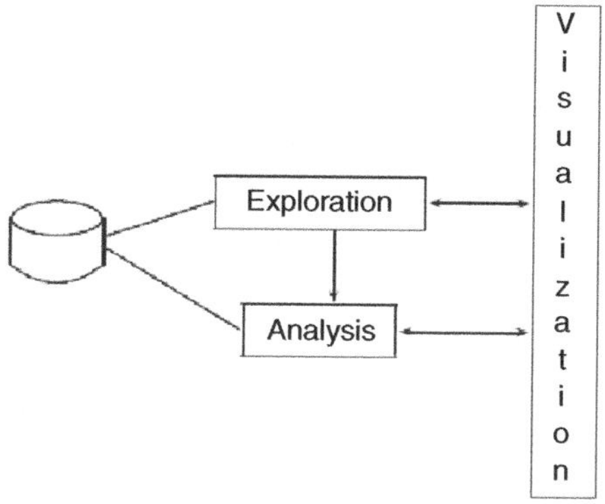

Figure 7.2 New sequence of treatments in data mining process.

visualization of results is much more easily understood than lengthy (verbal) descriptions. It facilitates mental representations of difficult concepts using images. We need good images and metaphors to ground our reasoning (Kuhn, 1996).

The sequence of steps in the data mining process is illustrated in the classical schema of Figure 7.1. This leads to the new schema of Figure 7.2 where visualization is the frame in which different treatments evolve: exploration or analysis. Such an architecture of the data mining process has also been suggested by Ankerst *et al.* (2003a, 2003b).

7.3 Choice of symbolic data representation

It is one thing to be convinced of the advantages visualization, but quite another to find a good visual representation. Once we tackle the question of the choice of representation, we have to specify what we want to represent and for what purpose.

In the exploratory phase, we want to visualize complex, multidimensional data. Symbolic data are multidimensional because, by definition, they involve many variables. They are complex because each variable may be given not only in the form of a single value, but also in the form of an interval or histogram. Chronological features can also be considered for some kinds of problems. Hierarchies of values can exist, as well as logical relations between variables.

The goal of the exploratory phase is first to better understand the data and generally to compare symbolic data in order to distinguish interesting patterns. Based on this preliminary analysis, a decision can be made on subsequent treatments. An exploratory analysis can also give some indication on relevant variables; the choice of variables will have some bearing on the choice of the following methods.

Before comparing objects, inspecting them one by one gives a better understanding of them as concepts, a better perception of their identity. Aberrant values can sometimes immediately be detected and appropriate action taken.

Once we have identified the data to be visualized and the goals, we can more easily specify the representation required. Nevertheless, as the problem is generally complex, there exists not one good representation but several. Each one highlights some aspect of the problem. The complementarity of representations gives rise to understanding and knowledge. It is important to make simple links between different views, using the interactive facilities provided by computers. In this way one can concentrate on mental tasks such as understanding and not on data manipulation or transformation tasks.

Several visualization methods have been suggested for representing multidimensional data, among them scatterplot matrices (Cleveland, 1993), parallel coordinates (Inselberg, 1985), icon-based techniques (Chernoff faces, stick figures, shape coding), pixel-oriented techniques, and worlds within worlds. Some interesting methods allow representation of numerous chronological data, for example pixel representation or calendar representation (Keim and Kriegel, 1994; Ankerst *et al.*, 2003a).

Few solutions exist for symbolic data, in particular because symbolic data analysis is still relatively new. We have been able to contribute to the domain thanks to the SODAS (Bock and Diday, 2000) and ISO-3D (Noirhomme-Fraiture, 2002) projects. We will develop our solution in the next section.

7.4 Visualisation with stars

We have already described the zoom star visualization in Noirhomme-Fraiture and Rouard (2000) and Noirhomme-Fraiture (2002). In this section, we will just recall the basic options of our symbolic object representation. As stated in the preceding section, we want to represent complex, multidimensional data in order to compare and understand symbolic objects. The variables can be of different types: quantitative, categorical, interval or modal. To take into account as many variables as possible, we represent one object by means of one graph, and not several objects in a space with a limited number of dimensions. In fact standard multidimensional representation enables visualization in two or three dimensions, sometimes more, but seldom more than five. Some solutions exist for representing distinct individuals with multiple variables: the iconic representation, parallel coordinates and radial coordinates. The iconic representation (auto-glyph, stick figure, colour icon) does not allow many variables to be taken into account: 5 to 10 would seem to be the maximum. Moreover, the variables must be categorical, or at least discrete. Parallel and radial coordinates are equivalent means in the sense that they dedicate one axis to one variable and allow us to represent as many variables as we want. The size and definition of the display is the only limitation, at least a priori, but of course human perception and mental load will impose natural limits on the number of axes. Instead of the now very well-known parallel coordinates (Inselberg, 1985), we have preferred a radial representation because it allows the early perception of a closed shape.

As stated in Section 7.1, humans construct visual objects mentally and are particularly adept at object recognition. A parallel coordinate representation (Inselberg, 1985) is not as easily identifiable as a circular representation, because the latter has a shape (Figure 7.3). Moreover, it optimizes the space occupied on the screen, at least on a PC.

To summarize our 2D zoom star representation, each variable is represented on a radial axis. For quantitative variables, values or interval limits are represented on standard graduated axes. Categorical or modal variables are equally distant dots, the size of which is proportional to the weight associated with the category. The limits of intervals and the dots of larger size are joined and the internal surface is coloured. If, in the description of the symbolic object, categorical or modal variables are in the majority, users usually prefer a 3D version where bar charts are directly visible on the axes. In this case, the form is lost but interactivity is greater because users can select the best viewing angle. They can also change the scale of the histograms (Figure 7.4).

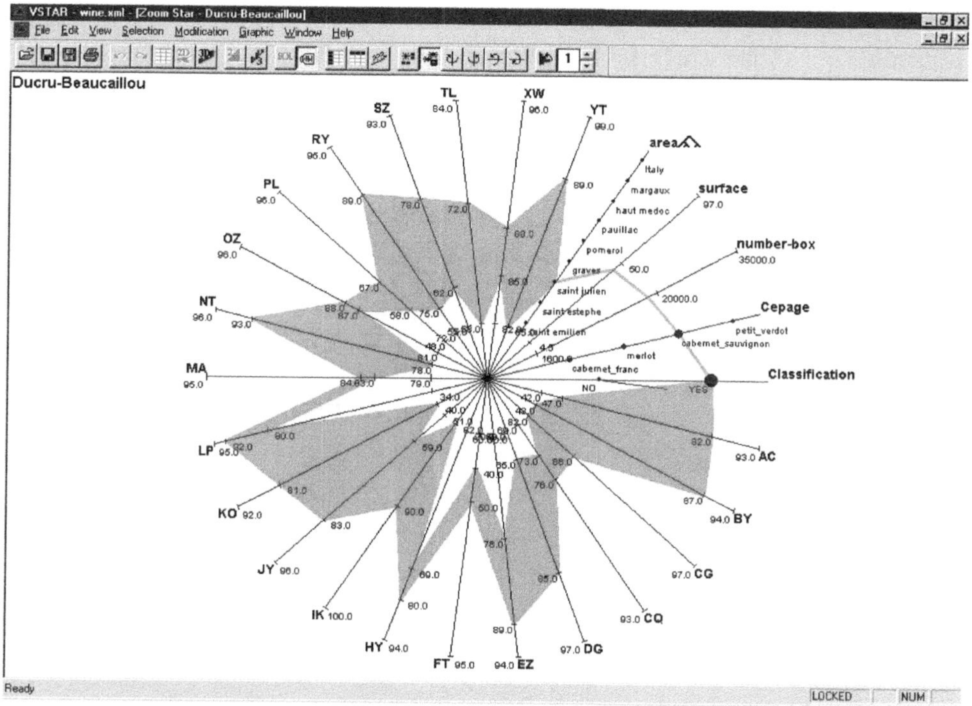

Figure 7.3 Two-dimensional zoom star with dependency.

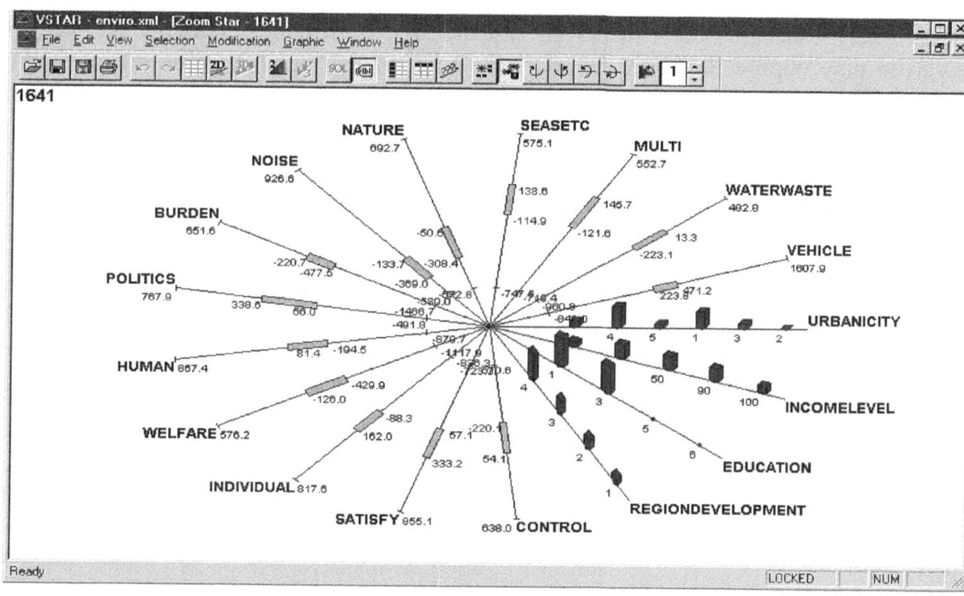

Figure 7.4 Three-dimensional zoom star.

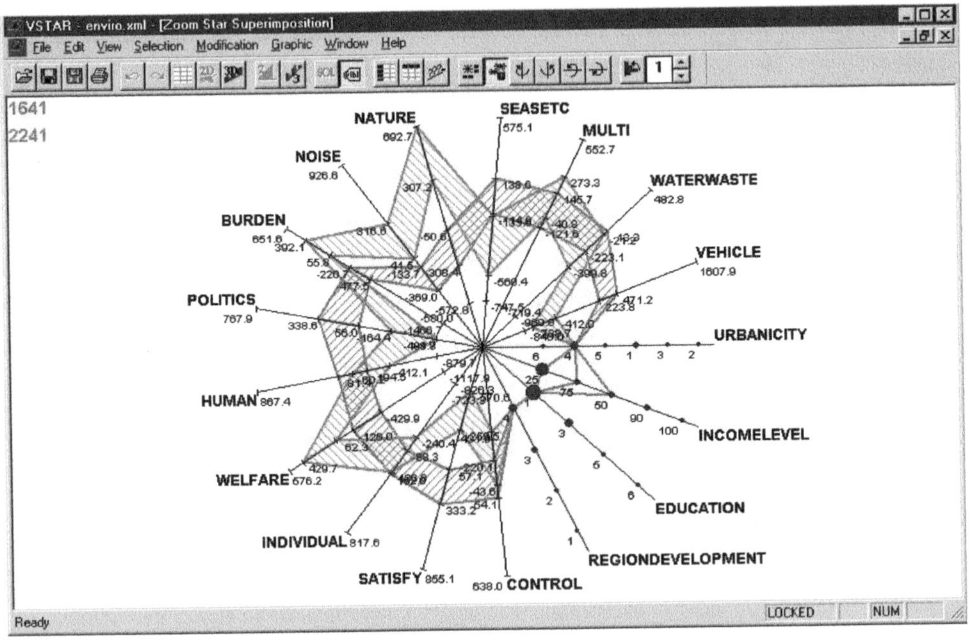

Figure 7.5 Superimposition of two symbolic objects.

7.5 Superimposition of stars

As explained in Noirhomme-Fraiture and Rouard (2002), it is possible to compare several objects when visualizing stars displayed side by side (tile mode), although the size of the screen may impose limits on the analysis. Another very helpful representation is the superimposition of figures (Figure 7.5). It is then much easier to detect differences between shapes. Fine-grained analysis of the detailed structure of the objects is provided. Of course, the number of objects is once again rather limited.

For superimposition, we prefer to use transparency for the surfaces, as in Figure 7.5. Unfortunately, the C++ library used in the SODAS software does not provide such a facility. We have thus used hatched surfaces. Otherwise, if we use the standard zoom star representation, one prominent object will hide the rest and prevent comparison between them.

7.6 Breakdown option

The breakdown option (Figure 7.6) is a way to build new symbolic objects or to retrieve existing ones. Thus this option could have been described in Chapter 5, on editing. But since, in SODAS2, breakdown is linked to object visualization, we have chosen to describe it here.

As explained in Chapter 1, a symbolic object may represent a concept but it can also be specified when aggregating a set of individuals. This is called its description in extension.

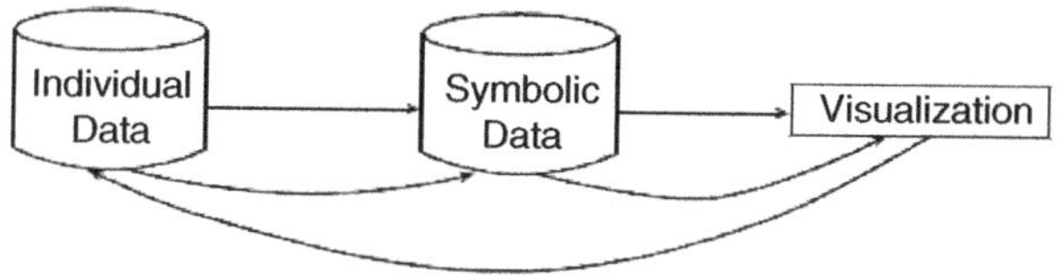

Figure 7.6 Visual breakdown.

7.6.1 Inclusion

Before describing the breakdown option, we have to make precise the notion of inclusion. When can we say that one symbolic object is included in another? Inclusion means that a concept described by a symbolic object is included in another, more general, described by another symbolic object. When both objects are described in extension, the problem is simpler because it concerns membership of individuals in a set.

Let S_1 and S_2 be two symbolic objects which respectively model two concepts C_1 and C_2. The inclusion between concepts or between symbolic objects will be denoted $C_1 \subset C_2$ or $S_1 \subset S_2$ and defined in two ways:

- *Definition by extension.* Let w be an individual. If both objects are known by extension, we can say

$$S_1 \subset S_2$$

if and only if

$$w \in \mathrm{Ext}_E(S_1) \implies w \in \mathrm{Ext}_E(S_2).$$

- *Definition by intension.* This is much more complex because it depends on the type of variables describing the symbolic objects: if

$$S_1 = [y_1 = V_{1,1}] \wedge [y_2 = V_{1,2}] \wedge \cdots \wedge [y_p = V_{1,p}],$$
$$S_2 = [y_1 = V_{2,1}] \wedge [y_2 = V_{2,2}] \wedge \cdots \wedge [y_p = V_{2,p}],$$

then $S_1 \subset S_2$ if and only if inclusion exists on all the variables. This means that

$$V_{1,j} \subset V_{2,j} \quad \text{for all } j = 1, \ldots, p.$$

Let us see what this means for the various types of variables. (We shall suppress the index j in what follows.)

For a quantitative single variable, the only description is of type constant:

$$[y = V] = [y = C].$$

Thus we must have

$$V_1 = V_2.$$

For an interval variable, $y = V$ can be written $y = [a, b]$.
Thus if, for S_1, $y = [a_1, b_1]$, and, for S_2, $y = [a_2, b_2]$, then we must have

$$[a_1, b_1] \subset [a_2, b_2].$$

For a categorical single variable, $y = V$ can be written $y = v_i$. We must have the same membership of a category for both objects:

$$V_1 = V_2 = [y = v_i].$$

For modal variables, let us first note that a categorical multiple variable can be considered as a modal one where weights are equal to 1 or 0. For a more general modal variable, we have

$$V = [m_1(p_1), m_2(p_2), \ldots, m_k(p_k)]$$

where $m_1, \ldots, m_k$ are the categories and $p_1, \ldots, p_k$ the related probabilities or weights. It is not possible here to say that one histogram is included in another. For inclusion, we need, in this case, an additional notion which is the weight of S_1 relative to S_2. Let us suppose that basic individuals belonging to S_1 and S_2 are known. Let n_1 be the number of individuals in S_1, n_1^* the number of individuals in the complement S_1^* of S_1 relative to S_2, and n_2 the number of individuals in S_2. We have

$$n_2 = n_1 + n_1^*.$$

The relative weight of S_1 and S_2 is

$$W = \frac{n_1}{n_2}.$$

We have the relation

$$p_{2,i} = \frac{p_{1,i} \star n_1 + p_{1,i}^* \star n_1^*}{n_1 + n_1^*}$$

where $p_{1,i}^*$ denotes the probability of category i in S_1^*, and thus

$$p_{1,i} \star \frac{n_1}{n_2} \leq p_{2,i}$$

or

$$p_{1,i} \star W \leq p_{2,i}.$$

If the relation holds for all i, then we can say that $S_1 \subset S_2$.

7.6.2 Hierarchies of symbolic objects

In practice, symbolic objects are often obtained by crossing factors in the basic data. We call the resulting objects *natural symbolic objects*. This operation leads in fact to hierarchies of symbolic objects.

For example, let us suppose that for each individual in the basic database, we know Gender, Age (three categories) and Educational level (two categories). At the first level, we can consider the object 'women' and the object 'men'. At the second level, crossing Age with Gender, we obtain the objects: (Women × Age 1), (Women × Age 2), (Women × Age 3), (Men × Age 1), (Men × Age 2), (Men × Age 3). At the third level, we can cross Gender × Age with Educational level, and so on. We obtain the hierarchy of objects shown in Figure 7.7. We summarize this hierarchy by writing

$$F_1 \rightarrow F_1 \times F_2 \rightarrow F_1 \times F_2 \times F_3.$$

At each level of the hierarchy, the objects are split into smaller ones.

Let x_i denote the category of F_1, y_j that of F_2, and z_k that of F_3, with $1 \leq i \leq 2$, $1 \leq j \leq 3$, $1 \leq k \leq 2$. An object of the hierarchy can be designated by (x_i, y_j, z_k) which means that, if it is the set of individuals with $F_1 = x_i$, $F_2 = y_j$ and $F_3 = z_k$, we have

$$(x_i, y_j, z_k) \subset (x_i, y_j) \subset x_i.$$

But, inverting the order of factors, we could have

$$F_2 \longrightarrow F_2 \times F_3 \longrightarrow F_2 \times F_3 \times F_1$$

$$(y_j, z_k, x_i)) \subset (y_j, z_k) \subset y_j.$$

We also have

$$(x_i, y_j) \subset x_i \quad \text{and} \quad (x_i, y_j) \subset y_j,$$

so that the hierarchy can exist in the hypercube $F_1 \times F_2 \times F_3$.

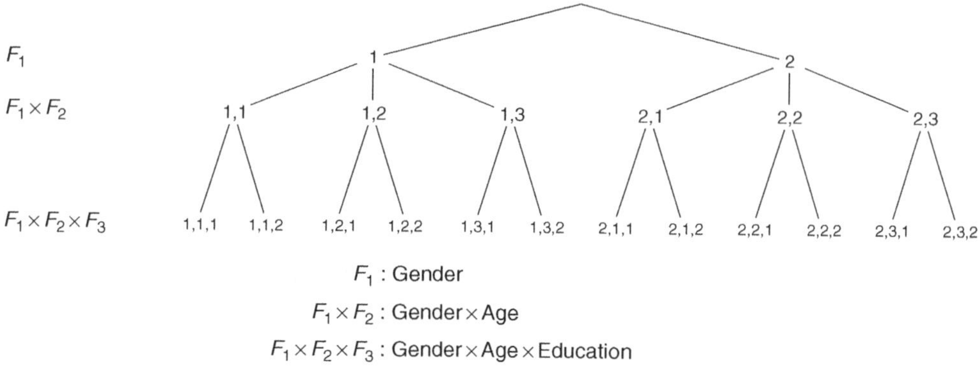

Figure 7.7 Sequence of treatments in symbolic data analysis.

7.6.3 Breakdown

When working with a collection of symbolic objects stored in a file, it is sometimes necessary to go to the lower level. For example, we work on level $F_1 \times F_2$ and want to go to level $F_1 \times F_2 \times F_3$. Two cases can occur:

1. The lower level in the hierarchy already exists, which means that the crossing has been done in the basic database and that the objects of this operation have been stored in a file. It is then only necessary to retrieve this file. If we want to work at the same time on objects of different levels ($F_1 \times F_2$ and $F_1 \times F_2 \times F_3$), we will first have to merge the two levels before being able to deal with the objects (see Chapter 5).

2. The lower level has not been aggregated, which implies that we have to go back to the basic database (if available) and carry out the task of symbolic object construction. This can be tedious if we have started a task of visualization or of other treatment.

7.6.4 Visual breakdown or drilldown

In SODAS2, we offer a facility for interactively constructing lower-level objects, in selecting the appropriate category on the display.

Breakdown or drilldown is a new interactive way to create a symbolic object inside a given class. When selecting a category on an axis of a zoom star, a search in the individuals

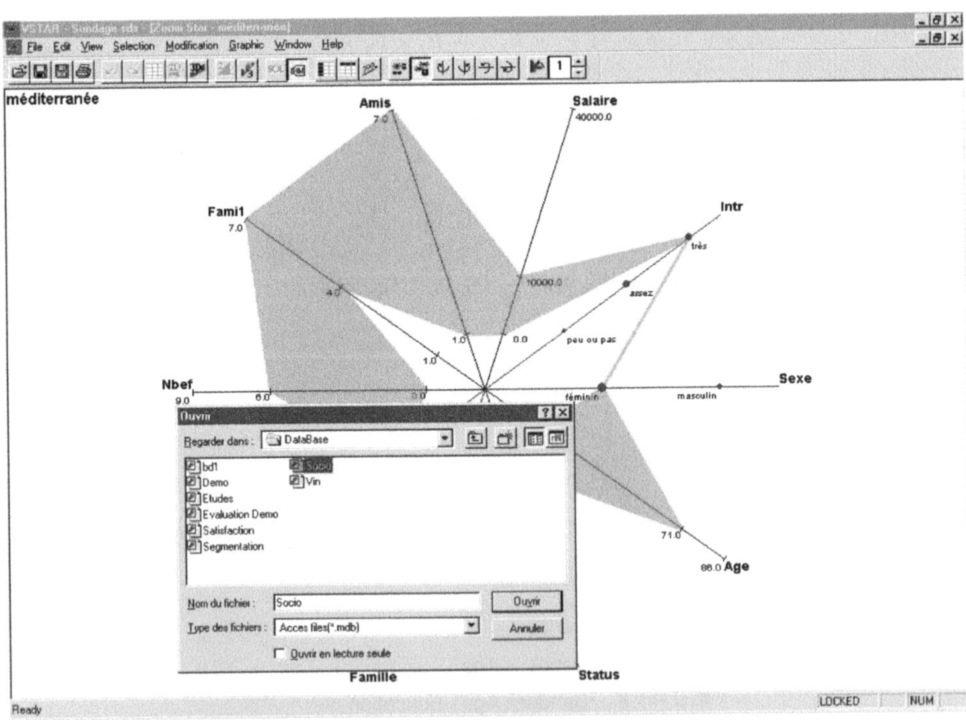

Figure 7.8 Original objects for breakdown.

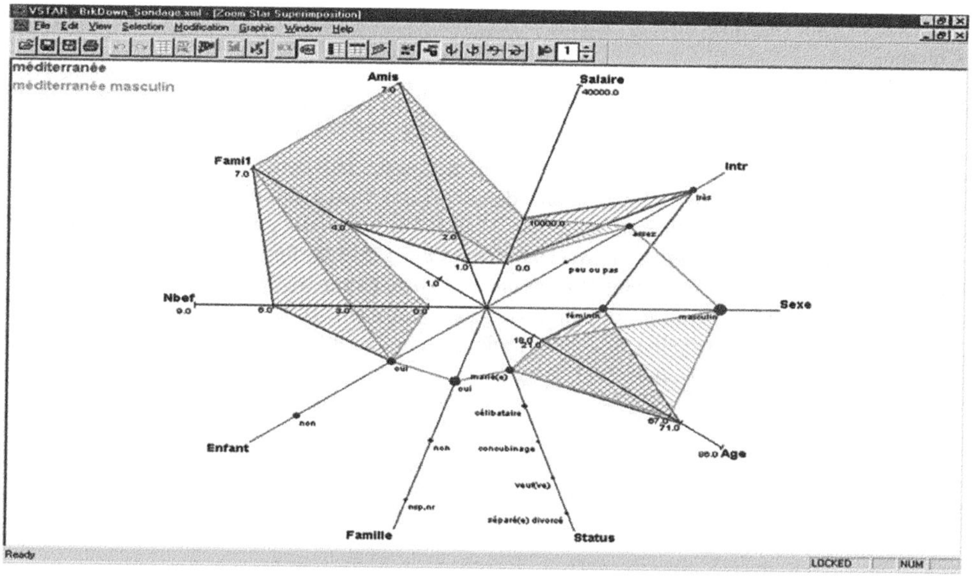

Figure 7.9 Breakdown superimposition of the new and the old objects.

database is generated and the corresponding object is automatically created. This represents the set of individuals who match the selected characteristic. For example, if we import data on sociological investigation, we can interactively create the objects corresponding to gender (women or men), see Figure 7.8, or the class of individuals corresponding to status (single person, widower, cohabitation, divorced). The two symbolic objects will be displayed on the same graph (with superimposition). Figure 7.9 shows the initial object ('Méditerranée') and the set of men inside it.

7.7 References

Ankerst, M., Elsen, C., Ester, M. and Kriegel, H.P. (2003a) Perception-based classification. *Informatica, An International Journal of Computing and Informatics*, 23(4), 493–499.

Ankerst, M., Jones, D., Kao, A. and Wang, C. (2003b) Data Jewel: Tightly integrating visualization with temporal data mining. In S. Simoff, M. Noirhomme, M. Böhlen and M. Ankerst (eds), *Third International Workshop on Visual Data Mining, Third IEEE International Conference on Data Mining*, 19–22 Nov. 2003, Melbourne, Florida.

Bock, H.-H. and Diday, E. (eds) (2000) *Analysis of Symbolic Data*. Springer-Verlag, Berlin.

Cleveland, W.S. (1993) *Visualizing Data*, AT&T Bell Laboratories, Murray Hill, NJ. Hobart Press, Summit, NJ.

Eysenck, M.W. (1993) *Principles of Cognitive Psychology*, Lawrence Erlbraum Associates, Hove.

Harris, N.G.E. (1999) Noticing. *Theory and Psychology*, 9(2), 147–164.

Inselberg, A. (1985) The plane with parallel coordinates. *The Visual Computer*, 1(4), 69–91.

Keim, D.A. and Kriegel, H.-P. (1994) VisDB: Database exploration using multidimensional visualization, *Computer Graphics & Applications*, September, 40–49.

Kuhn, T.S. (1996) *The Structure of Scientific Revolutions*, 3rd edn. University of Chicago Press, Chicago.

Lemaire, P. (1999) *Psychologie cognitive*. De Boeck, Brussels.

Miller, G.A. (1956) The magic number seven, plus or minus two: Some limits on our capacity for processing information. *Psychological Review*, 63, 81–93.

Noirhomme-Fraiture, M. (2002) Visualizing of large data sets: the zoom star solution, *E-Journal of Symbolic Data Analysis*, No. 0. http://www.jsda.unina2.it/newjsda/volumes/vol0/noirho.pdf

Noirhomme-Fraiture, M. and Rouard, M. (2000) Visualizing and editing symbolic objects. In H.-H. Bock and E. Diday (eds), *Analysis of Symbolic Data*, pp. 125–138. Springer-Verlag, Berlin.

Norman, D.A. and Draper, S.W. (1986) *User Centered System Design – New Perspectives on Human–Computer Interaction*. Lawrence Erlbaum Associates, Hillsdale, NJ.

Schneiderman, B. (2002) Inventing discovery tools: Combining information visualization with data mining. *Information Visualization*, 1(1), 5–12.

Part II
UNSUPERVISED METHODS

8

Dissimilarity and matching

Floriana Esposito, Donato Malerba and Annalisa Appice

8.1 Introduction

The aim of symbolic data analysis (SDA) is to investigate new theoretically sound techniques by generalizing some standard statistical data mining methods to the case of *second-order objects*, that is, generalizations of groups of individuals or classes, rather than single individuals (first-order objects).

In SDA, generalizations are typically represented by means of *set-valued* and *modal* variables (Bock and Diday, 2000). A variable Y is termed *set-valued* with domain $\mathcal{Y}$ if it takes its values in $P(\mathcal{X}) = \{U | U \subseteq \mathcal{Y}\}$, that is, the power set of $\mathcal{X}$. When $X(k)$ is finite for each k, then Y is called *multi-valued*, while when an order relation $\prec$ is defined on $\mathcal{Y}$, then the value returned by a set-valued variable is expressed by an interval $[\alpha, \beta]$ and Y is termed an *interval* variable. A modal variable Y is a multi-valued variable with *weights*, such that $Y(k)$ describes both multi-valued data $U(k)$ and associated weights $\pi(k)$.

Generalizations of different groups of individuals from the same population are described by the same set of symbolic variables. This leads to data tables, named *symbolic data tables*, which are more complex than those typically used in classical statistics. Indeed, the columns of a symbolic data table are called *symbolic variables*, while the rows correspond to distinct generalizations (or *symbolic descriptions*) describing a class of individuals that are in turn the partial or complete extent of a given concept.

Starting with a symbolic description, a *symbolic object* (SO) models the underlying concepts and provides a way to find at least the individuals of this class. In Bock and Diday (2000) and Diday and Esposito (2003), an SO is formally defined as a triple $s = (a, R, d)$, where R is a relation between descriptions (e.g., $R \in \{=, \equiv, \leq, \subseteq\}$ or R is an implication, a kind of matching), d is a description and a is a membership function defined by a set of

Symbolic Data Analysis and the SODAS Software Edited by E. Diday and M. Noirhomme-Fraiture

individuals Ω in a set L (e.g., $L = \{\text{true, false}\}$ or $L = [0, 1]$), such that a depends on R and d.

Many techniques for both the construction of SOs from records of individuals and the analysis of SOs are actually implemented in the ASSO Workbench. They deal with special classes of SOs, called *assertions*, where R is defined as $[d'Rd] = \wedge_{j=1,\dots,p}[d'_j R_j d_j]$, with $\wedge$ as the standard logical conjunction operator and a is defined as $a(w) = [y(w)Rd]$, with $y(w) = y_1(w), \dots, y_p(w)$ being the vector of variables describing w.

Dissimilarity and matching constitute one of the methods available in the ASSO Workbench. Several dissimilarity measures (DISS module) and matching functions (MATCH module) are implemented, enabling the user to compare symbolic data in a symbolic data table.

The DISS module implements the dissimilarity measures presented in Malerba *et al.* (2001, 2002). Henceforth, the dissimilarity measure d on a set of individuals E refers to a real-valued function on $E \times E$, such that $d_a^* = d(a, a) \le d(a, b) = d(b, a) < \infty$ for all a, $b \in E$. In contrast, a similarity measure s on a set of objects E is a real-valued function on $E \times E$ such that $s_a^* = s(a, a) \ge s(a, b) = s(b, a) \ge 0$ for all a, $b \in E$. Generally, $d_a^* = d^*$ and $s_a^* = s^*$ for each object a in E, and more specifically, $d^* = 1$ when $s^* = 0$ (Batagelj and Bren, 1995).

Since the similarity comparison can be derived by transforming a dissimilarity measure into a similarity one,[1] only dissimilarity measures are actually implemented in the DISS module.

The MATCH module performs a *directional* (or asymmetric) comparison between the SOs underlying symbolic descriptions stored in a symbolic data table, in order to discover their linkedness or differences. Notice that, while the dissimilarity measures are defined for symbolic descriptions and do not consider how the extent of corresponding SOs is effectively computed according to R, the matching comparison is performed at the level of SOs by interpreting R as a matching operator.

This matching comparison has a *referent* and a *subject* (Patterson, 1990). The former represents either a prototype or a description of a class of individuals, while the latter is either a variant of the prototype or an instance (individual) of a class of objects. In its simplest form, matching compares referent and subject only for equality, returning false when they fail in at least one aspect, and true otherwise. In more complex cases, the matching process performs the comparison between the description of a class C (subject of matching) and the description of some unit u (referent of matching) in order to establish whether the individual can be considered an instance of the class (inclusion requirement). However, this requirement of equality (*canonical matching*), even in terms of inclusion requirement, is restrictive in real-world problems because of the presence of noise, the imprecision of measuring instruments or the variability of the phenomenon described by the referent of matching. This makes it necessary to rely on a relaxed definition of matching that aims to compare two SOs in order to identify their similarities rather than to establish whether they are equal. The result is a *flexible matching* function with a range in the interval [0,1] that indicates the probability of precisely matching the subject with the referent, provided that some change is possibly made in the description of the referent. It is interesting to note

[1] This transformation is made possible by preserving properties defined with the induced quasi-ordering and imposing $d := \phi(s)$ and $s := \psi(d)$, respectively, where $\phi(\cdot)$ and $\psi(\cdot)$ are strictly decreasing functions with boundary conditions (e.g. $\phi(0) = 1$ and $\phi(1) = 0$, or $\phi(0) = 1$ and $\phi(\infty) = 0$).

that even flexible matching is not a dissimilarity measure, due to the non-symmetry of the matching function.

Both DISS and MATCH input a symbolic data table stored in a SODAS file and output a new SODAS file that includes the same input data, in addition to the matrix of the results of dissimilarity (or matching) computation.

More precisely, the dissimilarity (or matching) value computed between the ith symbolic description (SO) and the jth symbolic description (SO) taken from the input symbolic data table is written in the (i, j)th cell (entry) of the output dissimilarity (matching) matrix. The main difference is that the dissimilarity matrix is stored as a lower triangular matrix, since the upper values $(i < j)$ can be derived from the symmetry property of dissimilarity. Conversely, the matching matrix is stored as a sparse square matrix, due to the non-symmetry of the matching function.

Finally, the ASSO Workbench supports users in choosing the list of symbolic variables forming the symbolic descriptions (SOs) to be compared, the dissimilarity measure or matching function to be computed as well as some related parameters.

In this chapter, the use of DISS for the computation of dissimilarity measures is illustrated together with the VDISS module for the visualization of the dissimilarities by means of two-dimensional scatterplots and line graphs. The explanation of the outputs and the results of MATCH for the computation of the matching functions are given at the end of the chapter.

8.2 Input data

The main input of the dissimilarity and matching method is a SODAS file describing a symbolic data table, whose columns correspond to either set-valued or probabilistic symbolic variables. Each row represents the symbolic description d of an individual of E. The ASSO Workbench allows users to select one or more symbolic variables associated with the columns of the symbolic data table stored in the input SODAS file.

Statistical metadata concerning the source agencies of symbolic data, collection information, statistical populations, original variables and standards, symbolic descriptions and symbolic variables, logistical metadata and symbolic analysis previously performed on the data may be available in the metadata file, meta<SODAS file name>.xml, stored in the same directory as the input SODAS file. Notice that metadata information is not manipulated by the dissimilarity and matching method, but it is simply updated by recording the last dissimilarity measure or matching function computed on such data and the list of symbolic variables involved in the comparison.

8.3 Dissimilarity measures

Several methods have been reported in the literature for computing the dissimilarity between two symbolic descriptions d_a and d_b (Malerba *et al.*, 2001, 2002). In the following, we briefly describe the dissimilarity measures implemented in the DISS module for two kinds of symbolic descriptions, named *Boolean* and *probabilistic*. The former are described by set-valued variables, while the latter are described by probabilistic variables, that is, modal variables describing frequency distributions. SOs underlying Boolean and probabilistic symbolic descriptions are referred to as *Boolean symbolic objects* (BSOs) and *Probabilistic symbolic objects* (PSOs), respectively.

Mixed symbolic descriptions, that is, symbolic descriptions described by both set-valued and probabilistic variables, are treated by first separating the Boolean part from the probabilistic part and then computing dissimilarity values separately for these parts. Dissimilarity values obtained by comparing the Boolean and probabilistic parts respectively are then combined by sum or product.

8.3.1 Dissimilarity measures for Boolean symbolic descriptions

Let d_a and d_b be two symbolic descriptions described by m set-valued variables Y_i with domain $\mathcal{Y}_i$. Let $A_i(B_i)$ be the set of values (subset of Y_i) taken from Y_i in d_a (d_b). A class of dissimilarity measures between d_a and d_b is defined by aggregating dissimilarity values computed independently at the level of single variables Y_i (*componentwise dissimilarities*). A classical aggregation function is the Minkowski metric (or L_q distance) defined on $\mathbb{R}^m$. Another class of dissimilarity measures is based on the notion of *description potential* $\pi(d_a)$ of a symbolic description d_a, which corresponds to the *volume* of the Cartesian product $A_1 \times A_2 \times \cdots \times A_p$. For this class of measures no componentwise decomposition is necessary, so that no function is required to aggregate dissimilarities computed independently for each variable.

Dissimilarity measures implemented in DISS are reported in Table 8.1 together with their short identifier used in the ASSO Workbench. They are:

- Gowda and Diday's dissimilarity measure (U_1: Gowda and Diday, 1991);
- Ichino and Yaguchi's first formulation of a dissimilarity measure (U_2: Ichino and Yaguchi, 1994);
- Ichino and Yaguchi's normalized dissimilarity measure (U_3: Ichino and Yaguchi, 1994);
- Ichino and Yaguchi's normalized and weighted dissimilarity measure (U_4: Ichino and Yaguchi, 1994);
- de Carvalho's normalized dissimilarity measure for constrained[2] Boolean descriptions (C_1: de Carvalho, 1998);
- de Carvalho's dissimilarity measure (SO_1: de Carvalho, 1994);
- de Carvalho's extension of Ichino and Yaguchi's dissimilarity (SO_2: de Carvalho, 1994);
- de Carvalho's first dissimilarity measure based on description potential (SO_3: de Carvalho, 1998);
- de Carvalho's second dissimilarity measure based on description potential (SO_4: de Carvalho, 1998);

[2] The term *constrained Boolean descriptions* refers to the fact that some dependencies are defined between two symbolic variables X_i and X_j, namely *hierarchical dependencies* which establish conditions for some variables which are not measurable (not-applicable values), or *logical dependencies* which establish the set of possible values for a variable X_i conditioned by the set of values taken by the variable X_j. An investigation of the effect of constraints on the computation of dissimilarity measures is outside the scope of this paper, nevertheless it is always possible to apply the measures defined for constrained Boolean descriptions to unconstrained Boolean descriptions.

- de Carvalho's normalized dissimilarity measure based on description potential (SO_5: de Carvalho, 1998);
- a dissimilarity measure based on flexible matching between BSOs (SO_6).

The last measure (SO_6) differs from the others, since its definition is based on the notion of flexible matching (Esposito *et al.*, 2000), which is an asymmetric comparison. The dissimilarity measure is obtained by means of a symmetrization method that is common to measures defined for probabilistic symbolic descriptions.

The list of dissimilarity measures actually implemented in DISS cannot be considered complete. For instance, some clustering modules of the ASSO Workbench (e.g., NBCLUST and SCLUST; see Chapter 14) estimate the dissimilarity value between two boolean symbolic descriptions d_a and d_b as follows:

$$d(d_a, d_b) = \left(\sum_{i=1}^{m} \delta_i^2 (A_i, B_i) \right)^{1/2},$$

where δ_i denotes a dissimilarity index computed on each pair (A_i, B_i). In particular, if Y_i is an interval variable, we have that $A_i = [a_{i,\inf}, a_{i,\sup}]$ and $B_i = [b_{i,\inf}, b_{i,\sup}]$. In this case, $\delta_i(A_i, B_i)$ is computed in terms of:

- the Hausdorff distance defined by

$$\delta_i([a_{i,\inf}, a_{i,\sup}], [b_{i,\inf}, b_{i,\sup}]) = \max\{|a_{i,\inf} - b_{i,\inf}|, |a_{i,\sup} - b_{i,\sup}|\};$$

- the L_1 distance defined by

$$\delta_i([a_{i,\inf}, a_{i,\sup}], [b_{i,\inf}, b_{i,\sup}]) = |a_{i,\inf} - b_{i,\inf}| + |a_{i,\sup} - b_{i,\sup}|;$$

- the L_2 distance defined by

$$\delta_i([a_{i,\inf}, a_{i,\sup}], [b_{i,\inf}, b_{i,\sup}]) = (a_{i,\inf} - b_{i,\inf})^2 + (a_{i,\sup} - b_{i,\sup})^2.$$

On the other hand, if Y_i is a multi-valued variable that takes its values in the power set of Y_i (i.e., $\boldsymbol{P}(\mathcal{Y}_i) = \{U | U \subseteq \mathcal{Y}_i\}$), the dissimilarity index δ_i is computed by estimating the difference in the frequencies of each category value taken by Y_i.

Denoting by p_i the number of categories in Y_i ($p_i = |U_i|$, with U_i the range of Y_i), the frequency value $q_{i,U_i}(c_s)$ associated with the category value c_s ($s = 1, \ldots, p_i$) of the variable $Y_i = U_i (c_s \in U_i)$ is given by

$$q_{i,U_i}(c_s) = \begin{cases} \frac{1}{|U_i|}, & \text{if } c_s \in U_i, \\ 0, & \text{otherwise.} \end{cases}$$

Therefore, the symbolic descriptions d_a and d_b can be transformed as follows:

$$d_a = ((q_{1,A_1}(c_1), \ldots, q_{1,A_1}(c_{p_1})), \ldots, (q_{m,A_m}(c_1), \ldots, q_{m,A_m}(c_{p_m}))),$$
$$d_b = ((q_{1,B_1}(c_1), \ldots, q_{1,B_1}(c_{p_1})), \ldots, (q_{m,B_m}(c_1), \ldots, q_{m,B_m}(c_{p_m}))),$$

Table 8.1 Dissimilarity measures defined for Boolean symbolic descriptions.

Name	Componentwise dissimilarity measure	Objectwise dissimilarity measure
U_1	$D^{(i)}(A_i, B_i) = D_\pi(A_i, B_i) + D_s(A_i, B_i) + D_c(A_i, B_i)$ where $D_\pi(A_j, B_j)$ is due to *position*, $D_s(A_j, B_j)$ to *spanning* and $D_c(A_j, B_j)$ to *content*.	$\sum_{i=1}^{m} D^{(i)}(A_i, B_i)$
U_2	$\phi(A_i, B_i) = \|A_i \oplus B_i\| - \|A_i \otimes B_i\| + \gamma(2\|A_i \otimes B_i\| - \|A_i\| - \|B_i\|)$ with *meet* ($\otimes$) and *join* ($\oplus$) Cartesian operators.	$\sqrt[q]{\sum_{i=1}^{m} [\phi(A_i, B_i)]^q}$
U_3	$\psi(A_i, B_i) = \dfrac{\phi(A_i, B_i)}{\|X_i\|}$	$\sqrt[q]{\sum_{i=1}^{m} [\psi(A_i, B_i)]^q}$
U_4	$\psi(A_i, B_i) = \dfrac{\phi(A_i, B_i)}{\|X_i\|}$	$\sqrt[q]{\sum_{i=1}^{m} w_i [\psi(A_i, B_i)]^q}$
C_1	$D_1(A_i, B_i) = 1 - \alpha/(\alpha+\beta+\chi)$ $D_2(A_i, B_i) = 1 - 2\alpha/(2\alpha+\beta+\chi)$ $D_3(A_i, B_i) = 1 - \alpha/(\alpha+2\beta+2\chi)$ $D_4(A_i, B_i) = 1 - \frac{1}{2}\left(\frac{\alpha}{\alpha+\beta} + \frac{\alpha}{\alpha+\chi}\right)$ $D_5(A_i, B_i) =$ $1 - \alpha/\sqrt{(\alpha+\beta)(\alpha+\chi)}$ with $\chi = \mu(c(A_i) \cap B_i)$ $\alpha = \mu(A_i \cap B_i)$; $\beta = \mu(A_i \cap c(B_i))$	$\sqrt[q]{\dfrac{\sum_{i=1}^{m} [w_i D_r(A_i, B_i)]^q}{\sum_{i=1}^{m} \delta(i)}}$, where $\delta(i)$ is the indicator function

	For each subset $V_j \subseteq Y_i \mu(V_j) = \lvert V_j \rvert$ if Y_j is integer, nominal or ordinal and $\mu(V_j) = \lvert a - b \rvert$ if Y_j is continuous and $V_j = [a - b]$. $c(V_j)$ denotes the complementary set of V_j in the domain Y_i.	
SO_1		$\sqrt[q]{\sum_{i=1}^{m} [w_i D_r (A_i, B_i)]^q}$
SO_2	$\psi'(A_i, B_i) = \frac{\phi(A_i, B_i)}{\mu(A_i \oplus B_i)}$	$\sqrt[q]{\sum_{i=1}^{m} \frac{1}{m} [\psi'(A_i, B_i)]^q}$
SO_3	none	$\pi(d_a \oplus d_b) - \pi(d_a \otimes d_b)$ $+\gamma(2\pi(d_a \otimes d_b) - \pi(a) - \pi(b))$ where meet ($\otimes$) and join ($\oplus$) are Cartesian operators defined on BSOs.
SO_4	none	$\frac{\pi(d_a \oplus d_b) - \pi(d_a \otimes d_b) + \gamma(2\pi(d_a \otimes d_b) - \pi(d_a) - \pi(d_b))}{\pi(d_a^E)}$
SO_5	none	$\frac{\pi(d_a \oplus d_b) - \pi(d_a \otimes d_b) + \gamma(2\pi(d_a \otimes d_b) - \pi(d_a) - \pi(d_b))}{\pi(d_a \oplus d_b)}$
SO_6	none	$1 - [FlexMatch(a, b) + FlexMatch(b, a)]/2$ where *FlexMatch* denotes the flexible matching function, while a and b are the BSOs in the form of assertions underlying the descriptions d_a and d_b, respectively.

such that $\sum_{j=1}^{m_i} q_{i,A_i}(c_j) = 1$ and $\sum_{j=1}^{m_i} q_{i,B_i}(c_j) = 1$, for all $i \in \{1, \ldots, m\}$. Hence the dissimilarity index $\delta_i(A_i, B_i)$ is computed in terms of:

- the L_1 distance defined by

$$\delta_i(A_i, B_i) = \sum_{j=1}^{|Y_i|} |q_{i,A_i}(c_j) - q_{i,B_i}(c_j)|;$$

- the L_2 distance defined by

$$\delta_i(A_i, B_i) = \sum_{j=1}^{|Y_i|} (q_{i,A_i}(c_j) - q_{i,B_i}(c_j))^2;$$

- the de Carvalho distance defined by

$$\delta_i(A_i, B_i) = \sum_{j=1}^{|Y_i|} (\gamma q_{i,A_i}(c_j) + \gamma' q_{i,B_i}(c_j))^2,$$

where

$$\gamma = \begin{cases} 1, & \text{if } c_j \in A_i \wedge c_j \notin B_i, \\ 0, & \text{otherwise,} \end{cases}$$

$$\gamma' = \begin{cases} 1, & \text{if } c_j \notin A_i \wedge c_j \in B_i, \\ 0, & \text{otherwise.} \end{cases}$$

These dissimilarity measures will be implemented in an extended version of the DISS module.

8.3.2 Dissimilarity measures for probabilistic symbolic descriptions

Let d_a and d_b be two probabilistic symbolic descriptions and Y a multi-valued modal variable describing them. The sets of probabilistically weighted values taken by Y in d_a and d_b define two discrete probability distributions P and Q, whose comparison allows us to assess the dissimilarity between d_a and d_b on the basis of Y only. For instance, we may have: $P =$ (red:0.$\bar{3}$, white:0.$\bar{3}$, black:0.$\bar{3}$) and $Q =$ (red:0.1, white:0.2, black:0.7) when the domain of $\mathcal{Y}$ is $=$ {red, white, black}. Therefore, the dissimilarity between two probabilistic symbolic descriptions described by p symbolic probabilistic variables can be obtained by aggregating the dissimilarities defined on as many pairs of discrete probability distributions (componentwise dissimilarities). Before explaining how to aggregate them, some comparison functions $m(P, Q)$ for probability distributions are introduced.

Most of the comparison functions for probability distributions belong to the large family of 'convex likelihood-ratio expectations' introduced by both Csiszár (1967) and Ali and Silvey (1996). Some well-known members of this family are as follows:

- The *Kullback–Leibler* (KL) *divergence*, which is a measure of the difference between two probability distributions (Kullback and Leibler, 1951). This is defined as $m_{\text{KL}}(P, Q) := \Sigma_{x \in X} q(x) \log(q(x)/p(x))$ and measures to what extent the distribution P is an approximation of the distribution Q. It is asymmetric, that is, $m_{\text{KL}}(P, Q) \neq m_{\text{KL}}(Q, P)$ in general, and it is not defined when $p(x) = 0$. The KL divergence is generally greater than zero, and it is zero only when the two probability distributions are equal.
- The χ^2 *divergence*, defined as $m_{\chi^2}(P, Q) := \Sigma_{x \in X} |p(x) - q(x)|^2 / p(x)$, is strictly topologically stronger than the KL divergence, since the inequality $m_{\text{KL}}(P, Q) \leq m_{\chi^2}(P, Q)$ holds, i.e. the convergence in χ^2 divergence implies convergence in the KL divergence, but the converse is not true (Beirlant *et al.*, 2001). Similarly to the KL divergence, it is asymmetric and is not defined when $p(x) = 0$.
- The Hellinger coefficient is a similarity-like measure given by

$$m^{(s)}(P, Q) := \Sigma_{x \in X} q^s(x).p^{1-s}(x),$$

where s is a positive exponent with $0 < s < 1$. From this similarity-like measure *Chernoff's distance of order s* is derived as follows:

$$m_{\text{C}}^{(s)}(P, Q) := -\log m^{(s)}(P, Q).$$

This distance diverges only when the two distributions have zero overlap, that is, the intersection of their support is empty (Kang and Sompolinsky, 2001).

- *Rényi's divergence* (or *information gain*) of order s between two probability distributions P and Q is given by $m_{\text{R}}{}^{(s)}(P, Q) := -\log\ m^{(s)}(P, Q)/(s-1)$. It is noteworthy that, as $s \to 1$, Rényi's divergence approaches the KL divergence (Rached *et al.*, 2001).
- The *variation distance*, given by $m_1(P, Q) := \Sigma_{x \in X} |p(x) - q(x)|$, is also known as the *Manhattan distance* for the probability functions $p(x)$ and $q(x)$ and coincides with the *Hamming distance* when all features are binary. Similarly, it is possible to use *Minkowski* L_2 (or *Euclidean*) *distance* given by $m_2(P, Q) := \Sigma_{x \in X} |p(x) - q(x)|^2$ and, more generally, the Minkowski L_p distance with $p \in \{1, 2, 3, \ldots\}$. All measures $m_p(P, Q)$ satisfy the metric properties and in particular the symmetry property. The main difference between m_1 and m_p, $p > 1$, is that the former does not amplify the effect of single large differences (outliers). This property can be important when the distributions P and Q are estimated from noisy data.
- The *Kullback divergence* is given by $m_{\text{K}}(P, Q) := \Sigma_{x \in X} q(x) \log(q(x)/(1/2p(x) + 1/2q(x)))$ (Lin, 1991), which is an asymmetric measure. It has been proved that the Kullback divergence is upper bounded by the variation distance $m_1(P, Q) : m_{\text{K}}(P, Q) \leq m_1(P, Q) \leq 2$.

Some of the divergence coefficients defined above do not obey all the fundamental axioms that dissimilarities must satisfy. For instance, the KL divergence does not satisfy the symmetric property. Nevertheless, a symmetrized version, termed the *J-coefficient* (or *J-divergence*), can be defined as follows:

$$J(P, Q) := m_{\mathrm{KL}}(P, Q) + m_{\mathrm{KL}}(Q, P).$$

Alternatively, many authors have defined the *J-divergence* as the average rather than the sum $J(P, Q) := (m_{\mathrm{KL}}(P, Q) + m_{\mathrm{KL}}(Q, P))/2$. Generally speaking, for any (*possible*) non-symmetric divergence coefficient m there exists a symmetrized version $\underline{m}(P, Q) = m(Q, P) + m(P, Q)$ which fulfils all axioms for a dissimilarity measure, but typically not the triangle inequality. Obviously, in the case of Minkowski's L_p coefficient, which satisfies the properties of a dissimilarity measure and, more precisely, of a metric (triangular inequality), no symmetrization is required.

Given these componentwise dissimilarity measures, we can define the dissimilarity measure between two probabilistic symbolic descriptions d_a and d_b by aggregation through the generalized and weighted Minkowski metric:

$$d_p(d_a, d_b) = \sqrt[p]{\sum_{i=1}^{m} [c_i m(A_i, B_i)]^p},$$

where $\forall k \in \{1, \ldots, m\}, c_k > 0$ are weights with $\Sigma_{k=1\ldots m} c_k = 1$ and $m(A_i, B_i)$ is either the Minkowski L_p distance (LP) or a symmetrized version of the J-coefficient (J), χ^2 divergence (CHI2), Rényi's distance (REN), or Chernoff's distance (CHER). These are all variants of the dissimilarity measure denoted by P_1 in the ASSO Workbench. Notice that the Minkowski L_p distance, the J-coefficient, the χ^2 divergence, Rényi's distance and Chernoff's distance require no category of a probabilistic symbolic variable in a probabilistic symbolic description to be associated with a zero probability. To overcome these limitations, symbolic descriptions may be generated by using the *KT estimate* when estimating the probability distribution, in order to prevent the assignments of a zero probability to a category. This estimate is based on the idea that no category of a modal symbolic variable in a PSO can be associated with a zero probability. The KT estimate is computed as:

$$p(x) = \frac{(\text{No. of times } x \text{ occurs in } \{R_1, \ldots, R_M\}) + 1/2}{M + (K/2)},$$

where x is the category of the modal symbolic variable, $\{R_1, \ldots, R_M\}$ are sets of aggregated individuals, M is the number of individuals in the class, and K is the number of categories of the modal symbolic variable (Krichevsky and Trofimov, 1981).

The dissimilarity coefficients can also be aggregated through the product. Therefore, by adopting appropriate precautions and considering only Minkowski's L_p distance, we obtain the following normalized dissimilarity measure between probabilistic symbolic descriptions:

$$d'_p(d_a, d_b) = 1 - \frac{\prod_{i=1}^{m}\left(\sqrt[p]{2} - \sqrt[p]{\sum_{y_i} |p(x_i) - q(x_i)|^p}\right)}{\left(\sqrt[p]{2}\right)^m} = 1 - \frac{\prod_{i=1}^{m}\left(\sqrt[p]{2} - \sqrt[p]{L_p}\right)}{\left(\sqrt[p]{2}\right)^m},$$

where each x_i corresponds to a value of the ith variable domain.

Table 8.2 Dissimilarity measures defined for probabilistic symbolic descriptions.

Name	Componentwise dissimilarity measure	Objectwise dissimilarity measure
P_1	Either $m_p(P, Q)$ or a symmetrized version of $m_{KL}(P, Q)$, $m_\chi^2(P, Q)$, $m_C^{(s)}(P, Q)$, $m_R^{(s)}(P, Q)$	$\sqrt[p]{\sum_{i=1}^{m} [c_i m(A_i, B_i)]^p}$
P_2	$m_p(P, Q)$	$1 - \frac{\prod_{i=1}^{m}\left(\sqrt[p]{2} - \sqrt[p]{m_p(A_i, B_i)}\right)}{\left(\sqrt[p]{2}\right)^m}$
P_3	none	$1 - [FlexMatch(a, b) + FlexMatch(b, a)]/2$, where *FlexMatch* denotes the flexible matching function, while a and b are the PSOs in the form of assertions representing the descriptions d_a and d_b, respectively.

Note that this dissimilarity measure, denoted as P_2 in the ASSO Workbench, is symmetric and normalized in [0,1]. Obviously $d'_p(d_a, d_b) = 0$ if d_a and d_b are identical and $d'_p(d_a, d_b) = 1$ if the two symbolic descriptions are completely different.

Alternatively, the dissimilarity measure between two probabilistic dissimilarity descriptions d_a and d_b can be computed by estimating both the matching degree between the corresponding PSOs a and b and vice versa. The measure denoted as P_3 in the ASSO Workbench extends the SO_6 measure defined for BSOs. A summary of the three dissimilarity measures, defined on probabilistic symbolic descriptions, is reported in Table 8.2.

As already observed for the Boolean case, the list of dissimilarity measures implemented in DISS for PSOs is not exhaustive. Some clustering modules of the ASSO Workbench (e.g., NBCLUST and SCLUST; see Chapter 14) implement a further dissimilarity measure that estimates the dissimilarity between two probabilistic symbolic descriptions by composing the values of dissimilarity indices δ_i as follows:

$$d(d_a, d_b) = \left(\sum_{i=1}^{m} \delta_i^2((A_i, \pi_{A_i}), (B_i, \pi_{B_i}))\right)^{1/2}.$$

In this case, the dissimilarity index $\delta_i((A_i, \pi_{A_i}), (B_i, \pi_{B_i}))$ is computed in terms of:

- the L_1 distance defined by

$$\delta_i((A_i, \pi_{A_i}), (B_i, \pi_{B_i})) = \sum_{j=1}^{|Y_i|} |\pi_{A_i}(c_j) - \pi_{B_i}(c_j)|;$$

- the L_2 distance defined by

$$\delta_i(A_i, B_i) = \sum_{j=1}^{|Y_i|} (\pi_{A_i}(c_j) - \pi_{B_i}(c_j))^2;$$

- the de Carvalho distance defined by

$$\delta_i(A_i, B_i) = \sum_{j=1}^{|Y_i|} (\gamma\pi_{A_i}(c_j) + \gamma'\pi_{B_i}(c_j))^2,$$

  where γ and γ' are defined as before.

Also in this case we plan to implement these additional dissimilarity measures for PSOs in a new release of the DISS module.

8.4 Output of DISS and its visualization

The DISS module outputs a new SODAS file that includes both the input symbolic data table D and the dissimilarity matrix M resulting from the computation of the dissimilarity between each pair of symbolic descriptions from D. This means that $M(i, j)$ corresponds to the dissimilarity value computed between the ith symbolic description and the jth symbolic description taken from D. Since dissimilarity measures are defined as symmetric functions, M is a symmetric matrix with $M(i, j) = M(j, i)$. However, due to computation issues, M is effectively computed as a lower triangular matrix, where dissimilarity values are undefined for upper values $(i < j)$ of $M(i, j)$. In fact, upper values can be obtained without effort by exploiting the symmetry property of dissimilarity measures. In addition, DISS produces a report file that is a printer-formatted file describing both the input parameters and the matching matrix. When a metadata file is associated with the input SODAS file, DISS updates the metadata by recording the dissimilarity measure and the list of symbolic variables considered when computing the dissimilarity matrix in question.

Both the dissimilarity matrix and the dissimilarity metadata can be visualized by means of the SODAS2 module VDISS. More precisely, VDISS outputs the matrix M in either a *table format*, a two-dimensional *scatterplot* or *graphic* representation.

The table format visualization shows the dissimilarity matrix M as a symmetric matrix, where both rows and columns are associated with the individuals whose symbolic descriptions are stored in the symbolic data table stored in the input SODAS file. Although M is computed as a lower triangular matrix, undefined upper values of M ($M(i, j)$ with $i < j$), are now explicitly stated by imposing $M(i, j) = M(j, i)$.

Moreover, several properties can be checked on the dissimilarity matrix: the *definiteness* property,

$$M(i, j) = 0 \Rightarrow i = j, \quad \forall i, j = 1, \ldots, n;$$

the *evenness* property,

$$M(i, j) = 0 \Rightarrow M(i, k) = M(j, k), \quad \forall k = 1, \ldots, n;$$

the *pseudo-metric* or *semi-distance*,

$$M(i, j) \leq M(i, k) + M(k, j), \quad \forall i, j, k = 1, \ldots, n;$$

the *Robinsonian* property, by which, for each $k = 1, \ldots, n$, we have that

$$\begin{aligned} M(k, k) &\leq M(k, k+1) \leq \ldots \leq M(k, n-1) \leq M(k, n) \wedge M(k, k) \\ &\leq M(k, k-1) \leq \ldots \leq M(k, 2) \leq M(k, 1), \\ M(k, k) &\leq M(k+1, k) \leq \ldots \leq M(n-1, k) \leq M(n, k) \wedge M(k, k) \\ &\leq M(k-1, k) \leq \ldots \leq M(2, k) \leq M(1, k); \end{aligned}$$

Buneman's inequality,

$$M(i, j) + M(h, k) \leq \max\{M(i, h) + M(j, k), M(i, k) + M(j, h)\} \quad \forall i, j, h, k = 1, \ldots, n;$$

and, finally, the *ultrametric* property,

$$M(i, j) \leq \max\{M(i, k), M(k, j)\} \quad \forall i, j, k = 1, \ldots, n.$$

The two-dimensional scatterplot visualization is based on the non-linear mapping of symbolic descriptions stored in the input SODAS file and points of a two-dimensional space. This non-linear mapping is based on an extension of Sammon's algorithm (Sammon, 1969) that takes as input the dissimilarity matrix M and returns a collection of points in the two-dimensional space (visualization area), such that their Euclidean distances preserve the 'structure' of the original dissimilarity matrix.

Scatterplot visualization supports both scrolling operations (left, right, up or down) as well as zooming operations over the scatterplot area. For each point in the scatterplot area, the user can also display the (X, Y) coordinates as well as the name (or label) of the individual represented by the point.

The dissimilarity matrix M can also be output graphically in the form of a partial or total line, bar and pie chart. In line chart based output, dissimilarity values are reported along the vertical axis, while individual identifiers (labels or names) are reported on the horizontal axis. For each column j of M, a different line is drawn by connecting the set of points $P(i, j)$ associated with the $M(i, j)$ value. In particular, the (X, Y) coordinates of the point $P(i, j)$ represent the individual on the *ith* row of M and the dissimilarity value stored in the $M(i, j)$ cell, respectively. The main difference between a partial line chart and a total line chart is that the former treats M as a lower triangular matrix and draws a line for each column j of M by ignoring points whose ordinate value is undefined in M (i.e. $i < j$), while the latter treats M as a symmetric matrix and derives undefined values by exploiting the symmetry property of the dissimilarity measures.

Both partial and total line charts can be visualized in a two- or three-dimensional space. Dissimilarity values described with total line charts can also be output as bar or pie charts.

Finally, a report including the list of variables and the dissimilarity measures adopted when computing M can be output in a text box.

8.5 An Application of DISS

In this section, we show the use of both the DISS module for the computation of a dissimilarity matrix from a symbolic data table stored in an input SODAS file and the VDISS module for the visualization of dissimilarities by means of two-dimensional scatterplots and line graphs. For this purpose, we present a case study of the analysis of the symbolic data table stored in the SODAS file enviro.xml that contains symbolic descriptions of 14 individuals generated by DB2SO.

Symbolic data are extracted by generalizing the data derived from a survey conducted by Statistics Finland. The population under analysis is a sample of 2500 Finnish residents aged between 15 and 74 in December 2000. Data are collected by interview and missing values are imputed by logistic regression. The survey contains 46 questions, but only 17 questions representing both continuous and categorical variables are selected as independent variables for symbolic data generation. Symbolic data are constructed by Cartesian product among three categorical variables (grouping variables): gender (M, F), age ([15–24], [25–44], [45–64], [65–74]) and urbanicity (very urban and quite urban).[3] Statistical metadata concerning information about the sample Finnish population analysed for the survey, the original variables, the symbolic descriptions and the symbolic variables are stored in metaenviro.xml.

Starting from the enviro symbolic data, a new ASSO chain named enviro is created by selecting Chaining from the main (top-level) menu bar and clicking on New chaining or typing Ctrl-N. The base node is associated with the SODAS file enviro.xml and a new empty block is added to the running chain by right-clicking on the Base block and choosing Insert method from the pop-up menu. The DISS module is selected from Dissimilarity and Matching in the Methods drop-down list and dragged onto the empty block (see Figure 8.1).

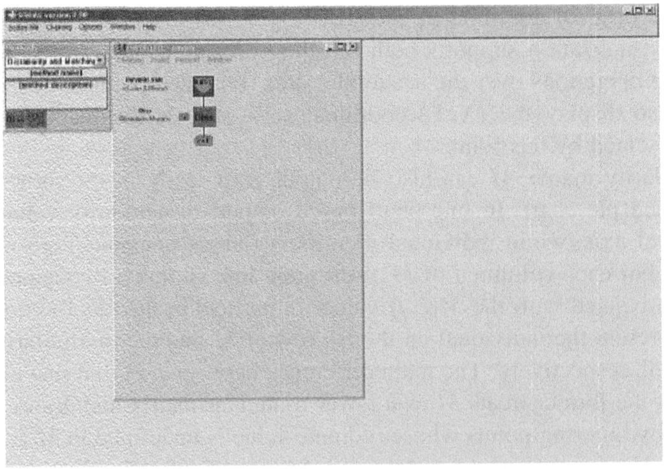

Figure 8.1 An example of the ASSO chain.

[3] The enviro.xml SODAS file contains the symbolic descriptions of only 14 individuals instead of 16. This is due to the fact that no enviro micro-data fall in two of the grouping sets obtained by DB2SO when aggregating enviro micro-data with respect to the 'gender–age–urbanicity' attributes.

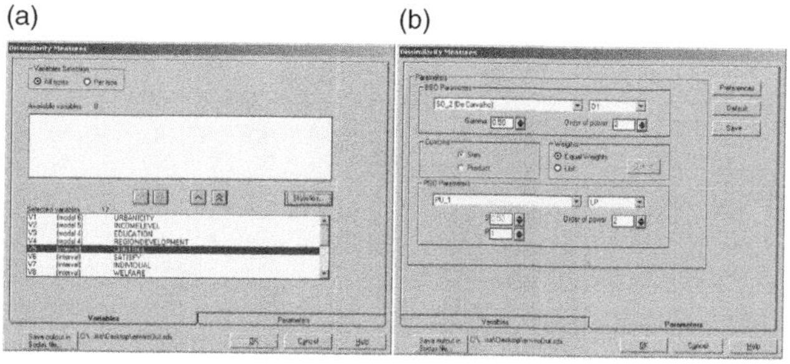

Figure 8.2 (a) List of variables and (b) dissimilarity measures selected when computing the dissimilarity matrix from the symbolic descriptions stored in the enviro SODAS file.

This chain is now able to compute and output the dissimilarity matrix representing dissimilarity values between each pair of symbolic descriptions stored in the enviro symbolic data table.

Before computing the dissimilarity matrix, users must choose the list of symbolic variables to be involved in computing dissimilarities, the dissimilarity measure(s) to be used, the name of the output SODAS file, and so on. Both the list of symbolic variables and the dissimilarity measures are chosen by selecting Parameters. . . from the pop-up menu associated with the DISS block in the chain. The list of symbolic variables taken from the input symbolic data table is shown in a list box and users choose the variables to be considered when computing dissimilarity (see Figure 8.2(a)).

For each symbolic variable some statistics can be output, namely, the minimum and maximum for continuous (single-valued or interval) variables and the frequency distribution of values for categorical (single-valued or multi-valued) variables.

By default, all the symbolic variables are available for selection without any restriction on the type. However, users may decide to output only a subset of the variables taken from the input symbolic data by filtering on the basis of the variable type. Whenever users select only set-valued (probabilistic) variables, symbolic descriptions to be compared are treated as Boolean (probabilistic) data. Conversely, when users select probabilistic variables in addition to set-valued variables, symbolic descriptions to be compared are treated as mixed data.

In this application, let us select all the symbolic variables (13 interval variables and four probabilistic variables) from the enviro data. This means that symbolic descriptions considered for dissimilarity computation are mixed data, where it is possible to separate the Boolean part from the probabilistic part. Users have to choose a dissimilarity measure for the Boolean part and a dissimilarity measure for the probabilistic part and to combine the result of computation by either sum or product (see Figure 8.2(b)).

In this example, we choose the dissimilarity measures SO_2 to compare the Boolean parts and P_1(LP) to compare the probabilistic parts of the enviro symbolic descriptions. Equal weights are associated with the set-valued variables, while dissimilarity values obtained by comparing Boolean parts and probabilistic parts are combined in an additive form.

When all these parameters are set, the dissimilarity matrix can be computed by choosing Run method from the pop-up menu associated with the DISS block in the running chain. DISS produces as output a new SODAS file that is stored in the user-defined path and

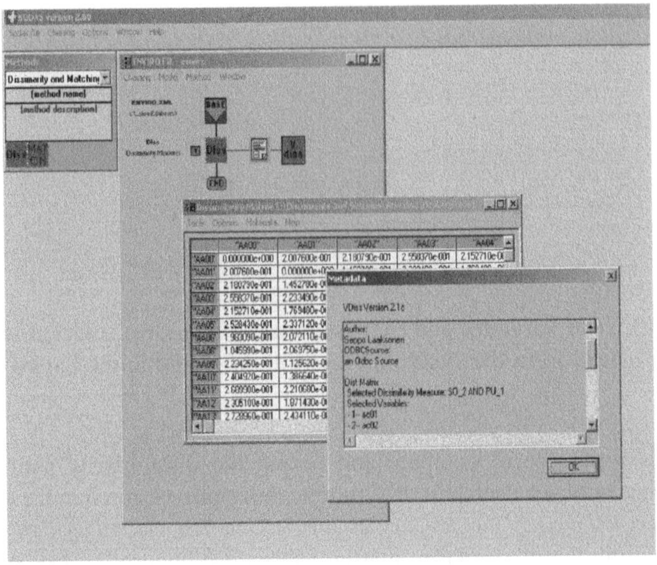

Figure 8.3 The table format output of the dissimilarity matrix and the dissimilarity metadata output of DISS on symbolic descriptions stored in the enviro.xml SODAS file.

includes both the symbolic data table stored in the input SODAS file and the dissimilarity matrix computed by DISS.

The metadata file is updated with information concerning the dissimilarity measures and the symbolic variables involved in the dissimilarity computation.

When the dissimilarity matrix is correctly constructed, the dissimilarity matrix is stored in the output SODAS file, in addition to the input symbolic data table. Moreover, the running chain is automatically extended with a yellow block (report block) that is directly connected to the DISS block. The report block allows users to output a report that describes the symbolic variables and the dissimilarity measures selected for the dissimilarity computation, as well as the lower triangular dissimilarity matrix computed by DISS. A pop-up menu associated with the report block allows users to either output this report as a printer-formatted file by selecting Open. . . and then View Result Report . . . or remove the results of the DISS computation from the running chain by selecting Delete results . . . from the menu in question.

Moreover, a red block connected to the yellow one is introduced in the running chaining. This block is automatically associated with the VDISS module and allows users to output both the dissimilarity matrix and the dissimilarity metadata (see Figure 8.3).

Table format output is shown by selecting Open. . . from the pop-up menu associated with the VDISS block of the running chain.

VDISS allows users to plot the symbolic descriptions taken from the enviro SODAS file as points on a two-dimensional scatterplot such that the Euclidean distance between the points preserves the dissimilarity values computed by DISS.

Notice that opting for a scatterplot highlights the existence of three clusters of similar symbolic descriptions (see Figure 8.4). In particular, symbolic descriptions labelled with AA00 and AA08 appear tightly close in the scatterplot area. This result is confirmed

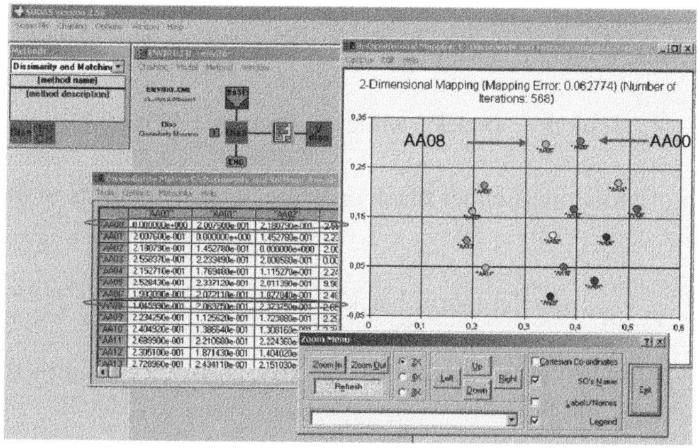

Figure 8.4 Scatterplot output of the symbolic descriptions stored in the enviro SODAS file, such that the Euclidean distance between the points preserves the dissimilarity values computed by the DISS module.

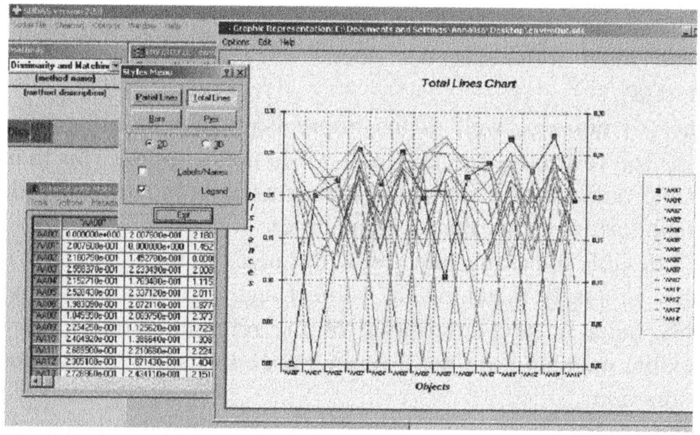

Figure 8.5 Line chart output of the dissimilarity matrix computed on the mixed symbolic data stored in the enviro SODAS file.

when visualizing line charts of the dissimilarity matrix in question (see Figure 8.5). This result suggests that the SOs underlying the symbolic descriptions AA00 and AA08 have a small dissimilarity (i.e., large similarity), that is, they identify 'homogeneous' classes of objects.

8.6 The matching functions

Matching comparison is a directional judgement involving a referent and a subject. In SDA, the referent is an SO representing a class description, while the subject is an SO that typically corresponds to the description of an individual.

The matching problem consists of establishing whether the individual described by the subject can be considered an instance of the referent. For instance, the SO

$$r = [\text{colour} = \{\text{black, white}\}] \wedge [\text{height} = [170, 200]]$$

describes a group of individuals either black or white, whose height is in the interval [170, 200], while the SO

$$s_1 = [\text{colour} = \text{black}] \wedge [\text{height} = 180]$$

corresponds to an individual in the extent of $r(s_1 \in \text{Ext}_E(r))$, since it fulfils the requirements stated in r. Conversely, the SO

$$s_2 = [\text{colour} = \text{black}] \wedge [\text{height} = 160]$$

does not correspond to an individual in the extent of $r(s_2 \notin \text{Ext}_E(r))$, since $160 \notin [170, 200]$. Thus, we can say that r matches s_1 but not s_2.

More formally, given two SOs, r and s, the former describes a class of individuals (referent of matching), the latter an individual (subject of matching) and matching checks whether s is an individual in the class described by r. Canonical matching returns either 0 (failure) or 1 (success).

The occurrence of noise as well as the imprecision of measuring instruments makes canonical matching too restrictive in many real-world applications. This makes it necessary to rely on a flexible definition of matching that aims at comparing two descriptions and identifying their similarities rather than the equalities.

The result is a flexible matching function with ranges in the interval [0,1] that indicates the probability of a precisely matching the subject against a referent, provided that some change is possibly made in the description of the referent. Notice that both canonical matching and flexible matching are not a resemblance measure, due to the non-symmetry of the matching function.

In the following, we briefly describe the matching operators implemented in the MATCH module for BSOs and PSOs. In the case of mixed SOs, matching values obtained by comparing the Boolean parts and the probabilistic parts are combined by product.

8.6.1 Matching functions for boolean symbolic objects

Let S be the space of BSOs in the form of assertions. The canonical matching between BSOs is defined as the function,

$$CanonicalMatch : S \times S \to \{0, 1\},$$

that assigns the value 1 or 0 to the matching of a referent $r \in S$ against a subject $s \in S$, where

$$r = [Y_1 \in w_1] \wedge [Y_2 \in w_2] \wedge \ldots \wedge [Y_p \in w_p],$$
$$s = [Y_1 \in v_1] \wedge [Y_2 \in v_2] \wedge \ldots \wedge [Y_p \in v_p].$$

More precisely, the canonical matching value is determined as follows:

$$CanonicalMatch(r, s) = \begin{cases} 1, & \text{if } v_j \subseteq w_j \forall j = 1, \ldots, p, \\ 0, & \text{otherwise.} \end{cases}$$

Conversely, the flexible matching between two BSOs is defined by

$$FlexMatch : S \times S \to [0, 1],$$

such that:

$$FlexMatch(r, s) = \begin{cases} 1, & \text{if } CanonicalMatch(r, s) = 1, \\ \in [0, 1], & \text{otherwise.} \end{cases}$$

Notice that the flexible matching yields 1 for an exact match.

In Esposito *et al.* (2000), the definition of flexible matching is based on probability theory in order to deal with chance and uncertainty. In this way, the result of flexible matching can be interpreted as the *probability* of r matching s, provided that a change is made in s.

More precisely, let

$$S(r) = \{s' \in S | CanonicalMatch(r, s') = 1\}$$

be the set of BSOs matched by r. Then the probabilistic view of flexible matching defines *FlexMatch* as the maximum conditional probability in $S(r)$, that is,

$$FlexMatch(r, s) = \max_{s' \in S(r)} P(s|s') \qquad \forall r, s \in S,$$

where $s(s')$ is the conjunction of simple BSOs (i.e., elementary events), that is, $s_1, \ldots, s_p(s'_1, \ldots, s'_p)$ such that each $s_i(s'_i)$ is in the form $[Y_i = v_i]([Y_i = v'_i])$. Then, under the assumption of conditional independence of the variables Y_j, the probability $\mathrm{P}(s|s')$ can be factorized as

$$P(s|s') = \prod_{i=1,\ldots,p} P(s_i|s') = \prod_{i=1,\ldots,p} P(s_i|s'_1 \wedge \ldots \wedge s'_p),$$

where $\mathrm{P}(s_i|s')$ denotes the probability of observing the event s_i given s'.

Suppose that s_i is an event in the form $[Y_i = v_i]$, that is, s describes an individual. If s' contains the event $[Y_i = v'_i]$, $P(s_i|s')$ is the probability that while we observed v_i, the true value was v'_i. By assuming that s_i depends exclusively on s'_i, we can write $P(s_i|s') = P(s_i|s'_i)$. This probability is interpreted as the similarity between the events $[Y_i = v_i]$ and $[Y_i = v'_i]$, in the sense that the more similar they are, the higher the probability:

$$P(s_i|s'_i) = P([Y_i = v_i]|[Y_i = v'_i]).$$

We denote by P the probability distribution of a random variable Y on the domain $\mathcal{Y}_i$ and δ_I a distance function such that $\delta_I : \mathcal{Y}_i \times \mathcal{Y}_i \to \mathbb{R}$. We obtain that

$$P(s_i|s'_i) = P([Y_i = v_i]|[Y_i = v'_i]) = P(\delta_I(v'_j, Y) \geq \delta_I(v'_j, v_j)).$$

Henceforth, we make some tacit assumptions on the distance δ_l as well as on the probability distribution P when they are not specified (Esposito *et al.*, 1991). In particular, we assume that the distance function δ_l for continuous-valued variables is the L_1 norm,

$$\delta_l(v, w) = |v - w|;$$

for nominal variables it is the binary distance,

$$\delta(v, w) = \begin{cases} 0, & \text{if } v = w, \\ 1, & \text{otherwise;} \end{cases}$$

while for ordinal variables it is

$$\delta_l(v, w) = |\text{ord}(v) - \text{ord}(w)|,$$

where ord(v) denotes the ordinal number assigned to the value v.

The probability distribution of Y_i on the domain $\mathcal{Y}_i$ is assumed to be the uniform distribution.

Example 8.1. We assume a nominal variable Y_i with a set $\mathcal{Y}_i$ of categories and a uniform distribution of Y on the domain $\mathcal{Y}_i$. If we use the binary distance, then

$$P([Y_i = v_i]|[Y_i = v_i']) = \frac{|\mathcal{Y}_i| - 1}{|\mathcal{Y}_i|},$$

where $|\mathcal{Y}_i|$ denotes the cardinality of $\mathcal{Y}_i$.

The definition of flexible matching can be generalized to the case of comparing any pair of BSOs and not necessarily comparing a BSO describing a class with a BSO describing an individual. In this case, we have that:

$$FlexMatch(r, s) = \max_{s' \in S(r)} \prod_{i=1,\dots,p} \sum_{j=1,\dots,q} \frac{1}{q} P(s_{ij}|s_i'),$$

when q is the number of categories for the variable j in the symbolic object s.

Example 8.2. (Flexible matching between BSOs). Let us consider a pair of BSOs r (referent of matching) and s (subject of matching) in the form of assertions, such that:

$$r = [R_1 \in \{\text{yellow, green, white}\}] \wedge [R_2 \in \{\text{Ford, Fiat, Mercedes}\}],$$
$$s = [S_1 \in \{\text{yellow, black}\}] \wedge [S_2 \in \{\text{Fiat, Audi}\}],$$

such that $\mathcal{Y}_1 =$ {yellow, red, green, white, black} is the domain of both R_1 and S_1 while $|\mathcal{Y}_1|$ is the cardinality of $\mathcal{Y}_1$ with $|\mathcal{Y}_1| = 5$. Similarly $\mathcal{Y}_2 =$ {Ford, Fiat, Mercedes, Audi, Peugeot,

Renault} is the domain of both R_2 and S_2 and $|\mathcal{Y}_2|$ is the cardinality of $\mathcal{Y}_2$ with $|\mathcal{Y}_2|=6$. We build the set S_r as follows:

$$\begin{aligned}
S(r) = \{s' \in S | CanonicalMatch\ (r, s') = 1\} = \{ \\
s'_1 = [S_1 = \text{yellow}] \wedge [S_2 = \text{Ford}]; \\
s'_2 = [S_1 = \text{yellow}] \wedge [S_2 = \text{Fiat}]; \\
s'_3 = [S_1 = \text{yellow}] \wedge [S_2 = \text{Mercedes}]; \\
s'_4 = [S_1 = \text{green}] \wedge [S_2 = \text{Ford}]; \\
s'_5 = [S_1 = \text{green}] \wedge [S_2 = \text{Fiat}]; \\
s'_6 = [S_1 = \text{green}] \wedge [S_2 = \text{Mercedes}]; \\
s'_7 = [S_1 = \text{white}] \wedge [S_2 = \text{Ford}]; \\
s'_8 = [S_1 = \text{white}] \wedge [S_2 = \text{Fiat}]; \\
s'_9 = [S_1 = \text{white}] \wedge [S_2 = \text{Mercedes}]\}.
\end{aligned}$$

When $s' = s'_1$, we obtain that:

$$\begin{aligned}
P(s_{11}|s'_{11}) &= P(S_1 = \text{yellow}|S_1 = \text{yellow}) = 1, \\
P(s_{12}|s'_{11}) &= P(S_1 = \text{black}|S_1 = \text{yellow}) = \frac{|\mathcal{Y}_1| - 1}{|\mathcal{Y}_i|} = \frac{4}{5}, \\
P(s_1|s'_1) &= 0.5(P(s_{11}|s'_{11}) + \text{P}(s_{12}|s'_{11})) = \frac{9}{10}, \\
P(s_{21}|s'_{12}) &= P(S_2 = \text{Fiat}|S_2 = \text{Ford}) = \frac{|\mathcal{Y}_2| - 1}{|\mathcal{Y}_2|} = \frac{5}{6}, \\
P(s_{22}|s'_{12}) &= \text{P}(S_2 = \text{Audi}|S_2 = \text{Ford}) = \frac{|\mathcal{Y}_2| - 1}{|\mathcal{Y}_2|} = \frac{5}{6}, \\
P(s_2|s'_1) &= 0.5(\text{P}(s_{21}|s'_{12}) + \text{P}(s_{22}|s'_{12})) = \frac{5}{6}.
\end{aligned}$$

Consequently, we have that $P(s_1|s'_1) \times P(s_2|s'_1) = \frac{3}{4}$. This means that $FlexMatch(r, s) \geq 0.75$.

8.6.2 Matching functions for probabilistic symbolic objects

The definition of the flexible matching function given for BSOs can be extended to the case of PSOs. If $r$and s are two PSOs, the flexible matching of r (referent of matching) against s (subject of matching) can be computed as follows:

$$FlexMatch(r, s) = \max_{s' \in S(r)} \prod_{i=1,\dots,p} P(s'_i) \sum_{j=1,\dots,q} P(s_{ij}) P(s_{ij}|s'_i).$$

Example 8.3. (Flexible matching between PSOs). Let us consider a pair of PSOs r and s, such that:

$$r = [R_1 \in \{\text{yellow}(0.2), \text{green}(0.5), \text{white}(0.3)\}]$$
$$\wedge [R_2 \in \{\text{Ford}(0.1), \text{Fiat}(0.5), \text{Mercedes}(0.4)\}]$$
$$s = [S_1 \in \{\text{white}(0.6), \text{green}(0.4)\}] \wedge [S_2 \in \{\text{Fiat}(0.3), \text{Audi}(0.7)\}],$$

such that $\mathcal{Y}_1 =$ {yellow, red, green, white, black} is the domain of both R_1 and S_1, while $\mathcal{Y}_2 =$ {Ford, Fiat, Mercedes, Audi, Peugeot, Renault} is the domain of both R_2 and S_2. We build the set S_r as follows:

$$S(r) = \{s' \in S | CanonicalMatch(r, s') = 1\} = \{$$
$$s'_1 = [S_1 = \text{yellow}] \wedge [S_2 = \text{Ford}];$$
$$s'_2 = [S_1 = \text{yellow}] \wedge [S_2 = \text{Fiat}];$$
$$s'_3 = [S_1 = \text{yellow}] \wedge [S_2 = \text{Mercedes}];$$
$$s'_4 = [S_1 = \text{green}] \wedge [S_2 = \text{Ford}];$$
$$s'_5 = [S_1 = \text{green}] \wedge [S_2 = \text{Fiat}];$$
$$s'_6 = [S_1 = \text{green}] \wedge [S_2 = \text{Mercedes}];$$
$$s'_7 = [S_1 = \text{white}] \wedge [S_2 = \text{Ford}];$$
$$s'_8 = [S_1 = \text{white}] \wedge [S_2 = \text{Fiat}];$$
$$s'_9 = [S_1 = \text{white}] \wedge [S_2 = \text{Mercedes}]\}.$$

When $s' = s'_1$, we obtain that:

$$P(s_{11}|s'_{11}) = P(S_1 = \text{white}|S_1 = \text{yellow}) = \frac{4}{5},$$

$$P(s_{11}) \times P(s_{11}|s'_{11}) = \frac{3}{5} \times \frac{4}{5} = \frac{12}{25},$$

$$P(s_{12}|b'_{11}) = P(S_1 = \text{green}|S_1 = \text{yellow}) = \frac{4}{5},$$

$$P(s_{12}) \times P(s_{11}|s'_{11}) = \frac{2}{5} \times \frac{4}{5} = \frac{8}{25},$$

$$P(s_{11}) \times P(s_{11}|s'_{11}) + P(s_{12}) \times P(s_{11}|s'_{11}) = \frac{12}{25} + \frac{8}{25} = \frac{4}{5},$$

$$P(s_{21}|s'_{12}) = P(S_2 = \text{Fiat}|S_2 = \text{Ford}) = \frac{5}{6},$$

$$P(s_{21}) \times P(s_{21}|s'_{12}) = \frac{3}{10} \times \frac{5}{6} = \frac{1}{4},$$

$$P(s_{22}|s'_{12}) = P(S_1 = \text{Audi}|S_2 = \text{Ford}) = \frac{5}{6},$$

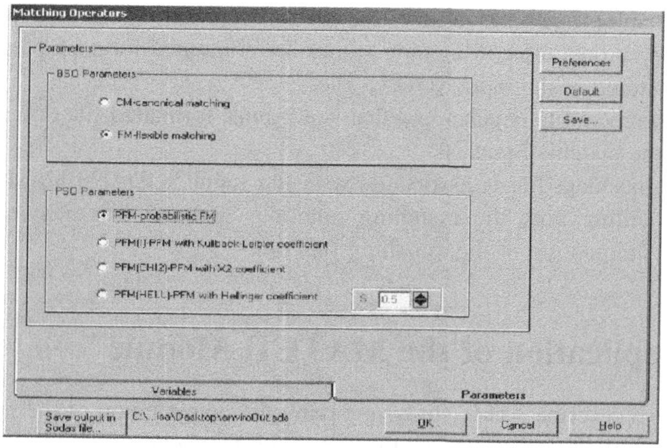

Figure 8.6 Setting the matching functions to compute the matching matrix.

$$P(s_{22}) \times P(s_{22}|s'_{12}) = \frac{7}{10} \times \frac{5}{6} = \frac{35}{60},$$

$$P(s_{21}) \times P(s_{21}|s'_{12}) + P(s_{22}) \times P(s_{22}|s'_{12}) = \frac{1}{4} + \frac{35}{60} = \frac{5}{6}.$$

Consequently, we have that $(P(s'_{11}) \times (P(s_{11}) \times P(s_{11}|s'_{11}) + P(s_{12}) \times P(s_{11}|s'_{11}))) \times (P(s'_{12}) \times (P(s_{21}) \times P(s_{21}|s'_{12}) + P(s_{22}) \times P(s_{22}|s'_{12}))) = (\frac{1}{5} \times \frac{4}{5}) \times (\frac{1}{10} \times \frac{5}{6}) = \frac{1}{75}$. This means that $FlexMatch(r, s) \geq \frac{1}{75}$.

Alternatively, the flexible matching of r against s can be performed by comparing each pair of probabilistic variables R_i and S_i, which take values on the same range $\mathcal{Y}_i$, according to some non-symmetric function f and aggregating the results by product, that is:

$$FlexMatch(r, s) = \prod_{i=1,\dots,p} f(R_i, S_i).$$

For this purpose, several comparison functions for probability distributions, such as the KL divergence, the χ^2 divergence and the Hellinger coefficient (see Section 8.3.2) have been implemented in the MATCH module. Notice that both the KL divergence and the χ^2 divergence are two dissimilarity coefficients. Therefore, they are not suitable for computing matching. However, they may be easily transformed into similarity coefficients by

$$f(P, Q) = \mathrm{e}^{-x},$$

where x denotes either the KL divergence value or the χ^2 divergence value.

8.7 Output of MATCH

The MATCH module outputs a new SODAS file that includes both the symbolic data table D stored in the input SODAS file and the matching matrix M such that $M(i, j)$ is the (canonical or flexible) matching value of the ith SO (referent) against the jth SO (subject) taken from the input data table. Matching values are computed for each pair of SOs whose descriptions are stored in the input SODAS file.

In addition, a report file is generated that is a printer-formatted file describing the input parameters and the matching matrix.

Finally, if a metadata file is associated with the input SODAS file, MATCH updates metadata by recording both the matching functions and the list of symbolic variables involved in the computation of the matching function.

8.8 An Application of the MATCH Module

In this section, we describe a case study involving the computation of the matching matrix from the SOs underlying as assertions the symbolic descriptions stored in enviro.xml. To this end, we create a new ASSO chain that includes a base block associated to the enviro.xml file. The running chain is then extended with a new block that is assigned to the MATCH module.

Before computing the matching matrix, users choose the list of symbolic variables to be involved in computing matching values, the matching functions to be computed, the name of the output SODAS file, and so on. Notice that in the current version of MATCH, users are not able to select a subset of SOs to be involved in matching computation. Conversely, all the SOs whose symbolic descriptions are stored in input SODAS file are processed to compute the matching matrix.

Both the list of variables and the matching functions are set by selecting Parameters... from the pop-up menu associated with the MATCH block in the running chain. The list of symbolic variables taken from the symbolic data table stored in the input SODAS file is shown in a list box and some statistics (e.g. minimum and maximum or frequency distribution) can be output for each variable.

Users choose symbolic variables to be considered when computing the matching matrix of the SOs taken from the enviro data. By default, all variables can be selected by users without any restriction on the type. However, users may decide to output only a subset of these variables (e.g. interval variables or probabilistic variables).

In this application, we decide to select all the symbolic variables (13 interval variables and four probabilistic variables) from the enviro data. This means that the SOs considered for matching computation are mixed SOs, where the Boolean part is separated from the probabilistic part. The matching values computed when comparing both the Boolean and probabilistic parts are then combined by product (see Figure 8.6).

When all the parameters are set, the matching matrix is built by choosing Run method from the pop-up menu associated with the MATCH block in the running chain.

If the matching matrix is correctly constructed, then MATCH produces as output a new SODAS file (e.g. enviroMatch.sds) that is associated with the current chain and is stored in the user-defined path and includes both the input symbolic data table and the matching matrix computed by MATCH.

A report file describing the input parameters (e.g. matching functions or symbolic variables) and the matching matrix is associated with a new block that is automatically introduced into the running chain and directly connected to the MATCH block (see Figure 8.7(a)).

(a) (b)

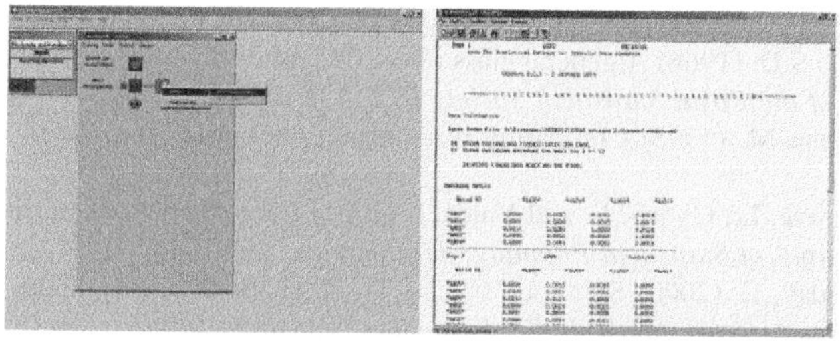

Figure 8.7 (a) Example of the ASSO chain including the Report block generated by MATCH and (b) the output of the report on the matching computation performed by MATCH.

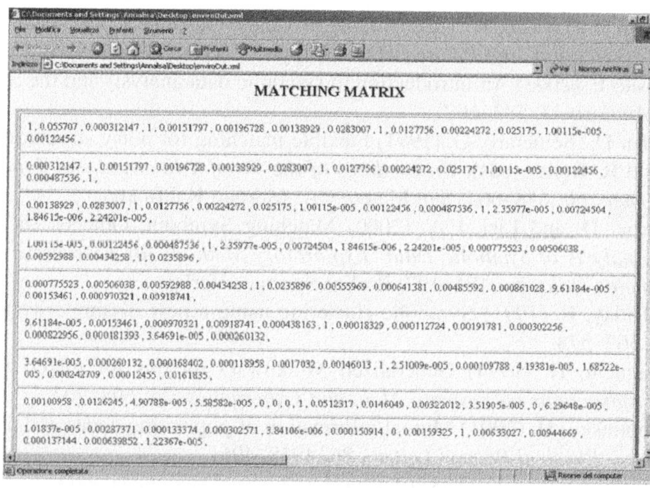

Figure 8.8 Matching matrix computed from the MATCH module on enviro data.

This report can be output as a printer-formatted file by selecting Open... and then View Result Report... from the pop-up menu associated with the report block (see Figure 8.7(b)). Alternatively, the report block can be removed from the running chain by selecting Delete Results... from the pop-up menu.

The matching matrix is stored in the output SODAS file that is associated with the current chain. Both the enviro symbolic data table and matching matrix are stored in XML format (see Figure 8.8). Such matching values are involved in the directional comparison of the SOs extracted from the enviro data in order to identify the set of SOs matched (i.e., covered) from each fixed SO.

Finally, the metadata file is updated by recording matching measures and symbolic variables involved in the matching comparison.

References

Ali, S.M. and Silvey, S.D. (1966) A general class of coefficient of divergence of one distribution from another. *Journal of the Royal Statistical Society B*, 2, 131–142.

Batagelj, V. and Bren, M. (1995) Comparing resemblance measures. *Journal of Classification*, 12, 73–90.

Beirlant, K. J., Devroye, L., Györfi, L. and Vajda, I. (2001) Large deviations of divergence measures on partitions. *Journal of Statistical Planning and Inference,* 93, 1–16.

Bock, H.-H. and Diday, E. (2000) Symbolic objects. In H.-H. Bock and E. Diday (eds), *Analysis of Symbolic Data: Exploratory Methods for Extracting Statistical Information from Complex Data*, pp. 54–77. Berlin: Springer-Verlag.

Csiszár, I. (1967) Information-type measures of difference of probability distributions and indirect observations. *Studia Scientiarum Mathematicarum Hungarica*, 2, 299–318.

de Carvalho, F.A.T. (1994) Proximity coefficients between Boolean symbolic objects. In E. Diday, Y. Lechevallier, M. Schader, P. Bertrand and B. Burtschy (eds), *New Approaches in Classification and Data Analysis*, pp. 387–394. Berlin: Springer-Verlag.

de Carvalho, F.A.T. (1998) Extension based proximity coefficients between constrained Boolean symbolic objects. In C. Hayashi, K. Yajima, H.-H. Bock, N. Ohsumi, Y. Tanaka and Y. Baba (eds), *Data Science, Classification, and Related Methods*, pp. 370–378. Tokyo: Springer-Verlag.

Diday, E. and Esposito F. (2003) An introduction to symbolic data analysis and the SODAS software. *Intelligent Data Analysis*, 7, 583–602.

Esposito, F., Malerba, D., Semeraro, G. (1991) Flexible matching for noisy structural descriptions. In J. Mylopoulos and R. Reiter (eds), *Proceedings of the Twelfth International Joint Conference on Artificial Intelligence*, pp. 658–664. San Mateo, CA: Morgan Kaufmann.

Esposito, F., Malerba, D. and Lisi, F.A. (2000) Matching Symbolic Objects. In H.-H. Bock and E. Diday (eds), *Analysis of Symbolic Data: Exploratory Methods for Extracting Statistical Information from Complex Data*, pp. 186–197. Berlin: Springer-Verlag.

Gowda, K.C. and Diday, E. (1991) Symbolic clustering using a new dissimilarity measure. *Pattern Recognition*, 24, 567–578.

Ichino, M. and Yaguchi, H. (1994) Generalized Minkowski metrics for mixed feature-type data analysis. *IEEE Transactions on Systems, Man, and Cybernetics*, 24, 698–707.

Kang, K. and Sompolinsky, H. (2001) Mutual information of population codes and distance measures in probability space. *Physical Review Letters*, 86, 4958–4961.

Krichevsky R.E. and Trofimov V.K. (1981) The performance of universal encoding. *IEEE Transaction Information Theory*, IT-27, 199–207.

Kullback, S. and Leibler, R.A. (1951) On information and sufficiency. *Annals of Mathematical Statistics*, 22, 76–86.

Lin, J. (1991) Divergence measures based on the Shannon entropy. *IEEE Transactions on Information Theory*, 37, 145–151.

Malerba, D., Esposito, F., Gioviale, V. and Tamma, V. (2001) Comparing dissimilarity measures for symbolic data analysis. In *Proceedings of Techniques and Technologies for Statistics – Exchange of Technology and Know-How*, Crete, 1, pp. 473–481. http://www.csc.liv.ac.uk/~valli/Papers/ntts-asso.pdf (accessed May 2007).

Malerba, D., Esposito, F. and Monopoli, M. (2002) Comparing dissimilarity measures for Probabilistic Symbolic Objects. In A. Zanasi, C.A. Brebbia, N.F.F. Ebecken and P. Melli (eds), *Data Mining III, Vol. 6: Series Management Information Systems*, pp. 31–40. Southampton: WIT Press.

Patterson, D.W. (1990) *Introduction to Artificial Intelligence and Expert Systems*. London: Prentice Hall.

Rached, Z., Alajaji, F. and Campbell, L.L. (2001) Rényi's divergence and entropy rates for finite alphabet Markov sources. *IEEE Transactions on Information Theory*, 47, 1553–1561.

Sammon, J.J.W. (1969) A nonlinear mapping for data structure analysis. *IEEE Transactions on Computers C*, 18, 401–409.

9

Unsupervised divisive classification

Jean-Paul Rasson, Jean-Yves Pirçon, Pascale Lallemand and Séverine Adans

9.1 Introduction

One of the most common tasks in data analysis is the detection and construction of *homogeneous* groups $C_1, C_2, \ldots$ of objects in a population Ω such that objects from the same group show a high similarity whereas objects from different groups are typically more dissimilar. Such groups are usually called *clusters* and must be constructed on the basis of the data which were recorded for the objects. This is commonly called the *clustering problem*.

The method discussed in this chapter is a divisive clustering method for a symbolic $n \times p$ data array $\underline{X}$. By the definition of a divisive clustering method, the algorithm starts with all objects in one cluster, and successively splits each cluster into (two) smaller ones until a suitable stopping rule prevents further divisions.

This algorithm proceeds in a monothetic way (Chavent, 1998). In other words, the algorithm assumes as input n data vectors and proceeds such that each split is carried out by using only a single variable (which is selected optimally).

The resulting classification structure is a k-partition $\mathcal{C} = (C_1, \ldots, C_k)$ such that:

- the created clusters are disjoint,

$$C_i \cap C_j = \emptyset, \qquad \forall i, j \in \{1, \ldots, k\}, i \neq j;$$

- each cluster is non-empty,

$$C_i \neq \emptyset, \qquad \forall i = 1, \ldots, k;$$

Symbolic Data Analysis and the SODAS Software Edited by E. Diday and M. Noirhomme-Fraiture

- the union of all classes is the whole set of objects,

$$\cup_{i=1}^{k} C_i = E,$$

where E is the initial set of objects.

9.2 Input data: interval data

This algorithm studies the case where n *symbolic objects* are described by p interval variables $y_1, \ldots, y_p$. The interval-valued variable y_j $(j = 1, \ldots, p)$ is measured for each element of the basic set $E = \{1, \ldots, n\}$. For each element $x \in \Omega$, we denote the interval $y_j(x)$ by $[\underline{y}_{jx}, \bar{y}_{jx}]$, thus $\underline{y}_{jx}$ $(\bar{y}_{jx})$ is the lower (upper) bound of the interval $y_j(x) \subseteq \mathbb{R}$.

A small example is given by Table 9.1. Here the fourth item is described by the fourth row of the table; i.e. the vector of intervals

$$x_4 = \begin{pmatrix} [9.0, 11.0] \\ [10.9, 12.5] \\ [7.1, 8.1] \end{pmatrix}.$$

9.3 The clustering tree method

We propose a recursive algorithm for organizing a given population of symbolic objects into classes. According to the clustering tree method, nodes are split recursively by choosing the best interval variable.

The original contribution of this method lies in the way a node is split. Indeed, the cut will be based on the sole assumption that the distribution of points can be modelled by a non-homogeneous Poisson process, where the intensity will be estimated by the kernel method. The cut will then be made so as to maximize the likelihood function.

9.3.1 Homogeneous and non-homogeneous Poisson process

A Poisson process is a natural point process which can be used on randomly and independently distributed data. This process is characterized by two elements: the variables counting

Table 9.1 A symbolic interval data table.

$x \backslash j$	y_1	y_2	y_3
1	[8.5, 10]	[13.0, 15.2]	[5.0, 8.2]
2	[6.3, 9.1]	[14.1, 16.0]	[6.3, 7.2]
3	[7.9, 11.8]	[11.6, 13.5]	[4.9, 6.5]
4	[9.0, 11.0]	[10.9, 12.5]	[7.1, 8.1]
5	[6.3, 7.2]	[12.9, 15.0]	[6.2, 7.4]
6	[7.1, 7.9]	[11.5, 12.9]	[4.8, 5.7]
7	[7.5, 9.4]	[13.2, 15.0]	[6.6, 8.1]
8	[6.6, 7.8]	[12.4, 13.2]	[5.7, 7.2]

the number of points into disjoint intervals are independent; the average number of points in every area A of the space is proportional to the Lebesgue measure $m(A)$ of this area.

In particular, consider a process N such that, $\forall A \subset \Omega$, $N(A)$ is a variable representing the cardinality of A. A process of n points is a *Poisson process of rate* λ *on* $\Omega \subset \mathbb{R}^p$, where $p \in \mathbb{N}$, if

1. $\forall A_1, \ldots, A_k \subset E$, $\forall i \neq j \in \{1, \ldots, k\}$, $A_i \cap A_j = \emptyset$:

$$N(A_i) \perp\!\!\!\perp N(A_j),$$

   i.e. the variables are independent.

2. $\forall A \subset \Omega$, and $k \in \mathbb{N}$, $k \geq 0$,

$$P(N(A) = k) = \mathrm{e}^{-\lambda m(A)} \frac{(\lambda m(A))^k}{k!},$$

   where $m(A)$ is the Lebesgue measure of A.

If the rate λ is constant, the Poisson process is a *homogeneous Poisson process*. If the rate λ is dependent on the points, it will be denoted a *non-homogeneous Poisson process*.

9.3.2 General hypothesis: non-homogeneous Poisson process

We consider a clustering problem where the observed points are independent and identically distributed. In particular, the observed points are generated by a non-homogeneous Poisson process with intensity $q(.)$ and are observed in E, where $E \subset \Omega$ is the union of k disjoint convex fields.

The likelihood function, for the observations $\underline{x} = (x_1, x_2, \ldots, x_n)$ with $x_i \in \mathbb{R}^p$, $i = 1, \ldots, n$, is

$$f_E(\underline{x}) = \frac{1}{(\rho(E))^n} \prod_{i=1}^{n} \mathbb{1}_E(x_i).q(x_i)$$

where $\rho(E) = \int_E q(x)\mathrm{d}x$ is the integrated intensity and $q(.)$ is the process intensity ($q(x) > 0\ \forall x$).

Consequently, if the intensity of the process is known, the maximum likelihood solution will correspond to k disjoint convex fields containing all the points for which the sum of the integrated intensities is minimal. When the intensity is unknown, it will be estimated.

9.3.3 Kernel method

To estimate the intensity of a non-homogeneous Poisson process, we will use a non-parametric method, the *kernel method*. Because this algorithm proceeds in a monothetic way, we do not need to extend formulae beyond one dimension.

The kernel estimator, which is a sum of 'bumps', each of which is placed on an observation, is defined by

$$\hat{q}(x) = \frac{1}{n}\sum_{i=1}^{n}\frac{1}{h}K\left(\frac{x - X_i}{h}\right)$$

where h is the window width, also called the *smoothing parameter*, which determines the width of the bumps, and K is the kernel with the following properties:

1. It is symetric.
2. It is continuous.
3. $\int K(x)\mathrm{d}x = 1$.
4. $K \geq 0$.
5. It determines the shape of the bumps.

The choice of smoothing parameter will be important. If it is too small, the estimator degenerates into a succession of peaks located at each point of the sample. If it is too large, the estimation approaches a uniform law and then we will have a loss of information. So, the question is how to determine this parameter.

9.3.4 Bumps and multi-modalities

Within the clustering context, Silverman (1981, 1986) has clearly distinguished the concept of *mode* from the concept of *bump*: a mode in a density f will be a local maximum, while a bump will be characterized by an interval in such way that the density is concave on this interval but not on a larger interval.

In the framework of density estimation by the kernel method, the number of modes will be determined by the smoothing parameter: for very large values of h, the density estimation will be unimodal; for h decreasing, the number of modes will increase. In Silverman's words: 'the number of modes is a decreasing function of the smoothing parameter h'.

This has been shown at least for the *normal kernel*. Consequently, to estimate the intensity of the non-homogeneous Poisson process, we will use the kernel method with this normal kernel, defined by

$$K_{\mathcal{N}}(t) = \frac{1}{\sqrt{2\pi}}\mathrm{e}^{-t^2/2}.$$

Since we use the normal kernel, there is a critical value h_{crit} of the smoothing parameter for which the estimation changes from unimodality to multi-modality. Our split criterion will look for this value.

9.3.5 Splitting criteria

For each variable, by a dichotomizing process, we find the greatest value of the parameter h, giving a number of modes of the associated intensities strictly larger than one. Once this h has been determined, we cut E into two convex disjoint fields E_1 and E_2, such that $E = E_1 \cup E_2$, for which the likelihood function

$$f_{E_1,E_2}(\underline{x}) = \frac{1}{(\rho(E_1)+\rho(E_2))^n} \prod_{i=1}^{n} 1\!\!1_{E_1 \cup E_2} . \hat{q}(x_i)$$

is maximum, i.e. for which the integrated density

$$\rho(E_1)+\rho(E_2)$$

is smallest. Since we are proceeding variable by variable, we will be able to select the best one, i.e. the one which generates the greatest likelihood function. This procedure is recursively performed until some *stopping rule* is satisfied: the number of points in a node must be below a cut-off value.

9.3.5.1 Set of binary questions for interval data

In the framework of the divisive clustering method, a node C is split on the basis of a single variable (suitably chosen) and answers (Chavent, 1992) to a specific binary question of the form 'Is $Y_j \leq c$?', where c is called the *cut value*.

An object $x \in C$ answers this question 'yes' or 'no' according to a binary function $q_c : E \rightarrow \{\text{true}, \text{false}\}$. The partition (C_1, C_2) of C induced by the binary question is as follows:

$$C_1 = \{x \in C \mid q_c(x) = \text{true}\}, \qquad C_2 = \{x \in C \mid q_c(x) = \text{false}\}.$$

Consider the case of interval variables. Let $Y_j(x) = [\alpha, \beta]$; the midpoint of $[\alpha, \beta]$ is $m_x = (\alpha + \beta)/2$.

1. The binary question is: Is $m_x \leq c$?
2. The function q_c is defined by

$$q_c(x) = \text{true if } m_x \leq c, \qquad q_c(x) = \text{false if } m_x > c.$$

3. The partition (C_1, C_2) of C induced by the binary question is

$$C_1 = \{x \in C \mid q_c(x) = \text{true}\}, \qquad C_2 = \{x \in C \mid q_c(x) = \text{false}\}.$$

9.3.6 Pruning method

At the end of the splitting process, we obtain a huge tree. The best subtree is then selected. Indeed, we have developed, under the hypothesis of a non-homogeneous Poisson process, a tree pruning method that takes the form of a classical hypothesis test, the gap test (Kubushishi, 1996; Rasson and Kubushishi, 1994).

Actually, we are testing each cut, i.e. we want to know if each cut is a good (gap test satisfied) or bad one (gap test not satisfied). In the case of two classes, D_1 and D_2, with $D_1 \cup D_2 = D$, the hypotheses are

$$H_0 : \text{there are } n = n_1 + n_2 \text{ points in } D_1 \cup D_2$$

versus

$$H_1 : \text{there are } n_1 \text{ points in } D_1 \text{ and } n_2 \text{ points in } D_2, \text{ with } D_1 \cap D_2 = \emptyset.$$

This pruning method crosses the tree branch by branch, from its root to its leaves, in order to index the good cuts and the bad cuts. The leaves for which there are only bad cuts are pruned.

9.3.7 Application to interval data

The current problem is to apply this new method to symbolic data of interval type (Bock and Diday, 2000; Chavent and Lechevallier, 2002). Define the interval space

$$I = \{[a_i, b_i], i = 1, \ldots, n, a_i \leq b_i\}.$$

We decide to represent each interval by its coordinates (midpoint, half-length), on the space $(M, L) \subseteq \mathbb{R} \times \mathbb{R}^+$.

As we use a divisive method, separations must respect the order of the classes centres, and thus we are led to consider, in the half-plane $\mathbb{R} \times \mathbb{R}^+$, only partitions invariant in relation to M.

We must minimize, in the most general case of a non-homogeneous Poisson process, the integrated intensity,

$$\int_{M_i}^{M_{i+1}} \rho_1(m)\mathrm{d}m + \int_{\min(L_i, L_{i+1})}^{\max(L_i, L_{i+1})} \rho_2(l)\mathrm{d}l, \tag{9.1}$$

and choose as our partition the one generated by any point being located inside the interval which maximizes (9.1).

9.3.8 Output data and results

After the tree-growing algorithm and the pruning procedure, we obtain the final clustering tree. The nodes of the tree represent the binary questions selected by the algorithm and the k leaves of the tree define the k-partition. Each cluster is characterized by a rule, that is, the path in the tree which provided it. The clusters therefore become new symbolic objects defined according to the binary questions leading from the root to the corresponding leaves.

9.4 Example

The above clustering method has been tested with Ichino's (1988) well-known oils data set. The data set is composed of eight oils described in terms of four interval variables, (Table 9.2). Our divisive algorithm yields the three-cluster partition shown in Figure 9.1.

Table 9.2 Table of oils and fats.

Sample	Specific gravity	Freezing point	Iodine value	Saponification value
linseed oil	[0.930, 0.935]	[−27, −18]	[170, 204]	[118, 196]
perilla oil	[0.930, 0.937]	[−5, −4]	[192, 208]	[188, 197]
cottonseed oil	[0.916, 0.918]	[−6, −1]	[99, 113]	[189, 198]
sesame oil	[0.920, 0.926]	[−6, −4]	[104, 116]	[187, 193]
camellia oil	[0.916, 0.917]	[−21, −15]	[80, 82]	[189, 193]
olive oil	[0.914, 0.919]	[0, 6]	[79, 90]	[187, 196]
beef tallow	[0.860, 0.870]	[30, 38]	[40, 48]	[190, 199]
hog fat	[0.858, 0.864]	[22, 32]	[53, 77]	[190, 202]

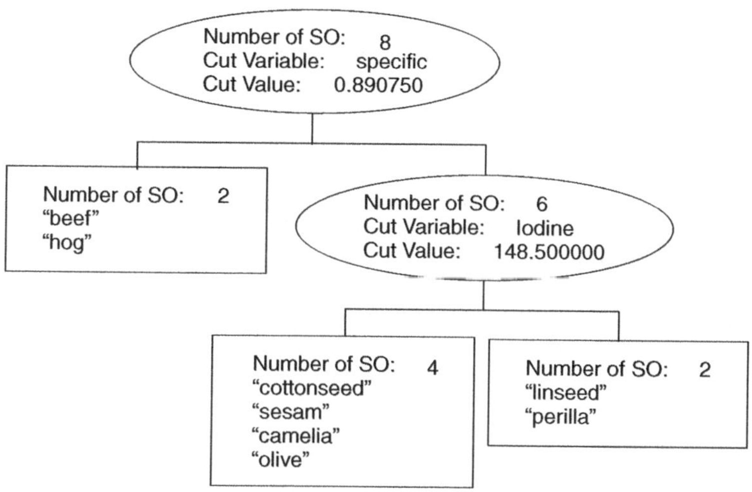

Figure 9.1 Classification of oils into three clusters (each terminal node corresponds to a cluster).

Two binary questions correspond to two binary functions $E \to \{\text{true}, \ \text{false}\}$, given by

$$q_1 = [\text{Spec.Grav.}(x) \leq 0.89075], \qquad q_2 = [\text{Iod.Val.}(x) \leq 148.5].$$

Each cluster corresponds to a symbolic object, e.g. a query assertion:

$$C_1 = [\text{Spec. Grav.}(x) \leq 0.89075],$$
$$C_2 = [\text{Spec. Grav.}(x) > 0.89075] \wedge [\text{Iod. Val.}(x) \leq 148.5],$$
$$C_3 = [\text{Iod. Val.}(x) > 148.5].$$

Then the resulting three-cluster partition is:

$$C_1 = \{\text{beef}, \ \text{hog}\},$$

$$C_2 = \{\text{cottonseed, sesam, camelia, olive}\},$$
$$C_3 = \{\text{linseed, perilla}\}.$$

For instance, the object 'linseed' will be in the cluster C_3 because

- Spec.Grav.(linseed) = $[0.930, 0.935]$ and $(0.930 + 0.935)/2 > 0.890\,75$,
- Iod.Val.(linseed) = $[170, 204]$ and $(170 + 204)/2 > 148.5$.

References

Bock, H.-H. and Diday, E. (2000) *Analysis of Symbolic Data: Exploratory Methods for Extracting Statistical Information from Complex Data.* Berlin: Springer-Verlag.

Chavent, M. (1992) Analyse des données symboliques: Une méthode divisive de classification. Doctoral thesis, Université Paris IX-Dauphine.

Chavent, M. (1998) A montothetic clustering method. *Pattern Recognition Letters*, 19: 989–996.

Chavent, M. and Lechevallier, Y. (2002) Dynamical clustering of interval data: Optimization of an adequacy criterion based on Hausdorff distance. In K. Jajuga, A. Sokolowski and H.-H. Bock (eds), *Classification, Clustering, and Data Analysis*, pp. 53–60. Berlin: Springer-Verlag.

Ichino, M. (1988) General metrics for mixed features – the cartesian space theory for pattern recognition. In *Proceedings of the 1988 IEEE International Conference on Systems, Man and Cybernetics*, Volume 1, pp. 494–497. Beijing: International Academic Publishers.

Kubushishi, T. (1996) On some applications of the point process theory in cluster analysis and pattern recognition. Doctoral thesis, Facultés Universitaires Notre-Dame de la Paix.

Rasson, J.P. and Kubushishi, T. (1994) The gap test: an optimal method for determining the number of natural classes in cluster analysis. In E. Diday, Y. Lechevallier, M. Schader, P. Bertrand and B. Burtschy (eds), *New Approaches in Classification and Data Analysis*, pp. 186–193. Berlin: Springer-Verlag.

Silverman, B. (1981) Using kernel density estimates to investigate multimodality. *Journal of Royal Statistic Society B*, 43: 97–99.

Silverman, B. (1986) *Density Estimation for Statistics and Data Analysis*. London: Chapman & Hall.

10

Hierarchical and pyramidal clustering

Paula Brito and Francisco de A.T. de Carvalho

10.1 Introduction

This chapter addresses the problem of clustering symbolic data, using the hierarchical and the pyramidal models. Pyramids (Diday, 1984, 1986; Bertrand and Diday, 1985) extend the hierarchical clustering model by allowing overlapping clusters which are not nested, but they impose a linear order on the set of individuals to be clustered, and all clusters formed are intervals of this order. Pyramidal clustering produces a structure that is richer than a hierarchy, in that it allows for the formation of a larger number of clusters and provides a seriation of the given data set.

A symbolic clustering method has been proposed (Brito and Diday, 1990; Brito, 1991, 1994), using the hierarchical and pyramidal models and allowing the clustering of multi-valued data. This method was further developed in order to allow for categorical variables for which a frequency or probability distribution is given (modal variables) (Brito, 1998), defining appropriate operators for clustering modal symbolic data. Later, Brito and de Carvalho extended this work so as to allow for the existence of hierarchical rules between multi-valued categorical variables (Brito and de carvalho, 1999) and between modal variables (Brito and de carvalho 2002), by suitably defining the generalization operators and the generality measures for this case. The consideration of hierarchical rules in symbolic data analysis has also been widely studied by Vignes (1991), de Carvalho (1994) and Csernel and de Carvalho (1999). The general clustering method falls within the framework of conceptual clustering, since each cluster formed is associated with a conjunction of properties in the input variables, which constitutes a necessary and sufficient condition for cluster membership. Clusters are thus 'concepts', described both extensionally, by the set of their members, and intentionally, by a symbolic object expressing their properties.

Symbolic Data Analysis and the SODAS Software Edited by E. Diday and M. Noirhomme-Fraiture

Many other hierarchical clustering methods have since been proposed for symbolic data, which differ in the type of symbolic variables they work on and/or the criteria considered. An agglomerative approach has been introduced which forms composite symbolic objects using a join operator whenever mutual pairs of symbolic objects are selected for agglomeration based on minimum dissimilarity (Gowda and Diday, 1999) or maximum similarity (Gowda and Diday 1992). Ichino and Yaguchi (1994) define generalized Minkowski metrics for mixed feature variables and present dendrograms obtained from the application of standard linkage methods for data sets containing numeric and symbolic feature values. Gowda and Ravi (1995a, 1995b) have presented divisive and agglomerative algorithms for symbolic data based on the combined usage of similarity and dissimilarity measures. Chavent (1998) has proposed a divisive clustering method for symbolic data which simultaneously provides a hierarchy for the symbolic data set and a characterization of each cluster in the hierarchy by necessary and sufficient descriptions. El-Sonbaty and Ismail (1998) have introduced an on-line agglomerative hierarchical technique based on the concept of a single-linkage method for clustering both symbolic and numerical data. Gowda and Ravi (1999) have presented a hierarchical clustering algorithm for symbolic objects based on the gravitational approach, which is inspired by the movement of particles in space due to their mutual gravitational attraction.

In Section 10.2, we describe the general symbolic hierarchical and pyramidal clustering method, starting by recalling the hierarchical and pyramidal clustering models, and then detailing the generalization procedure and the computation of generality measures for the different types of variables, and the clustering algorithm. Section 10.3 addresses the problem of symbolic clustering in the presence of hierarchical rules, both for categorical and for modal data. In Section 10.5, the HIPYR module of the SODAS software is presented and its use explained. Finally, in Section 10.6 an application illustrates the method.

10.2 Symbolic hierarchical and pyramidal clustering: the basic method

10.2.1 Hierarchical and pyramidal models

The general aim of a clustering method is to aggregate the elements of a set E into homogeneous clusters. In the case of hierarchical or pyramidal clustering, the clusters formed are organized in a tree-like structure. In a bottom-up approach, the most similar objects are merged together, then similar clusters are merged, until a unique cluster, containing all the elements in E, is formed. In the case of a hierarchy, each level corresponds to a partition; in the case of a pyramid we get, at each level, a family of overlapping clusters, but all clusters are intervals of a total linear order on E.

Formally, a hierarchy on a set E is a family H of non-empty subsets $h, h', \ldots$ of E, such that:

- $\forall w \in E : \{w\} \in H$,
- $E \in H$,
- $\forall h, h' \in H : h \cap h' = \emptyset$ or $h \subseteq h'$ or $h' \subseteq h$.

Often, a non-negative real value is associated with each cluster, characterizing its heterogeneity. An indexed hierarchy or dendrogram is a pair (H, f), where H is a hierarchy and f a mapping $f: H \to \mathrm{R}^+$ such that:

- $f(h) = 0 \Leftrightarrow \mathrm{card}(h) = 1$ (where $\mathrm{card}(h)$ stands for the cardinality of h),
- $h \subset h' \Rightarrow f(h) \leq f(h')$.

A cluster $h \in H$ is said to be a successor of a cluster $h' \in H$ if $h \subseteq h'$ and there does not exist a cluster $h'' \in H$, $h'' \neq h$, $h'' \neq h'$, such that $h \subset h'' \subset h'$; h' is then said to be a predecessor of h. In a hierarchy, each cluster has at most one predecessor.

The pyramidal model (Diday, 1984, 1986, Bertrand and Diday, 1985) generalizes hierarchies by allowing non-disjoint classes at each level, but it imposes the existence of a linear order on E such that all classes are intervals of this order. Figure 10.1 shows a hierarchy and a pyramid on a five-element set.

Formally, a pyramid is defined as a family P of non-empty subsets $p, p', \ldots,$ of E, such that:

- $\forall w \in E: \{w\} \in P$,
- $E \in P$,
- $\forall p, p' \in P: p \cap p' = \emptyset$ or $p \cap p' \in P$,
- there exists an order θ on E such that each element of P is an interval of θ.

A pyramid indexed in the broad sense is a pair (P, f), where P is a pyramid and f a mapping $f: P \to \mathrm{R}^+$ such that

- $f(p) = 0 \Leftrightarrow \mathrm{card}(p) = 1$,
- $p \subset p' \Rightarrow f(p) \leq f(p')$,
- $f(p) = f(p')$ with $p \subset p'$ and $p \neq p' \Rightarrow \exists p_1 \neq p, p_2 \neq p$ such that $p = p_1 \cap p_2$.

The notions of successor and predecessor, defined for hierarchies, also apply to the case of pyramids. It is easy to prove that in a pyramid each cluster admits up to two predecessors. Hence, a pyramid provides both a clustering and a seriation on the data. The pyramidal model, leading to a system of clusters which is richer than that produced by hierarchical

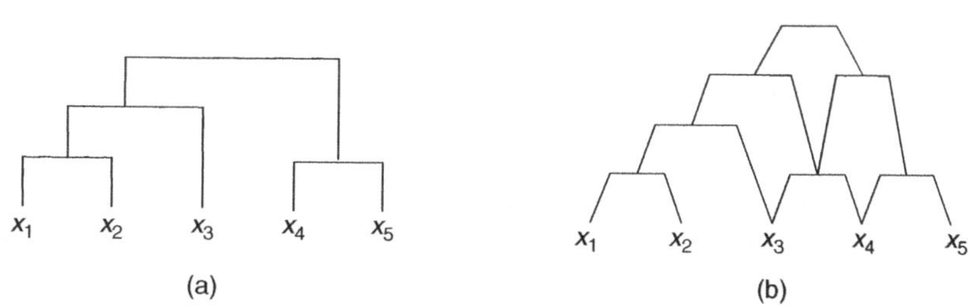

Figure 10.1 (a) Hierarchy, (b) pyramid.

clustering, allows for the identification of clusters that the hierarchical model would not identify; the existence of a compatible order on the objects leads, however, to a structure which is much simpler than lattices.

Pyramidal clustering algorithms have been developed by suitably adapting the ascending hierarchical clustering algorithm (Bertrand & Diday 1985; Rodríguez et al., 2001).

10.2.2 Clustering criterion: complete objects

The symbolic hierarchical/pyramidal clustering method (Brito and Diday, 1990; Brito,1991, 1994) takes as its input a symbolic data set, where each element is described by possibly mixed kinds of symbolic variables, $y_j : \Omega \to O_j$, $j = 1, \ldots, p$, Ω being the population under study. Let $O = O_1 \times \cdots \times O_p$ be the observation space. Let $E = \{w_1, \ldots, w_n\}$ be the set of units we wish to cluster, and $s_i = (a_i, R_i, y(w_i))$ where

$$a_i = \bigwedge_{j=1}^{p} [y_j R_j y_j(w_i)]$$

is the assertion associated with w_i, $i = 1, \ldots, n$.

Recall that the *extent* of a symbolic object $s = (a, R, y(w))$, where $a = \bigwedge_{j=1}^{p} [y_j R_j d_j]$, in E is defined as $\mathrm{Ext}_E(s) = \{w \in E : y_j(w) R_j d_j, j = 1, \ldots, p\}$. The *virtual extent* of s is defined as $\mathrm{Ext}_O(s) = d_1 \times \cdots \times d_p \subseteq O$.

A symbolic object s is said to be *more general* than a symbolic object s' if $\mathrm{Ext}_O(s') \subseteq \mathrm{Ext}_O(s)$. In this case, s' is said to be *more specific* than s.

The *intent* of a set $C \subseteq E$, denoted $\mathrm{int}(C)$, is the most specific symbolic object s such that $\forall w \in C, w \in \mathrm{Ext}_E(s)$.

A symbolic object s is said to be *complete* if the intent of its extent is the object itself: s complete $\Rightarrow \mathrm{int}(\mathrm{Ext}_E(s)) = s$. That is, s is complete if it is the most specific object that exhaustively describes its extent. A *concept* is a pair (C, s), where C is a subset of E and s a complete symbolic object, such that s is the intent of C and C is the extent of s: $\mathrm{int}(C) = s$, $\mathrm{Ext}_E(s) = C$.

The initial set of concepts is $\{(\{w_1\}, s_1), \ldots, (\{w_n\}, s_n)\}$: it is assumed that all $(\{w_i\}, s_i)$, $i = 1, \ldots, n$, are concepts.

The criterion that guides cluster formation is the intent–extent duality: each cluster of the hierarchy or pyramid should correspond to a concept, that is, each cluster, which is a part of E, is represented by a complete symbolic object whose extent is the cluster itself. This means that each cluster is by construction associated with a symbolic object that generalizes its members, and that no element outside the cluster meets the description given by this symbolic object.

Clusters, and the corresponding concepts, are formed recursively: at each step a new concept (C, s) is formed by merging suitable and previously constructed concepts (C_α, s_α) and (C_β, s_β), with $C = C_\alpha \cup C_\beta$, and providing that s, the generalization of s_α and s_β, is such that $s = s_\alpha \cup s_\beta = \mathrm{int}(C)$. The generalization procedure for the different variable types is detailed below.

An additional criterion must then be considered for choosing among the possible aggregations. The principle is that clusters associated with less general symbolic objects should be formed first. Since the generality relation between symbolic objects is just a partial order relation, a numerical criterion has been defined that allows the generality of a symbolic object to be evaluated: this is the 'generality degree', which is detailed in Section 10.2.3.

The generalization procedure

When forming a new concept by merging previously formed concepts, it is necessary to determine its intent, in the form of a symbolic object. The symbolic object to be formed should generalize the symbolic objects associated with the concepts that are merged.

Generalizing two symbolic objects s and s' means determining a symbolic object s'' such that s'' is more general than both s and s', and is the least general symbolic object fulfilling this condition. Generalization is performed variable-wise and the procedure differs according to the variable type. Let $s_1, \ldots, s_q$ be a set of symbolic objects we wish to generalize.

Interval-valued variables In the presence of an interval-valued variable y, the symbolic objects s_h, $h = 1, \ldots, q$, with $s_h = (a_h, R_h, d_h)$, use events of the form $e_h = [y \subseteq I_h]$, with $I_h = [l_h, u_h]$, $h = 1, \ldots, q$. In the generalized object, the event corresponding to this variable will be

$$e = [y \subseteq [\min\{l_h\}, \max\{u_h\}]]$$

since this is the smallest interval that contains all the intervals $I_h = [l_h, u_h]$, $h = 1, \ldots, q$.

Example
Consider the symbolic objects s_{Alfa} and s_{Porsche} defined on variables Price and Engine Capacity, describing the car models Alfa 145 and Porsche, respectively, associated with the following assertions:

$$a_{\text{Alfa}} = [\text{Price} \subseteq [27\,806, 33\,596]] \wedge [\text{Engine Capacity} \subseteq [1000, 3000]],$$

$$a_{\text{Porsche}} = [\text{Price} \subseteq [147\,704, 246\,412]] \wedge [\text{Engine Capacity} \subseteq [3387, 3600]].$$

The generalized object is associated with the assertion:

$$a = [\text{Price} \subseteq [27\,806, 246\,412]] \wedge [\text{Engine Capacity} \subseteq [1000, 3600]].$$

Categorical single and categorical multi-valued variables For a categorical single- or multi-valued variable y with domain $O = \{m_1, \ldots, m_k\}$, the events in the symbolic objects $s_1, \ldots, s_q$ take the form $e_h = [y = v_h]$ with $v_h \in O$ (if y is a classical categorical single-valued variable), or $e_h = [y \subseteq V_h]$ with $V_h \subseteq O$ (if y is a multi-valued variable). For generalization purposes, $e_h = [y = v_h]$ is equivalent to $e_h = [y \subseteq \{v_h\}]$, so both cases can be considered together. Then, generalization is made by applying the union operator to the sets associated with the corresponding events; in the generalized object, the event corresponding to the variable in question will be $e = [y \subseteq V]$, with $V = \bigcup_{h=1}^{q} V_h$ since V is the smallest set that contains all the V_h, $h = 1, \ldots, q$.

Example
Consider the symbolic objects z_{Alfa} and z_{Porsche}, defined on variables Fuel and Category, describing car models Alfa 145 and Porsche, respectively, associated with the assertions:

$$b_{\text{Alfa}} = [\text{Fuel} \subseteq \{\text{Petrol, Diesel}\}] \wedge [\text{Category} \subseteq \{\text{Utilitarian}\}],$$

$$b_{\text{Porsche}} = [\text{Fuel} \subseteq \{\text{Petrol}\}] \wedge [\text{Category} \subseteq \{\text{Sporting}\}].$$

The generalized object is associated with the assertion:

$$b = [\text{Fuel} \subseteq \{\text{Petrol, Diesel}\}] \wedge [\text{Category} \subseteq \{\text{Sporting, Utilitarian}\}].$$

Modal Variables For a modal variable y, with underlying domain $O_y = \{m_1, \ldots, m_k\}$, we have $y(w) = \{m_1(p_1^w), \ldots, m_k(p_k^w)\}$. We shall consider events of the form $[y(w) R \{m_1(p_1), \ldots, m_k(p_k)\}]$, with the following relations:

'$=$' such that $[y(w) = \{m_1(p_1), \ldots, m_k(p_k)\}]$ is true iff $p_\ell^w = p_\ell$, $\ell = 1, \ldots, k$,

'$\leq$' such that $[y(w) \leq \{m_1(p_1), \ldots, m_k(p_k)\}]$ is true iff $p_\ell^w \leq p_\ell$, $\ell = 1, \ldots, k$,

'$\geq$' such that $[y(w) \geq \{m_1(p_1), \ldots, m_k(p_k)\}]$ is true iff $p_\ell^w \geq p_\ell$, $\ell = 1, \ldots, k$.

Generalization can then be defined in two ways – by the maximum and by the minimum – each with a particular semantics.

Generalization of events *by the maximum* is done by taking, for each category m_ℓ, the maximum of its probabilities/frequencies. The generalization of $e_1 = [yR\{m_1(p_1^1), \ldots, m_k(p_k^1)\}], \ldots, e_q = [yR\{m_1(p_1^q), \ldots, m_k(p_k^q)\}]$ with $R \in \{=, \leq\}$ is $e = [y \leq \{m_1(p_1), \ldots, m_k(p_k)\}]$ where $p_\ell = \max\{p_\ell^h, h = 1, \ldots, q\}$, $\ell = 1, \ldots, k$. Notice that the generalized event is typically not probabilistic, in the sense that the obtained distribution is not a probability distribution, since $p_1 + \cdots + p_k > 1$ is possible.

Example
Consider events e_{Alfa} and e_{Porsche} describing the colour distribution of the car models Alfa 145 and Porsche, respectively. Let

$$e_{\text{Alfa}} = [\text{Colour} = \{\text{red}(30\%), \text{black}(70\%)\}],$$

$$e_{\text{Porsche}} = [\text{Colour} = \{\text{red}(60\%), \text{black}(20\%), \text{white}(20\%)\}].$$

The generalized event is:

$$e = [\text{Colour} \leq \{\text{red}(60\%), \text{black}(70\%), \text{white}(20\%)\}].$$

Generalization of events *by the minimum* is done by taking, for each category, the minimum of its probabilities/frequencies. The generalization of $e_1 = [yR\{m_1(p_1^1), \ldots, m_k(p_k^1)\}], \ldots, e_q = [yR\{m_1(p_1^q), \ldots, m_k(p_k^q)\}]$ with $R \in \{=, \geq\}$ is $e = [y \geq \{m_1(p_1), \ldots, m_k(p_k)\}]$ where $p_\ell = \min\{p_\ell^h, h = 1, \ldots, q\}$, $\ell = 1, \ldots, k$. The generalized event is typically not probabilistic, since, in this case, $p_1 + \cdots + p_k < 1$ is possible.

Example
Define e_{Alfa} and e_{Porsche} as in the previous example. The generalized event is now:

$$e' = [\text{Colour} \geq \{\text{red}(30\%), \text{black}(20\%)\}].$$

Ordinal variables Let $y : \Omega \longrightarrow O, i = 1, \ldots, p$. y is an ordinal variable if its observation space is an ordered set $(O, \leq)$. Let $O = \{m_1 \leq m_2 \leq \cdots \leq m_k\}$. An interval of O is a set $I \subseteq O$ such that if $m_i, m_j \in I$ then $m_\ell \in I, \forall m_\ell$ such that $m_i \leq m_\ell \leq m_j$. An interval of an ordinal variable will be denoted $I = [\min(I), \max(I)]$. The principle as concerns ordinal variables is that the descriptions should only admit intervals, that is, an event corresponding to an ordinal variable is of the form $e = [y \subseteq I]$ with I an interval of O.

In the presence of an ordinal variable y, the assertions a_h associated with symbolic objects $s_h = (a_h, R_h, d_h), h = 1, \ldots, q$, contain events of the form $e_h = [y \subseteq I_h]$, with $I_h = [l_h, u_h], h = 1, \ldots, q$. As in the case of interval-valued variables, in the generalized object, the event corresponding to this variable will be $e = [y \subseteq [\min\{l_h\}, \max\{u_h\}]$, since this is the smallest interval that contains all the intervals $I_h = [l_h, u_h], h = 1, \ldots, q$.

Taxonomic variables Let $y : \Omega \longrightarrow O, i = 1, \ldots, p$. y is a taxonomic variable if O, the observation space of y, is ordered into a tree structure. Taxonomies may be taken into account in generalizing: first, we proceed as in the case of categorical multi-valued variables, then each set of values of O is replaced by the lowest value in the taxonomy covering all the values of the given set. We choose to go up to level h when at least two successors of h are present:

$$[\text{Form} = \text{triangle, rectangle}] \longrightarrow [\text{Form} = \text{polygon}].$$

10.2.3 The generality degree

The *generality degree* allows us to express the generality of a symbolic object, and hence to choose among the possible aggregations. For interval and categorical multi-valued variables, it expresses the proportion of the underlying domain that is covered by the assertion associated with a symbolic object; for modal variables, it expresses how much the given distributions are close to uniform distributions. Let $s = (a, R, d)$ be the symbolic object associated with the assertion $a = \bigwedge_{j=1}^{p}[y_j \in V_j], V_j \subseteq O_j$. We allow O_j to be bounded, $1 \leq j \leq p$. Then, the generality degree of s is

$$G(s) = \prod_{j=1}^{p} \frac{c(V_j)}{c(O_j)} = \prod_{j=1}^{p} G(e_j), \qquad 0 < G(s) \leq 1. \tag{10.1}$$

The generality degree is computed variable-wise; for each variable y_j, corresponding to event e_j, a factor $G(e_j) \in [0, 1]$ is computed, and these factors are then combined to get a measure of the variability of the symbolic object. The definition of the operator $c(\cdot)$ depends on the type of variables; it is a measure of the size of the underlying sets, and it should be increasing with respect to set inclusion and hence with respect to generality (the larger the extent of e_j, the larger the value of $c(V_j)$).

For interval-valued variables, $c(\cdot)$ is defined as the length of the interval; for categorical single, categorical multi-valued and ordinal variables, $c(\cdot)$ is defined as the cardinality.

For modal variables with k categories, $m_1, \ldots, m_k$, on which we have a probability/frequency distribution $p_1, \ldots, p_k$, and $e = [yR\{m_1(p_1), \ldots, m_k(p_k)\}]$, with $R \in \{=, \leq, \geq\}$, two measures have been proposed, according to the generalization operator used:

- If $R \in \{=, \leq\}$ and generalization is performed by the maximum of the probability/frequency values,

$$G_1(e) = \sum_{j=1}^{k} \sqrt{p_j \times \frac{1}{k}} = \frac{\sum_{j=1}^{k} \sqrt{p_j}}{\sqrt{k}}. \tag{10.2}$$

- If $R \in \{=, \geq\}$ and generalization is performed by the minimum of the probability/frequency values,

$$G_2(e) = \sum_{j=1}^{k} \sqrt{(1-p_j) \times \frac{1}{k(k-1)}} = \frac{\sum_{j=1}^{k} \sqrt{1-p_j}}{\sqrt{k(k-1)}}. \tag{10.3}$$

These expressions evaluate, in each case, the similarity between the given distribution and the uniform distribution. $G_1(e)$ is the affinity coefficient (Matusita, 1951) between $(p_1, \ldots, p_k)$ and the uniform distribution; it is maximum when the given distribution is uniform: $p_j = 1/k, j = 1, \ldots, k$. $G_2(e)$ is the affinity coefficient between $(1-p_1, \ldots, 1-p_k)$ and $(k^{-1}(k-1)^{-1}, \ldots, k^{-1}(k-1)^{-1})$; again, if $p_1 + \cdots + p_k = 1$ for the modal variable Y, $G_2(e)$ is maximum, and $G_2(e) = 1$ when the given distribution is uniform. Thus, the more similar the corresponding distributions are to uniform distributions, the more general we consider an object.

Table 10.1 summarizes the generalization procedures and the corresponding operators to compute the generality degree, for the different variable types.

10.2.4 Algorithm

Let $E = \{w_1, \ldots, w_n\}$ be the set of units we wish to cluster, and s_i the symbolic object associated with w_i, , $i = 1, \ldots, n$. The initial set of concepts is $\{(\{w_1\}, s_1), \ldots, (\{w_n\}, s_n)\}$: it is assumed that all $(\{w_i\}, s_i), i = 1, \ldots, n$, are concepts. Let $G(s)$ be the generality degree of a symbolic object s. Let (C_α, s_α) and (C_β, s_β) be candidates for merging, $C = C_\alpha \cup C_\beta$ and $s = s_\alpha \cup s_\beta$. To be merged, the concepts (C_α, s_α) and (C_β, s_β) should fulfil the following conditions:

(a) C_α and C_β can be merged together according to the desired clustering structure.

(b) $\text{Ext}_E(s) = C$, i.e., no element of E outside C belongs to the extent of s (i.e. fulfils the conditions expressed by s).

(c) A numerical criterion is minimum. This criterion may be the generality degree of the resulting symbolic object s, $G(s)$ or the increase in the generality degree.

Table 10.1 Generalization operators and corresponding measures.

Variables	Generalization	$c(\cdot)$
Interval	covering interval	interval length
categorical single and multi-valued	set union	cardinal
modal (prob./freq. distr.)	maximum	G_1
modal (prob./freq. distr.)	minimum	G_2

Then the concept corresponding to the new cluster is (C, s). If no pair of concepts (C_α, s_α), (C_β, s_β) fulfils conditions (a) and (b), the algorithm proceeds by trying to merge more than two concepts at a time (with suitable adaptation of the merging conditions). By using the minimum generality degree criterion to choose among the pairs of concepts (C_α, s_α), (C_β, s_β), fulfilling conditions (a) and (b), the algorithm forms clusters first which are associated with less general symbolic objects.

The new cluster C is indexed by $f(C) = G(s)$, the value of the generality degree of s. Notice that $f(C) = G(s)$ is monotone increasing as regards cluster inclusion; that is, given two concepts (C_α, s_α), (C_β, s_β), if $C_\alpha \subseteq C_\beta$ then $G(s_\alpha) \leq G(s_\beta)$. This ensures the non-existence of inversions in the graphical representation.

The algorithm stops when the cluster E is formed which corresponds to a concept (E, s), for a suitable s.

The following algorithm constructs an indexed hierarchy or a pyramid indexed in the broad sense, such that each cluster formed corresponds to a concept. Let P_t denote the set of clusters formed after step t, Q_t the corresponding set of concepts, and $S_t \subseteq P_t \times P_t$ the set of pairs of elements of P_t that may be merged at step $t+1$, according to the chosen model. For the sake of simplicity of presentation, we shall assume that $S_t \neq \emptyset$ in all steps.

Initialization: $P_0 = E$, $Q_0 = \{(\{w_1\}, s_1), \ldots, (\{w_n\}, s_n)\}$, $S_0 = P_0 \times P_0$, $C_i = \{w_i\}$, $f(C_i) = 0$,
$i = 1, \ldots, n$. Aggregation/generalization:
After step t: $P_t = \{C_h, h = 1, \ldots, m\}$, $Q_t = \{(C_h, s_h), h = 1, \ldots, m\}$,
$S_t - \{(C_h, C_{h'}) \subseteq P_t \times P_t : C_h$ may be merged with $C_{h'}\}$
While $E \notin P_t$:
1 Let $(\alpha, \beta) : G(s_\alpha \cup s_\beta) = \min\{G(s_h \cup s_{h'})\text{for}(C_h, C_{h'}) \in S_t\}$
If $\text{Ext}_E(s_\alpha \cup s_\beta) = C_\alpha \cup C_\beta$
Then
$C_{m+1} = C_\alpha \cup C_\beta$
$s_{m+1} = s_\alpha \cup s_\beta$
$f(C_{m+1}) = G(s_\alpha \cup s_\beta)$
$P_{t+1} = P_t \cup \{C_{m+1}\}$
$Q_{t+1} = Q_t \cup \{(C_{m+1}, s_{m+1})\}$
Else
$S_t = S_t \setminus (C_\alpha, C_\beta)$
Go to 1

10.3 Symbolic clustering in the presence of rules

The symbolic clustering method presented in the previous section may also be applied in the case where the data present some dependence rules between variables. The general algorithm remains the same, provided that suitable adaptations are introduced in the generalization procedure and in the computation of the generality degree.

We shall start by clearly defining hierarchical dependence, which is the type of dependence that is allowed for, and then detail the adaptations to be introduced in the generalization procedure and in the computation of the generality degree, both in the case of rules between categorical single- or categorical multi-valued variables, and in the case of rules between modal variables.

10.3.1 Hierarchical dependence

A variable z is said to be dependent on another variable y if the range of values for z depends on the value recorded for y (Bock and Diday, 2000). A variable z is said to be *hierarchically dependent* on a variable y if z makes no sense for some categories of y, and hence becomes 'non-applicable'.

Let O_y be the range of y and O_z the range of z, and $O_y' \subseteq O_y$ the set of values for which z makes no sense. Then the hierarchical dependence may be expressed by the rule (de Carvalho, 1998):

$$y \text{ takes values in } O_y' \Longleftrightarrow z \text{ is not applicable.}$$

When such a dependence exists, the special code NA (non-applicable) is used to express the fact that no value can be recorded for z. This corresponds to considering an enlarged domain which contains NA as a category, $O_z' = O_z \cup \{\text{NA}\}$.

Example
Let $y =$ gender, $O_y = \{\text{male, female}\}$ and $z =$ number of pregnancies, $O_z = \{0, 1, 2, 3, \ldots\}$. Clearly, the applicability of z depends on the value of y: if $y(w) = \text{male}$ for some w then Z is not applicable for w, $z(w) = \text{NA}$.

10.3.2 Rules between non-modal variables

In this section we will only consider hierarchical dependence between categorical single- or categorical multi-valued variables.

The generalization procedure

As concerns generalization, in the case of existence of hierarchical rules, NA is treated like any other category of a categorical single- or categorical multi-valued variable. Suppose z is hierarchically dependent on y. Then if $a_1 = [y \in V_1] \wedge [z \in W_1]$ and $a_2 = [y \in V_2] \wedge [z \in W_2]$, we have

$$a_1 \cup a_2 = [y \in V_1 \cup V_2] \wedge [z \in W_1 \cup W_2],$$

where NA may belong to either W_1 or W_2 (or both).

The generality degree

In the presence of hierarchical rules $r_1, \ldots, r_t$, the generality degree must be adjusted by subtracting the *cardinal* corresponding to the non-coherent descriptions.

Consider again a variable z hierarchically dependent on a variable y:

$$y \text{ takes values in } O_y' \Leftrightarrow z \text{ is not applicable}$$

and $O_z' = O_z \cup \{NA\}$. Let $s = (a, R, d)$ where $a = [y \in O_y' \cup O_y''] \wedge [z \in O_z'' \cup \{NA\}]$, where $O_y'' \subseteq O_y \setminus O_y'$ and $O_z'' \subseteq O_z$. Then

$$G(s \mid r) = \frac{c(O_y' \cup O_y'') \times c(O_z'' \cup \{NA\}) - c(O_y') \times c(O_z'') - c(O_y'') \times c(\{NA\})}{c(O_y) \times c(O_z') - c(O_y') \times c(O_z) - c(O_y \setminus O_y') \times c(\{NA\})}. \quad (10.4)$$

Example
As in the previous example, let $y =$ gender, $O_y = \{\text{male, female}\}$ and $z =$ number of pregnancies. We have

$$O_z{}' = \{0, 1, 2, 3, 4, \text{ more, NA}\}, y(w) = \text{male} \Rightarrow z(w) = NA.$$

Consider a symbolic object s defined by the assertion

$$a = [y \in \{\text{male, female}\}] \wedge [z \in \{0, 1, 2, \text{ NA}\}].$$

Then

$$G(s) = \frac{(2 \times 4) - (1 \times 3 + 1 \times 1)}{(2 \times 6) - (1 \times 5 + 1 \times 1)} = \frac{4}{6} = 0.6667.$$

10.3.3 Rules between modal variables

When we are in the presence of hierarchically dependent variables, for which probability/frequency distributions are known, we say that we have *constrained probabilistic data*. In such cases, the distributions possess some special features, which follow from the imposition of coherence of the descriptions.

Example
Again let $y =$ gender and $z =$ number of pregnancies. An example of an assertion associated with a probabilistic constrained object is:

$$[y = \{\text{female}(0.7), \text{male}(0.3)\}] \wedge [z = \{0(0.3), 1(0.3), 2(0.1), \text{NA}(0.3)\}].$$

Notice that the frequency of NA for the dependent variable z must equal the frequency of the category 'male' for variable y, which triggered off the rule. In fact, if in a sample there are 30% of males then, of course, for 30% of the individuals the variable z is not applicable.

In general, if $y(w) \in O_y{}'$ implies that z is not applicable, then the frequency of the NA category must equal the sum of the frequencies of all categories in $O_y{}'$. Another alternative would be to use conditional probabilities in the description of the dependent variable z. In this work, taking into account the generalization operators used and the corresponding interpretation, we shall use absolute probabilities.

The generalization procedure

For modal variables, two possibilities have been presented:

Generalization by the maximum In this case, we take for each category the maximum of its frequencies. If there are hierarchical rules of the form

$$y(w) \in O_y{}' \subset O_y \Rightarrow z(w) = \text{NA},$$

the maximum is computed for all categories except NA and then the weight of NA in the generalized object is given by $\min\{1, \sum_{m_i \in O_y{}'} p_i\}$.

Example
Let y = Education, with O_y = {basic, secondary, superior} and z = University diploma, with O_z = {BSc, MSc, PhD}. Clearly, if $y(w)$ = basic or $y(w)$ = secondary, for some w, then z is not applicable for w, $z(w)$ = NA.

Let us consider a town whose education profile may be described by the assertion

$$\begin{aligned} a_1 = &[y = \{\text{basic}(0.3), \text{secondary}(0.5), \text{superior}(0.2)\}] \\ &\wedge [z = \{\text{BSc}(0.1), \text{MSc}(0.1), \text{PhD}(0.0), \text{NA}(0.8)\}] \end{aligned}$$

and another town with education profile given by

$$\begin{aligned} a_2 = &[y = \{\text{basic}(0.1), \text{secondary}(0.8), \text{superior}(0.1)\}] \\ &\wedge [z = \{\text{BSc}(0.0), \text{MSc}(0.05), \text{PhD}(0.05), \text{NA}(0.9)\}]. \end{aligned}$$

The generalized profile is described by the assertion

$$\begin{aligned} a_1 \cup a_2 = &[y \leq \{\text{basic}(0.3), \text{secondary}(0.8), \text{superior}(0.2)\}] \\ &\wedge [z \leq \{\text{BSc}(0.1), \text{MSc}(0.1), \text{PhD}(0.05), \text{NA}(1.0)\}]. \end{aligned}$$

The extent of the corresponding symbolic object is composed by towns with *at most* 30% people with basic education, *at most* 80% people with secondary education, *at most* 20% people with superior education, etc., and so z may be non-applicable in up to 100% of the cases.

Generalization by the minimum In this case, we generalize by taking for each category the minimum of its frequencies. In the presence of hierarchical rules, the minimum is computed for all categories except NA and then the weight of NA in the generalized object is given by $\sum_{m_i \in O_y'} p_i$.

Example
Consider again the previous example. The generalized profile is now described by the assertion

$$\begin{aligned} a_1 \cup a_2 = &[y \geq \{\text{basic}(0.1), \text{secondary}(0.5), \text{superior}(0.1)\}] \\ &\wedge [z \geq \{\text{BSc}(0.0), \text{MSc}(0.05), \text{PhD}(0.0), \text{NA}(0.6)\}]. \end{aligned}$$

The extent of the corresponding symbolic object is composed by towns with *at least* 10% people with basic education, *at least* 50% people with secondary education, *at least* 10% people with superior education, etc., and so z is non-applicable in *at least* 60% of the cases.

The generality degree

For modal variables with dependence rules, generality measures G_1 and G_2, described in Section 10.2.3, are used. However, in the presence of hierarchical rules $r_1, \ldots, r_t$, the generality degree has to be adjusted by subtracting the generality of the non-coherent descriptions to the value of generality computed without taking rules into account. Let $G(s \mid r_1 \wedge \ldots \wedge r_t)$ be the generality of a symbolic object taking into account rules $r_1, \ldots, r_t$. Then

$$G(s \mid r_1 \wedge \ldots \wedge r_t) = G(s) - G(s \mid \neg(r_1 \wedge \ldots \wedge r_t)). \tag{10.5}$$

Example
Let us again consider the town whose education profile is described by $s = (a, R, d)$ associated with the assertion

$$\begin{aligned} a = &[y = \{\text{basic}(0.3), \text{secondary}(0.5), \text{superior}(0.2)\}] \\ &\wedge [z = \{\text{BSc}(0.1), \text{MSc}(0.1), \text{PhD}(0.0), \text{NA}(0.8)\}]. \end{aligned}$$

In this example, the hierarchical dependence between Education and University diploma is expressed by the rule $r: y(w) \in \{\text{basic, secondary}\} \Longleftrightarrow z(w) = \text{NA}$. The non-coherent descriptions are those for which there is a contradiction with the rule r, that is, those for which $y(w) =$ basic or secondary and $z(w) =$ BSc, MSc or PhD, or $y(w) =$ superior and $z(w) =$ NA. Then,

$$G_1(s) = \left(\frac{\sqrt{0.3} + \sqrt{0.5} + \sqrt{0.2}}{\sqrt{3}}\right)\left(\frac{\sqrt{0.1} + \sqrt{0.1} + \sqrt{0} + \sqrt{0.8}}{\sqrt{4}}\right) = 0.75$$

$$G_1(s \mid \neg r) = \left(\frac{\sqrt{0.3} + \sqrt{0.5}}{\sqrt{3}}\right)\left(\frac{\sqrt{0.1} + \sqrt{0.1} + \sqrt{0}}{\sqrt{4}}\right) + \left(\frac{\sqrt{0.2}}{\sqrt{3}}\right)\left(\frac{\sqrt{0.8}}{\sqrt{4}}\right) = 0.34$$

is the generality corresponding to the non-coherent descriptions ({basic, secondary}, {BSc, MSc, PhD}) and ({superior}, NA). Therefore,

$$G_1(s \mid r) = G_1(s) - G_1(s \mid \neg r) = 0.41.$$

10.4 Postprocessing

10.4.1 Pruning

Pruning a hierarchy or pyramid involves suppressing clusters with the aim of obtaining a structure which is easier to interpret, without important loss of information. Let C be a cluster with a single predecessor C' (recall that in the case of pyramids a cluster may have up to two predecessors). Let f be the index function. Suppose that $f(C') - f(C) < \epsilon$, where $\epsilon > 0$ is a fixed threshold. We can then suppress C from the structure, with no great loss (see Figure 10.2). The choice of ϵ depends on the degree of simplification we wish to achieve: too small an ϵ will scarcely change the structure, while a large ϵ will considerably simplify it, and remove a lot of clusters; usually ϵ will be a suitable percentage of the maximum height. Once the graphical representation is displayed, the user may ask for a simplification of the structure, choosing the pruning parameter ϵ.

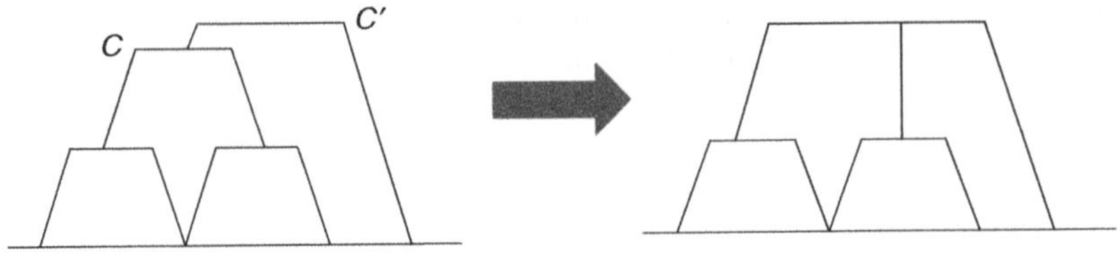

Figure 10.2 Pruning a pyramid.

10.4.2 Rule generation

If the hierarchy/pyramid is built from a symbolic data table, we obtain an inheritance structure, in the sense that each cluster inherits the properties associated with its predecessors. This fact allows us to generate rules between clusters. Two methods are considered:

- *Fusion* method (pyramids only). Let (p_1, s_1), (p_2, s_2) be two concepts in the pyramid, and (p, s) be another concept such that $p = p_1 \cap p_2$ (see Figure 10.3). We can then write the rule:

$$s_1 \wedge s_2 \Rightarrow s.$$

- *Splitting* method (hierarchies and pyramids). Let (p, s) be a concept obtained by merging (p_1, s_1) with (p_2, s_2), that is, $p = p_1 \cup p_2$, $s = s_1 \cup s_2$ (see Figure 10.4). Then,

$$s \Rightarrow s_1 \vee s_2.$$

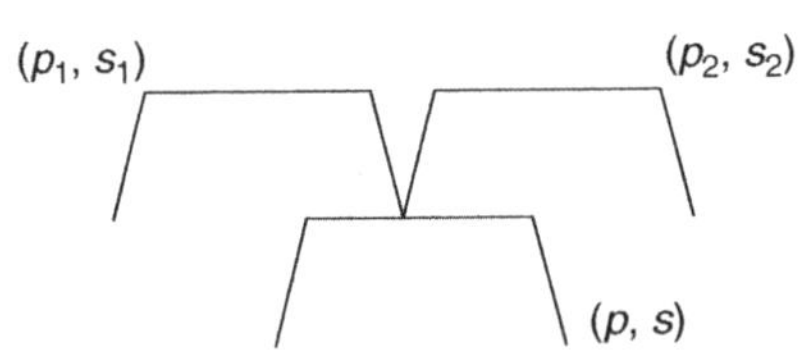

Figure 10.3 Fusion rule.

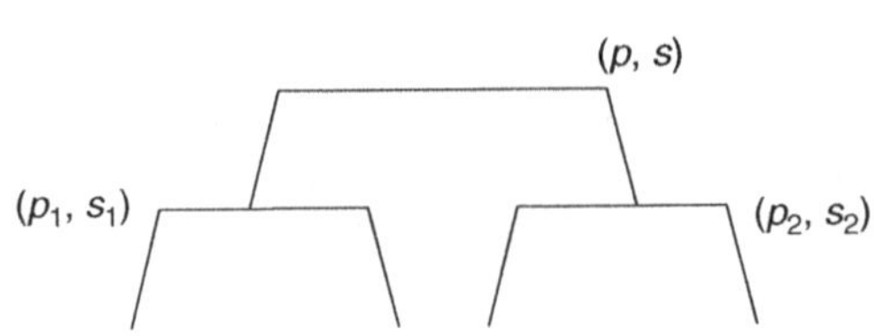

Figure 10.4 Fission rule.

10.5 Hierarchical and pyramidal clustering with the SODAS software

10.5.1 Objectives

The HIPYR module allows a hierarchy or a pyramid to be constructedon a set of symbolic data. HIPYR assumes that you have a set of entities, hereafter called 'individuals' (persons, institutions, cities, objects, etc.), that you wish to organize in a nested clustering structure. Individuals, associated with symbolic objects, are described by quantitative single, interval, categorical single, categorical multi-valued and/or modal variables; mixed types are allowed. If a dissimilarity matrix is available, it may be used for numerical clustering. Taxonomies and hierarchical dependencies defined on variables may be taken into account.

HIPYR constructs the cluster structure based either on the individual-variable data set or on dissimilarity values between the elements of the data set. In the former case, the symbolic hiearchical/pyramidal clustering method described in Sections 10.2 and 10.3 is applied. The description of each cluster is then given by the associated symbolic object and each cluster is defined by the set of its members together with its description (as a complete symbolic object that generalizes its members).

If the clustering is based on a dissimilarity matrix between the elements of E, then dissimilarites must previously be computed by the DISS module and the input of HIPYR is a file containing a triangular dissimilarity matrix. Then, a classical hierarchical or pyramidal clustering algorithm is applied. The most similar clusters are aggregated at each step and aggregation measures, which express the dissimilarity between clusters, are used. In HIPYR the following aggregation measures may be used: complete linkage, single linkage, average linkage and diameter. In this case, the clusters are not automatically associated with a discriminant description, only dissimilarity values are available, so no description is given for the clusters.

Once the structure has been completed, a comparison measure between individuals may be obtained. This is an induced dissimilarity measure: for any two individuals, the induced dissimilarity is equal to the height of the cluster where they first appear together when exploring the structure bottom-up. This induced dissimilarity may then be compared with the dissimilarity or generality degree values directly obtained from the data.

Let d_o be the initial comparison measure (dissimilarity or generality) and d_I the induced measure. The evaluation value ev is defined as

$$ev = \frac{\sum_{i=1}^{n-1}\sum_{j=i+1}^{n}\left(d_o(w_i, w_j) - d_I(w_i, w_j)\right)^2}{\sum_{i=1}^{n-1}\sum_{j=i+1}^{n} d_o(w_i, w_j)}. \tag{10.6}$$

The program also allows for an automatic selection of 'interesting' clusters. Here, a cluster is considered to be 'interesting' if its index value (i.e., its height) differs significantly from the height of its predecessors.

The program starts by computing the differences between the index values of the clusters and those of their predecessors. Let $D(C)$ be the value of this difference for cluster C. Then, the mean $\bar{D}$ and the standard deviation SD_D of those differences are computed, and standardized values $D'(C)$ are computed for all clusters:

$$D'(C) = \frac{D(C) - \bar{D}}{SD_D}. \tag{10.7}$$

Clusters C for which $|D'(C)| > 2$ are selected.

Visualization of the hierarchy or pyramid is performed by VPYR. Pruning may be performed from the graphical representation.

10.5.2 Options and parameters

HIPYR provides several different choices for clustering data; therefore, the user must specify the values of a variety of parameters which characterize special options. On the other hand, it is also possible to run HIPYR in a semi-automatic way, i.e., without worrying about the correct choice of all parameters, since HIPYR provides default (standard) values for all parameters. Figure 10.5 shows the options dialogue box for the HIPYR module. The options are explained below in the same order as they appear in the dialogue box.

- *Hierarchy/pyramid.* The user must choose to form either a hierarchy or a pyramid and click on the corresponding button. By default, a hierarchy is formed.
- *Data source.* Here, the user must choose whether the clustering is to be done on the basis of a dissimilarity matrix (numerical clustering) or using the symbolic objects (symbolic clustering).
- *Aggregation function.* The user must select the aggregation function to be used by the clustering algorithm. For numerical clustering (data source: dissimilarity matrix) the following options are available: maximum (complete linkage), minimum (single linkage), average linkage and diameter. For symbolic clustering (data source: symbolic objects) there are two options: generality degree and increment of the generality degree. By default, the maximum is selected for numerical clustering and the generality degree for symbolic clustering.

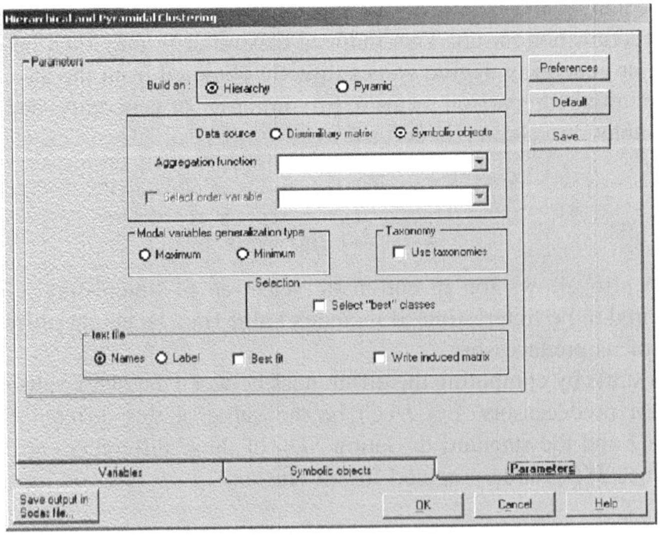

Figure 10.5 HIPYR module options dialogue box.

- *Select order variable*. If the user wants the objects of the data array to be ordered in a given way (thus fixing the basis of the pyramid), this button must be checked. Then the user must indicate the variable which contains this order. Only quantitative single or ordinal variables are displayed. By default, no order variable is selected.
- *Modal variables generalization type*. If modal variables are present, the user may choose whether to perform generalization (in symbolic clustering) by the maximum or by the minimum. Only one of these options may be chosen; by default, generalization is performed by the maximum.
- *Taxonomy*. If taxonomies are defined for some variables, HIPYR allows the user to use them in the generalization process (symbolic clustering). If this option is checked, taxonomies are used, otherwise they are not. By default, taxonomies are not used.
- *Selection*. If this option is checked, the program will automatically produce a set of 'interesting' formed clusters. By default, the selection is not made.
- *Text file identification*. Here the user must choose whether to use names or labels in the text output file. A 'best fit' option is also available, which chooses the best solution depending on the length of the labels. By default, names are used.
- *Write induced matrix*. Here the user must indicate whether he wants the induced dissimilarity matrix to be written in the output listing. This is an $n \times n$ triangular matrix. By default, it is not written in the listing.
- *Save output file*. Here the user must indicate the name of the output file. This file contains the initial data array – top left $n \times p$ array – and the description of the clusters formed in the basis of the initial variables (one row for each cluster). Some columns are added to the initial variables: index value, a quantitative single variable that indicates the height of the cluster (equals 0 for all single individuals); list of cluster members, a categorical multi-valued variable.

10.5.3 Output results

Listing

The HIPYR algorithm produces as output a SODAS file containing both the input data and a new set of symbolic objects corresponding to the clusters of the hierarchy or pyramid. Supplementary results are provided in a text file:

- File and options used.
- List of clusters, and, for each cluster, the clusters merged to form the present cluster, membership list, index value, and symbolic object – if the hierarchy or pyramid was constructed by a symbolic clustering method, then this symbolic description constitutes a necessary and sufficient condition for cluster membership.
- Evaluation value: this gives the fit between the obtained structure and the original data – the lower its value the better the fit.

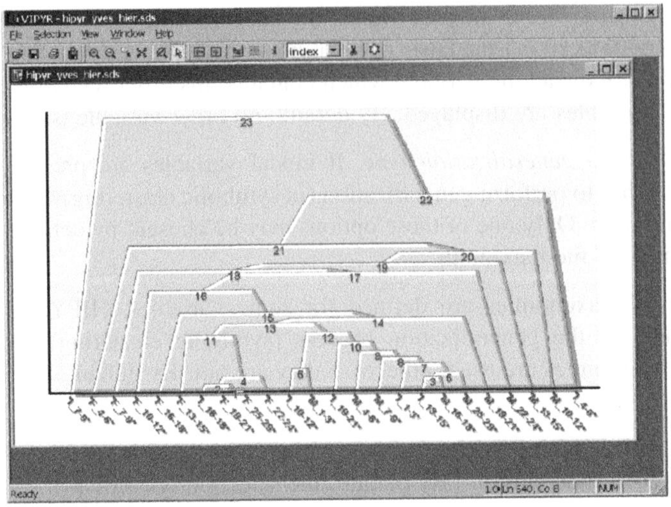

Figure 10.6 Graphical representation of a hierarchy.

- List of the selected 'interesting' clusters, if requested.
- The induced dissimilarity matrix, if requested.

Graphical representation

A graphical output is provided by VPYR (see Figure 10.6). Several options are then available to explore the structure (hierarchy or pyramid):

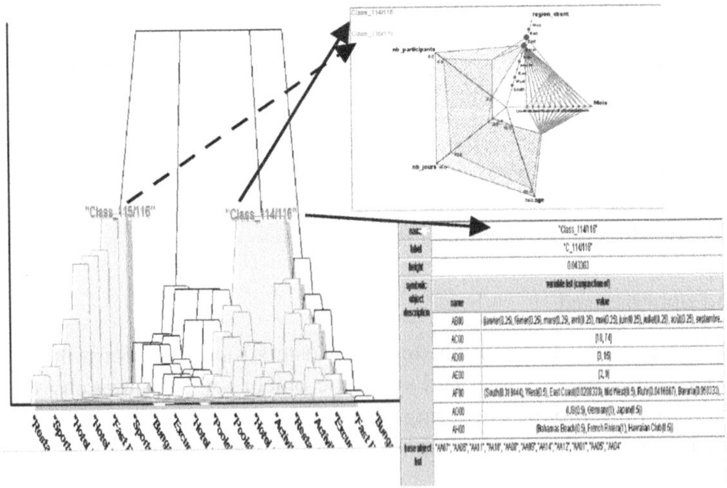

Figure 10.7 Interpretation tools on the graphical representation.

- A cluster is selected by clicking on it. Then the user may get the description of the cluster in terms of a list of chosen variables or its representation by a zoom star (Figure 10.7); see Noirhomme-Fraiture and Rouard (2000), and Chapter 7 of the present volume.
- The user can prune the hierarchy or pyramid using the aggregation heights as a criterion. First, the rate of simplification must be selected, then the user must click on the simplification button in the menu; this launches a new graphic window with a simplified graphic.
- If the hierarchy/pyramid is built from a symbolic data table, rules may be generated and saved in a specified file.
- Should the user be interested in a particular cluster, he may obtain a window with the structure restricted to this cluster and its successors.

10.6 Example: cultural survey data

This data set was obtained from a survey, carried out in five Portuguese towns in 1997, on participation in cultural and leisure activities. The 1409 individuals in the survey database were aggregated according to gender, age group and qualifications, leading to a data set of 35 symbolic objects describing the distribution of each variable in each group. Among the large number of variables in the database, we have chosen those indicated in Table 10.2, with the corresponding categories.

We performed symbolic hierarchical and pyramidal clusterings on these data. Figure 10.8 shows the symbolic hierarchy obtained, when generalization is done by the maximum and using the generality degree as criterion for cluster formation. Figure 10.9 shows the zoom star comparing the descriptions of clusters 32 and 33. Figure 10.10 shows the symbolic pyramid obtained on these data, for the same parameters, and where the automatically selected clusters are indicated. Figure 10.11 shows the same pyramid after pruning at a 20% rate.

Table 10.2 Variables used in the cultural survey application.

Variables	Categories
Frequency watching TV	everyday from time to time almost never never
Radio	yes, no
Art museum	yes, no
Books	yes, no
Cinema	yes, no
Football	yes, no
Daily newspapers	yes, no
Weekly newspapers	yes, no

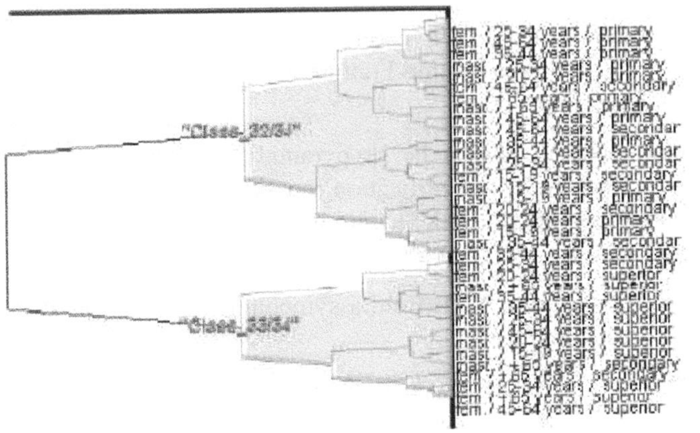

Figure 10.8 Symbolic hierarchy for the cultural survey data.

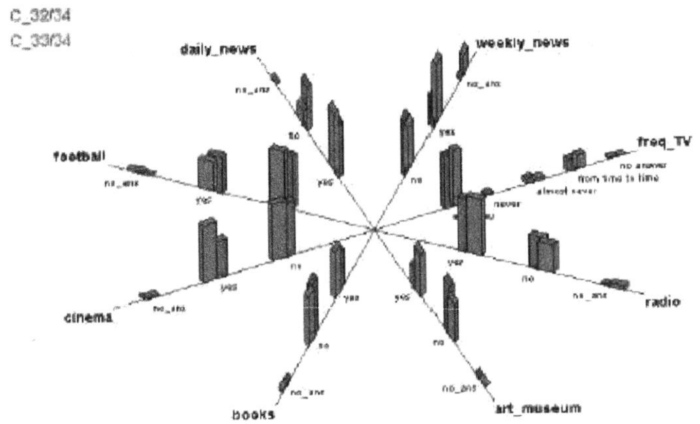

Figure 10.9 Zoom star for clusters 32 and 33 of the symbolic hierarchy for the cultural survey data.

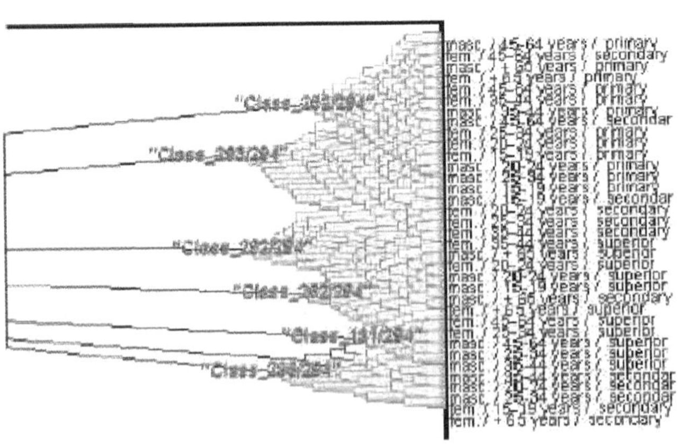

Figure 10.10 Symbolic pyramid for the cultural survey data.

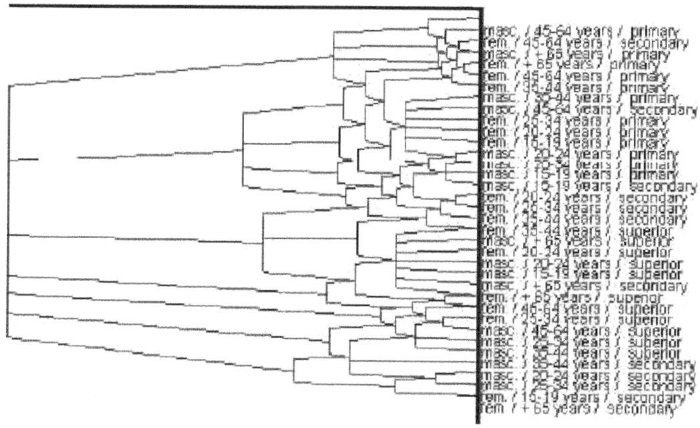

Figure 10.11 Pyramid pruned at a 20% rate.

For instance, cluster 181 of the pyramid, composed of {fem._25_34 years _ superior, fem._45_64 years_superior, masc._45_64 years_superior, masc._25_34 years_superior, masc._35_44years_superior, masc._35_44 years_secondary}, is associated with the following assertion:

$$[\text{freq\_TV} = \{\text{everyday} \leq 0.916667, \text{almost never} \leq 0.0681818, \text{from time to time} \leq 0.2, \text{no\_answer} \leq 0.030303\}]$$
$$\wedge [\text{radio} = \{\text{yes} \leq 0.873239, \text{no} \leq 0.5, \text{no\_ans} \leq 0.08\}]$$
$$\wedge [\text{art\_museum} = \{\text{yes} \leq 0.666667, \text{no} \leq 0.757576, \text{no\_ans} \leq 0.0333333\}]$$
$$\wedge [\text{books} = \{\text{yes} \leq 0.583333, \text{no} \leq 0.787879, \text{no\_ans} \leq 0.04\}]$$
$$\wedge [\text{cinema} = \{\text{no} \leq 0.757576, \text{yes} \leq 0.605634\}]$$
$$\wedge [\text{football} = \{\text{no} \leq 1, \text{yes} \leq 0.484848, \text{no\_ans} \leq 0.0333333\}]$$
$$\wedge [\text{daily\_news} = \{\text{yes} \leq 0.939394, \text{no} \leq 0.25\}]$$
$$\wedge [\text{weekly\_news} = \{\text{no} \leq 0.454545, \text{yes} \leq 0.746479\}].$$

We can see that this group of people who have superior education or are men between 35 and 44 years old and have secondary education is characterized by high values of the 'yes' category for most cultural activities and not so high a value for attending football matches.

Pruning the pyramid at a 20% rate allowed for a much more simplified clustering structure, where the most important clusters are kept and intermediate clusters are pruned.

References

Bertrand, P. and Diday, E. (1985) A visual representation of the compatibility between an order and a dissimilarity index: The pyramids. *Computational Statistics Quarterly*, 2(1), 31–42.

Bock, H.-H. and Diday, E. (eds) (2000) *Analysis of Symbolic Data: Exploratory Methods for Extracting Statistical Information from Complex Data*. Berlin: Springer-Verlag.

Brito, P. (1991) Analyse de données symboliques. Pyramides d'héritage. Doctoral thesis, Univ. Paris IX Dauphine.

Brito, P. (1994) Use of pyramids in symbolic data analysis. In E. Diday, Y. Lechevallier, M. Schader, P. Bertrand and B. Burtschy (eds), *New Approaches in Classification and Data Analysis*, pp. 378–386. Berlin: Springer-Verlag.

Brito, P. (1995) Symbolic objects: Order structure and pyramidal clustering. *Annals of Operations Research*, 55, 277–297.

Brito, P. (1998) Symbolic clustering of probabilistic data. In A. Rizzi, M. Vichi and H.-H. Bock (eds), *Advances in Data Science and Classification*. Berlin: Springer-Verlag.

Brito, P. (2000) Hierarchical and pyramidal clustering with complete symbolic objects. In H.-H. Bock and E. Diday (eds) *Analysis of Symbolic Data: Exploratory Methods for Extracting Statistical Information from Complex Data*, pp. 312–324. Berlin: Springer-Verlag.

Brito, P. and de Carvalho, F.A.T. (1999) Symbolic clustering in the presence of hierarchical rules. In *Studies and Research, Proceedings of the Conference on Knowledge Extraction and Symbolic Data Analysis (KESDA'98)*, pp. 119–128. Luxembourg: Office for Official Publications of the European Communities.

Brito, P. and de Carvalho, F.A.T. (2002) Symbolic clustering of constrained probabilistic data. In M. Schwaiger and O. Opitz (eds), *Exploratory Data Analysis in Empirical Research*, pp. 12–21. Berlin: Springer-Verlag.

Brito, P. and Diday, E. (1990) Pyramidal representation of symbolic objects. In M. Schader and W. Gaul (eds),*Knowledge, Data and Computer-Assisted Decisions*. Berlin: Springer-Verlag.

Chavent, M. (1998) A monothetic clustering method. *Pattern Recognition Letters*, 19, 989–996.

Csernel, M. and de Carvalho, F.A.T. (1999) Usual operations with symbolic data under normal symbolic form. *Applied Stochastic Models in Business and Industry*, 15, 241–257.

de Carvalho, F.A.T. (1994) Proximity coefficients between Boolean symbolic objects. In E. Diday, Y. Lechevallier, M. Schader, P. Bertrand and B. Burtschy (eds), *New Approaches in Classification and Data Analysis*, pp. 387–394. Berlin: Springer-Verlag.

de Carvalho, F.A.T. (1998) Extension based proximity coefficients between constrained Boolean symbolic objects. In C. Hayashi, K. Yajima, H.-H. Bock, N. Ohsumi, Y. Tanaka and Y. Baba (eds), *Data Science, Classification, and Related Methods*, pp. 370–378. Tokyo: Springer-Verlag.

Diday, E. (1984) Une représentation visuelle des classes empiétantes: Les pyramides. Research Report 291. INRIA, Rocquencourt, Le Chesnay, France.

Diday, E. (1986) Orders and overlapping clusters by pyramids. In J. de Leeuw, J. Heiser, J. Meulman and F. Critchley (eds), *Multidimensional Data Analysis*, pp. 201–234. Leiden: DSWO Press.

El-Sonbaty, Y. and Ismail, M.A. (1998) On-line hierarchical clustering. *Pattern Recognition Letters*, 19, 1285–1291.

Gowda, K.C. and Diday, E. (1991) Symbolic clustering using a new dissimilarity measure. *Pattern Recognition*, 24(6), 567–578.

Gowda, K.C. and Diday, E. (1992) Symbolic clustering using a new similarity measure. *IEEE Transactions on Systems, Man and Cybernetics*, 22, 368–378.

Gowda, K.C. and Ravi, T.R. (1995a) Divisive clustering of symbolic objects using the concepts of both similarity and dissimilarity. *Pattern Recognition*, 28(8), 1277–1282.

Gowda, K.C. and Ravi, T.R. (1995b) Agglomerative clustering of symbolic objects using the concepts of both similarity and dissimilarity. *Pattern Recognition Letters*, 16, 647–652.

Gowda, K.C. and Ravi, T.R. (1999) Clustering of symbolic objects using gravitational approach. *IEEE Transactions on Systems, Man and Cybernetics*, 29(6), 888–894.

Ichino, M. and Yaguchi, H. (1994) Generalized Minkowski metrics for mixed feature type data analysis. *IEEE Transactions on Systems, Man and Cybernetics*, 24(4), 698–708.

Matusita, K. (1951) Decision rules based on distance for problems of fit, two samples and estimation. *Annals of Mathematical Statistics*, 3, 1–30.

Noirhomme-Fraiture, M. and Rouard, M. (2000) Visualising and editing symbolic objects. In H.-H. Bock and E. Diday (eds), *Analysis of Symbolic Data: Exploratory Methods for Extracting Statistical Information from Complex Data*, pp. 125–138. Berlin: Springer-Verlag.

Rodríguez, O., Brito, P. and Diday, E. (2001) Algoritmos para la classificación piramidal simbólica. *Revista de Matemática: Teoría y Aplicaciones*, 7(1–2), 23–42.

Vignes, R. (1991) Caractérisation automatique de groupes biologiques. Doctoral thesis, Université Paris VI.

11

Clustering methods in symbolic data analysis

Francisco de A.T. de Carvalho, Yves Lechevallier and Rosanna Verde

11.1 Introduction

This chapter deals with methods for clustering concepts modelled by symbolic objects (SOs) into a predefined number of homogeneous classes. The classes are suitably interpreted and represented by class prototypes. The proposed partitioning algorithms are based on a generalization of the classical dynamic clustering method. The general optimization criterion is a measure of the best fit between the partition and the representation of the classes. The prototype is a model of a class, and its representation may be an element of the same space of representation of the concepts to be clustered which generalizes the characteristics of the elements belonging to the class. Therefore, the prototype is a concept itself and is also modelled as an SO.

We can extend the algorithm to cluster a set of individuals E having a symbolic description. In this context, the prototypes are symbolic descriptions of the classes of individuals.

The allocation function for the assignment of the elements to the classes depends on the nature of the variables which describe the SOs. The choice of the allocation function must be related to the particular type of prototype taken as the representation model of the a class.

Finally, we distinguish between two main approaches to dynamic clustering: the symbolic clustering algorithm (SCLUST) and the symbolic clustering algorithm on distance tables (DCLUST). The former method has as input a set of concepts modelled by SOs, while the latter has as input a distance matrix. The distances between the SOs are computed using suitable functions which measure the proximity between SOs according to their descriptions.

Symbolic Data Analysis and the SODAS Software Edited by E. Diday and M. Noirhomme-Fraiture

Moreover, the choice of these measures must be consistent with the type of prototypes that represent the classes of the partition.

What distinguishes the symbolic clustering method is the interpretation of the classes as concepts. Modelling the concepts by prototypes, defined as SOs, makes it possible to give a symbolic meaning to the elements of the partition at each step of the algorithm and not only at the end of the procedure.

Tools for the interpretation of clusters and for the evaluation of the quality of the partition are presented in Section 11.5. Two applications on artificial and on real data, illustrated in Section 11.6, serve to corroborate the procedure.

11.1.1 Generality of the dynamic clustering method

Let F be a set of elements to be clustered. The aim of the dynamic clustering algorithm (Diday, 1971; Diday and Simon, 1976; Celeux *et al.*, 1989) is to divide a set of elements $x \in F$ into k homogeneous clusters. The general idea is to construct a suitable representation for each cluster by means of a description of the elements of F belonging to this cluster in order to allocate new elements to the clusters obtained. The algorithm is based on the following principles:

- the number k of classes must be fixed, but a different number of classes can be requested in order to look for the best partition into k classes. This can be done by moving k between two and a selected maximum number of classes.
- the algorithm is performed in two steps. The *representation step* describes the k classes $(P_1, \ldots, P_i, \ldots, P_k)$ of the partition P of the set F by a vector $L = (G_1, \ldots, G_i, \ldots, G_k)$ of k prototypes. The *allocation step* assigns the elements of F to the classes, according to their proximity to the prototypes. This proximity is a value given by an allocation function ψ defined on $F \times \Lambda$, where Λ is a set of prototypes, such that $G_i \in \Lambda$ and $L \in \Lambda^k$.

The algorithm looks for the partition $P \in \mathcal{P}_k$ of F into k classes, among all the possible partitions P_k, and the vector $L \in \Lambda^k$ of k prototypes representing the classes in P, such that a criterion Δ is minimized:

$$\Delta(P^*, L^*) = \min\{\Delta(P, L) | P \in \mathcal{P}_{k||}, \mathcal{L} \in \Lambda^{k||}\}.$$

Moreover, the algorithm can be reinitialized in order to improve the final partition and the value of the optimization criterion.

The convergence of the algorithm to a stationary value of the criterion function Δ is guaranteed by the optimal fit between the type of representation of the classes by prototypes and the allocation function $\psi(\cdot)$.

11.1.2 Existence and uniqueness conditions in the dynamic clustering process

Usually, the criterion $\Delta(P, L)$ is defined as

$$\Delta(P, L) = \sum_{i=1}^{k} \sum_{x \in P_i} \psi(x, G_i).$$

This criterion being additive, convergence can be guaranteed by the existence and uniqueness of a $G_i \in \Lambda$. Given a predefined and arbitrary index on the vector $L = (G_1, \ldots, G_k)$, we define the argmin function on L by $i = \text{argmin}\{\psi(x, G_l)/l = 1, \ldots, k\}$, for $x \in F$, as the index of the element G_i of the vector L which satisfies the following two properties: $\psi(x, G_i) = \min\{\psi(x, G_l)/l = 1, \ldots, k\}$ and the index of G_i is minimal.

The definition of the argmin function guarantees the decreasing of the criterion during the allocation step, whereas the decreasing of the criterion during the representation step is guaranteed by the existence and the uniqueness of $G_i \in \Lambda$ for all the k classes P_i of the partition P, i.e.,

$$\forall P, \forall P_i \in P, \exists G_i \in \Lambda : \sum_{x \in P_i} \psi(x, G_i) < \sum_{x \in P_i} \psi(x, G),$$

where $G \in \Lambda$ and $G \neq G_i$.

11.2 Symbolic dynamic clustering approaches

The proposed dynamic clustering method generalizes the standard clustering method in order to partition a set of individuals E, modelled by symbolic descriptions, or a set of concepts C, modelled by SOs, into k classes. The algorithm is based on the choice of prototypes for representing the classes, and the choice of a proximity function to allocate the concepts to classes at each step. The clustering criterion to be optimized is a measure of best fit between the partition of the set of concepts and the prototype descriptions.

11.2.1 The input data

Like other symbolic data analysis (SDA) techniques, the clustering methods proposed here run on a symbolic data table, denoted X (Bock and Diday, 2000, Chapter 3). In the SODAS software, this matrix is furnished in .sds or .xml format. The columns of the input data table are associated with *symbolic variables* $y_1, \ldots, y_j, \ldots, y_p$ and the rows contain the descriptions $d_c = (d_c^1, \ldots, d_c^p) \in D$ of the concepts c of C, which are modelled by SOs $s_c = (a_c, R, d_c)$, where R is a binary or fuzzy relation between descriptions, and a_c is the mapping from a set of individuals Ω to the interval [0,1] or to the set {0,1} (Bock and Diday, 2000, Chapter 1).

The prototype g_i of a class $P_i \in P$ is modelled as an SO $g_i = (a_{g_i}, R, G_i)$. We denote by Λ the space of the prototypes $g \in \Lambda$ and by Γ the space of their descriptions.

If the space D of descriptions of the elements of C is the same as the space Γ of the descriptions of the prototypes g_i then we have $\psi(c, g_i) = R(y(c), y(g_i)) = [d_c R G_i]$.

The symbolic dynamic clustering algorithm can be applied to symbolic data described by any type of symbolic variable – interval, categorical multi-valued and modal. It considers the classical quantitative and categorical single-valued variables as a particular case of, respectively, interval-valued and categorical multi-valued symbolic variables.

According to the nature of the symbolic variables which describe the SOs and the prototypes, different proximity measures or distances are proposed to allocate the SOs associated with the elements of C to classes. Classical L_2 and χ^2 distances are used when the SOs are described by real-valued or categorical variables, while the Hausdroff distance, based on the L_1-norm, is used to compute the proximity between SOs described by interval variables, and a context-dependent proximity, or alternatively a ϕ^2 function, is proposed when the SOs are described by categorical multi-valued or modal variables.

Generally the proximity measures are computed on the symbolic descriptions of the SOs associated with the concepts or on the symbolic descriptions of the individuals. Nevertheless, we remark how, in this context, rules, taxonomies and metadata, that moreover characterize SOs, can be taken into account in the computation of the dissimilarity between concepts.

Finally, tools for evaluating the partition obtained are proposed in Section 11.4.

11.2.2 Symbolic interpretation

The following tools give a measure of the quality of the partition as well as of the classes generalizing the inertia criterion to symbolic data contained in the symbolic data table X.

The classical decomposition of the total sum of squares (TSS) into the within sum of squares (WSS) and the between sum of squares (BSS) is given by

$$\underbrace{\sum_{h=1}^{n} \delta^2(x_h, G)}_{TSS} = \underbrace{\sum_{i=1}^{k} \sum_{h \in P_i} \delta^2(x_h, G_i)}_{WSS} + \underbrace{\sum_{i=1}^{k} n_i \delta^2(G_i, G)}_{BSS}, \qquad (11.1)$$

where x_h (for $h = 1, \ldots, n$) is the hth row of X which contains the description d_h. Furthermore, δ^2 is the Euclidean distance, G is the mean of the n points $x_h \in \mathbb{R}^p$, G_i is the mean of the points x_h associated with the elements of the class P_i, and $n_i = \text{card}(P_i)$.

As is well known, the mean of a cluster P_i is the point $G_i \in \mathbb{R}^p$ which minimizes the adequacy criterion

$$f_{P_i}(G_i) = \sum_{h \in P_i} \delta^2(x_h, G_i).$$

In the clustering methods proposed in this chapter, we have generalized the concept of the mean of a cluster P_i by a prototype g_i which minimizes the adequacy criterion:

$$f_{P_i}(g_i) = \sum_{c \in P_i} \psi(c, g_i).$$

The allocation function ψ measures the degree of proximity between a concept c and the SOs g_i. We remark that when the set of elements to be clustered are elements of E or C, the function ψ measures the degree of proximity between an individual and a suitable symbolic description of the class.

TSS (the total inertia) and WSS (the within-cluster inertia) can be generalized by using the previously defined prototypes and comparison functions. Specifically, the generalization of WSS is the criterion $\Delta(P, L)$ with $P = (P_1, \ldots, P_k)$ and $L = (G_1, \ldots, G_k)$ performed by the dynamic cluster algorithm; and the generalization of TSS is the adequacy criterion $f_C(G_C)$, where G_C is the prototype of the whole set of n concepts in C.

Of course, equation (11.1) no longer holds after generalization to symbolic data.

11.3 Symbolic dynamic clustering algorithm (SCLUST)

The symbolic dynamic clustering method is an generalization of the standard dynamic clustering method to cluster a set of concepts $c \in C$ into k homogeneous classes. The procedure is based on the same principles as the classical one:

- fixing the number k of classes;
- a *representation step*, describing the k classes $(P_1, \ldots, P_i, \ldots, P_k)$ of the partition P of the set C by a vector of prototypes $g = (g_1, \ldots, g_i, \ldots, g_k)$ $g \in \Lambda$;
- an *allocation step*, defining an allocation function to assign the elements of C to the classes of the partition P.

The criterion Δ minimized by the algorithm is assumed additive and is based on the function $\psi(.)$ Therefore, the convergence of the algorithm to an optimum value of the function Δ is guaranteed by the consistency between the representation of the classes by prototypes and the allocation function $\psi(.)$.

11.3.1 The general scheme of the symbolic dynamical clustering algorithm

- *Initialization*. Let $P^{(0)} = \{P_1^{(0)}, \ldots, P_k^{(0)}\}$ be the initial random partition of C in k classes.
- *Representation step t*. For $i = 1, \ldots, k$, compute a prototype $g_i^{(t)}$ as the SO representing the class $P_i \in P^{(t)}$.
- *Allocation step t*. Any concept $c \in C$ is assigned to the class P_i if and only if $\psi(c, g_i)$ is a minimum:

$$P_i^{(t+1)} = \{c \in C \mid i = \text{argmin}\{\psi(c, g_l)/l = 1, \ldots, k\}.$$

- *Stopping rule or stability*. If $P^{(t+1)} = P^{(t)}$ then stop, else go to the representation step.

Notice that the algorithm can be initialized by a random set of k concepts of C. The SOs associated to these concepts are assumed as the starting prototypes $g_1^{(0)}, \ldots, g_k^{(0)}$. Then the initial partition is obtained by the allocation step.

The criterion $\Delta(P, L)$ optimized by the dynamic clustering algorithm is additive with respect to the data descriptors. We propose to define the criterion $\Delta(P, L)$ as the sum of the allocation function $\psi(c, g_i)$ for each concept c belonging to a class $P_i \in P$ and the prototype $g_i \in \Lambda$:

$$\Delta(P, L) = \sum_{i=1}^{k} \sum_{c \in P_i} \psi(c, g_i).$$

An alternative general scheme, based on the optimization of the criterion $\Delta(P, L)$, is a *k-means-like* one (MacQueen, 1967). This approach performs the allocation step before the representation step:

- *Initialization*. Let $g^{(0)} = (g_1^{(0)}, \ldots, g_k^{(0)})$ be the initial prototypes. These prototypes are obtained by random symbolic objects associated of the concepts of C.
- *Step t*. For all $c \in C$:

- *Allocation step t*. An object c is assigned to the class P_i^t, if and only if $\psi(c, g_i^{(t)})$ is minimum.
- *Representation step t*. The prototype $g_i^{(t+1)}$ is updated by including the last allocated element.

- *Stopping rule or stability*. If $P^{(t+1)} = P^{(t)}$ then stop, else go to previous step.

In the first algorithm, the description G_i of each prototype g_i changes when all the objects have been assigned to the class. In the second, it is modified after the assignment of each object to a new class P_i of the partition.

Partitioning of classical data

Whenever input data are described by classical variables y_j $(j = 1, \ldots, p)$, even of different types, that is, quantitative and categorical single-valued, the prototypes of the classes can be described by both classical and symbolic variables. Different kinds of proximity or dissimilarity measures can be proposed as allocation functions (see Table 11.1). All distances used in this chapter are described in Chapter 8 of Bock and Diday (2000).

Partitioning of Boolean symbolic data

The partitioning method, illustrated here, is performed on symbolic data, described by symbolic variables of two types: categorical multi-valued and interval variables. This method allows a partition of symbolic objects to be constructed using dissimilarity or distance measures defined for both types of variables (Table 11.2). The class prototypes are also described by symbolic variables.

We can distinguish two main ways of representing a class: a prototype expressed by a single element of the class (e.g., the element at the minimum average distance from all the elements of the class as well as by the element which minimizes the criterion function); or a prototype chosen as a summarizing function of the elements of the class. In the latter case,

Table 11.1 Default parameters for classical input data.

SO descriptors	G_i	$\delta_j(x_h, G_i)$
quantitative	quantitative (real-valued)	L_2
nominal	nominal (categories)	χ^2
quantitative	interval	Hausdorff distance
nominal	categorical multi-valued	de Carvalho
nominal	modal (frequency distr.)	de Carvalho

Table 11.2 Definitions for Boolean symbolic input data.

SO descriptors	G_i described by	$\delta_j(x_h, G_i)$
interval	interval	Hausdorff distance
categorical multi-valued	modal	de Carvalho

the prototype can be suitably modelled by a modal SO (Bock and Diday, 2000, Chapter 3). The description of a modal SO is given by frequency (or probability) distributions associated with the categories of the p categorical descriptors.

In the *representation step*, depending on the nature of the descriptors of the set C of SOs we distinguish different cases:

(i) all the SO descriptors are intervals;

(ii) all the SO descriptors are multi-valued categorical variables.

In the first case, Chavent (1997), Chavent and Lechevallies (2002) and Chavent *et al.* (2003) demonstrated that is possible to represent a class by an interval chosen as the one at minimum average distance from all the other intervals, elements of the class. The optimal solution was found analytically, assuming as distance a suitable measure defined on intervals: the *Hausdorff distance*.

The second case can be considered according to two different approaches:

(i) the prototype is expressed by the most representative element of the class (or by a virtual element v_b) according to the allocation function;

(ii) the prototype is a high-order SO, described by the distribution function associated with the multinominal variables domains.

In particular, in the first approach the prototype is selected as the neighbour of all the elements of the class. Given a suitable allocation function $\psi(c_h, g_i)$ the prototype g_i of the class P_i is chosen as the SO associated with the concept c_h, where $h = \text{argmin}\{\sum_{h' \in P_i} \psi(c_{h''}, c_{h'}) : c_{h''} \in P_i\}$. Similarly, a prototype g of P_i can be constructed considering all the descriptions of the SOs $c_h \in P_i$ and associating with them a set of descriptions corresponding to the most representatives among the elements of P_i, such that $\sum_{h \in P_i} \psi(c_h, g) = \min\left\{\sum_{h \in P_i} \psi(c_h, g') : g' \in \Gamma\right\}$. A similar criterion has been used by Chavent and Lechavallies (2002) and Chavent *et al.* (2003) in the choice of the description of the prototypes as interval data, according to the Hausdorff distance.

Nevertheless, we observe that if the virtual prototype g is not an SO associated with a concept of C, its description may be inconsistent with the conceptual meaning of a symbolic object. Thus, instead of taking g to represent the class P_i, it is more appropriate to choose the nearest SO of $c_h \in P_i$, according to the allocation function value.

This choice is a generalization of the nearest-neighbour algorithm criterion in the dynamic clustering. However, it respects the numerical criterion of the minimum dissimilarity measure and guarantees coherence with the allocation function.

In the second approach, we can also assume a uniform distribution associated with multi-categorical variables to transform SO descriptors into *modal* ones. The prototype g_i, representing the cluster P_i, is described by the minimum generalization of the descriptions of the elements belonging to the class P_i, for all the categorical multi-valued variables (de Carvalho *et al.*, 1999).

In the *allocation step* the coherence between the prototype and the allocation function guarantees the convergence of the partitioning algorithm. Thus, we distinguish two different situations: (i) the description space of SOs associated with $c_h \in C$ and prototypes is the same; (ii) the prototypes are represented as modal SOs.

The first case corresponds to the situation where the prototypes and SOs are modelled by vectors of intervals for interval variables, as well as by sets of categories for categorical multi-valued variables. Finally, they are also in the same space of representation whenever both are described by *modal* variables.

The second case corresponds to the situation where the prototypes are modelled by modal variables, while the SOs are described by interval and/or categorical multi-valued variables.

Hausdorff distance for interval data As noted above, Chavent (1997) proposed the Hausdorff distance in the SO clustering procedure. The Hausdorff distance is usually proposed as a distance between two sets A and B and is defined as follows:

$$\delta_H(A, B) = \max\{\sup_{a\in A}\inf_{b\in B} \delta(a, b), \sup_{b\in B}\inf_{a\in A} \delta(a, b)\},$$

where $\delta_H(A, B)$ is a distance between two elements a and b. When A and B are two intervals, $[\underline{a}, \overline{a}]$, $[\underline{b}, \overline{b}] \subset \mathbb{R}$, the Hausdroff distance is

$$\delta_H(A, B) = \max\{|\underline{a} - \underline{b}|, |\overline{a} - \overline{b}|\},$$

which is a L_∞-distance.

It is known that the Hausdorff distance $\delta(A, B)$ between two intervals can be decomposed as the sum of the distances between the middle points, μ_A and μ_B, and between the radii λ_A and λ_B (semi-length) of A and B:

$$\delta_H(A, B) = |\mu_A - \mu_B| + |\lambda_A - \lambda_B|.$$

Chavent and Lechevallier (2002) and Chavent *et al.* (2003) found the best representation of clusters of interval-valued data, which is obtained by minimizing the Hausdorff average distance between all the intervals describing the objects of the class and the more suitable interval $[\underline{\theta}, \overline{\theta}]$ representing the class. Because the Hausdorff distance is additive, the criterion function (for all classes P_i, $i = 1, \ldots, k$) is

$$f(G_h) = \sum_{h\in P_i} \delta(x_h, G_i) = \sum_{h\in P_i}\sum_{j=1}^{J} \delta(x_h^j, G_i^j),$$

where J is the number of interval variables. Thus, the 'best' interval can be found for each interval variable y_j ($j = 1, \ldots, J$) by minimizing

$$f(G_i^j) = \sum_{h\in P_i} \delta(x_h^j, G_i^j) = \max \sum_{h\in P_i} \{|\underline{\theta}^j - \underline{a}_h^j|, |\overline{\theta^j} - \overline{a_h^j}|\}.$$

A dissimilarity function to compare multinominal data Among the dissimilarity functions proposed in the SDA context to compare SOs, we can assume as an *allocation function* in the clustering algorithm the most general *dissimilarity function* (Gower, 1966; Ichino and Yaguchi, 1994; Gordon, 1999) when the SOs of C are all described by interval data or by categorical multi-valued variables. In the case of interval descriptions, this measure is an alternative to the Hausdorff distance.

Ichino and Yaguchi proposed this measure as a suitable comparison function between pairs of objects. In our context of analysis, this is the measure consistent with the fitting criterion of the algorithm when the representation prototype is in the same space of representation of the elements of the class.

To define this *dissimilarity* we need to introduce the *description potential* function $\eta(.)$, proposed by de Carvalho and Souza (1999):

- $\eta(d_j^h)$ is the length of the interval d_j^h if y_j is an interval variable;
- $\eta(d_j^h)$ is the cardinality of the set of values included in d_j^h if y_j is multi-categorical.

Given two concepts $c_h, c_{h'} \in C$, the Ichino–Yaguchi measure is expressed by the following formula:

$$\delta_j(x_h, x_{h'}) = \eta(d_j^h \oplus d_j^{h'}) - \eta(d_j^h \cap d_j^{h'}) + \gamma(2\eta(d_j^h \cap d_j^{h'}) - \eta(d_j^h) - \eta(d_j^{h'}))$$

where γ is between 0 and 0.5 and $\oplus$ is the *Cartesian join* which is defined as:

$$d_j^h \oplus d_j^{h'} = \begin{cases} [\min\{d_j^h, d_j^{h'}\}, \max\{d_j^h, d_j^{h'}\}] & \text{if } y_j \text{ is a quantitative or a ordinal variable,} \\ d_j^h \cup d_j^{h'}, & \text{if } y_j \text{ is a nominal variable.} \end{cases}$$

When γ is set to 0.5, the Ichino–Yaguchi function becomes

$$\delta_j(x_h, x_{h'}) = \eta(d_j^h \oplus d_j^{h'}) - \frac{\eta(d_j^h) + \eta(d_j^{h'})}{2}.$$

Starting from this measure, different kinds of normalization of δ can be considered: by the *description potential* $\eta(D_j)$ of the domain D_j (Ichino and Yaguchi, 1994) of y_j or by the *description potential* of the combination set of the two symbolic descriptions, $\eta(d_j^h \oplus d_j^{h'})$ (de Carvalho and Souza, 1999). Further values for the parameter γ in [0, 0.5] were also considered. Then, the aggregation function is usually chosen according to a Minkowski metric (with parameter r):

$$\delta(x_h, x_{h'}) = p^{-1}\left[\sum_{j=1}^{p}(\delta_j(x_h, x_{h'}))^r\right]^{1/r} \qquad (r \geq 1).$$

The prototypes of the classes, described by interval data, are chosen as the neighbouring elements from all the other ones, according to the dissimilarity measure selected. This means that the prototype of the class is the SO belonging to this class such that the sum of the dissimilarities between this so and all the other SO belonging to this class is the smallest one. In this context, the *virtual* intervals, equal to the median intervals of the elements of each class, are consistent for the Hausdorff distance but are not consistent with respect to this different type of dissimilarity measure.

Two-component distance as allocation matching function When the SO concept and prototype are in two different description spaces D and Γ, we consider a suitable *context-dependent proximity function* δ (de Carvalho and Souza, 1999), as a matching function, to compare the objects to prototypes.

In this case the description space of the concepts c_h to be clustered and the description space Γ are not homogeneous because the symbolic object of c_h is described by categorical multi-valued or interval variables, and the prototypes are described by *modal* ones. To retrieve the two configurations of data in the same description space, we consider the transformation on the SO description by a uniform distribution (de Carvalho and Souza, 1999) and we estimate the empirical distribution of the prototypes. Note that the distributions can be assumed non-uniform on the basis of external information on the SO descriptions.

Therefore, the allocation function ψ can be considered as a suitable comparison function which measures the degree of matching between the concept c_h and the prototype g_i, according to their description G_i and d_h. The particular allocation function that we consider is based on a function of two additive components (de Carvalho *et al.*, 1999; Verde *et al.*, 2000).

Partitioning for modal symbolic data

When the set of elements to be clustered are modal objects and $D=\Gamma$, suitable dissimilarity measures can be proposed as allocation function.

When both concepts and prototypes are modelled by distributions, a classical ϕ^2 distance can be proposed as allocation function (see Table 11.3). As noted above, modal data can be derived by imposing a system of weights (pseudo-frequencies, probabilities, beliefs) on the domain of categorical multi-valued or interval descriptors. These transformations of the SO description space are requested wherever prototypes have been chosen as modal SOs.

In this way, concepts $c_h \in C$ are allocated to the classes according to their minimum distance from the prototype. Because ϕ^2 is also an additive measure, the distance of x_h to G_i can be expressed as

$$\phi^2(x_h, G_i) = \sum_{j=1}^{J} \sum_{v=1}^{L_j} \frac{1}{q_{v\cdot}} (q^h_{vj} - q^i_{vj})^2,$$

where $v=1, \ldots, L_j$ is the set of the L_j categories constituting the domain of a categorical multi-valued variable y_j or a set of elementary intervals of an interval variable y_j (Chavent *et al.*, 2003, 2006). The q^i_{vj} is obtained as the average profile of the distribution $q^h_j = (q^h_{1j}, \ldots, q^h_{L_j j})$ associated with $c_h \in P_i$ for the modal variable y_j. Thus, simple Euclidean distances between profiles can also be used.

Partitioning for mixed symbolic data

All the proposed distance functions for p variables are determined by sums of dissimilarities corresponding to the univariate symbolic component descriptors y_j. The most appropriate dissimilarity functions are listed in the tables above according to the type of variables.

In practice, however, symbolic data to be clustered are typically described by different types of variables. In order to compute an overall dissimilarity measure two approaches are proposed here:

Table 11.3 Definitions for symbolic modal input data.

SO descriptors	G_i	$\delta_j(x_h, G_i)$
modal	modal	ϕ^2

1. *Weighted linear combination of dissimilarity measures.* If the SO associated with c is described by a different type of variable the overall dissimilarity between c and g_i is obtained by a linear combination of the proximity measures computed with respect to the different (classical or symbolic) variables:

$$\psi(c, g_i) = \sum_{k=1}^{4} \lambda_k \sum_{j \in V_k} \xi_j \delta_k(x_h^j, G_i^j), \tag{11.2}$$

where V_k is the set of variables of a priori type k ($k = 1$ for quantitative, $k = 2$ for categorical, $k = 3$ for interval, and $k = 4$ for modal), λ_k is a normalizing factor for the set of variables of type k, and ξ_j is a normalizing factor for the jth variable y_j.

2. *Categorization (discretization, segmentation, ...).* In this case, all the variables y_j are transformed to the same type.

11.3.2 Output of SCLUST in SODAS software

The result of SCLUST is a new set of concepts g_i modelled by SOs. The SODAS software produces as output a file in .sds or .xml format containing a description of this set of symbolic objects. At the symbolic input data table X, the program adds a categorical variable in order to classify the SO according to the clusters given by SCLUST.

The following supplementary output is furnished by SODAS in a text file:

- a list of classes (membership list, binary membership vector, class summary);
- a description vector associated with the prototypes;
- the relation for each variable,
- the extension mapping.

The visualization of the prototypes is performed by the VSTAR and VPLOT modules. Tools for interpreting the clusters of the partition obtained are also available (see Section 11.6).

11.4 Clustering algorithm on distance tables (DCLUST)

The aim of the clustering method on distance tables (DCLUST) is to partition a set C into a (fixed) number k of homogeneous classes on the basis of the proximities between pairs of concepts or individuals of C. The criterion which is optimized by DCLUST is based on the sum of the dissimilarities (or the sum of squares of the distances) between elements belonging to the same cluster. In the SODAS software the proximity table is provided by the DISS module (or by DI or DIM). Alternatively, it can be given directly by proximity functions which are able to take into account dependencies (hierarchical and logical) between variables. Therefore, the use of proximity functions allows logical dependencies and hierarchical dependencies to be taken into account. Thus, input proximity between two elements is obtained by context-free proximity functions or by a dissimilarity table.

11.4.1 Generality of the method

The clustering process aims to group the objects of a set C into k homogeneous clusters on the basis of a dissimilarity table. The proposed approach is an application of the dynamical clustering algorithm applied to a dissimilarity table (Lechevallier, 1974). The algorithm follows the main principles of the method.

The number k of classes must be fixed, but a different number of classes can be requested in order to look for the best partition in k classes. This can be done by moving k between 2 classes and a selected maximum number of classes.

Moreover, the algorithm can be reinitialized in order to improve the final partition and the value of the optimization criterion.

The DCLUST algorithm is as follows:

- *Initialization*. The initial vector of prototypes, $q_{(0)} = \{g_1^{(0)}, \ldots, g_k^{(0)}\}$ contains random concepts or elements of C.
- *Allocation step t*. An object c_h is assigned to the class $P_i^{(t)}$ if and only if $i = \text{argmin}\{\psi(c, g_l^{(t-1)})/l = 1, \ldots, k\}$.
- *Representation step t*. For $i = 1, \ldots, k$, the prototype $g_i^{(t)}$ representing class $P_i^{(t)} \in P^{(t)}$ is the SO of the concept where the index is equal to

$$h = \text{argmin}\left\{ \sum_{c_l \in P_i^{(t)}} \psi(c_l, s_m)/m = 1, \ldots, |C| \right\}.$$

  Then the prototype $g_i^{(t)}$ is equal to the SO s_h of the concept c_h.
- *Stopping rule or stability*. If $P^{(t)} = P^{(t-1)}$ then stop, else go to the allocation step.

11.4.2 Input data

DCLUST can be used on dissimilarity matrices on a set of concepts or a set of individuals. If DCLUST is used on a set of individuals, the prototypes are only defined by the symbolic descriptions of the elements of E.

With the SODAS software the dissimilarity table must be given by a file in .sds or .xml format containing the data table between the individuals. This data table is obtained by one of the following strategies: a SODAS file (e.g., using the procedure DISS in the new version of the SODAS software or DI or DIM in the old version); or using functions (taking into account hierarchical and logical dependencies).

11.4.3 Output data and results

SODAS produces as output a file in .sds or .xml format containing an informatics representation of the set of concepts or individuals associated with the classes. If the input is just a dissimilarity table, only the alternative algorithm DCLUST can be used. A symbolic description of the classes is not available.

If the input consists of both a symbolic data table and a proximity table, the basic DCLUST algorithm produces as output a .sds or .xml file containing a new set of symbolic objects: the clusters of the final partition, described by a vector of the description of the prototypes. In the DCLUST algorithm, an indicator variable is added to the original symbolic data table containing the index of the classes of the final partition.

Supplementary results are furnished in a text file as described in Section 11.3.2

If the input is just a given dissimilarity table it is not possible to construct a symbolic description of the classes.

If the symbolic data table is furnished and the basic algorithm is applied, the visualization of the prototypes will be performed by VSTAR.

11.5 Cluster interpretative aids

This section deals with several indices to measure and interpret the quality of the partition or the quality of the clusters of the partition. The classical interpretative aids are usually based on the inertia criterion. In our approach we propose to generalize this criterion to symbolic data, where the centre of gravity is replaced by the symbolic description of the prototypes. Our interpretative aids are only based on the symbolic description of the prototypes; the relation R and the set of the mapping functions are not used.

As in the classical method (Celeux *et al.*, 1989), the proposed indices to describe a partition are based on the decomposition of the total sum of squares (*TSS*) into the within-cluster (*WSS*) and between-clusters (*BSS*) sum of squares.

In order to simplify our presentation we assume that all the elements of C have the same weight equal to 1.

The gain in inertia obtained by replacing the description prototype G of C by the k description prototypes $(G_1, \ldots, G_k)$ of the partition P is no longer the between-cluster inertia (*BSS* in (11.1)).

Considering that by construction of the prototypes we have

$$\underbrace{\sum_{i=1}^{k} f_{C_i}(G_i)}_{\Delta(P,L)} \leq \underbrace{\sum_{i=1}^{k} f_{C_i}(G_C)}_{f_C(G_C)}, \tag{11.3}$$

we define the gain in homogeneity obtained by replacing the n objects by the k prototypes, as the difference between $f_C(G_C)$ and $\Delta(P, L)$. In other words, the gain can be interpreted as the 'difference' between the null hypothesis 'no structure = partition in one cluster' and the k-cluster partition obtained by minimization of the adequacy criterion Δ. Thus, the indices used to interpret a partition, as well as its clusters, are:

- $f_{C_i}(G_i)$, which is a measure of homogeneity of the cluster C_i;
- $\Delta(P, L)$, which is a measure of within-cluster homogeneity of the partition P;
- $f_C(G_C)$, which is a measure of total homogeneity of the set of concepts C.

We consider as homogeneity measures $f_{C_i}(G_i)$ the allocation function used in the clustering approach.

11.5.1 Interpretation of the partition

The *quality index* $Q(P)$ of a partition P can be interpreted as the gain between the null hypothesis 'no structure = partition into one cluster' and the partition into k clusters obtained by optimizing the fitting criterion between the partition P and the corresponding vector of prototypes, normalized by the total homogeneity $f_C(G_C)$:

$$Q(P) = 1 - \frac{\Delta(P, L)}{f_C(G_C)}. \tag{11.4}$$

This index takes values between 0 and 1. It is equal to 1 when all the clusters are reduced to a single object or to identical objects. It is equal to 0 for the one-cluster partition of C. Because this criterion decreases with the number of clusters it can only be used to compare two partitions having the same number of clusters. Because one k-cluster partition is better than another if the criterion $\Delta(P, L)$ is smaller, this partition will be better if $Q(P)$ is larger.

For classical quantitative data,

$$Q(P) = \frac{BSS}{TSS} = 1 - \frac{WSS}{TSS}$$

is that part of the inertia of C explained by P. In our context $Q(P)$ measures, in the same way, the part of the homogeneity of C explained by P.

For each variable j, we similarly define the quality of the partition on the variable j:

$$Q_j(P) = 1 - \frac{\sum_{i=1}^{k} \tilde{f}_{C_i}(G_i^j)}{f_C(G_C^j)} \tag{11.5}$$

which is the part of the homogeneity of the classes C_i of P due by the variable j. This criterion measures the *power of discrimination* of the variable j on the partition P. Moreover, because the quality of $Q(P)$ is a weighted average of the values $Q_j(P)$,

$$Q(P) = \sum_{j=1}^{p} \frac{f_C(G_C^j)}{f_C(G_C)} Q_j(P), \tag{11.6}$$

this index measures even the importance of the variable j in the building of the partition. Finally, a relative index $Q_j(P)$ is obtained by normalizing it by $Q(P)$.

11.5.2 Interpretation of the clusters

The quality of a cluster C_i is defined by:

$$Q(C_i) = 1 - \frac{f_{C_i}(G_i)}{f_{C_i}(G_C)}. \tag{11.7}$$

If the value of $Q(C_i)$ is near 1 then the elements of the clusters are very different from the general prototype G_C.

A *contribution measure* is given as the ratio between the homogeneity value computed for each cluster (or variable) and the global homogeneity criterion of the partition $\Delta(P, G)$.

The sum of all the contributions is equal to 1. We can measure the *contribution of the cluster* C_i to the global variability by

$$K(C_i) = \frac{f_{C_i}(G_i)}{\Delta(P, L)}. \tag{11.8}$$

The sum of the k contributions is obviously 1.

A final criterion useful in interpreting a cluster according to a variable j is:

$$Q_j(C_i) = 1 - \frac{f_{C_i}(G_i^j)}{f_{C_i}(G_C^j)}. \tag{11.9}$$

If this value is near to 1 then the description prototype of the cluster C_i is very far from G_C. In other words, this criterion helps to find the variables which characterize the cluster C_i. Because the quality of the cluster, $Q(C_i)$, is a weighted average of the values $Q_j(C_i)$,

$$Q(C_i) = \sum_{j=1}^{p} \frac{\tilde{f}_{C_i}(G_C^j)}{f_{C_i}(G_C)} Q_j(C_i), \tag{11.10}$$

the values of the criterion $Q_j(C_i)$ have to be interpreted by comparison with the value $Q(C_i)$. In other words, we will consider that a variable j characterizes the cluster C_i if $Q_j(C_i) > Q(C_i)$.

11.6 Applications

11.6.1 Application to micro-organism data

This simple example is designed to show the main differences between dynamical clustering for symbolic objects and numerical clustering methods. The data set is a recoding set proposed by Michalski *et al.* (1981) in a conceptual classification context. The components of the problem are: the set of objects (micro-organisms) to be clustered (see Figure 11.1); the set of categorical variables selected for describing micro-organisms shown in Table 11.4.

A dynamic classification has been carried out on this data set. The ten symbolic objects are classified into $k = 3$ classes according to their characteristics by using as allocation function the context-dependent proximity measure. The set of descriptors consists of four categorical multi-valued variables with the following categories:

body spots = {top, bottom, right, left},

body part = {triangle, rectangle, circle},

tail type = {top, bottom, right, nobody},

texture = {black, white}.

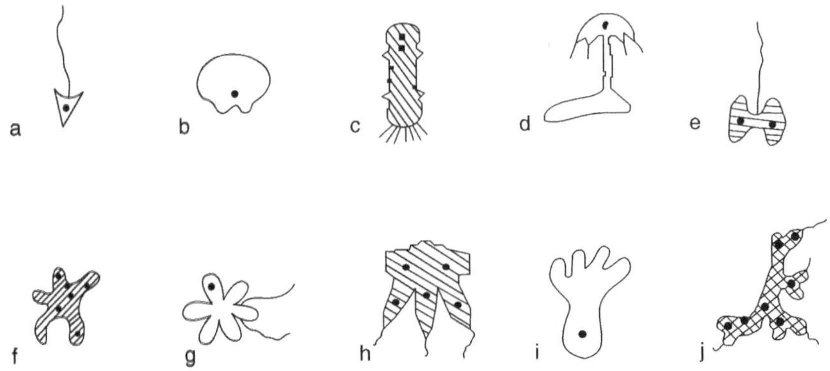

Figure 11.1 Micro-organisms.

Table 11.4 Categorical descriptions of micro-organisms.

Micro-organism	Body parts	Body spots	Texture	Tail type
a	1	one	blank	single
b	1	one	blank	none
c	1	many	striped	multiple
d	2	one	blank	multiple
e	2	many	striped	single
f	many	many	striped	none
g	many	one	blank	multiple
h	many	many	striped	multiple
i	many	one	blank	none
j	many	many	crosshatched	multiple

Table 11.5 Symbolic descriptions of micro-organisms.

Micro-organism	Body spots	Body part	Tail type	Texture
a	bottom	triangle	top	white
b	bottom	circle	nobody	white
c	top	rectangle, circle	bottom	black, white
d	top	rectangle, circle	top, bottom	white
e	bottom, right, left	circle	top	black, white
f	top, right, left	nobody	striped	black, white
g	top	circle	right	black
h	top, bottom, right, left	triangle, rectangle	bottom	black, white
i	bottom	rectangle, circle	bottom, right	white
j	top, bottom, right, left	rectangle, circle	bottom, right	black, white

Table 11.6 Step 1: symbolic descriptions of prototypes.

Prototypes	Body spots	Body part	Tail type	Texture
G_1	top(1/4), bottom(31/60) right(7/30), left(7/30)	triangle(1/5), rectangle(1/5) circle(3/5)	top(2/5), bottom(3/10) right(1/10), nobody(1/5)	black(3/10) white(7/10)
G_2	top(3/4), bottom(1/12) right(1/12), left(1/12)	triangle(1/6), rectangle(1/3) circle(1/2)	top(1/6), bottom(1/2) right(1/3)	black(1/6) white(5/6)
G_3	top(1/6), bottom(1/2) right(1/6), left(1/6)	rectangle(1/4) circle(3/4)	nobody(1)	black(1/4) white(3/4)

Table 11.7 Step 1: distance between objects and prototypes.

Prototypes	$\delta(\cdot, G_1)$	$\delta(\cdot, G_2)$	$\delta(\cdot, G_3)$	Partition
a	2.18	2.75	4.75	C_1
b	1.98	3.58	1.0	C_3
c	1.65	0.92	2.83	C_2
d	1.55	0.92	3.08	C_2
e	1.25	2.08	2.42	C_1
f	1.72	2.58	0.75	C_3
g	2.35	1.58	3.33	C_2
h	1.3	1.0	3.25	C_2
i	1.78	3.25	0.75	C_3
j	0.8	0.33	2.0	C_2

Table 11.5 shows the symbolic description of the micro-organisms.

Denoting the 10 SOs by $a, b, \ldots, j$ the algorithm initializes by clustering the objects as $C_1 = \{a, b, c, e, j\}$, $C_2 = \{d, g, h\}$ and $C_3 = \{f, i\}$ and representing the classes by prototypes shown in Table 11.6.

The allocation function of each object to a class is given by Table 11.7, the final column of which shows the assigned class of the object.

According to the minimum value of the allocation function between the object x and the set of prototypes G_i, we obtain the following partition $P^{(1)}$ of the set C: $C_1 = \{a, e\}$, $C_2 = \{c, d, g, h, j\}$ and $C_3 = \{b, f, i\}$. This differs from the previous one for the elements b, c, j that move from C_1 to C_2 and C_3.

The partial criteria computing with respect to each cluster are respectively $\Delta_1^{(1)} = 3.43$, $\Delta_2^{(1)} = 4.75$, $\Delta_3^{(1)} = 2.5$. Thus, the global criterion $\Delta^{(1)}$ is equal to 10.68. The representation by prototype G_i of the new clusters C_i, at step 2, is shown in Table 11.8.

The partition $P^{(2)}$ of the set C obtained at step 2, $C_1 = \{a, e\}$, $C_2 = \{c, d, g, h, j\}$ and $C_3 = \{b, f, i\}$ (see Table 11.9), is coincident with that obtained at the previous step. The partial criterion values according to this partition are $\Delta_1^{(2)} = 2.0$, $\Delta_2^{(2)} = 4.7$, $\Delta_3^{(2)} = 1.58$ and the global criterion value is equal to 8.28, less than $\Delta^{(1)}$.

Table 11.8 Step 2: symbolic descriptions of prototypes.

Prototypes	Body spots	Body part	Tail type	Texture
G_1	bottom(4/6) right(1/6), left(1/6)	triangle(1/2) circle(1/2)	top(1)	black(3/4) white(1/4)
G_2	top(7/10), bottom(1/10) right(1/10), left(1/10)	triangle(1/10), rectangle(2/5) circle(1/2)	top(1/10), bottom(3/5) right(3/10)	black(3/10) white(7/10)
G_3	top(1/9), bottom(4/6) right(1/9), left(1/9)	rectangle(1/6) circle(5/6)	nobody(1)	black(1/6) white(5/6)

Table 11.9 Step 2: distance between objects and prototypes.

Prototypes	$\delta(\cdot, G_1)$	$\delta(\cdot, G_2)$	$\delta(\cdot, G_3)$	Partition
a	1.08	3	4.5	C_1
b	3.08	3.7	0.67	C_3
c	5	0.8	2.89	C_2
d	3.75	1.0	3.05	C_2
e	0.5	2.1	2.28	C_1
f	3.5	2.6	0.83	C_3
g	4.75	1.8	3.22	C_2
h	3.25	0.9	3.33	C_2
i	3.59	3.3	0.5	C_3
j	3.25	0.2	2	C_2

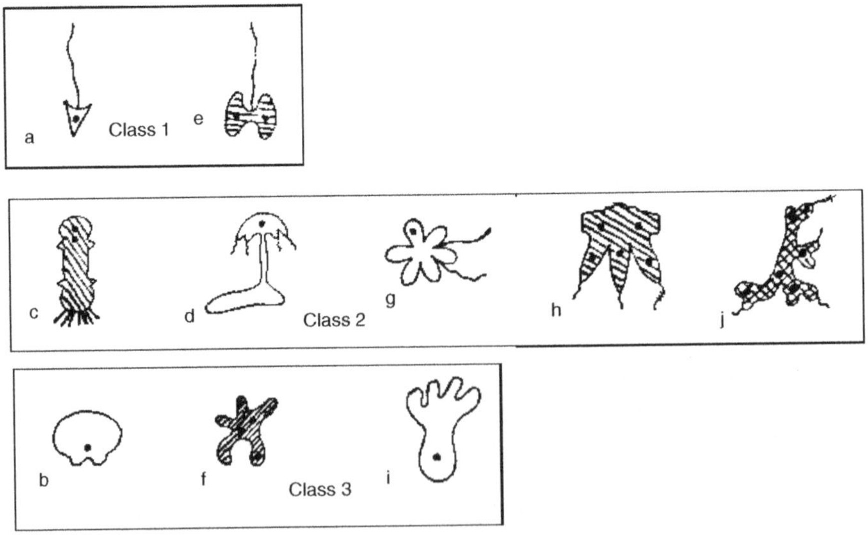

Figure 11.2 Partition.

The algorithm is stopped at this best partition $P^{(2)}$ of the set C whose clusters are optimally represented by the prototype G_i (Figure 11.2).

The partition we have obtained with our algorithm coincides with the best partition given by PAF in Michalski *et al.* (1981) on the same objects but described by classical nominal variables. Such partition correspond well to the human solution (Michalski *et al.* (1981, p. 52).

11.6.2 Application to 60 meteorological stations in China

In this subsection we partition a set of data from the Long-Term Instrumental Climatic Data Base of the People's Republic of China (http://dss.ucar.edu/datasets/ds578,5/data).

This set of data contains the monthly (maximum and minimum) temperatures observed in 60 meteorological stations in China. For our example we have considered the temperatures of the year 1988 and we have constructed a table of dimension 60 rows × 12 columns, corresponding to the number of stations and the number of months of the year (see Table 11.10). The different quality and contribution indices have been computed on an example of partition into five clusters obtained by a dynamical clustering algorithm on symbolic data described by interval variables. For instance, the ChangSha station is described by the 12 intervals of the monthly temperatures which is the symbolic description used by our method:

$$\begin{aligned}
&[\texttt{January} = [2.7:7.4]] \wedge [\texttt{February} = [3.1:7.7]] \\
&\wedge [\texttt{March} = [6.5:12.6]] \wedge [\texttt{April} = [12.9:22.9]] \\
&\wedge [\texttt{May} = [19.2:26.8]] \wedge [\texttt{June} = [21.9:31]] \\
&\wedge [\texttt{July} = [25.7:34.8]] \wedge [\texttt{August} = [24.4:32]] \\
&\wedge [\texttt{September} = [20:27]] \wedge [\texttt{October} = [15.3:22.8]] \\
&\wedge [\texttt{November} = [7.6:19.6]] \wedge [\texttt{December} = [4.1:13.3]].
\end{aligned}$$

The symbolic data table contains the descriptions of the 60 meteorological stations.

Table 11.10 Minima and maxima monthly temperatures recorded by the 60 meteorological stations.

Meteorological station	January	February	...	December
AnQing	[1.8, 7.1]	[5.2, 11.2]	...	[4.3, 11.8]
BaoDing	[−7.1, 1.7]	[−5.3, 4.8]	...	[−3.9, 5.2]
BeiJing	[−7.2, 2.1]	[−5.3, 4.8]	...	[−4.4, 4.7]
...	...	...	...	...
ChangChun	[−16.9, −6.7]	[−17.6, −6.8]	...	[−15.9, −7.2]
ChangSha	[2.7, 7.4]	[3.1, 7.7]	...	[4.1, 13.3]
ZhiJiang	[2.7, 8.2]	[2.7, 8.7]	...	[5.1, 13.3]

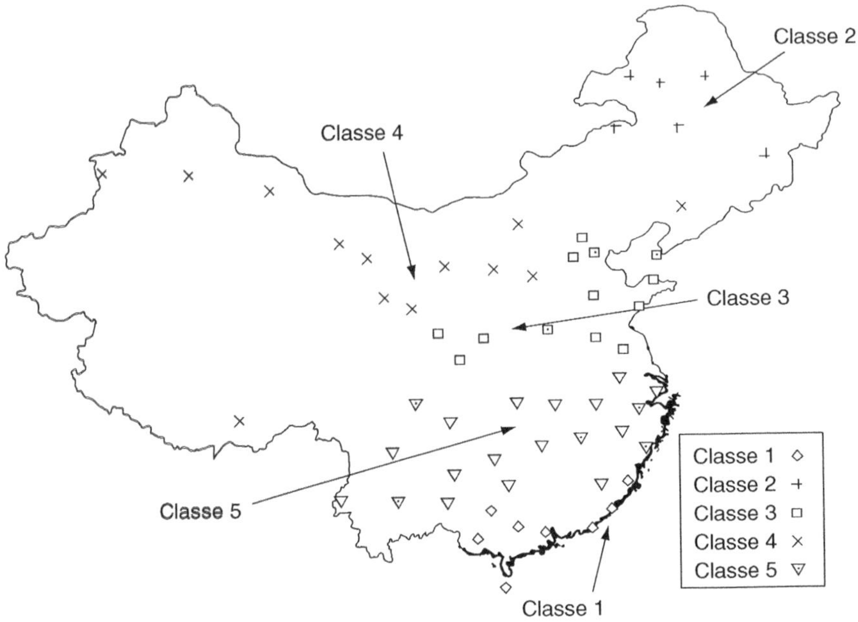

Figure 11.3 The five clusters of the partition of the 60 meteorological stations.

Fixing the number of clusters to 5, the algorithm is reiterated 50 times and the best solution is found for the minimum value of the criterion equal to $D = 3848.97$. It is worth noting that the obtained partition of the 60 elements conforms to follows the geographical contiguity of the stations (see Figure 11.3).

Table 11.11 Quality and contribution measures (×100) of the intervals of temperatures observed in the 12 months to the partition of the stations into five clusters.

Variable	Quality	Contribution with P	Contribution with E
January	**69.50**	**13.76**	12.74
February	**66.18**	**12.63**	12.28
March	**64.52**	9.30	9.27
April	**64.36**	6.74	6.73
May	61.68	6.15	6.42
June	53.36	4.56	5.50
July	46.31	4.05	5.63
August	47.19	3.73	5.08
September	**61.10**	6.05	6.37
October	**70.41**	8.97	8.19
November	**70.63**	**10.79**	9.83
December	**71.33**	**13.26**	11.96

According to the kind of representation of the clusters by intervals proposed in the partitioning algorithm on interval data, the prototype of each cluster is the interval which minimizes the Hausdorff distances from all the elements belonging to the cluster.

In Table 11.11 we have indicated the values of the different indices of quality and contribution proposed here. We can observe that the winter months are more discriminant of the cluster (high value of the quality index and contribution index) than the summer ones. In Figure 11.4 the prototypes of the clusters computed on the interval values of January and December are much more separated than the ones in Figure 11.5 corresponding to the prototype of temperatures of June and September. For the months January, February, November and December the values of the quality and the contribution are similar, while for the months April and October, even if the quality measure assumes quite the same value of the winter months, the contribution value is lower, which means the discriminant power of these months is weaker.

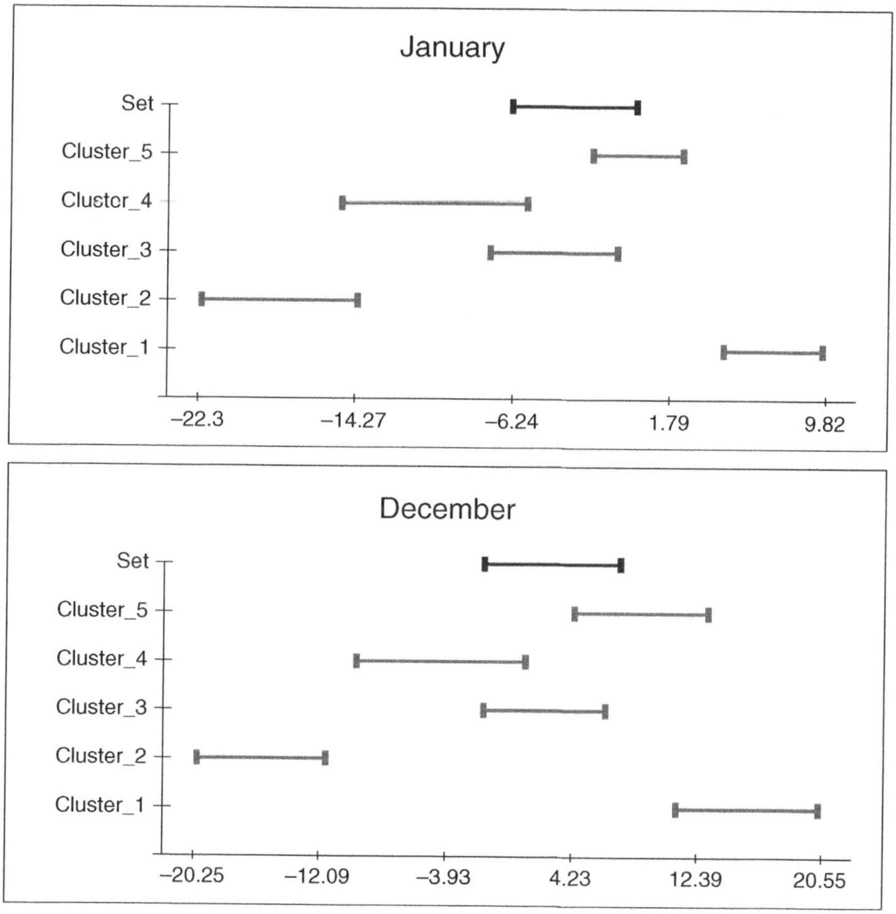

Figure 11.4 Representation of the prototypes of the clusters for the months December and January.

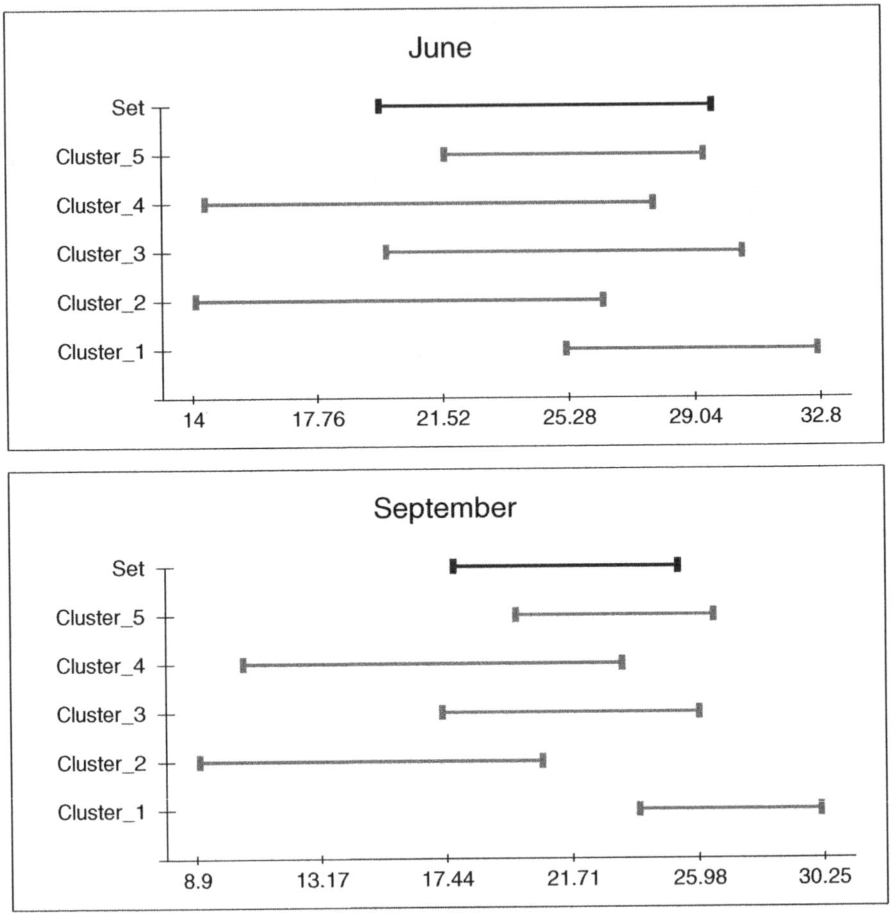

Figure 11.5 Representation of the prototypes of the clusters for the months June and September.

References

Bock, H.-H. and Diday, E. (eds), (2000) *Analysis of Symbolic Data*. Springer Verlag. Berlin.

Celeux, G., Diday, E., Govaert, G., Lechevallier, Y. and Ralambondrainy, H. (1988) Classification Automatique des Données. Bordas, Paris.

Chavent, M. (1997) Analyse des données symboliques. Une méthode divisive de classification. Thesis, Université Paris IX Dauphine.

Chavent, M. and Lechevallier, Y. (2002) Dynamical clustering algorithm of interval data: optimization of an adequacy criterion based on Hausdorff distance. In A. Sokolowski and H.-H. Bock (eds), *Classification, Clustering and Data Analysis*, pp. 53–59. Springer-Verlag, Berlin.

Chavent, M., de Carvalho, F.A.T., Lechevallier, Y. and Verde, R. (2003) Trois nouvelles méthodes de classification automatique des données symboliques de type intervalle. *Revue de Statistique Appliquée*, 51(4), 5–29.

Chavent, M., de Carvalho, F.A.T., Lechevallier, Y. and Verde, R. (2006) New clustering methods for interval data. *Computational Statistics*, 21, 211–230.

de Carvalho, F.A.T. (1992) Méthodes descriptives en analyse des données symboliques. Doctoral thesis, Université Paris IX Dauphine.

de Carvalho, F.A.T. (1994) Proximity coefficients between Boolean symbolic objects. In E. Diday, Y. Lechevallier, M. Schader, P. Bertrand and B. Bustschy (eds), *New Approaches in Classification and Data Analysis*, pp. 387–394. Springer-Verlag, Berlin.

de Carvalho, F.A.T. and Souza, R.M.C. (1998) Statistical proximity functions of Boolean symbolic objects based on histograms. In A. Rizzi, M. Vichi and H.-H. Bock (eds), *Advances in Data Science and Classification*, pp. 391–396. Springer-Verlag, Berlin.

de Carvalho, F.A.T. and Souza, R.M.C. (1999) New metrics for constrained Boolean symbolic objects. In *Studies and Research: Proceedings of the Conference on Knowledge Extraction and Symbolic Data Analysis (KESDA'98)*. pp. 175–187. Office for Official Publications of the European Communities, Luxembourg.

de Carvalho, F.A.T, Verde, R. and Lechevallier, Y. (1999) A dynamical clustering of symbolic objects based on a context dependent proximity measure. In H. Bacelar-Nicolau, F.C. Nicolau and J. Janssen (eds), *Proc. IX International Symposium – ASMDA'99*, pp. 237–242. LEAD, University of Lisbon.

Diday, E. (1971) La méthode des nuées dynamiques *Revue de Statistique Appliquée*, 19(2), 19–34.

Gordon, A.D. (1999) *Classification*. Chapman and Hall/CRC, Boca Raton, FL.

Gower, J.C. (1966) Some distance properties of latent root and vector methods using multivariate analysis. *Biometrika*, 53, 325–338.

Ichino, M. and Yaguchi, H. (1994). Generalized Minkowski metrics for mixed feature type data analysis. *IEEE Transactions on Systems, Man and Cybernetics*, 24, 698–708.

Lechevallier, Y. (1974) *Optimisation de quelques critères en classification automatique et application à l'étude des modifications des protéines sériques en pathologie clinique*. Thesis, Université Paris VI.

MacQueen, J. (1967) Some methods for classification and analysis of multivariate observations. *Proceedings of the Fifth Berkeley Symposium on Mathematical Statistics and Probability*, Vol. 1, pp. 281–297. University of California Press, Berkeley.

Michalski, R.S., Stepp, R.E. and Diday, E. (1981) A recent advance in data analysis: Clustering objects into classes characterized by conjunctive concepts. In L.N. Kanal and A. Rosenfeld A. (eds), *Progress in Pattern Recognition*, pp. 33–56. North-Holland, Amsterdam.

Verde, R., de Carvalho, F.A.T. and Lechevallier, Y. (2000) A dynamical clustering algorithm for multi-nominal data. In H.A.L. Kiers, J.-P. Rasson, P.J.F. Groenen and M. Schader (eds), *Data Analysis, Classification, and Related Methods*, pp. 387–394. Springer-Verlag, Berlin.

12

Visualizing symbolic data by Kohonen maps

Hans-Hermann Bock

12.1 Introduction

Visualizing data in the form of illustrative diagrams and searching such diagrams for structures, clusters, trends, dependencies or anomalies (e.g., outliers) is one of the main motives and strategies in exploratory statistics, data analysis and data mining. In the case of symbolic data, powerful methods are provided by suitable 'symbolic' adaptations of classical methods such as symbolic principal component analysis (SPCA) and symbolic generalized canonical (or factor) analysis (SGCA) that produce, for example, two-dimensional displays in the space of the first and second 'factors'.

An alternative visualization of symbolic data is obtained by constructing a *Kohonen map*: instead of displaying the individual items $k = 1, \ldots, n$ (data points, data rectangles, etc.) as n points or rectangles in a two-dimensional 'factor space' as in SPCA, the n items are first clustered into a (smaller) number m of 'mini-clusters' (this is essentially a data reduction step), and these mini-clusters are then assigned to (and represented by) the vertices of a fixed, prespecified rectangular lattice $\mathcal{L}$ of points (or cells) in the plane such that 'similar' clusters (in the original data space) are represented by neighbouring vertices in $\mathcal{L}$ (see Figures 12.2, 12.4 and 12.12).

In SODAS such a map is constructed by the SYKSOM module, which provides several options for the clustering process. In order to visualize the properties of the clusters, SYKSOM determines, during the cluster construction process, suitable cluster representatives or prototypes. These prototypes are visualized, in a third step, by clicking on the corresponding vertex of the screen, or by displaying all cluster prototypes simultaneously, as illustrated by Figures 12.4–12.7. Such a display is called a *Kohonen map*. It reveals the properties of the data units and clusters in a 'landscape' (a map, a factor space) where

Symbolic Data Analysis and the SODAS Software Edited by E. Diday and M. Noirhomme-Fraiture

the multidimensional structure of the data is unfolded in two dimensions and 'similar' data items are located in the same region of the landscape.

SYKSOM assumes a data table for n items or individuals (rows) that are described by p interval-type variables (columns; see Table 12.1). The visualization of clusters is obtained by the VMAP, VIEW and VPLOT modules (accessible directly from the output screen), and by other SODAS modules for displaying the properties of clusters (including also, for example, additional categorical variables).

Why and when should Kohonen maps be used for analysing data? Three arguments must be considered in this context. First, the implicit dimension reduction process (from p to 2 dimensions) operates here in a *non-linear* way and therefore also works for data (data rectangles) that are located along complex, non-linear, and possibly even bent manifolds in $\mathbb{R}^p$ (in contrast, for example, to SPCA; see Figure 12.2). Second, the Kohonen strategy is well suited to the analysis of a large number n of items since the implicitly underlying clustering process reduces the n items to a relatively small number m of clusters (or vertices) whose properties can be more easily visualized than those of the original n data (where SPCA or SGCA might yield unwieldy displays). Third, we will see that in the Kohonen approach both *clustering* and *class prototype construction* are conducted in a simultaneous process where the data are entered *sequentially* and current results (clusters, prototypes) are *progressively updated* by considering further data points.[1] It is this sequential approach that enables the analysis of thousands of data items, in contrast to other clustering or visualization methods that operate bulkwise, that is, simultaneously with all data, and can therefore run into computational difficulties.

This chapter is organized as follows: In Section 12.2 we describe the basic ideas that underlie the Kohonen and SYKSOM approach, in particular the sequential cluster construction process and the subsequent visualization methods. We mention three versions of the SYKSOM approach whose details will be explained later on. Section 12.3 specifies various notations and concepts not explained in Section 12.2 – dissimilarity measures, cluster prototypes, kernel functions, etc. – and describes various optional steps (initialization, learning, cooling). Finally, in Section 12.4, we detail the three variants of the SYKSOM approach – the StochApprox, MacQueen1, and MacQueen2 algorithms.

12.2 Kohonen maps

12.2.1 The linear projection paradigm in PCA and SPCA

A classical data visualization method is provided by *principal component analysis* (PCA). We are given n p-dimensional data vectors $x_1, \ldots, x_n \in \mathbb{R}^p$ which describe the properties of n objects $k = 1, \ldots, n$ (individuals, items, ...) in terms of p real-valued variables (components). As illustrated in Figure 12.1, in order to visualize the (often high-dimensional) data, PCA selects an appropriate (linear!) hyperplane H of a given low dimension s in the Euclidean space $\mathbb{R}^p$ (typically $s = 2$) and projects the data points $x_1, \ldots, x_n$ onto H. The resulting configuration of projected points $y_1, \ldots, y_n \in H$ is then displayed on the screen and can reveal interesting features of the data set, for example, clustering tendencies and

[1] This 'self-organizing' structure explains the acronym SYKSOM that combines the SYmbolic context with the classical terminology 'Kohonen's Self-Organizing Map'.

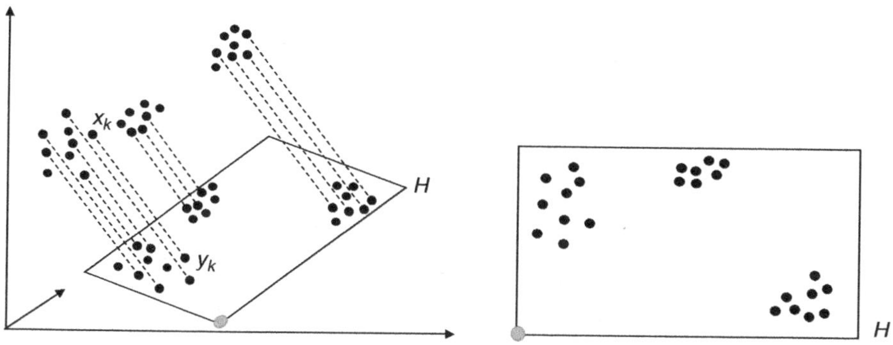

Figure 12.1 PCA of n data points $x_k \in \mathrm{R}^3$, with two-dimensional on-screen display.

trends. A suitable generalization to symbolic interval-type data is described in Chapter 9 of Bock and Diday (2000) and implemented by the SPCA module in the SODAS software.

From their construction principles, both PCA and SPCA are oriented towards a situation where the data $x_1, \ldots, x_n$ are essentially concentrated in the neighbourhood of an s-dimensional *linear* manifold of $\mathbb{R}^p$. Unfortunately, this linearity assumption is rarely fulfilled in practice since the cloud of data points is often concentrated near a *non-linear* manifold F of $\mathbb{R}^p$ which may be bent down or back (see Figure 12.2). Then PCA (and SPCA) may return an erroneous, artificial, or non-interpretable configuration, or even a methodological artefact.

12.2.2 The Kohonen approach for visualizing symbolic data

In order to overcome this difficulty, Kohonen (1982, 1995) proposed an alternative visual representation which does not use a linear hyperplane H in $\mathbb{R}^p$ (that is displayed on the screen), but represents the data (specifically, m data clusters) by the vertices $P_1, \ldots, P_m$ of a *rectangular lattice* $\mathcal{L}$ with b rows and a columns, such that each vertex P_i of this lattice represents a homogeneous *cluster* C_i of objects, and also a *prototype* z_i describing the overall properties of this cluster (Figure 12.2). The construction of clusters and prototypes is conducted in a way such that, in verbal terms, the neigbourhood structure of the vertices P_i in $\mathcal{L}$ approximates the neighbourhood structure of the corresponding clusters C_i or prototypes z_i in $\mathbb{R}^p$ and thus of a non-linear manifold F on which the data may be located (*topological correctness*).

While the classical Kohonen approach is designed for data *points* $x_k = (x_{k1}, \ldots, x_{kp})' \in \mathbb{R}^p$, the SYKSOM module provides a generalization to the case of symbolic *interval-type data*, that is, n vectors of intervals $x_1, \ldots, x_n \subset \mathbb{R}^p$ with

$$x_k = \begin{pmatrix} [a_{k1}, b_{k1}] \\ \vdots \\ [a_{kp}, b_{kp}] \end{pmatrix},$$

Table 12.1 A data table for $n = 8$ items (cities, $k = 1, \ldots, 8$) and $p = 3$ interval-type variables (price ranges for three brands of whisky, $j = 1, 2, 3$).

$k \setminus j$	Var. 1	Var. 2	Var. 3
1	[8.5, 10.0]	[13.0, 15.2]	[5.0, 8.2]
2	[6.3, 9.1]	[14.1, 16.0]	[6.3, 7.2]
3	[7.9, 11.8]	[11.6, 13.5]	[4.9, 6.5]
4	[9.0, 11.0]	[10.9, 12.5]	[7.1, 8.1]
5	[6.3, 7.2]	[12.9, 15.0]	[6.2, 7.4]
6	[7.1, 7.9]	[11.5, 12.9]	[4.8, 5.7]
7	[7.5, 9.4]	[13.2, 15.0]	[6.6, 8.1]
8	[6.6, 7.8]	[12.4, 13.2]	[5.7, 7.2]

for example

$$x_4 = \begin{pmatrix} [\ 9.0, 11.0] \\ [10.9, 12.5] \\ [\ 7.1,\ \ 8.1] \end{pmatrix}, \tag{12.1}$$

where x_k describes the properties of the kth item and may, equivalently, be interpreted as a rectangle (hypercube)

$$\boxed{Q_k = [a_k, b_k] \subset \mathbb{R}^p}$$

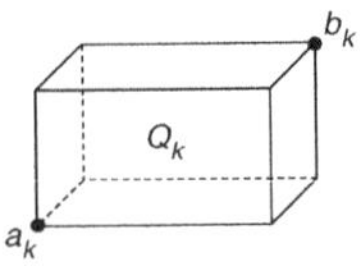

in the p-dimensional space $\mathbb{R}^p$ with component-specific sides $[a_{kj}, b_{kj}]$ for $j = 1, \ldots, p$. Here

$$a_k = \begin{pmatrix} a_{k1} \\ \vdots \\ a_{kp} \end{pmatrix} \quad \text{and} \quad b_k = \begin{pmatrix} b_{k1} \\ \vdots \\ b_{kp} \end{pmatrix} \tag{12.2}$$

are the 'left lower vertex' and the 'right upper vertex' of Q_k. We will often identify the interval-type *data vector* x_k with the corresponding hypercube $Q_k \in \mathbb{R}^p$.

Table 12.1 provides an example of an interval-type data matrix for $n = 8$ items with $p = 3$ interval-type variables, where x_k is given by the kth row, and Figure 12.2 illustrates a case where the data hypercubes $Q_1, \ldots, Q_n$ are concentrated near a bent hypersurface F (such that SPCA would be an inappropriate tool).

The symbolic Kohonen approach, implemented by the SODAS module SYKSOM, essentially combines the following steps (see also Bock, 1997, 2003, 2004):

1. The n data hypercubes are clustered into m non-overlapping clusters $C_1, \ldots, C_m \subset \{1, \ldots, n\}$ of (items with) 'similarly located' hypercubes (where $m = b \cdot a$ is the number of vertices of the selected $b \times a$ lattice $\mathcal{L}$). This is essentially a data aggregation step where the size of the constructed clusters is typically small in relation to the total number n of individuals. We will refer to these clusters as 'mini-clusters'.

2. Each mini-cluster C_i is characterized by a *prototype hypercube* z_i in $\mathbb{R}^p$ (as a class representative).

3. Each mini-cluster C_i, and thus each prototype z_i, is assigned to a vertex P_{ν_i} of the lattice $\mathcal{L}$ (with $\nu_i \in \{1, \ldots, n\}$ for all i).

4. This assignment process is conducted in a way such that any two prototypes z_i, z_j (clusters C_i and C_j) which are *neighbouring in* $\mathbb{R}^p$ (i.e., which have a small dissimilarity to or distance from each other) are typically assigned to two vertices P_{ν_i}, P_{ν_j} of $\mathcal{L}$ which are *neighbouring in* $\mathcal{L}$ (i.e., such that the 'path distance' between the vertices P_{ν_i} and P_{ν_j} along the edges of $\mathcal{L}$ is small).[2]

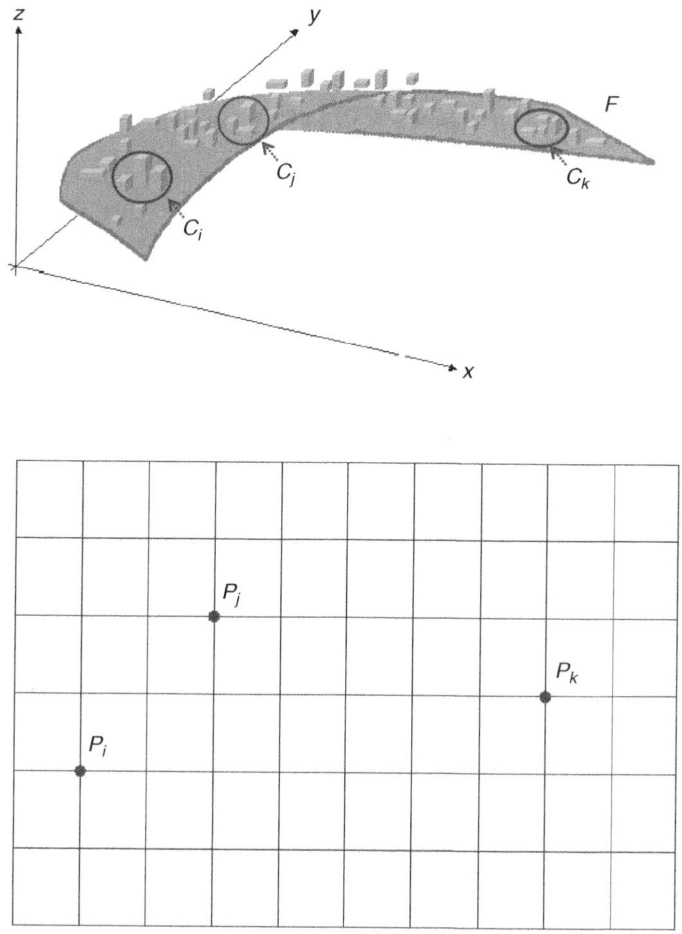

Figure 12.2 A set of hypercubes Q_k in R^3 concentrated near a non-linear manifold F with clusters C_i to be displayed by the $m = b \cdot a = 7 \cdot 11 = 77$ vertices P_i in the rectangular lattice $\mathcal{L}$ in a topologically correct form.

---

[2] In the following, we use a labelling of clusters and vertices such that C_i and z_i are assigned to the vertex P_i (i.e., with $\nu_i \equiv i$).

After a series of iterative steps, the SYKSOM algorithm yields a final partition $(C_1, \ldots, C_m)$ of objects and describes each mini-cluster C_i by a 'typical' hypercube $z_i \subset \mathbb{R}^p$ that is called the 'prototype' of C_i.

It is expected that, in the end, the lattice $\mathcal{L}$ together with the m mini-clusters and cluster prototypes provides an illustrative 'flat representation' of the (unknown) bent and twisted surface F which is visualized by the Kohonen map on the screen (see Figure 12.4) and may be interpreted by using the graphical modules VMAP, VPLOT, VIEW, etc. (see Figures 12.5–12.8).

The SYKSOM approach comprises three different but similar methods for building the mini-clusters and their cluster representatives from n interval-type data $x_1, \ldots, x_n$: StochApprox, MacQueen1 and MacQueen2. All three follow the same lines, described in the next subsection.

12.2.3 The basic steps of the SYKSOM algorithms

Kohonen map approaches are often considered in the context of *neural networks* and *sequential learning*. This is due to the fact that the underlying algorithms (and also SYKSOM) enter the data in their *sequential order* $x_1, x_2, x_3, \ldots$ (e.g., time order), and not simultaneously as a bulk $\{x_1, \ldots, x_n\}$ of n data vectors. Here we briefly describe the basic steps of this sequential process. A detailed presentation is deferred to Section 12.4, after various concepts and definitions have been explained in Section 12.3.

The SYKSOM approach proceeds in a series of steps $t = 0, 1, 2, \ldots$. At stage t we have included the first t data hypercubes $x_1, \ldots, x_t$ and obtained, as a preliminary result, m mini-clusters $C_1^{(t)}, \ldots, C_m^{(t)}$ which form a partition of the set $\{1, \ldots, t\}$ of the first t items (these classes are assigned to the vertices $P_1, \ldots, P_m$ of $\mathcal{L}$) and also m class prototypes $z_1^{(t)}, \ldots, z_m^{(t)}$ (hypercubes in $\mathbb{R}^p$). Then, at the next step $t+1$, we include the $(t+1)$th data rectangle x_{t+1} and update the previous clusters and prototypes in some suitable way. This is iterated until a stopping criterion is fulfilled.

A basic version of the algorithm proceeds as follows.

12.2.3.1 Initialization

When starting at $t=0$ we define an initial set of $m = b \cdot a$ empty classes $C_1^{(0)} = \emptyset, \ldots, C_m^{(0)} = \emptyset$ of items and m cluster prototypes $z_1^{(0)}, \ldots, z_m^{(0)}$, that is, hypercubes in $\mathbb{R}^p$ (various options exist; see Section 12.3.1). For all $i = 1, \ldots, m$, the class $C_i^{(0)}$ is assigned to the vertex P_i of the lattice.

12.2.3.2 Iteration step

At the end of step t, we have processed the first t data hypercubes $x_1 = [a_1, b_1], \ldots, x_t = [a_t, b_t]$ and obtained

- an m-partition $\mathcal{C}^{(t)} = (C_1^{(t)}, \ldots, C_m^{(t)})$ of the set $\{1, \ldots, t\}$ of the first t objects;
- a system $\mathcal{Z}^{(t)} = (z_1^{(t)}, \ldots, z_m^{(t)})$ of m interval-type prototypes where $z_i^{(t)} = [u_i^{(t)}, v_i^{(t)}]$ is a hypercube in $\mathbb{R}^p$ with lower-left vertex $u_i^{(t)}$ and upper-right vertex $v_i^{(t)}$ $(i = 1, \ldots, m)$.

In the next step $t+1$, the algorithm considers the next data vector $x_{t+1} = [a_{t+1}, b_{t+1}]$ (a hypercube Q_{t+1} in $\mathbb{R}^p$) and proceeds as follows.

Minimum-distance assignment First, it determines, from the m currently available class prototypes $z_1^{(t)}, \ldots, z_m^{(t)}$, the prototype $z_{i^*}^{(t)}$ which is closest to x_{t+1} in the following sense:

$$d(x_{t+1}, z_{i^*}^{(t)}) = \min_{j=1,\ldots,m} d(x_{t+1}, z_j^{(t)}). \tag{12.3}$$

Here $d(\cdot,\cdot)$ is a dissimilarity measure between the data hypercube $x_{t+1} = [a_{t+1}, b_{t+1}]$ and the prototype $z_j^{(t)} = [u_j^{(t)}, v_j^{(t)}]$. This measure has to be specified by the user. In the case of the StochApprox algorithm, SYKSOM provides three options:

- vertex-type distance;
- Hausdorff-type L_1-distance;
- Hausdorff-type L_2-distance.

For details, see Section 12.3.2. For the MacQueen1 algorithm, see (ii*) in Section 12.4.2.2 (vertex-type distance); and for MacQueen2, see the specifications in (12.50) and (12.51) of Section 12.4.2.4.

The new object $t+1$ is then assigned to the class $C_{i^*}^{(t)}$,

$$C_{i^*}^{(t+1)} := C_{i^*}^{(t)} \cup \{t+1\}, \tag{12.4}$$

whereas all other classes remain unchanged:

$$C_i^{(t+1)} := C_i^{(t)}, \qquad \text{for all } i \text{ with } i \neq i^*.$$

Updating the class prototypes Having included the $(t+1)$th item in the class $C_{i^*}^{(t)}$, all m cluster prototypes $z_1^{(t)}, \ldots, z_j^{(t)}$ are updated (not only $z_{i^*}^{(t)}$!). Each prototype $z_j^{(t)} = [u_j^{(t)}, v_j^{(t)}]$ is shifted towards x_{t+1}, but the amount of shifting $z_j^{(t)}$ depends on the distance $\delta(P_j, P_{i^*})$ between those vertices P_j and P_{i^*} in the lattice $\mathcal{L}$ which correspond to the classes C_j and C_{i^*}, respectively. In fact, it is this 'weighting' scheme which finally guarantees that the topological ordering of the vertices $P_1, \ldots, P_m$ in the lattice $\mathcal{L}$ resembles the geometrical configuration of the obtained classes or prototypes $z_1^{(t)}, \ldots, z_m^{(t)}$ in $\mathbb{R}^p$.

Shifting rectangles is described by two update formulae for the lower-left and the upper-right vertices of the hypercube $z_j^{(t+1)} = [u_j^{(t+1)}, v_j^{(t+1)}]$ as follows:

$$\begin{aligned} u_j^{(t+1)} &= u_j^{(t)} + \alpha_{t+1} \cdot K_{i^*j} \cdot (a_{t+1} - u_j^{(t)}), \\ v_j^{(t+1)} &= v_j^{(t)} + \alpha_{t+1} \cdot K_{i^*j} \cdot (b_{t+1} - v_j^{(t)}) \end{aligned} \tag{12.5}$$

$(j = 1, \ldots, m,\ t = 0, 1, \ldots)$. In this formula:

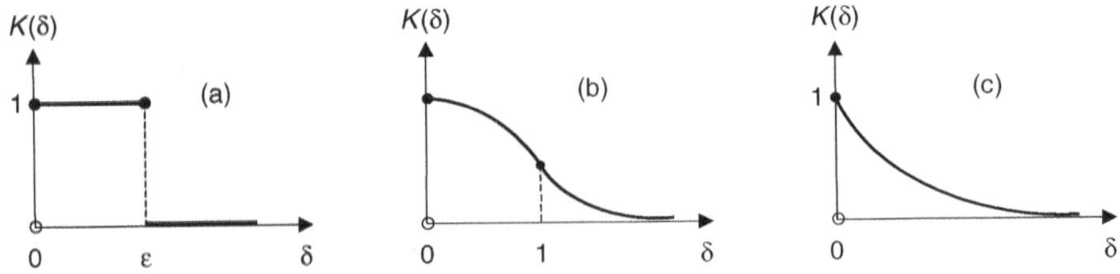

Figure 12.3 Three common kernel functions: (a) threshold kernel, $K(\delta) = 1$ and 0 if $\delta \le \epsilon$ and $\delta > \epsilon$, respectively (with a threshold $\epsilon > 0$); (b) Gaussian kernel, $K(\delta) = e^{-\delta^2/2}$; and (c) exponential kernel, $K(\delta) = e^{-\delta}$ for $\delta > 0$.

- $(\alpha_1, \alpha_2, \ldots)$ denotes a decreasing sequence of 'learning factors' $\alpha_t > 0$. In the case of StochApprox they are to be specified by the user (see Sections 12.3.4 and 12.4.1). In the case of the MacQueen algorithms these factors depend on the data and are automatically calculated by the algorithm (see (12.43) and (12.54) below).
- The 'weights' K_{i^*j} are defined as functions of the path distance $\delta(P_{i^*}, P_j)$:

$$K_{i^*j} := K(\delta(P_{i^*}, P_j)) = K_{ji^*} \tag{12.6}$$

where $K(\delta)$ is, for $\delta \ge 0$, a typically decreasing 'weighting' or 'kernel' function (see Figure 12.3). This kernel must be selected by the user (see Sections 12.3.5, 12.3.6, and 12.4.1).

12.2.3.3 Three different construction schemes for a Kohonen map

SYKSOM provides three different versions of the update formulae (12.5), yielding essentially three different, but related algorithms:

- *Stochastic approximation (StochApprox)*, which is a direct generalization of Kohonen's original self-organizing map method for classical data vectors and uses a prespecified sequence (α_t) of learning factors (see Section 12.4.1).
- Two generalized clustering schemes due to MacQueen, *MacQueen1* and *MacQueen2*, which use, for each class, its 'optimal' class prototype – as in the classical clustering method of MacQueen (1967) – with the consequence that the 'learning factors' are automatically determined by the data (see Section 12.4.2).

12.2.3.4 Iteration cycles and stopping

The previous steps are iterated for $t = 1, 2, \ldots, n$ such that all n available data rectangles $x_1, \ldots, x_n$ are processed. This first cycle results in an m-partition $\mathcal{C}^{(n)} = (C_1^{(n)}, \ldots, C_m^{(n)})$ of $\{1, \ldots, n\}$ and a system $\mathcal{Z}^{(n)} = (z_1^{(n)}, \ldots, z_m^{(n)})$ of m class prototypes.

After this first cycle ($\ell = 1$) the algorithm repeats, in a series of further cycles ($\ell = 2, 3, \ldots$), the same procedure for the same n data in the same order. The cluster prototypes obtained at the end of the ℓth cycle are used as the initial cluster prototypes for the $(\ell + 1)$th cycle. Cycling stops after a prespecified number c of cycles, or if the class prototypes have attained a stationary state. A detailed formulation is as follows.

Stopping rule The algorithm is stopped after the first full cycle $\ell = 1, 2, \ldots$ (i.e., after $t = \ell \cdot n$ updates) for which either $\ell \geq c$ or the difference between the current prototypes $z_i^{(\ell \cdot n)}$ and the prototypes $z_i^{((\ell-1) \cdot n)}$ at the end of the previous cycle $(\ell - 1)$ is sufficiently small in the sense that

$$\Delta_l := \frac{\sum_{i=1}^{m} \|z_i^{(l \cdot n)} - z_i^{(l-1) \cdot n}\|^2}{\sum_{i=1}^{m} \|z_i^{(l \cdot n)}\|^2} < \delta, \tag{12.7}$$

where δ is a prespecified accuracy threshold. Otherwise a new cycle $(\ell + 1)$ with n updates is performed. The initial configuration $\mathcal{Z}^{(0)}$ of this new cycle is given either

(a) by the prototype system $\mathcal{Z}^{(\ell \cdot n)} = (z_1^{(\ell \cdot n)}, \ldots, z_m^{(\ell \cdot n)})$ obtained in the end of the ℓth cycle (default; ‘no centre replacement’),

(b) or by the system $\tilde{\mathcal{Z}}^{(0)}$ of the most typical prototypes for the classes $C_1^{(\ell \cdot n)}, \ldots, C_m^{(\ell \cdot n)} \subset \{1, \ldots, n\}$ in the sense of the option which has been selected from Section 12.3.3 below (‘centre replacement’).

In both cases, the algorithm starts the new cycle with m empty classes and proceeds as described in Section 12.2.3.2.

Remark 12.1. The SYKSOM module conducts always at least 20 cycles even if c is fixed below 20 or if the selected accuracy level is attained before.

Remark 12.2. When the algorithm finally stops we obtain an m-partition $\mathcal{C} = (C_1, \ldots, C_m)$ of the set $\{1, \ldots, n\}$ of all n items with clusters $C_1, \ldots, C_m$, together with m cluster prototypes (hypercubes) $z_1, \ldots, z_m$ in $\mathbb{R}^p$ which can be saved by the options *Save partition in SODAS file...* and *Save prototypes in SODAS file...* in the *Parameters* dialogue box of SYKSOM.

12.2.4 Visualizing the SYKSOM output by means of Kohonen maps

Neither the (mini-)clusters $C_1, \ldots, C_m$ nor the class prototypes $z_1, \ldots, z_m$ constructed by SYKSOM can be directly seen in the lattice $\mathcal{L}$. The usefulness of the Kohonen approach lies in the fact that there exist various tools for visualizing the mini-clusters, their properties and similarities, and thus the data. The SODAS software provides, in particular, the following three modules:

- VMAP, a module for displaying the lattice $\mathcal{L}$ with icons, zoom stars, and diagrams describing the properties of the classes in a (topologically ordered) map;
- VIEW, a module for displaying the class prototypes, for example, by zoom stars;
- VPLOT, a module for displaying the (projection of the) class prototypes in the two-dimensional Euclidean space spanned by two variables which can be selected by the user.

In the following we will briefly illustrate these options.

12.2.4.1 The VMAP module

The VMAP module displays the Kohonen map in the form of Figure 12.4. Each cell (vertex of $\mathcal{L}$) corresponds to a mini-cluster C_i and contains two icons, a circle and a square. The circle indicates the size $|C_i|$ of a cluster C_i (with area proportional to $|C_i|$). The square displays essentially the volume of the corresponding prototype hypercube $z_i = [u_i, v_i]$: its area is proportional to the geometric mean $[\prod_{j=1}^{p}(v_{ij} - u_{ij})]^{1/p}$ of the p side lengths $v_{ij} - u_{ij}$ $(j = 1, \ldots, p)$ of the rectangle z_i. Up to the exponent $1/p$, this is the volume of z_i (standardization is such that the maximum volume is 1).

Left-clicking on a cell yields the member list of the corresponding class. Right-clicking on a cell gives a detailed description of the corresponding class prototype either in the form of a zoom star (*Polaire*) or a bar diagram (*Parallèle*). The latter displays, for the selected mini-cluster C_i, the p sides $[u_{ij}, v_{ij}]$ of the corresponding prototype rectangle $z_i = [u_i, v_i] = ([u_{i1}, v_{i1}], \ldots, u_{ip}, v_{ip}])$ dependent on the variables coded by $j = 1, \ldots, p$ (see Figure 12.5).

The *Proximité* option provides a (smoothed) version of the map where the dissimilarities between all classes and a selected class are displayed by the grey level. The *Toutes* option displays all three displays.

The user can also generate these displays for two or several clusters simultaneously. This option makes it easy to compare the properties of clusters.

12.2.4.2 VIEW: visualizing the class prototypes

While the VMAP module provides a class description for one or more mini-clusters C_i, the user can also display the properties of *all m classes simultaneously* just by saving the

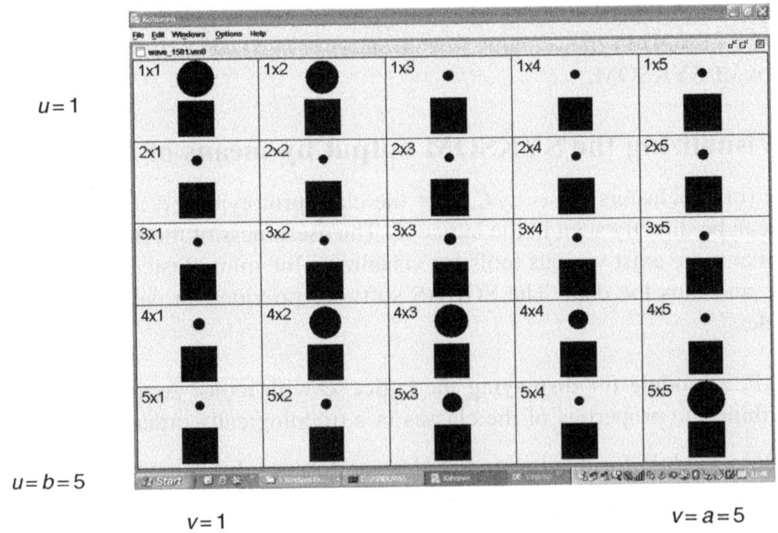

Figure 12.4 Display of a 5 × 5 lattice $\mathcal{L}$ (the 25 cells of the display correspond to the 25 vertices of $\mathcal{L}$). Circles represent the cluster size $|C_i|$, squares the geometric mean of the p side lengths of the prototype hypercubes.

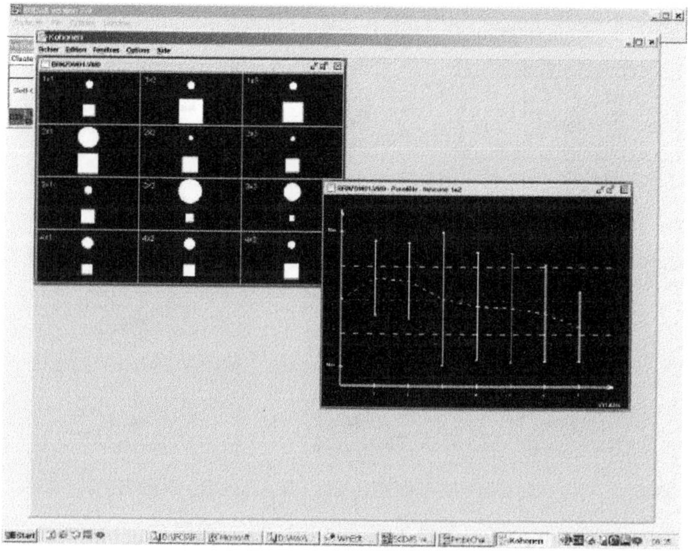

Figure 12.5 A Kohonen map of size 4×3 with the bar diagram of the mini-cluster C_2 represented by the vertex $P_2 \triangleq 1 \times 2$.

prototypes $z_1, \ldots, z_m$ in a 'prototype file' (by choosing some parameter in a SYKSOM window) and utilizing the VIEW module.

This is illustrated by Figures 12.6 and 12.7, where $p = 12$ interval-type variables (monthly temperatures) were considered with a lattice of size 3×4 and 4×4, respectively. The first uses bar diagrams to display the class properties, while the second uses zoom stars.

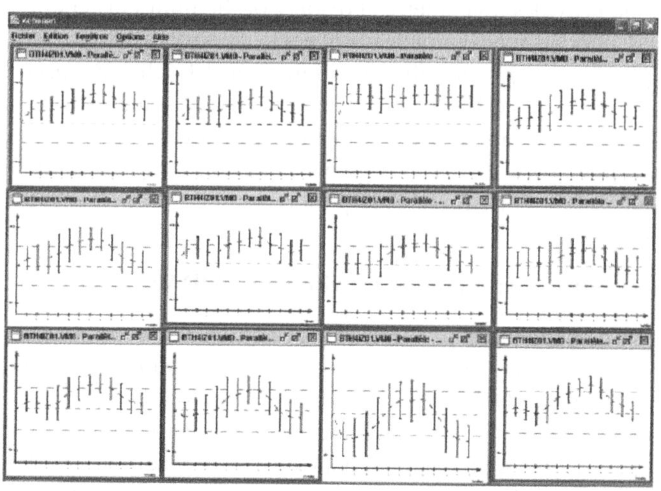

Figure 12.6 Bar diagram of the sides of the class-specific prototype rectangles z_i in the case of $p = 12$ interval-type variables (monthly temperatures), simultaneously for all $m = 12$ mini-clusters $C_1, \ldots, C_{12}$.

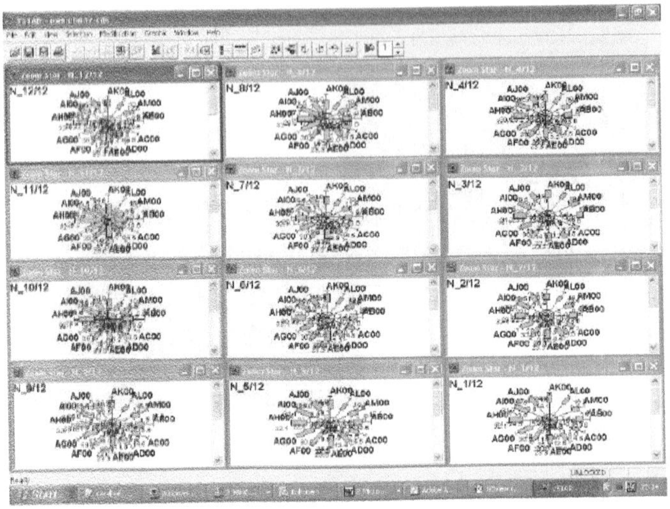

Figure 12.7 Zoom star diagrams of class-specific prototype rectangles z_i in the case of $p = 12$ interval-type variables, simultaneously for all $m = 16$ mini-clusters $C_1, \ldots, C_{16}$.

In principle, it is to be expected that zoom stars in the same region of the map are similar to each other, whereas distant vertices should show marked differences in the display.

12.2.4.3 The VPLOT display

The VPLOT module provides a geometrical display of the mini-clusters in the space of two arbitrarily selected interval-type variables $j, j' \in \{1, \ldots, p\}$: it displays the projection of the class prototypes $z_1, \ldots, z_m$ onto the two-dimensional Euclidean space spanned by the selected variables j and j' (see Figure 12.8). This provides some idea of how clusters are distributed in this two-dimensional space and (by trying several different pairs of variables j, j') even in the whole space $\mathbb{R}^p$. Such displays can also reveal the amount of overlap and separation that may exist among the cluster prototypes.

12.2.4.4 Other visualization methods provided by SODAS

A further approach considers the partition $(C_1, \ldots, C_m)$ of the original set $\{1, \ldots, n\}$ of items as a new (categorical) variable and stores for each item its class membership as a $(p+1)$th variable (the SYKSOM module provides such an option). To these $p+1$ variables we may also add other (quantitative, qualitative, or modal) variables which are available for the n items and thereby obtain an enlarged data table $\tilde{X}$ of size $n \times (p+1+q)$, say, which includes the original $n \times p$ table X of interval-type data that was used for the construction of the mini-clusters. Describing the m mini-clusters now by the m corresponding symbolic class descriptions (symbolic objects), we may analyse and visualize this class-specific behaviour of all recorded variables in $\tilde{X}$ by using other modules in the SODAS software.

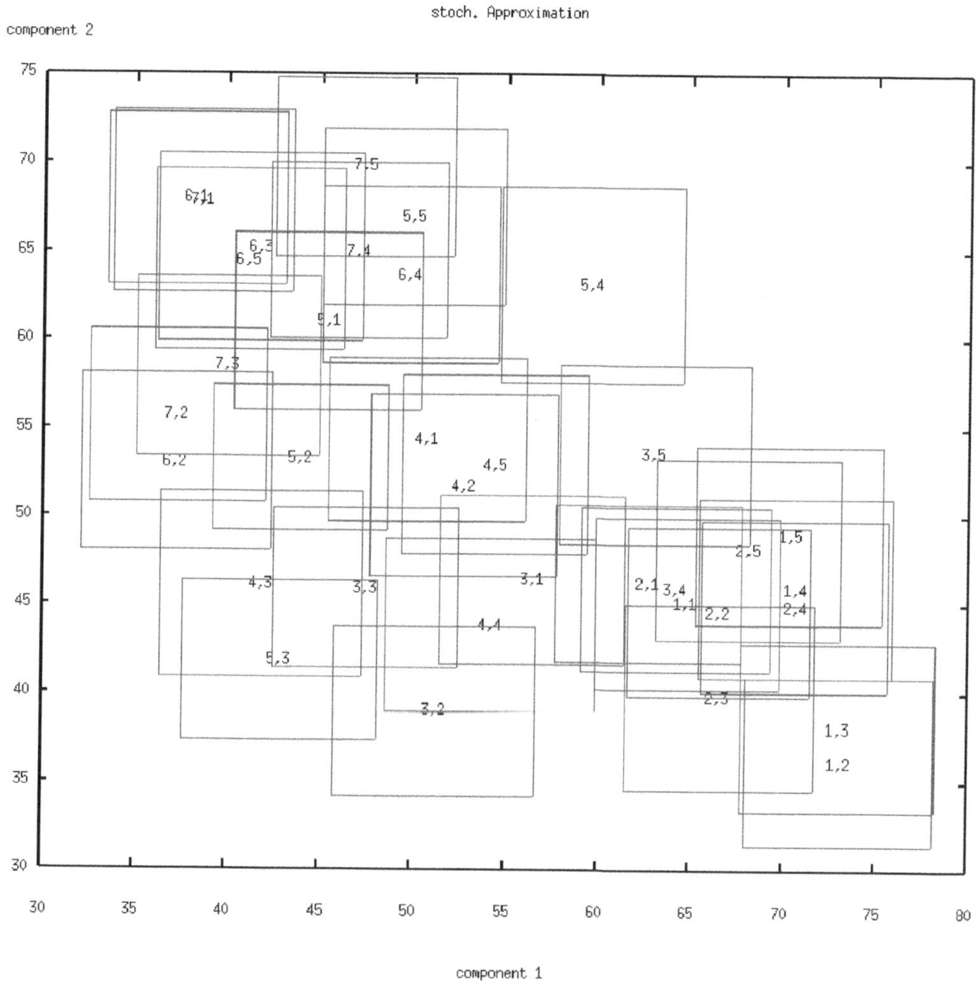

Figure 12.8 Projection of the class prototype rectangles $z_1, \ldots, z_m$ (here $m = 7 \cdot 5 = 35$) onto the space of two selected variables $j = 1$ and $j' = 2$ (the underlying sample consists of 10 000 randomly simulated rectangles in a square of $\mathbb{R}^5$).

12.3 Technical definitions and methodological options

In this section we define a range of theoretical concepts that are used in the iterative cluster construction process and thus specify options that are provided by the SYKSOM approach.

12.3.1 Initial configuration of class prototypes

As described above, the SYKSOM algorithm starts from an initial configuration $\mathcal{Z}^{(0)} = (z_1^{(0)}, \ldots, z_m^{(0)})$ of m class prototypes in the form of m hypercubes in $\mathbb{R}^p$. SYKSOM provides two standard options for defining $\mathcal{Z}^{(0)}$.

12.3.1.1 Random single-point prototypes

For a given number n of data rectangles $x_k = [a_k, b_k]$, let

$$amin_j := \min\{a_{kj} \mid k = 1, \ldots, n\},$$
$$bmax_j := \max\{b_{kj} \mid k = 1, \ldots, n\}$$

be the minimum and maximum boundary of n intervals for the jth variable ($j = 1, \ldots, p$). Let $Z = (Z^{(1)}, \ldots, Z^{(p)})$ be a random vector of independent real-valued variables where each $Z^{(j)}$ has a uniform distribution in the interval $[amin_j, bmax_j]$ such that Z has a uniform distribution in the p-dimensional hypercube $Q_{\max} = [amin_1, bmax_1] \times \cdots \times [amin_p, bmax_p]$.

The initial configuration $\mathcal{Z}^{(0)}$ is constructed by simulating m independent realizations $z_1, \ldots, z_m \in \mathbb{R}^p$ of Z and defining the m class prototypes by

$$z_1^{(0)} := [z_1, z_1] \ldots, z_m^{(0)} := [z_m, z_m],$$

that is, as single-point hypercubes.

In the course of the subsequent map construction process, these single-point prototypes $z_i^{(0)} = [z_i, z_i]$ of volume 0 will be blown up into 'regular' rectangles, that is, with a positive volume.

12.3.1.2 First-data method

This classical initialization method uses the first m data hypercubes $x_1, \ldots, x_m$ as the m initial prototypes:

$$z_1^{(0)} := x_1, \quad \ldots, \quad z_m^{(0)} := x_m.$$

This method may be used whenever it is plausible that the first m data rectangles are a 'random' choice over all n rectangles. This does not necessarily hold for all applications.

Remark 12.3. Under this option, the recursive update process has to start with $x_{m+1}, x_{m+2}, \ldots$. Therefore the update formulae below necessitate, in principle, a shift by m in the indices of the data x_k.

12.3.2 Metrics and distances for hypercubes

SYKSOM provides three distance measures $d(Q, Q')$ for measuring the dissimilarity between two p-dimensional rectangles $Q = [a, b]$, $Q' = [a', b']$ in $\mathbb{R}^p$ with 'lower left' and 'upper right' vertices $a = (a_1, \ldots, a_p)$ and $b = (b_1, \ldots, b_p)$, respectively, and similarly for $a', b' \in \mathbb{R}^p$ with $a \le b$ and $a' \le b'$ (componentwise).

12.3.2.1 The vertex-type distance

Here $d(Q, Q')$ is defined as the sum of the squared Euclidean distances between the 2^p vertices $v_\epsilon = a + \epsilon \cdot (b - a)$ and $v'_\epsilon = a' + \epsilon \cdot (b' - a')$ of the hypercubes Q and Q', where $\epsilon = (\epsilon_1, \ldots, \epsilon_p) \in \{0, 1\}^p$ is a p-vector of zeros and ones, and $\epsilon \cdot u = (\epsilon_1 u_1, \ldots, \epsilon_p u_p)$ denotes, for any vector $u \in \mathbb{R}^p$, the vector of componentwise products (see Figure 12.9(a)). The corresponding formula

$$\begin{aligned} d(Q, Q') &:= \sum_{\epsilon \in \{0,1\}^p} \|v_\epsilon - v'_\epsilon\|^2 = \sum_{\epsilon \in \{0,1\}^p} \|a - a' + \epsilon \cdot (b - b' - a + a')\|^2 \\ &= \sum_{\epsilon \in \{0,1\}^p} \sum_{j=1}^{p} [(1 - \epsilon_j)(a_j - a'_j) + \epsilon_j (b_j - b'_j)]^2 \\ &= 2^{p-1} \cdot (\|a - a'\|^2 + \|b - b'\|^2) = 2^{p-1} \left\| \begin{pmatrix} a \\ b \end{pmatrix} - \begin{pmatrix} a' \\ b' \end{pmatrix} \right\|^2 \end{aligned} \tag{12.8}$$

shows that $d(Q', Q')$ is, up to the factor 2^{p-1} (which will be neglected in all subsequent formulae), identical to

- either the sum of squared (p-dimensional) Euclidean distances $\|a - a'\|^2$ and $\|b - b'\|^2$ of the lower and upper vertices of Q and Q', respectively (see Figure 12.9(a)),
- or, equivalently, the squared ($2p$-dimensional) Euclidean distance of the vectors

$$Q \mathrel{\hat{=}} \begin{pmatrix} a \\ b \end{pmatrix}, \qquad Q' \mathrel{\hat{=}} \begin{pmatrix} a' \\ b' \end{pmatrix} \tag{12.9}$$

in R^{2p} which represent the two hypercubes Q and Q'.

12.3.2.2 Hausdorff-type distance measures

The *Hausdorff distance* between two sets Q and Q' of $\mathbb{R}^p$ is defined by

$$\begin{aligned} d(Q, Q') &:= \max\{ \max_{\alpha \in Q}\{\min_{\beta \in Q'} \|\alpha - \beta\|\}, \max_{\beta \in Q'}\{\min_{\alpha \in Q} \|\alpha - \beta\|\}\} \\ &= \max\{\|\alpha_1 - \beta_1\|\}, \|\alpha_2 - \beta_2\|\}\} \end{aligned} \tag{12.10}$$

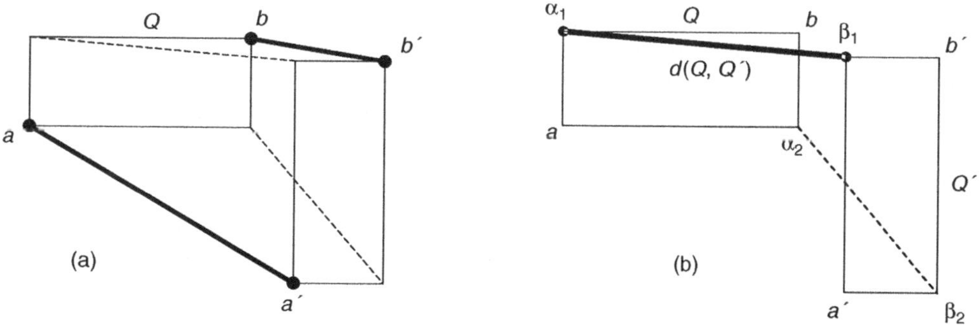

Figure 12.9 (a) Vertex-type distance; (b) Hausdorff distance.

(see Figure 12.9(b)). For the one-dimensional case ($p = 1$) where $\|\alpha - \beta\| = |\alpha - \beta|$ is the absolute difference between $\alpha, \beta \in \mathbb{R}^1$ (L_1-distance), we obtain for two intervals $Q = [a, b]$, $Q' = [a', b']$ in $\mathbb{R}^1$ the Hausdorff distance formula

$$d(Q, Q') = d_1([a, b], [a', b']) := \max\{|a - a'|, |b - b'|\}. \tag{12.11}$$

In the p-dimensional case the determination of the Hausdorff distance (12.10) is computationally more demanding (an algorithm is provided in Bock, 2005). For this case, SYKSOM uses the following modification:

$$\begin{aligned} d(Q, Q') = d_p([a, b], [a', b']) &:= \left(\sum_{j=1}^{p} d_1([a_j, b_j], [a'_j, b'_j])^2 \right)^{1/2} \\ &= \left(\sum_{j=1}^{p} [\max\{|a_j - a'_j|, |b_j - b'_j|\}]^2 \right)^{1/2}, \end{aligned} \tag{12.12}$$

which is called a *Hausdorff-type* L_2*-distance* here. It combines the p one-dimensional, coordinatewise Hausdorff distances $d_1([a_j, b_j], [a'_j, b'_j])$ in a way which is similar to the definition of Euclidean (L_2) distance in $\mathbb{R}^p$.

A *Hausdorff-type* L_1*-distance* results in the alternative definition

$$\tilde{d}(Q, Q') := \sum_{j=1}^{p} d_1([a_j, b_j], [a'_j, b'_j]) = \sum_{j=1}^{p} \max\{|a_j - a'_j|, |b_j - b'_j|\} \tag{12.13}$$

which is analoguous to the L_1-distance in $\mathbb{R}^p$.

Remark 12.4. Various other dissimilarity measures for interval data have been proposed. They introduce suitable weights for the variables $j = 1, \ldots, p$, or they use the range of those variables (in the data set) and the length of the overlap between the intervals $[a_j, b_j]$ and $[a'_j, b'_j]$ in order to find an empirically attractive *relative* weighting for the distances in the jth coordinate (see Sections 8.3 and 11.2.2 in Bock and Diday, 2000).

12.3.3 Cluster prototypes

In the case of interval-type data $x_1, x_2, \ldots$ (p-dimensional hypercubes), the prototype of a class $C \subset \{1, \ldots, n\}$ of items will be assumed to be a suitable p-dimensional rectangle $G = G(C) = [u, v]$, with two vertices $u, v \in \mathbb{R}^p$ where $u \leq v$. Since 'suitability' can be defined in various ways, SYKSOM proposes three different options for choosing a class prototype $G(C)$. For ease of presentation, we denote this class by $C = \{1, \ldots, n\}$ with n interval-type data vectors $x_1, \ldots, x_n$ where x_k corresponds to the hypercube $Q_k = [a_k, b_k] \subset \mathbb{R}^p$.

12.3.3.1 The envelope-type prototype

A first choice for the prototype $G = G(C) = [u, v]$ is the *smallest rectangle which includes all rectangles* Q_k *from* C (Figure 12.10(a)). The lower and upper vertices $u = (u_1, \ldots, u_p)$, $v = (v_1, \ldots, v_p)$ of G are then given by

$$u_j := \min_{k \in C} a_{kj}, \quad v_j := \max_{k \in C} b_{kj}, \qquad j = 1, \ldots, m. \tag{12.14}$$

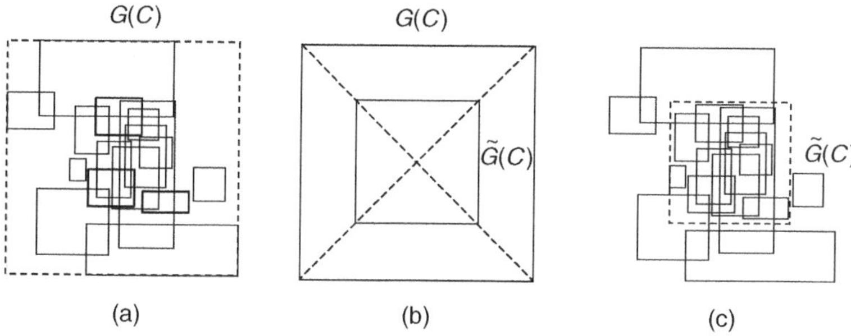

Figure 12.10 Prototype of a class C of rectangles in $\mathbb{R}^2$: (a) envelope-type; (b) truncation process; (c) α-truncated envelope-type $\tilde{G}(C)$, here with $\alpha = 2/3$.

In the terminology of Ichino and Yaguchi (1994), G is the *join* of all Q_k from C and denoted by $G = \bigotimes_{k \in C} Q_k$ (see also Section 8.3 in Bock and Diday, 2000).

12.3.3.2 The α-truncated envelope-type prototype

The envelope rectangle $G = [u, v]$ from Section 12.3.3.1 is obviously sensitive to outliers in C. Therefore it may be preferable to truncate G in some way and to use a smaller rectangle $\tilde{G} = \tilde{G}(C) = [\tilde{u}, \tilde{v}]$ in $G = [u, v]$ as the class prototype for C. We postulate that:

(i) $\tilde{G}$ should be homothetic to the envelope-type rectangle $G = [u, v]$ from Section 12.3.3.1;

(ii) $\tilde{G}$ has the same midpoint $(u + v)/2$ as G;

(iii) $\tilde{G}$ should contain at least a given percentage α (with $0.7 \leq \alpha < 1$, say) of rectangles Q_k from C.

Such a rectangle $\tilde{G}$ is given by the lower and upper vertices defined by

$$\tilde{u} = (0.5 - \gamma)v + (0.5 + \gamma)u,$$
$$\tilde{v} = (0.5 + \gamma)v + (0.5 - \gamma)u,$$

where $0 < \gamma < 1/2$ is a suitable (class-specific) scaling factor which guarantees the α-overlap condition (iii) for the class C (see Figures 12.10(b) and 12.10(c)).

12.3.3.3 The most typical rectangle of a cluster

A general approach for defining a prototype $G \subset \mathbb{R}^p$ for a given cluster C proceeds by first defining, for a given rectangle G, a *measure of typicality* with respect to the n hypercubes in C, and then choosing a rectangle with *maximum typicality* (see Bock, 1974, pp. 100–103). A suitable measure of typicality is provided by the sum of all dissimilarities $d(Q_k, G)$ between G and the hypercubes Q_k from C which has to be minimized with respect to G. Then $G = G(C)$ is the rectangle with 'minimum average deviation' in the sense

$$g(C, G) := \sum_{k \in C} d(Q_k, G) \quad \longrightarrow \min_G . \qquad (12.15)$$

For a general distance measure d the solution of this optimization problem may be computationally difficult or even prohibitive. However, for the vertex-type distance and the Hausdorff-type L_1-distance we obtain neat and explicit formulae (see also Bock, 2005).

Average rectangle for vertex-type distance If in (12.15) we use the vertex-type distance $d(Q_k, G) = \|a_k - u\|^2 + \|b_k - v\|^2$ between $Q_k = [a_k, b_k]$ and $G = [u, v]$, this criterion is minimized by the rectangle $G = [\overline{u}, \overline{v}]$ with the 'average' lower left and upper right vertices given by:

$$\overline{u} := \overline{a}_C := \frac{1}{|C|} \cdot \sum_{k \in C} a_k, \qquad \overline{v} := \overline{b}_C := \frac{1}{|C|} \cdot \sum_{k \in C} b_k. \tag{12.16}$$

In fact, here the optimization problem (12.15) is equivalent to the classical sum-of-squares minimization problem in $\mathbb{R}^{2p}$:

$$\begin{aligned} g(C, G) &= \sum_{k \in C} d(Q_k, G) = \sum_{k \in C} \|a_k - u\|^2 + \sum_{k \in C} \|b_k - v\|^2 \\ &= \sum_{k \in C} \left\| \begin{pmatrix} a_k \\ b_k \end{pmatrix} - \begin{pmatrix} u \\ v \end{pmatrix} \right\|^2 \longrightarrow \min_{\binom{u}{v} \in \mathbb{R}^{2p}}, \end{aligned} \tag{12.17}$$

whose solution $u = \overline{u}$, $v = \overline{v}$ automatically satisfies the condition $\overline{u} \leq \overline{v}$ as required.

Remark 12.5. If one more element $n+1$ is added to the class $C = \{1, \ldots, n\}$ with a data vector $x_{n+1} = [a_{n+1}, b_{n+1}]$, the resulting class $\widehat{C} := C + \{n+1\} = \{1, \ldots, n, n+1\}$ has the prototype $\widehat{Q} = [\widehat{u}, \widehat{v}] = [\overline{a}_{\widehat{C}}, \overline{b}_{\widehat{C}}]$. It appears that its lower left and upper right vertices $\widehat{u}, \widehat{v}$ can be calculated from $Q = [\overline{a}_C, \overline{b}_C]$ by a simple recursion:

$$\begin{aligned} \widehat{u} = \overline{a}_{\widehat{C}} = \frac{1}{|C|+1} \left(\sum_{k \in C} a_k + a_{n+1} \right) &= \overline{a}_C + \frac{1}{|C|+1} (a_{n+1} - \overline{a}_C), \\ \widehat{v} = \overline{b}_{\widehat{C}} = \frac{1}{|C|+1} \left(\sum_{k \in C} b_k + b_{n+1} \right) &= \overline{b}_C + \frac{1}{|C|+1} (b_{n+1} - \overline{b}_C). \end{aligned} \tag{12.18}$$

Median prototype for Hausdorff-type L_1-distance If we use for d the Hausdorff-type L_1-distance $\tilde{d}(Q_k, G)$ between the rectangles Q_k and G in $\mathbb{R}^p$, see (12.13), then the criterion (12.15) is minimized by a rectangle $G = [\tilde{u}, \tilde{v}]$ which is obtained from the medians of the centres and of the lengths of the n coordinatewise intervals $[\alpha_{kj}, b_{kj}]\mathbb{R}^1$, $k = 1, \ldots, n$ (for each $j = 1, \ldots, p$).

More specifically, let

$$m_{kj} := \frac{a_{kj} + b_{kj}}{2}, \qquad \ell_{kj} := \frac{b_{kj} - a_{kj}}{2} \tag{12.19}$$

be the midpoint and the half-length of the interval $[a_{kj}, b_{kj}] = [m_{kj} - \ell_{kj}, m_{kj} + \ell_{kj}]$ that is the jth coordinate in Q_k (or x_k). We write a prototype G in the analogous form

$$G = [u, v] = [\mu - \lambda, \mu + \lambda]$$

where $\mu = (\mu_1, \ldots, \mu_p)' := (v + u)/2$ is the midpoint of G and the vector $\lambda = (\lambda_1, \ldots, \lambda_p)' := (v - u)/2$ comprises the half-lengths $\lambda_1, \ldots, \lambda_p$ of the coordinatewise intervals $[u_1, v_1], \ldots, [u_p, v_p]$ of G. Then the most typical rectangle for C in the sense of (12.15) is obtained from the coordinatewise medians (Chavent and Lechevallier, 2002): the centre of G is given by

$$\tilde{\mu}_j := \text{median}\{m_{1j}, \ldots, m_{nj}\}, \qquad j = 1, \ldots, p,$$

and the half side-lengths of G are given by

$$\tilde{\lambda}_j := \text{median}\{\ell_{1j}, \ldots, \ell_{nj}\}, \qquad j = 1, \ldots, p.$$

We call this optimum rectangle $G = G(C)$ the *median prototype*. Formally, it is given by

$$G(C) = [\tilde{u}, \tilde{v}] := \begin{pmatrix} [\tilde{\mu}_1 - \tilde{\lambda}_1, \tilde{\mu}_1 + \tilde{\lambda}_1] \\ \vdots \\ [\tilde{\mu}_p - \tilde{\lambda}_p, \tilde{\mu}_p + \tilde{\lambda}_p] \end{pmatrix}. \tag{12.20}$$

12.3.4 The learning factors

In the update formulae (12.5) the *learning factors* $\alpha_1, \alpha_2, \ldots$ determine the amount by which the prototypes are changed in the steps $t = 1, 2, \ldots$ of the SYKSOM algorithm. Whereas these factors are automatically calculated by the *MacQueen1* and *MacQueen2* methods (see (12.43) and (12.54) with (12.49) below), they must be chosen by the user in the case of the *StochApprox* algorithm. In order to attain convergence, the general theory of stochastic approximation suggests the use of a sequence (α_t) with

$$\sum_{t=1}^{\infty} \alpha_t = \infty \quad \text{and} \quad \sum_{t=1}^{\infty} \alpha_t^2 < \infty. \tag{12.21}$$

SYKSOM provides two alternatives:

- *Classical.* The fixed standard choice for these learning factors is the arithmetic sequence

$$\alpha_t = \frac{1}{t}, \qquad \text{for} \quad t = 1, 2, \ldots, n, n+1, n+2, \ldots. \tag{12.22}$$

- *Cyclic repetition.* The first n elements of the previous sequence are repeated at the begining of each cycle of n data rectangles, that is,

$$\alpha_t = \frac{1}{t \bmod n} \qquad \text{for} \quad t = 1, 2, \ldots, \tag{12.23}$$

which yields the series $1/1, 1/2, \ldots, 1/n, 1/1, 1/2, \ldots, 1/n, 1/1, \ldots$. The latter alternative makes sense only for a large number n of objects.

12.3.5 The weighting kernel function

As shown by the updating formulae (12.5), the amount by which a class prototype $z_j^{(t)}$ is shifted 'towards the new data x_{t+1}' in step t is different for different classes $C_j^{(t)}$, due to the factor $K_{i^*j} = K(\delta(P_{i^*}, P_j))$ which depends on the path distance $\delta(P_{i^*}, P_j)$ between the vertices P_{i^*}, P_j in the lattice $\mathcal{L}$ (recall that i^* denotes the class prototype $z_{i^*}^{(t)}$ which is closest to x_{t+1} in $\mathbb{R}^p$; see (12.3)).

It is a major trick of Kohonen approaches to arrange the shifting process such that a prototype $z_j^{(t)}$ is shifted by a larger amount if P_j is *close* to P_{i^*} in $\mathcal{L}$ (e.g., is in an ϵ-neighbourhood of P_{i^*} in terms of the path distance δ) than in the case where P_j is *far away* from P_{i^*} in $\mathcal{L}$.

This shifting idea is modelled by the formula

$$K_{ij} := K(\delta(P_i, P_j)), \qquad i, j = 1, \ldots, m, \tag{12.24}$$

where $K(\delta) \geq 0$ is a non-increasing *kernel function* of the path distance $\delta \geq 0$, typically with $K(0) = 1$. Often K is identical (or proportional) to a common probability distribution density. SYKSOM uses three alternative options for the kernel function $K(\cdot)$ (see Figures 12.3 and 12.11):

(i) The *threshold kernel* which defines the ϵ-neighbourhood of a vertex P_i in $\mathcal{L}$:

$$K(\delta) := K_\epsilon(\delta) := \begin{cases} 1, & \text{for } 0 \leq \delta \leq \epsilon, \\ 0, & \text{for } \epsilon < \delta < \infty \end{cases} \tag{12.25}$$

(see Figure 12.3(a)). Adopting this kernel, no centre $z_j^{(t)}$ with a distance $\delta(P_{i^*}, P_j) > \epsilon$ is shifted in the updating step t.

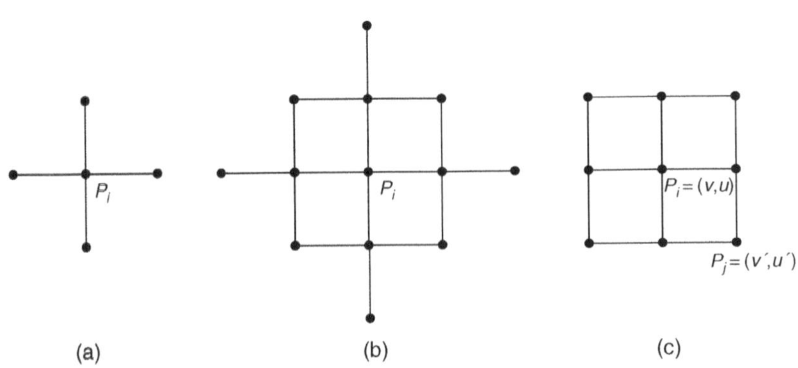

Figure 12.11 (a), (b) ϵ-neighbourhood of P_i for $\epsilon = 1$ and $\epsilon = 2$ (c) 8-point neighbourhood of P_i.

Remark 12.6. SYKSOM provides the alternatives $\epsilon = 0, 1, 2, 3$ and 4 (with $\epsilon = 1$ as default). For $\epsilon = 1$ and $\epsilon = 2$ the corresponding neighbourhoods (in terms of the path distance $\delta = \delta(P_i, P_j)$) are illustrated in Figure 12.11.

(ii) The *Gaussian distribution kernel* with

$$K(\delta) := \mathrm{e}^{-\delta^2/2}, \qquad \text{for } \delta \geq 0 \tag{12.26}$$

(see Figure 12.3(b)).

(iii) The *exponential distribution kernel* with

$$K(\delta) := \mathrm{e}^{-\delta}, \qquad \text{for } \delta \geq 0 \tag{12.27}$$

(see Figure 12.3(c)).

Note that in order to limit computation time, SYKSOM sets $K_{ij} := 0$ whenever $\delta(P_i, P_j) > 4$. This corresponds to replacing the original weighting scheme (12.24) by

$$K_{ij} := K(\delta(P_i, P_j)) \cdot K_4(\delta(P_i, P_j)). \tag{12.28}$$

Since the weight values computed from (12.24) will typically be negligible for the truncated pairs (P_i, P_j), there is no much difference between the truncated and non-truncated versions.

(iv) A fourth option uses the Euclidean distance $\Delta(P_i, P_j) = ((v - v')^2 + (u - u')^2)^{1/2}$ of the lattice points $P_i \hat{=} (v, u)$ and $P_j \hat{=} (v', u')$ which are here described by their Cartesian coordinates in $\mathbb{R}^2$ (see Figure 12.12). For this option the weights are specified by

$$K_{ij} = K_\epsilon(\Delta(P_i, P_j)). \tag{12.29}$$

SYKSOM uses only the value $\epsilon = 1.5$ which corresponds to the *eight-point neighbourhood* displayed in Figure 12.11(c)

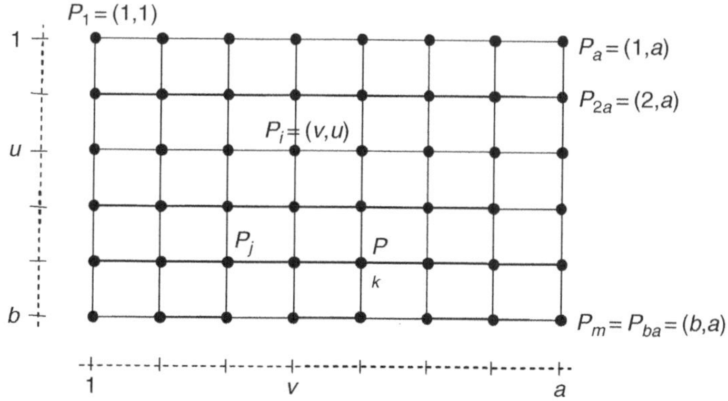

Figure 12.12 Coordinates in a rectangular lattice $\mathcal{L}$ with $b = 6$ rows, $a = 8$ columns and $m = b \cdot a = 6 \cdot 8 = 48$ vertices corresponding to 48 mini-clusters.

12.3.6 Cooling and the cooling parameter $T(t)$

Practical as well as theoretical considerations lead to the idea of shrinking the 'effective neighbourhood' of a lattice point P_i more and more during the sequential updating process such that finally only those class prototypes are updated that are very close to the 'winning' prototype z_{i*} in (12.3). For example, start with a large ϵ-neighbourhood and decrease ϵ after each cycle of size n, and simultaneously strengthen those weights K_{ij} in (12.24) that correspond to very close pairs of vertices $P_i, P_j \in \mathcal{L}$, whereas these weights are more and more reduced if P_i, P_j are far away from each other in $\mathcal{L}$.

Technically this idea is formalized by introducing a positive factor $T = T(t)$ which is usually called the 'temperature' and depends on the step t of the algorithm, and by replacing the transformation formula (12.24) by

$$K_{ij} := K^{T(t)}(\delta(P_i, P_j)), \tag{12.30}$$

where K^T denotes the rescaled kernel

$$K^T(\delta) := K\left(\frac{\delta}{T}\right), \qquad \delta \geq 0. \tag{12.31}$$

In fact, if $K(\delta)$ is the distribution density of a random variable X, $K(\delta/T)/T$ provides the density of the rescaled variable $\tilde{X} := T \cdot X$ (for any fixed temperature $T > 0$). For $t = 1, 2, \ldots$ the function $T = T(t)$ is decreasing from a value $T_{\max}$ to a value $T_{\min}$.

SYKSOM adopts three specifications for $T(t)$: no cooling, single-step linear cooling, and batchwise linear cooling. Recall that the algorithm starts with the data $x_1, \ldots, x_n$ in the first cycle, then uses the same data (now denoted by $x_{n+1}, \ldots, x_{2n}$) in the second cycle and so on, until we stop after (maximally) c cycles, say. Thus we have the steps $t = 0, 1, 2, \ldots, cn - 1$ where in step t the observation x_{t+1} is observed.

1. *No cooling*. Here $T(t) = 1$ for all t. This option will automatically be used with the threshold kernel (12.25) and the 8-point neighbourhood kernel (see (iv) in Section 12.3.5 and Fig. 12.11(c)). For the Gaussian and exponential kernels, it is the default option in SYKSOM.

2. *Single-step linear cooling*. $T(t)$ is *linearly* decreasing from $T_{\max}$ to a smaller value $T_{\min}$ for $t = 0, 1, \ldots, nc - 1$. This leads to the formula

$$T(t) = T_{\max} - \frac{T_{\max} - T_{\min}}{nc - 1} \cdot T, \qquad t = 0, 1, \ldots, nc - 1. \tag{12.32}$$

3. *Batchwise linear cooling*. Here $T(t)$ is constant in each cycle and decreases linearly from cycle 1 to cycle c. Since the rth cycle comprises the steps $t = (r-1)n, \ldots, rn{-}1$, this amounts to

$$T(t) = T_{\max} - \frac{(T_{\max} - T_{\min})}{c} \cdot r, \qquad \text{for} \quad t = (r-1)n, \ldots, rn - 1; \ r = 1, \ldots, c. \tag{12.33}$$

The parameters $0 < T_{\min} < T_{\max}$ are automatically fixed by SYKSOM as described in the following remark.

Table 12.2 Weights $K_{ij} = K^T(\delta(P_i, P_j))$ for various values of $\delta = \delta(P_i, P_j)$ in the case of the Gaussian and the exponential kernel (for $\tilde{\epsilon} = 4$).

δ	Gaussian kernel		Exponential kernel	
	$T_{max} = 0.466\tilde{\epsilon} = 1.864$	$T_{min} = 0.915$	$T_{max} = 0.434\tilde{\epsilon} = 1.737$	$T_{min} = 0.558$
0	1.000	1.000	1.000	1.000
1	0.750	0.303	0.562	0.167
2	0.325	0.008	0.316	0.028
3	0.075	0.000	0.178	0.005
4	0.010	0.000	0.100	0.001

Remark 12.7. In the case of the Gaussian and the exponential kernel, (12.26) and (12.27) respectively, SYKSOM determines T_{max} and T_{min} from a distance threshold $\tilde{\epsilon}$ which determines the maximum neighbourhood from a vertex to be considered in the cooling method (note that $\tilde{\epsilon}$ has a different meaning here than when defining a threshold kernel (i) in Section 12.3.5). The user must specify some value $\tilde{\epsilon} = 1, 2, 3, 4$. The default value is $\tilde{\epsilon} = 4$.

- The initial value $T(0) = T_{max}$ is determined from the condition $K^{T(0)}(0)/K^{T(0)}(\tilde{\epsilon}) = k_1 := 10$. This means that the centre $P_j = P_i$ of an $\tilde{\epsilon}$-neighbourhood of P_i gets a weight $K^{T(0)}(0) = K^{T(0)}(\delta(P_i, P_i))$ 10 times as large as a vertex P_j with path distance $\delta(P_i, P_j) = \tilde{\epsilon}$ from P_i, that is, on the boundary of this neighbourhood.
- The final value T_{min} is determined from the condition $K^T(1)/K^T(2) = k_2 := 6$. This means that in the $\tilde{\epsilon}$-neighbourhood of P_i all points with path distance $\delta = 1$ have a weight 6 times as large as those with path distance $\delta = 2$.

These rules lead, in case of the Gaussian kernel, to the specifications

$$T(0) = T_{max} = \sqrt{\frac{\tilde{\epsilon}^2}{2 \ln k_1}}, \qquad T_{min} = \sqrt{\frac{3}{2 \ln k_2}}, \tag{12.34}$$

and for the exponential kernel to

$$T(0) = T_{max} = \frac{\tilde{\epsilon}}{\ln k_1}, \qquad T_{min} = \frac{1}{\ln k_2}. \tag{12.35}$$

In order to get some idea about the range of absolute weight values, we display in Table 12.2 the weights obtained for the threshold parameter $\tilde{\epsilon} = 4$.

12.4 The StochApprox and MacQueen algorithms

SYKSOM provides essentially two recursive algorithms for building a Kohonen map: *StochApprox* and *MacQueen*. There are two versions, *MacQueen1* and *MacQueen2*, of the latter. For all these algorithms, the basic skeleton is the same and has been described in Section 12.2.3.

In the following we motivate and summarize these procedures as far as this has not yet been done in Sections 12.2.3 and 12.3. Essentially, we have only to specify the update formula (12.5), but we also include, as remarks, some guidelines for running the SYKSOM module.

12.4.1 Generalized stochastic approximation

The *StochApprox* option provides a direct generalization of Kohonen's classical self-organizing algorithm for constructing a topologically ordered map of data in a lattice $\mathcal{L}$ in $\mathbb{R}^2$. Essentially, it is based on classical methods for stochastic approximation (see Bock, 1998, 1999).

Before running the StochApprox algorithm the user must make some choices:

1. Open the data file and select p interval-type variables for construction.
2. Select the size of the lattice $\mathcal{L}$, i.e., the numbers b and a of rows and columns, respectively.
3. Select a dissimilarity measure $d(Q, Q')$ for comparing two rectangles Q, Q' in $\mathbb{R}^p$ (see Section 12.3.2).
4. Specify the initial system of class prototypes (see Section 12.3.1).
5. Select the type of cluster prototypes (see Section 12.3.3).
6. Choose the weights K_{ij} by selecting
   (a) the kernel $K(\delta)$ (see Section 12.3.5),
   (b) the neighbourhood size $\epsilon \in \{1, 2, 3, 4\}$ (in case of the threshold kernel (12.26)), and
   (c) the cooling method (no cooling, single-step or batchwise linear cooling).
7. Select the maximum number c of cycles and the precision threshold δ (see Section 12.2.3.4).

After these preparatory specifications, the *StochApprox* algorithm proceeds as described in Section 12.2.3 by including sequentially all data hypercubes $x_1, \ldots, x_n$ in $\mathbb{R}^p$. Formulae (12.5) for adapting the m current class prototypes (hypercubes) $z_1^{(t)} = [u_1^{(t)}, v_1^{(t)}], \ldots, z_m^{(t)} = [u_m^{(t)}, v_m^{(t)}]$ in $\mathbb{R}^p$ to the new data hypercube $x_{t+1} = [a_{t+1}, b_{t+1}]$ (assigned to cluster $C_{i^*}^{(t+1)}$) are in this case as follows:

$$\begin{aligned} u_j^{(t+1)} &= u_j^{(t)} + \alpha_{t+1} K_{i^* j} \cdot (a_{t+1} - u_j^{(t)}), \\ v_j^{(t+1)} &= v_j^{(t)} + \alpha_{t+1} K_{i^* j} \cdot (b_{t+1} - v_j^{(t)}), \end{aligned} \qquad (12.36)$$

for $j = 1, \ldots, m$, with weights $K_{i^* j} = K^{T(t)}(\delta(P_{i^*}, P_j))$ which depend on the path distance $\delta(P_{i^*}, P_j)$ between the vertices P_{i^*}, P_j in the lattice $\mathcal{L}$.

12.4.2 The MacQueen1 and MacQueen2 algorithms

In this subsection we first describe the sequential clustering method of MacQueen designed for classical data points and then generalize it to the case of data hypercubes and the construction of Kohonen maps.

12.4.2.1 MacQueen's classical clustering method for data points

For the case of single-valued data points $x_1, .x_2, ... \in \mathbb{R}^p$, MacQueen (1967) proposed a sequential clustering method that proceeds with steps $t = 0, 1, 2, ...$ and is based on the following principles (see also Bock, 1974, pp. 279–297):

(i) At the end of step t we have incorporated the data points $x_1, \ldots, x_t$ and obtained a partition $\mathcal{C}^{(t)} = (C_1^{(t)}, \ldots, C_m^{(t)})$ of the set $\{1, \ldots, t\}$ of the first t items and a system $\mathcal{Z}^{(t)}$ of class centroids $z_1^{(t)}, \ldots, z_m^{(t)}$ (points in $\mathbb{R}^p$). In the next step $(t+1)$, a new data point x_{t+1} is observed.

(ii) The data point x_{t+1} (i.e., the item $t+1$) is assigned to the cluster $C_{i^*}^{(t)}$ which is closest to x_{t+1} in the sense that

$$d(x_{t+1}, z_{i^*}^{(t)}) = \|x_{t+1} - z_{i^*}^{(t)}\|^2 = \min_{j=1,\ldots,m} \{\|x_{t+1} - z_j^{(t)}\|^2\}, \qquad (12.37)$$

such that $C_{i^*}^{(t+1)} = C_{i^*}^{(t)} \cup \{t+1\}$ and $C_i^{(t+1)} = C_i^{(t)}$ for all $i \neq i^*$ (see also (12.3) and (12.4)).

(iii) The prototype of each cluster $C_i^{(t+1)}$ is defined as the cluster centroid

$$z_i^{(t+1)} := \bar{x}_{C_i^{(t+1)}} = \frac{1}{|C_i^{(t+1)}|} \sum_{k \in C_i^{(t+1)}} x_k, \qquad \text{for } i = 1, \ldots, m, \qquad (12.38)$$

that is, the most typical point of $C_i^{(t+n)}$ in $\mathbb{R}^p$ in the classical sum-of-squares sense.

(iv) Steps (i) to (iii) are iterated until a stopping criterion is fulfilled.

Remark 12.8. In the previously described step $(t+1)$, only the i^*th class and its centroid are changed whereas all other classes and prototypes are maintained. As a matter of fact, and with a view to the update formulas below, the new prototype $z_{i^*}^{(t+1)}$ can be calculated by a simple recursion formula instead of using the definition formula (12.38) that involves all data points in $C_{i^*}^{(t+1)}$. This recursion formula is given by

$$z_{i^*}^{(t+1)} = \bar{x}_{C_{i^*}^{(t+1)}} = z_{i^*}^{(t)} + \frac{1}{|C_{i^*}^{(t)}| + 1} (x_{t+1} - z_{i^*}^{(t)}), \qquad (12.39)$$

while $z_i^{(t+1)} = z_i^{(t)}$ for all $i \neq i^*$. This formula resembles (12.5), but involves a learning factor $\alpha_{t+1} = 1/(|C_{i^*}^{(t)}| + 1)$ which is dependent on class size and the data.

12.4.2.2 A symbolic version of MacQueen's clustering method for interval data

In the case of interval-type data with hypercubes $x_k = [a_k, b_k]$ (instead of data points) in $\mathbb{R}^p$, we can easily formulate a 'symbolic' version of MacQueen's algorithm just by applying the previous update formula to the vertices of a prototype rectangle. This yields the following *symbolic sequential clustering approach* (where we have italized the modifications to MacQueen's original method (i) to (iv)):

(i*) At the end of step t we have incorporated the data *hypercubes* $x_1, \ldots, x_t$ and obtained a partition $\mathcal{C}^{(t)} = (C_1^{(t)}, \ldots, C_m^{(t)})$ of the set $\{1, \ldots, t\}$ of the first t items and a system $\mathcal{Z}^{(t)}$ of class *prototypes* $z_1^{(t)} = [u_1^{(t)}, v_1^{(t)}], \ldots, z_m^{(t)} = [u_m^{(t)}, v_m^{(t)}]$ (i.e., *hypercubes* in $\mathbb{R}^p$). In the next step $(t+1)$, a new data *hypercube* $x_{t+1} = [a_{t+1}, b_{t+1}]$ is observed.

(ii*) The data *hypercube* x_{t+1} (i.e., the item $t+1$) is assigned to the cluster $C_{i^*}^{(t)}$ which has the closest prototype $z_i^{(t)}$ when using the vertex-type distance (12.8), i.e., with

$$
\begin{aligned}
d(x_{t+1}, z_{i^*}^{(t)}) &= ||a_{t+1} - u_{i^*}^{(t)}||^2 + ||b_{t+1} - v_{i^*}^{(t)}||^2 \\
&= \min_{j=1,\ldots,m} \{||a_{t+1} - u_j^{(t)}||^2 + ||b_{t+1} - v_j^{(t)}||^2\}
\end{aligned}
\tag{12.40}
$$

whereas all other classes remain unchanged: $C_{i^*}^{(t+1)} = C_{i^*}^{(t)} \cup \{t+1\}$ and $C_i^{(t+1)} = C_i^{(t)}$ for all $i \neq i^*$.

(iii*) The prototype of each cluster $C_i^{(t+1)}$ is defined as the most typical *hypercube* in $\mathbb{R}^p$ in the sense of (12.17) with the average rectangles from (12.16):

$$
z_i^{(t+1)} = [u_i^{(t+1)}, v_i^{(t+1)}] = [\overline{a}_{C_i^{(t+1)}}, \overline{b}_{C_i^{(t+1)}}]. \tag{12.41}
$$

Thus all class prototypes except for $z_{i^*}^{(t+1)}$ remain unchanged.

(iv*) Steps (i*) to (iii*) are iterated until a stopping criterion is fulfilled.

Remark 12.9. Similarly as in Remark 12.8, there exists a recursion formula for the lower and upper boundaries (averages) of the prototype $z_i^{(t+1)} = [u_i^{(t+1)}, v_i^{(t+1)}]$ of the new class $C_{i^*}^{(t+1)}$:

$$
\begin{aligned}
u_{i^*}^{(t+1)} &= u_{i^*}^{(t)} + \frac{1}{|C_{i^*}^{(t)}|+1}(a_{t+1} - u_{i^*}^{(t)}), \\
v_{i^*}^{(t+1)} &= v_{i^*}^{(t)} + \frac{1}{|C_{i^*}^{(t)}|+1}(b_{t+1} - v_{i^*}^{(t)}).
\end{aligned}
\tag{12.42}
$$

12.4.2.3 MacQueen1

In the SODAS framework, the previously described symbolic MacQueen method is generalized so as to produce a Kohonen map such that the neighbourhood topology of the class prototypes $z_1^{(t)}, \ldots, z_m^{(t)}$ in $\mathbb{R}^p$ is finally approximated by the neighbourhood topology of the vertices $P_1, \ldots, P_m$ in the rectangular lattice $\mathcal{L}$ as formulated in Sections 12.1 and 12.2.

This new method, *MacQueen1*, is obtained (a) by using the update formula (12.42) for *all* m class prototypes $z_j^{(t+1)}$ (and not only for the new class $C_{i^*}^{(t+1)}$) and (b) by introducing the weights $K_{i^*j} = K(\delta(P_{i^*}, P_j))$ defined in Section 12.2.3.2 in order to differentiate among the classes.

This new symbolic MacQueen-type construction of a Kohonen map proceeds as described in the previous section with (i*), (ii*) and (iv*), but the update formula (iii*) is replaced by the following:

$$\begin{aligned} u_j^{(t+1)} &= u_j^{(t)} + \frac{K_{i^*j}}{|C_j^{(t)}|+1}(a_{t+1} - u_j^{(t)}), \\ v_j^{(t+1)} &= v_j^{(t)} + \frac{K_{i^*j}}{|C_j^{(t)}|+1}(b_{t+1} - v_j^{(t)}), \end{aligned} \tag{12.43}$$

for $j = 1, \ldots, m$, where the index i^* characterizes the class $C_{i^*}^{(t)}$ which has the smallest *vertex-type* distance $d(x_{t+1}, z_i^{(t)})$ from x_{t+1} (see (12.40)).

12.4.2.4 MacQueen2

The alternative method, *MacQueen2*, provides another way to define class prototypes and is based on an optimality criterion (*K-criterion*) formulated by Bock (1999) and Anouar *et al.* (1997) in order to optimize Kohonen maps in the case of *n data points* $x_1, \ldots, x_n \in \mathbb{R}^p$. A discrete version of this criterion is given by

$$g_n(\mathcal{C}, \mathcal{Z}) := \sum_{i=1}^{m} \sum_{k \in C_i} \left[\sum_{j=1}^{m} K_{ij} \, \|x_k - z_j\|^2 \right] \longrightarrow \min_{\mathcal{C}, Z} \tag{12.44}$$

and must be minimized with respect to all systems $\mathcal{Z} = (z_1, \ldots, z_m)$ of m class centres (points) in $\mathbb{R}^p$ and all m-partitions $\mathcal{C} = (C_1, \ldots, C_m)$ of the given items $\{1, \ldots, n\}$. Note that the inner sum can be interpreted as a dissimilarity between the class C_i and the hypercube x_k.

Considering here only the set $\{1, \ldots, t\}$ of the first t items, it can be shown that for a fixed m-partition $\mathcal{C} = \mathcal{C}^{(t)} = (C_1^{(t)}, \ldots, C_m^{(t)})$ of these items the optimum centre system $\mathcal{Z}^{(t)} = (z_1^{(t)}, \ldots, z_m^{(t)})$ is given by

$$z_i^{(t)} \; := \; \sum_{j=1}^{m} w_{ij}^{(t)} \, \overline{x}_{C_j^{(t)}}, \qquad i = 1, \ldots, m, \tag{12.45}$$

with the class centroids $\overline{x}_{C_j^{(t)}}$ (see also (12.38)) and non-negative weights

$$w_{ij}^{(t)} \; := \; |C_j^{(t)}| \cdot K_{ij} / w_i^{(t)}, \qquad i, j = 1, \ldots, m, \tag{12.46}$$

with class-specific norming factors

$$w_i^{(t)} := \sum_{j=1}^{m} |C_j^{(t)}| \cdot K_{ij}, \tag{12.47}$$

such that $\sum_{j=1}^{m} w_{ij}^{(t)} = 1$ for all i. Before generalizing this method to the case of interval-type data, the following remark, is in order.

Remark 12.10. If (only) one of the classes of $\mathcal{C}^{(t)}$ is modified by adding a new element x_{t+1}, for example, $C_{i^*}^{(t+1)} = C_{i^*}^{(t)} \cup \{t+1\}$ while $C_i^{(t+1)} = C_i^{(t)}$ for all $i \neq i^*$, the new weights and the optimum centres can be obtained by a recursion: since (12.47) implies

$$w_i^{(t+1)} = w_i^{(t)} + K_{ii^*}, \qquad i = 1, \ldots, m,$$

we obtain the recursion formula

$$\begin{aligned} z_i^{(t+1)} &= \sum_{j=1}^{m} w_{ij}^{(t+1)} \bar{x}_{C_j^{(t+1)}}, \qquad i = 1, \ldots, m, \\ &= z_i^{(t)} + \beta_i^{(t)} (x_{t+1} - z_i^{(t)}), \end{aligned} \tag{12.48}$$

where the coefficients $\beta_i^{(t)}$ are defined by

$$\beta_i^{(t)} := \frac{K_{ii^*}}{w_i^{(t+1)}} = \frac{K_{ii^*}}{w_i^{(t)} + K_{ii^*}}. \tag{12.49}$$

The *MacQueen2* method generalizes the previous approach to the case of *interval-type data* with data hypercubes $x_k = [a_k, b_k]$ in $\mathbb{R}^p$ $(k = 1, \ldots, n)$, essentially by considering the hypercubes as $2p$-dimensional data points $\tilde{x}_k = (a_k, b_k)$ in $\mathbb{R}^{2p}$ and then introducing the update formulae from Remark 12.10 into the algorithmic steps (i*) to (iv*) of the *MacQueen1* algorithm instead of (12.43). The resulting algorithm is as follows:

(i**) At the end of step t we have incorporated the data *hypercubes* $x_1, \ldots, x_t$ and obtained a partition $\mathcal{C}^{(t)} = (C_1^{(t)}, \ldots, C_m^{(t)})$ of the set $\{1, \ldots, t\}$ of the first t items and a system $\mathcal{Z}^{(t)}$ of class *prototypes* $z_1^{(t)} = [u_1^{(t)}, v_1^{(t)}], \ldots, z_m^{(t)} = [u_m^{(t)}, v_m^{(t)}]$ (i.e., *hypercubes* in $\mathbb{R}^p$). In the next step $(t+1)$, a new data *hypercube* $x_{t+1} = [a_{t+1}, b_{t+1}]$ is observed.

(ii**) The data hypercube $x_{t+1} = [a_{t+1}, b_{t+1}]$ (i.e., the item $t+1$) is assigned to the cluster $C_{i^*}^{(t)}$ which has the closest prototype $z_i^{(t)}$ (minimum-distance assignment, see (12.3)). But in contrast to *MacQueen1*, *MacQueen2* provides two alternatives for measuring dissimilarity: in addition to

- Option 1: Vertex-type (as in *MacQueen1*, see (iii*) or (12.17)):

$$d(x_{t+1}, z_i^{(t)}) := \|a_{t+1} - u_i^{(t)}\|^2 + \|b_{t+1} - v_i^{(t)}\|^2 \quad \longrightarrow \min_i, \tag{12.50}$$

MacQueen2 also includes

- Option 2: Average-type (default) (based on criterion (12.44)):

$$d(x_{t+1}, z_i^{(t)}) := \sum_{j=1}^{m} K_{ij} \left(\|a_{t+1} - u_j^{(t)}\|^2 + \|b_{t+1} - v_j^{(t)}\|^2 \right) \quad \longrightarrow \min_i. \tag{12.51}$$

All other classes remain unchanged: $C_{i^*}^{(t+1)} = C_{i^*}^{(t)} + \{t+1\}$ and $C_i^{(t+1)} = C_i^{(t)}$ for all $i \neq i^*$.

(iii**) For both options, the new prototype system $\mathcal{Z}^{(t+1)} = (z_1^{(t+1)}, \ldots, z_m^{(t+1)})$ is defined to be the solution of (12.44) (with the fixed partition $\mathcal{C} = \mathcal{C}^{(t+1)}$) and is given by the hypercubes $z_i^{(t+1)} = [u_i^{(t+1)}, v_i^{(t+1)}]$ with

$$u_i^{(t+1)} = \sum_{j=1}^{m} w_{ij}^{(t+1)} \overline{u}_{C_j^{(t+1)}}, \tag{12.52}$$

$$v_i^{(t+1)} = \sum_{j=1}^{m} w_{ij}^{(t+1)} \overline{v}_{C_j^{(t+1)}}, \tag{12.53}$$

for $i = 1, \ldots, m$. By analogy with (12.48), this yields the update formulae

$$\begin{aligned} u_i^{(t+1)} &= u_i^{(t)} + \beta_i^{(t)} (a_{t+1} - u_i^{(t)}), \\ v_i^{(t+1)} &= v_i^{(t)} + \beta_i^{(t)} (b_{t+1} - v_i^{(t)}), \end{aligned} \tag{12.54}$$

with coefficients $\beta_i^{(t)}$ as before.

(iv**) The steps (i**) to (iii**) are iterated until the stopping criterion is fulfilled (see Section 12.2.3.4).

Comparing the *StochApprox* method with *MacQueen1* and *MacQueen2*, it is evident that all proceed in a very similar way, but that with *StochApprox* the learning factors α_t must be specified (somewhat arbitrarily), whereas MacQueen derives them from a general 'most typical cluster prototype' definition, dependent on the data (and without additional cooling). It should be noted, however, that the clusters and centres produced by Option 2 of *MacQueen2* do *not* in general share the classical properties and do *not* typically have a spherical, ellipsoidal or at least a convex form. Thus the clusters obtained with Option 2 must always be checked for interpretability (even if they are derived by a general optimality criterion).

Remark 12.11. The alert reader may wonder why Option 1 of *MacQueen2* (identical to (ii*) of *MacQueen1*) is combined with the update formula (12.54) and not with the former update (12.43) of *MacQueen1*. In fact, this is the same inconsistency as in Kohonen's original self-organizing map method (Kohonen, 1982). For details, see Bock (1999).

References

Anouar, F., Badran, F. and Thiria, S. (1997) Cartes topologiques et nuées dynamiques. In S. Thiria, Y. Lechevallier, O. Gascuel and S. Canu (eds), *Statistique et méthodes neuronales*, pp. 190–206. Dunod, Paris.

Bock, H.-H. (1974) *Automatische Klassifikation. Theoretische und praktische Methoden zur Strukturierung von Daten (Clusteranalyse)*. Vandenhoeck & Ruprecht, Göttingen.

Bock, H.-H. (1997) Simultaneous visualization and clustering methods as an alternative to Kohonen maps. In G. Della Riccia, H.-J. Lenz and R. Kruse (eds), *Learning, Networks and Statistics*, pp. 67–85. Springer-Verlag, Vienna.

Bock, H.-H. (1998) Clustering and neural networks. In A. Rizzi, M. Vichi and H.-H. Bock (eds), *Advances in Data Science and Classification*, pp. 265–277. Springer-Verlag, Berlin.

Bock, H.-H. (1999) Clustering and neural network approaches. In W. Gaul and H. Locarek-Junge (eds), *Classification in the Information Age*, pp. 42–57. Springer-Verlag, Berlin.

Bock, H.-H. (2003) Clustering algorithms and Kohonen maps for symbolic data. *Journal of the Japanese Society of Computational Statistics*, 15, 217–229.

Bock, H.-H. (2004) Visualizing symbolic data tables by Kohonen maps: The SODAS module SYKSOM. In *User Manual for* SODAS2 *Software*, public deliverable D3.4b of ASSO project (IST-2000-25161). http://www.assoproject.be/sodaslink/.

Bock, H.-H. (2005) Optimization in symbolic data analysis: dissimilarities, class centers, and clustering. In D. Baier, R. Decker, L. Schmidt-Thieme (eds), *Data Analysis and Decision Support*, pp. 3–10. Springer-Verlag, Berlin.

Bock, H.-H. and Diday, E. (eds) (2000) *Analysis of Symbolic Data. Exploratory Methods for Extracting Statistical Information from Complex Data.* Springer-Verlag, Berlin.

Chavent, M. and Lechevallier, Y. (2002) Dynamic clustering of interval data. In K. Jajuga, A. Sokolowski and H.-H. Bock (eds), *Classification, Clustering, and Data Analysis – Recent Advances and Applications*, pp. 53–60. Springer-Verlag, Berlin.

Ichino, M. and Yaguchi, H. (1994) Generalized Minkowski metrics for mixed feature type data analysis. *IEEE Transactions on Systems, Man and Cybernetics*, 24(4), 698–708.

Kohonen, T. (1982) Self-organized formation of topologically correct feature maps. *Biological Cybernetics*, 43, 59–69.

Kohonen, T. (1995) *Self-Organizing Maps.* Springer-Verlag, New York.

MacQueen, J. (1967) Some methods for classification and analysis of multivariate observations. In L. LeCam and J. Neyman (eds), *Proceedings of the Fifth Berkeley Symposium on Mathematical Statistics and Probability*, vol. 1, pp. 281–297. University of California Press, Berkeley.

13

Validation of clustering structure: determination of the number of clusters

André Hardy

13.1 Introduction

The aim of cluster analysis is to identify structure within a data set. When hierarchical algorithms are used, an important problem is then to choose one solution in the nested sequence of partitions of the hierarchy. On the other hand, optimization methods for cluster analysis usually demand the a priori specification of the number of groups. So most clustering procedures require the user to fix the number of clusters, or to determine it in the final solution.

Some studies have been proposed to compare procedures for the determination of the number of clusters. For example, Milligan and Cooper, (1985) conducted a Monte Carlo evaluation of 30 indices for determining the number of clusters. Hardy (1996) compared three methods based on the hypervolumes clustering criterion with four other methods available in the Clustan software. Gordon (1998) modified the five stopping rules whose performance was best in the Milligan and Cooper study in order to detect when several different, widely separated values of c, the number of clusters, would be appropriate, that is, when a structure is detectable at several different scales.

In this chapter we consider two hypothesis tests for the number of clusters based on the hypervolumes clustering criterion: the hypervolumes test and the gap test. These statistical methods are based on the assumption that the points we observe are generated by a homogeneous Poisson process (Karr, 1991) in k disjoint convex sets. We consider the five criteria for the number of clusters analysed by Milligan and Cooper (1985). We show

Symbolic Data Analysis and the SODAS Software Edited by E. Diday and M. Noirhomme-Fraiture

how these methods can be extended in order to be applied to symbolic objects described by interval-valued, multi-valued and modal variables (Bock and Diday, 2000).

In this chapter, then, we are interested in what is usually considered as the central problem of cluster validation: the determination of the number of clusters for symbolic objects described by interval-valued, multi-valued and modal variables.

13.2 The clustering problem

The clustering problem we are interested in is the following. Let $E = \{w_1, w_2, \ldots, w_n\}$ be a set of units. On each of the n units we measure the value of p variables $y_1, y_2, \ldots, y_p$. The objective is to find a 'natural' partition $P = \{C_1, C_2, \ldots, C_k\}$ of the set E into k clusters.

13.3 Classical criteria for the number of clusters

Many different methods for the determination of the number of clusters have been published in the scientific literature. The most detailed and complete comparative study was undertaken by Milligan and Cooper. They analysed and classified 30 indices for the determination of the number of clusters, and investigated the extent to which these indices were able to detect the correct number of clusters in a series of simulated data sets containing a known structure. The five best rules investigated in this study are defined below in the case of classical quantitative data, in the order in which they were ranked in Milligan and Cooper's investigation (see also Hardy and André, 1998). In Section 13.9 we extend these methods to symbolic objects described by interval-valued, multi-valued and modal variables.

13.3.1 The Calinski and Harabasz method

The Calinski and Harabasz (1974) index is given by

$$CH = \frac{B/(c-1)}{W/(n-c)}$$

where n is the total number of units, and c the number of clusters in the partition. B and W denote, the total between-clusters sum of squared distances (about the centroids) and the total within-cluster sum of squared distances, respectively. The maximum value of the index is used to indicate the true number of clusters in the data set.

13.3.2 The J-index

Duda and Hart (1973) proposed a hypothesis test for deciding whether a cluster should be subdivided into two sub-clusters. The test statistic is based on the comparison between W_1, the within-cluster sum of squared distances, and W_2, the sum of within-cluster sums of squared distances when the cluster is optimally partitioned into two clusters. The null hypothesis of a single cluster is rejected if

$$J = \left(-\frac{W_2}{W_1} + 1 - \frac{2}{\pi p}\right)\left(\frac{2(1-8/\pi^2 p)}{np}\right)^{-1/2} > z_{1-\alpha}$$

where p denotes the dimensionality of the data, n the number of objects in the cluster being investigated, and $z_{1-\alpha}$ a standard normal deviate specifying the significance level of the test. Several values for the standard score were tested by Milligan and Cooper (1985), Gordon (1998) and Hardy and André (1998). The best results were obtained when the value was set to 3.20 (Milligan and Cooper) or 4 (Gordon; Hardy and André).

13.3.3 The C-index

This index requires the computation of V, the sum of the within-cluster pairwise dissimilarities. If the partition has r such dissimilarities, we denote by $V_{\min}$ ($V_{\max}$) the sum of the r smallest (largest) pairwise dissimilarities. The C-index (Hubert and Levin, 1976) is then defined by

$$C = \frac{V - V_{\min}}{V_{\max} - V_{\min}}.$$

The minimum value of the index across the partitions into ℓ clusters ($\ell = 1, \ldots, K$) is used to indicate the optimal number of clusters where K is a reasonably large integer fixed by the user. The best minimal value is 0. This absolute minimum is attained when a partition is such that the largest within-cluster dissimilarity is less than the smallest between-clusters dissimilarity.

13.3.4 The Γ-index

Here comparisons are made between all within-cluster pairwise dissimilarities and all between-clusters pairwise dissimilarities. A comparison is defined as consistent (inconsistent) if a within-cluster dissimilarity is strictly smaller (greater) than a between-clusters dissimilarity. The Γ-index (Baker and Hubert, 1975) is computed as

$$\Gamma = \frac{\Gamma_+ - \Gamma_-}{\Gamma_+ + \Gamma_-}$$

where Γ_+ (Γ_-) represents the number of consistent (inconsistent) comparisons. The maximum value of the Γ-index indicates the correct number of clusters. We observe that the absolute maximum value of the index is 1.

13.3.5 The Beale test

Beale (1969) proposed a hypothesis test in order to decide the existence of k_2 versus k_1 clusters in the data ($k_2 > k_1$). The test statistic is based on the increase in the mean square deviation from the cluster centroids as one moves from k_2 to k_1 clusters against the mean square deviation when k_2 clusters were present. The test is applied at each level of a hierarchy, in order to test whether a cluster should be divided into two clusters. In this chapter, $k_1 = 1$ and $k_2 = 2$.

The test involves the statistic

$$\frac{(W_1 - W_2)/W_2}{(n-1)(n-2)^{-1}2^{2/p} - 1}.$$

Under the null hypothesis that the cluster should not be subdivided, this statistic has an F distribution with p and $(n-2)p$ degrees of freedom, where p is the number of variables and n is the sample size. Milligan and Cooper (1985) found that the 0.005 significance level gave the best results for the data they analysed.

13.4 Symbolic variables

Symbolic data analysis is concerned with the extension of classical data analysis and statistical methods to more complex data called symbolic data. Here we are interested in symbolic objects described by interval-valued, multi-valued and modal variables. Consider the classical situation with a set of units $E = \{w_1, \ldots, w_n\}$ and a series of p variables $y_1, \ldots, y_p$. This chapter is based on the following definitions (Bock and Diday, 2000).

13.4.1 Multi-valued and interval-valued variables

A variable y is termed *set-valued* with the domain Y, if, for all $w_k \in E$,

$$y : E \rightarrow D$$
$$w_k \longmapsto y(w_k)$$

where the description set D is defined by $D = \mathcal{P}(Y) = \{U \neq \emptyset \mid U \subseteq Y\}$.

A set-valued variable y is called *multi-valued* if its description set D_c is the set of all finite subsets of the underlying domain Y; so $|y(w_k)| < \infty$, for all units $w_k \in E$.

A set-valued variable y is called *categorical multi-valued* if it has a finite range Y of categories and *quantitative multi-valued* if the values $y(w_k)$ are finite sets of real numbers.

A set-valued variable y is called *interval-valued* if its description set D_I is the set of intervals of R.

13.4.2 Modal variables

A modal variable y on a set $E = \{w_1, \ldots, w_n\}$ of objects with domain Y is a mapping

$$y(w_k) = (U(w_k), \pi_k), \qquad \text{for all } w_k \in E,$$

where π_k is a measure or a (frequency, probability or weight) distribution on the domain Y of possible observation values (completed by a suitable σ-field), and $U(w_k) \subseteq Y$ is the support of π_k in the domain Y. The description set of a modal variable is denoted by D_m.

13.5 Dissimilarity measures for symbolic objects

Clustering algorithms and methods for the determination of the number of clusters usually require a dissimilarity matrix which reflects the similarity structure of the n units. In this section we present distance measures on the set $E = \{w_1, \ldots, w_n\}$ in order to determine an $n \times n$ distance matrix on E.

13.5.1 Interval-valued variables

Let $E = \{w_1, \ldots, w_n\}$ be a set of n units described by p interval-valued variables $y_1, \ldots, y_p$. p dissimilarity indices $\delta_1, \ldots, \delta_p$ defined on the description sets D_{Ij} are used to form a global dissimilarity measure on E,

$$\delta_j : D_{Ij} \times D_{Ij} \to \mathbb{R}^+$$
$$(w_{kj}, w_{\ell j}) \longmapsto \delta_j(w_{kj}, w_{\ell j}).$$

If $w_{kj} = [\alpha_{kj}, \beta_{kj}]$ and $w_{\ell j} = [\alpha_{\ell j}, \beta_{\ell j}]$, three distances are defined for interval-valued variables: the Hausdorff distance,

$$\delta_j\,(w_{kj}, w_{\ell j}) = \max\{\,|\,\alpha_{kj} - \alpha_{\ell j}\,|, |\,\beta_{kj} - \beta_{\ell j}\,|\,\};$$

the L_1 distance,

$$\delta_j\,(w_{kj}, w_{\ell j}) = \,|\,\alpha_{kj} - \alpha_{\ell j}\,| + |\,\beta_{kj} - \beta_{\ell j}\,|;$$

and the L_2 distance,

$$\delta_j\,(w_{kj}, w_{\ell j}) = \,(\alpha_{kj} - \alpha_{\ell j})^2 + (\beta_{kj} - \beta_{\ell j})^2.$$

The p dissimilarity indices $\delta_1, \ldots, \delta_p$ defined on the description sets D_{I_j} are combined to form a global dissimilarity measure on E,

$$d : E \times E \longrightarrow \mathbb{R}^+$$
$$(w_k, w_\ell) \longmapsto d(w_k, w_\ell) = \left(\sum_{j=1}^{p} \delta_j^2(w_{kj}, w_{\ell j}) \right)^{1/2},$$

where δ_j is one of the dissimilarity measures defined above.

13.5.2 Multi-valued variables

Let $E = \{w_1, \ldots, w_n\}$ be a set of n units described by p multi-valued variables $y_1, \ldots, y_p$ with domains $Y_1, \ldots, Y_p$ respectively. Let m_j denote the number of categories of y_j. The frequency value $q_{j,w_k}(c_s)$ associated with the category c_s $(s = 1, \ldots, m_j)$ of the variable y_j is given by

$$q_{j,w_k}(c_s) = \begin{cases} \frac{1}{|y_j(w_k)|}, & \text{if } c_s \in y_j(w_k), \\ 0, & \text{otherwise.} \end{cases}$$

The symbolic description of the unit $w_k \in E$ is an $(m_1 + \cdots + m_p)$-dimensional vector given by

$$w_k = ((q_{1,w_k}(c_1), \ldots, q_{1,w_k}(c_{m_1})), \ldots, (q_{p,w_k}(c_1), \ldots, q_{p,w_k}(c_{m_p}))).$$

So the original data matrix $\underline{X} = (y_j(w_k))$ is transformed into a frequency matrix $\tilde{X}$

	y_1			...	y_p		
	1	...	m_1	...	1	...	m_p
w_1	$q_{1,w_1}(c_1)$	...	$q_{1,w_1}(c_{m_1})$	...	$q_{p,w_1}(c_1)$	...	$q_{p,w_1}(c_{m_p})$
$\vdots$	$\vdots$		$\vdots$		$\vdots$		$\vdots$
w_k	$q_{1,w_k}(c_1)$	...	$q_{1,w_k}(c_{m_1})$	...	$q_{p,w_k}(c_1)$	...	$q_{p,w_k}(c_{m_p})$
$\vdots$	$\vdots$		$\vdots$		$\vdots$		$\vdots$
w_n	$q_{1,w_n}(c_1)$	...	$q_{1,w_n}(c_{m_1})$	...	$q_{p,w_n}(c_1)$	...	$q_{p,w_n}(c_{m_p})$

where, for all $w_k \in E$ and for all $j \in \{1, \ldots, p\}$, $\sum_{i=1}^{m_j} q_{j,w_k}(c_i) = 1$.

Let δ_j be a distance function defined on D_{cj}:

$$\delta_j : D_{cj} \times D_{cj} \to \mathbb{R}^+$$
$$(w_{kj}, w_{\ell j}) \longmapsto \delta_j(w_{kj}, w_{\ell j}).$$

The L_1 and L_2 distances on D_{cj} are respectively defined by

$$\delta_j(w_{kj}, w_{\ell j}) = \sum_{i=1}^{|Y_j|} |q_{j,w_k}(c_i) - q_{j,w_\ell}(c_i)| \quad \text{and} \quad \delta_j(w_{kj}, w_{\ell j}) = \sum_{i=1}^{|Y_j|} (q_{j,w_k}(c_i) - q_{j,w_\ell}(c_i))^2,$$

and the de Carvalho distance by

$$\delta_j(w_{kj}, w_{\ell j}) = \sum_{i=1}^{|Y_j|} (\gamma q_{j,w_k}(c_i) + \gamma' q_{j,w_\ell}(c_i)),$$

where

$$\gamma = \begin{cases} 1, & \text{if } c_i \in y_j(w_k) \text{ and } c_i \notin y_j(w_\ell), \\ 0, & \text{otherwise,} \end{cases}$$

$$\gamma' = \begin{cases} 1, & \text{if } c_i \notin y_j(w_k) \text{ and } c_i \in y_j(w_\ell), \\ 0, & \text{otherwise.} \end{cases}$$

The p dissimilarity indices $\delta_1, \ldots, \delta_p$ defined on the sets D_{cj} are combined to form a global dissimilarity measure on E,

$$d : E \times E \longrightarrow \mathbb{R}^+$$
$$(w_k, w_\ell) \longmapsto d(w_k, w_\ell) = \left(\sum_{j=1}^{p} \delta_j^2(w_{kj}, w_{\ell j}) \right)^{1/2},$$

where δ_j is one of the dissimilarity measures defined above.

13.5.3 Modal variables

The case of modal variables is similar to that of multi-valued variables. The frequencies $q_{j,w_k}(c_s)$ are simply replaced by the values of the distribution $\pi_{j,k}$ associated with the categories of $y_j(w_k)$.

13.6 Symbolic clustering procedures

13.6.1 Introduction

In order to generate partitions, several symbolic clustering methods are considered. SCLUST (Verde *et al.*, 2000), see also Chapter 11, is a partitioning clustering method; it is a symbolic extension of the well-known dynamical clustering method (Diday, 1972; Diday and Simon, 1976). DIV (Chavent, 1998) is a symbolic hierarchical monothetic divisive clustering procedure based on the extension of the within-class sum-of-squares criterion. SCLASS (Pirçon, 2004), see also Chapter 9, is a symbolic hierarchical monothetic divisive method based on the generalized hypervolumes clustering criterion (Section 13.8.2).

The hierarchical part of HIPYR (Brito, 2000, 2002), see also Chapter 10, is a module including the symbolic extensions of four hierarchical clustering methods: the single linkage, the complete linkage, the average linkage and the diameter. The corresponding aggregation measures are based on a dissimilarity matrix computed by the DISS module of the SODAS2 software (Chapter 8).

Hardy (2004) has developped a module called SHICLUST containing the symbolic extensions of four well-known classic hierarchical clustering methods: the single linkage, complete linkage, centroid and Ward methods.

These agglomerative methods execute a complete sequence of fusions. In the first step, the two closest objects are joined, leaving $n-1$ clusters, one of which contains two objects while the rest have only one. In each successive step, the two closest clusters are merged. The aggregation measures used are defined below.

13.6.2 Aggregation indices

Let $\mathcal{P}(E)$ be the set of all subsets of E. An *aggregation index* between groups of objects h and h' is a mapping

$$d: \quad \mathcal{P}(E) \times \mathcal{P}(E) \quad \longrightarrow \quad \mathbb{R}$$

such that, for all $h, h' \in \mathcal{P}(E)$, $d(h, h') \geq 0$ (positivity) and $d(h, h') = d(h', h)$ (symmetry).

13.6.3 Four classic hierarchical methods

The single linkage method

Groups initialy consisting of single individuals are fused according to the distance between their nearest members, the groups with the smallest distance being fused. The aggregation measure is defined by

$$d_1(h, h') = \min_{\substack{x \in h \\ y \in h'}} \delta(x, y),$$

where δ is any dissimilarity measure between objects x and y.

The complete linkage method

This method is the opposite of the single linkage method. The distance between groups is now defined as the distance between their most remote pair of individuals. The complete linkage method is based on

$$d_2(h, h') = \max_{\substack{x \in h \\ y \in h'}} \delta(x, y),$$

where δ is any dissimilarity measure between objects x and y.

The centroid method

The distance between groups is defined as the distance between the group centroids. The procedure fuses groups according to the distance between their centroids, the groups with the smallest distance being fused first. Thus,

$$d_3(h, h') = \delta(g^{(h)}, g^{(h')}),$$

where δ is any dissimilarity measure between objects x and y and $g^{(h)}$ the centroid of group h.

The centroid of group h is a p-dimensional vector

$$g^{(h)} = (g_1^{(h)}, \ldots, g_p^{(h)}),$$

where p is the number of variables, $|h|$ the number of objects in h and

$$g_j^{(h)} = \frac{\sum_{x \in h} x_j}{|h|}, \qquad (j = 1, \ldots, p).$$

In the case of interval-valued variables, we have

$$Y_j : E \to \mathcal{B}_j : x_i \mapsto Y_j(x_i) = x_{ij} = [\alpha_{ij}, \beta_{ij}] \subset \mathbb{R}, \qquad j = 1, \ldots, p.$$

The centroid of group h is the hyperrectangle of gravity of h defined by

$$g^{(h)} = \left(\left[\frac{1}{n_h} \sum_{x_i \in h} \alpha_{i1}, \frac{1}{n_h} \sum_{x_i \in h} \beta_{i1} \right], \ldots, \left[\frac{1}{n_h} \sum_{x_i \in h} \alpha_{ip}, \frac{1}{n_h} \sum_{x_i \in h} \beta_{ip} \right] \right)$$

where n_h is the number of objects in class h.

In the case of multi-valued variables, we have

$$Y_j : E \to \mathcal{B}_j : x_i \mapsto Y_j(x_i) = x_{ij} = \{c_1, \ldots, c_{m_j}\}, \qquad j = 1, \ldots, p.$$

The frequency $q_{j,x_i}(c_s)$ associated with category c_s $(s = 1, \ldots, m_j)$ of $Y_j(x_i)$ is given by

$$q_{j,x_i}(c_s) = \begin{cases} \frac{1}{|Y_j(x_i)|}, & \text{if} \quad c_s \in Y_j(x_i) \\ 0, & \text{otherwise.} \end{cases}$$

The centroid of group h is defined by

$$g^{(h)} = \frac{1}{n_h} \sum_{x_i \in h} ((q_{1,x_i}(c_1), \ldots, q_{1,x_i}(c_{m_1})), \ldots, (q_{p,x_i}(c_1), \ldots, q_{p,x_i}(c_{m_p}))).$$

The case of modal variables is similar to the case of multi-valued variables. The frequencies $q_{j,x_i}(c_s)$ are simply replaced by the values of the distribution $\pi_{j,i}$ associated with the categories of $y_j(x_i)$.

The Ward method

The aggregation measure is based on the computation of the total sum of squared deviations of every point from the mean of the cluster to which it belongs. At each step of the procedure, the union of every possible pair of clusters is considered and the two clusters whose fusion results in the minimum increase in the error sum of squares are combined. The method is thus based on

$$d_4(h, h') = I(h \cup h') - I(h) - I(h')$$

where $I(h)$ is the inter-cluster inertia of cluster h, defined by

$$I(h) = \sum_{x \in h} \delta^2(x, g^{(h)}),$$

in which δ is any dissimilarity measure between objects x and y and $g^{(h)}$ is the centroid of group h.

13.6.4 Symbolic extensions of the four classic hierarchical methods: SHICLUST

All these aggregation indices are based on a dissimilarity measure between the elements of E. In this chapter we use the L_1, L_2, Hausdorff and de Carvalho dissimilarity measures. The corresponding methods are included in the SHICLUST module. But we can also consider the dissimilarity measures included in the DISS module of SODAS2 (Chapter 8). Thanks to these aggregation indices and dissimilarity measures, we are able to classify symbolic objects described by interval-valued, multi-valued and modal variables.

13.7 Determination of the number of clusters for symbolic objects

The five best methods for the determination of the number of clusters from the Milligan and Cooper (1985) study are based on a dissimilarity matrix. Such a dissimilarity matrix can be computed for symbolic objects described by interval-valued, multi-valued and modal variables. Consequently, these five methods can be used in order to determine the structure of symbolic data. They have been included in a module called NBCLUST (Hardy *et al.*, 2002; Hardy and Lallemand, 2004), integrated in the SODAS2 software. The five methods of NBCLUST are computed at each level of the four hierarchies of SHICLUST.

Concerning SCLUST, we select the best partition into ℓ clusters, for each value of ℓ ($\ell = 1, \ldots, K$, where K is a reasonably large integer fixed by the user), and we compute the three indices available for non-hierarchical classification (Calinski–Harabasz, the C-index and the Γ-index). The analysis of these indices should provide the "best" number of clusters.

13.8 Statistical models based on the Poisson processes

13.8.1 The hypervolumes clustering method

13.8.1.1 Definition: homogeneous (or stationary) Poisson process

N is a Poisson process with intensity q ($q \in \mathbb{R}$) on a set $D \subset \mathbb{R}^p$ $(0 < m(D) < \infty)$ if the following hold (Cox and Isham, 1980):

- For all $A_1, \ldots, A_k \subset D$, for all $i \neq j \in \{1, \ldots, k\}$ with $A_i \cap A_j = \emptyset$,

$$N(A_i) \perp\!\!\!\perp N(A_j).$$

  The random variables $N(A_i)$ and $N(A_j)$ that count the number of points in disjoint regions of the space are independent.

- For all $A \subset D$, for all $k > 0$,

$$P(N(A) = k) = \frac{(q\,m(A))^k}{k!}\, \mathrm{e}^{-q\,m(A)}.$$

  The random variable $N(A)$ has a Poisson distribution with mean $m(A)$ where $m(\cdot)$ is the multidimensional Lebesgue measure.

13.8.1.2 Conditional uniformity property for the homogeneous Poisson process

If $N(D) = n$ is finite, then the n points are independently and uniformly distributed on D. This conditional uniformity property allows us to write the density function associated with the homogeneous Poisson process

$$f(x) = \frac{1}{m(D)}\, I_D(x)$$

and thus, if $x = (x_1, \ldots, x_n)$, the likelihood function L_D takes the form

$$L_D(x) = \frac{1}{(m(D))^n} \prod_{i=1}^{n} I_D(x_i)$$

where I_D denotes the indicator function of the set D.

13.8.1.3 Starting problem: the estimation of a convex set

The starting point of this approach is the following problem: given a realization of a homogeneous Poisson process N with an unknown intensity q over a convex compact domain D, find D (using inferential statistical methods).

The convex hull of the points belonging to D is both a sufficient statistic and the maximum likelihood estimate of the domain D. An unbiaised estimator can be obtained by taking a dilation of the convex hull of the points from its centroid (Ripley and Rasson, 1977; Moore, 1984).

13.8.1.4 The hypervolumes clustering criterion

The hypervolumes clustering method (Hardy and Rasson, 1982; Hardy, 1983) assumes that the n p-dimensional observation points $x_1, \ldots, x_n$ are a random sample of a homogeneous Poisson process N in a set D included in the Euclidean space $\mathbb{R}^p$ (with $0 < m(D) < \infty$). The set D is supposed to be the union of k disjoint convex compact domains $D_1, D_2, \ldots, D_k$. The problem is then to estimate the unknown domains D_i in which the points were generated. We denote by $C_i \subset \{x_1, \ldots, x_n\}$ the subset of the points belonging to D_i $(1 \leq i \leq k)$.

The likelihood function L_D takes the form given in Section 13.8.1.2. The maximization of L_D is equivalent to the minimization of the hypervolumes clustering criterion (Hardy, 1983)

$$\max_{D_1,\ldots,D_k} L_D(x) \iff \min_{P \in \mathcal{P}_k} \sum_{i=1}^{k} m(H(C_i)),$$

where $\mathcal{P}_k$ is the set of all the partitions of C into k clusters. The maximum likelihood estimates of the k unknown domains $D_1, D_2, \ldots, D_k$ are the k convex hulls $H(C_i)$ of the k subgroups of points C_i such that the sum of the Lebesgue measures of the disjoint convex hulls $H(C_i)$ is minimal.

The hypervolumes clustering criterion is defined by

$$W_k = \sum_{i=1}^{k} m(H(C_i)) = \sum_{i=1}^{k} \int_{H(C_i)} m(\mathrm{d}x)$$

where $H(C_i)$ is the convex hull of the points belonging to C_i and $m(H(C_i))$ is the multidimensional Lebesgue measure of that convex hull. This clustering criterion has to be minimized over the set of all the partitions of the observed sample into k clusters. In the context of a clustering problem, we try to find the partition P^* such that

$$P^* = \arg \min_{P_k \in \mathcal{P}_k} \sum_{i=1}^{k} \int_{H(C_i)} m(\mathrm{d}x).$$

13.8.2 The generalized hypervolumes clustering method

13.8.2.1 Definition: non-homogeneous (or non-stationary) Poisson process

The non-homogeneous Poisson process N with intensity $q(\cdot)$ on the domain $D \subset \mathbb{R}^p$ $(0 < m(D) < \infty)$ is characterized by the two following properties:

- For all $A \subset D$, $N(A)$ has a Poisson distribution with mean $\int_A q(x)\, m(\mathrm{d}x)$.
- Conditional property: if $N(A) = n$, then the n points are independently distributed in A, with a density function proportional to $q(x)$.

13.8.2.2 The generalized hypervolumes clustering criterion

The generalized hypervolumes clustering method (Kubushishi, 1996; Rasson and Granville, 1996) assumes that the n p-dimensional points $x_1, \ldots, x_n$ are generated by a non-homogeneous Poisson process N with intensity $q(\cdot)$ in a set $D \subset \mathbb{R}^p$ $(0 < m(D) < \infty)$ where D is the union of k disjoint convex domains $D_1, \ldots, D_k$. The problem is then to estimate the unknown domains D_i in which the points were generated.

Thanks to the conditional property for the non-homogeneous Poisson process, we can write the density of the process as

$$f(x) = \frac{q(x)\, I_D(x)}{\int_D q(t)\, m(dt)} = \frac{q(x)\, I_D(x)}{\rho(D)}$$

where $\rho(D) = \int_D q(t)\, m(dt)$ is called the integrated intensity of the process on D.

The likelihood function can be written as $(x = (x_1, \ldots, x_n))$

$$L_D(x) = \prod_{i=1}^{n} f_X(x_i) = \frac{1}{(\rho(D))^n} \prod_{i=1}^{n} I_D(x_i)\, q(x_i).$$

The generalized hypervolumes clustering criterion is deduced from this statistical model using maximum likelihood estimation.

Let $\mathcal{P}_k$ be the set of all the partitions of C into k clusters. If the intensity $q(\cdot)$ of the non-homogeneous Poisson process is known, the maximization of the likelihood function L_D is equivalent to the minimization of the generalized hypervolumes criterion W_k^*:

$$\max_{D_1,\ldots,D_k} L_D(x) \iff \min_{P \in \mathcal{P}_k} \sum_{i=1}^{k} \rho(H(C_i)) \iff \min_{P \in \mathcal{P}_k} W_k^*.$$

The generalized hypervolumes criterion is defined by

$$W_k^* = \sum_{i=1}^{k} \rho(H(C_i)) = \sum_{i=1}^{k} \int_{H(C_i)} q(x) m(dx)$$

where $q(\cdot)$ is the intensity of the nonhomogeneous Poisson process and $H(C_i)$ is the convex hull of the points belonging to C_i. In the context of a clustering problem, we try to find the partition P^* such that

$$P^* = \arg \min_{P_k \in \mathcal{P}_k} \sum_{i=1}^{k} \int_{H(C_i)} q(x)\, m(dx).$$

If $q(\cdot)$ is not known, it must be estimated.

13.8.2.3 Estimation of the intensity of the non-homogeneous Poisson process

In order to estimate q, we use a non-parametric method: the kernel method. The kernel estimate $\hat{q}$ of q is defined by

$$\hat{q}(x) = \frac{1}{n} \sum_{i=1}^{n} \frac{1}{h} K\left(\frac{x - x_i}{h}\right)$$

where the kernel K has the following properties:

- K is symmetric and continuous;
- $K \geq 0$;
- $\int K(x)\, dx = 1$.

The parameter h is the window width also called the 'smoothing parameter'. The kernel estimate is a sum of 'bumps' around the observations x_i. The kernel function K determines the shapes of the bumps while the window width h determines their width. In order to estimate h, we distinguish the notions of 'bump' and 'mode'. A mode of a density q is a local maximum of that density. A bump is characterized by an interval such that the density is concave on that interval, but not on a larger interval.

Due to its properties, we use a normal kernel defined by

$$K_N(x) = \frac{1}{\sqrt{2\pi}}\, e^{-x^2/2}.$$

Silverman (1981, 1986) proved that the number of modes for a normal kernel is a decreasing function of the smoothing parameter h. So for practical purposes we estimate h by choosing the largest value of h such that $\hat{q}$ remains multimodal.

13.9 Statistical tests for the number of clusters based on the homogeneous Poisson point process

13.9.1 The hypervolumes test

The statistical model based on the homogeneous Poisson process allows us to define a likelihood ratio test for the number of clusters (Hardy, 1996). Let us denote by $C = \{C_1, C_2, \ldots, C_\ell\}$ the optimal partition of the sample into ℓ clusters and $B = \{B_1, B_2, \ldots, B_{\ell-1}\}$ the optimal partition into $\ell - 1$ clusters. We test the hypothesis H_0: $t = \ell$ against the alternative H_1: $t = \ell - 1$, where t represents the number of 'natural' clusters ($\ell \geq 2$). The test statistic is defined by

$$S(x) = \frac{W_\ell}{W_{\ell-1}}$$

where W_i is the value of the hypervolumes clustering criterion associated with the best partition into i clusters.

Unfortunately the sampling distribution of S is not known. But $S(x)$ belongs to $[0, 1]$. Consequently, for practical purposes, we can use the following decision rule: reject H_0 if S is close to 1. We apply the test in a sequential way: if ℓ_0 is the smallest value of $\ell \geq 2$ for which we reject H_0, we choose $\ell_0 - 1$ as the best number of 'natural' clusters.

13.9.2 The gap test

The gap test (Kubushishi, 1996; Rasson and Kubushishi, 1994) is based on the same statistical model (homogeneous Poisson process). We test H_0: the $n = n_1 + n_2$ observed points are a realization of a Poisson process in D against H_1: n_1 points are a realization of a homogeneous Poisson process in D_1 and n_2 points in D_2 where $D_1 \cap D_2 = \emptyset$. The sets D, D_1, D_2 are unknown. Let us denote by C (C_1, C_2) the set of points belonging to D (D_1, D_2). The test statistic is given by

$$Q(x) = \left(1 - \frac{m(\Delta)}{m(H(C))}\right)^n,$$

where $H(C)$ is the convex hull of the points belonging to C, $\Delta = H(C) \setminus (H(C_1) \cup H(C_2))$ is the 'gap space' between the clusters and m is the multidimensional Lebesgue measure. The test statistic is the Lebesgue measure of the gap space between the clusters.

The decision rule is the following (Kubushishi, 1996). We reject H_0, at level α, if (the asymptotic distribution)

$$\frac{nm(\Delta)}{m(H(C))} - \log n - (p-1)\log\log n \geq -\log(-\log(1-\alpha)).$$

13.9.3 From the homogeneous Poisson process to the non-homogeneous Poisson process

By a change of variables (Cox and Isham, 1980) it is possible to transform a non-homogeneous Poisson process into a homogeneous Poisson process. Let $(x_1, \ldots, x_n)$ be a realization of a non-homogeneous Poisson process with intensity $q(x)$.

We use the following change of variables:

$$\tau(x) = \int_0^x q(t)\, m(\mathrm{d}t) \text{ on } \mathbb{R}^p.$$

We obtain new data $(\tau_1, \ldots, \tau_n)$, where $\tau_i = \tau(x_i) \in \mathbb{R}^p$ $(i = 1, \ldots, n)$. These data are distributed according to a homogeneous Poisson process.

The hypervolumes test and the gap test were initially developed with the homogeneous Poisson process. Thanks to this transformation, they can be used with the non-homogeneous Poisson process.

13.9.4 Application to interval-valued data

The hypervolumes test and the gap test are not based on the existence of a dissimilarity matrix. They require the computation of convex hulls of points. In order to extend these tests to interval-valued data, we proceed as follows. An interval is summarized by two numbers: its centre and its length ((C, L) modelling). Each object can thus be represented as a point in a $2p$-dimensional space where p is the number of interval-valued variables measured on each object.

For practical purposes, the hypervolumes test and the gap test are applied to the points obtained from (C, L) modelling, in the two-dimensional spaces associated with each of the p interval-valued variables.

When the hypervolumes test is applied to the hierarchies of partitions generated by each of the four hierarchical methods included in SHICLUST, it computes, in each of the p (C, L) representations, the areas of the convex hulls corresponding to the partitions generated, at the corresponding level of the hierarchy. Consequently, the number of clusters obtained with one interval-valued variable may be different from the number of clusters obtained with another interval-valued variable.

In order to solve this problem we select from the set of all the variables the most discriminant one and we apply the hypervolumes test and the gap test in the two-dimensional (C, L) space associated with that variable.

The total inertia (Celeux *et al.*, 1989) of the set E of objects is defined by

$$T = \sum_{i=1}^{n} (x_i - g)' \ (x_i - g).$$

It is possible to compute the contribution of the class C_ℓ or of the jth variable to the total inertia T:

$$T = \sum_{i=1}^{n} (x_i - g)' \ (x_i - g) = \sum_{\ell=1}^{k} \sum_{j=1}^{p} \sum_{x_i \in C_\ell} (x_{ij} - g_j)^2 = \sum_{\ell=1}^{k} \sum_{j=1}^{p} T_j^{(\ell)} = \sum_{\ell=1}^{k} T^{(\ell)},$$

where $T^{(\ell)} = \sum_{j=1}^{p} T_j^{(\ell)}$. We also have

$$T = \sum_{i=1}^{n} (x_i - g)' \ (x_i - g) = \sum_{\ell=1}^{k} \sum_{j=1}^{p} \sum_{x_i \in C_\ell} (x_{ij} - g_j)^2 = \sum_{\ell=1}^{k} \sum_{j=1}^{p} T_j^{(\ell)} = \sum_{j=1}^{p} T_j,$$

where $T_j = \sum_{\ell=1}^{k} T_j^{(\ell)}$. So $T^{(\ell)}$ is the contribution of class C_ℓ to the total inertia T. T_j is the contribution of variable j to the total inertia T.

We have a similar decomposition for the inter-class inertia B:

$$B = \sum_{\ell=1}^{k} n_\ell \ (g^{(\ell)} - g)' \ (g^{(\ell)} - g) = \sum_{\ell=1}^{k} \sum_{j=1}^{p} n_\ell \ (g_j^{(\ell)} - g_j)^2 = \sum_{\ell=1}^{k} \sum_{j=1}^{p} B_j^{(\ell)} = \sum_{\ell=1}^{k} B^{(\ell)},$$

where $B^{(\ell)} = \sum_{j=1}^{p} B_j^{(\ell)}$. We also have

$$B = \sum_{\ell=1}^{k} n_\ell \ (g^{(\ell)} - g)' \ (g^{(\ell)} - g) = \sum_{\ell=1}^{k} \sum_{j=1}^{p} n_\ell \ (g_j^{(\ell)} - g_j)^2 = \sum_{\ell=1}^{k} \sum_{j=1}^{p} B_j^{(\ell)} = \sum_{j=1}^{p} B_j,$$

where $B_j = \sum_{\ell=1}^{k} B_j^{(\ell)}$. So $B^{(\ell)}$ is the contribution of class C_ℓ to the inter-class inertia B and B_j is the contribution of the jth variable to the inter-class inertia B.

Thanks to these decompositions, we can determine the most discriminant variable by the following indices:

$$\text{cor}(j) = 100.\frac{B_j}{T_j}, \qquad \text{ctr}(j) = 100.\frac{B_j}{B}.$$

We proceed as follows. The symbolic clustering method SCLUST is applied to the original interval-valued data. We consider the successive partitions of E into ℓ clusters ($\ell = 1, \ldots, K$, where K is a reasonably large integer fixed by the user). We transform the symbolic objects into classical data using (C, L) modelling. We apply the hypervolumes test and the gap test in the two-dimensional space associated with the most discriminant interval-valued variable.

When cor and ctr do not produce the same choice of the best variable, we recommend analysing the best partitions given by each in order to validate one of them.

13.10 Examples

13.10.1 Symbolic artificial data set

The artificial data set in Table 13.1 consists of 30 objects described by two interval-valued variables. Thus each object can be represented by a rectangle in the two-dimensional space

Table 13.1 Artificial data.

Objects	Variable 1	Variable 2
1	[3.37, 6.28]	[0.35, 1.10]
2	[0.81, 3.17]	[3.35, 5.11]
3	[2.19, 4.69]	[1.57, 1.72]
4	[1.06, 3.59]	[1.92, 3.55]
5	[2.33, 5.31]	[2.72, 3.95]
6	[0.63, 2.64]	[1.21, 3.04]
7	[0.43, 2.59]	[2.96, 4.28]
8	[2.10, 5.08]	[−0.63, 1.01]
9	[1.38, 5.13]	[−0.04, 1.53]
10	[1.05, 3.42]	[1.91, 3.37]
11	[13.21, 16.12]	[2.63, 3.78]
12	[10.07, 12.43]	[3.66, 5.58]
13	[10.46, 12.96]	[0.58, 1.66]
14	[11.05, 13.58]	[1.93, 3.87]
15	[10.66, 13.64]	[3.64, 5.35]
16	[13.79, 15.81]	[−0.30, 0.87]
17	[10.60, 12.77]	[2.78, 4.04]
18	[11.63, 14.62]	[0.95, 2.00]
19	[11.77, 13.93]	[−0.44, 1.14]
20	[9.02, 11.39]	[1.00, 2.68]
21	[13.27, 16.18]	[0.64, 2.03]
22	[10.64, 13.00]	[3.13, 4.50]
23	[19.68, 22.07]	[4.30, 5.59]
24	[21.39, 23.84]	[0.63, 1.68]
25	[20.14, 22.99]	[1.71, 3.30]
26	[19.64, 22.07]	[3.81, 5.07]
27	[19.01, 21.44]	[2.32, 4.12]
28	[23.21, 26.17]	[0.43, 2.38]
29	[19.48, 22.32]	[0.24, 1.99]
30	[19.90, 22.37]	[1.80, 3.37]

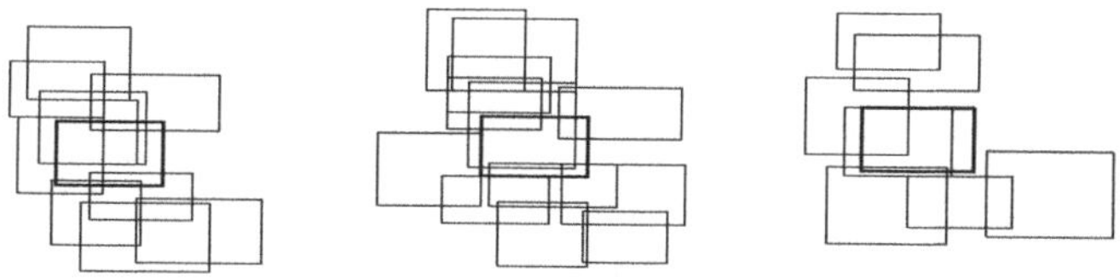

Figure 13.1 Visualization of the interval-valued data of Table 13.1.

$\mathbb{R}^2$. The data were generated in order to present a well-defined structure into three clusters (Figure 13.1). The three rectangles printed in bold in Figure 13.1 are the prototypes of the three clusters, that is, the hyper-rectangles of gravity of the three natural clusters.

13.10.1.1 SHICLUST

NBCLUST is applied to the hierarchies of partitions given by the four hierarchical methods included in SHICLUST. The five criteria for the number of classes give the correct number of natural clusters. For example, Table 13.2 presents the values of the indices given by NBCLUST applied to the hierarchy of partitions given by the centroid clustering method. The maximum value of the indices for the Calinski and Harabasz method and the Γ-index is obtained for $k = 3$. The minimum value of the C-index is also obtained for $k = 3$. If we choose the best levels for the statistical tests proposed by Milligan and Cooper (1985), the J-index and the Beale test also detect the existence of the same three natural clusters in the data.

13.10.1.2 SCLUST

The SCLUST procedure is applied to the original interval-valued data. We consider the best partitions into ℓ clusters ($\ell = 1, \ldots, K$). The three number-of-clusters criteria of NBCLUST available for non-hierarchical clustering methods are then implemented. The results are presented in Table 13.3. We conclude that the natural structure contains three clusters.

For the hypervolumes test and the gap test, we determine the most discriminant variable using cor(j) and ctr(j) from Section 13.9.4. The values of these indices are presented in Table 13.4, showing that the first variable is the most discriminant one.

Table 13.2 Centroid and NBCLUST.

Centroid	Calinski and Harabasz	J-index	C-index	Γ-index	Beale test
$k=8$	147.46524	0.93037	0.00214	0.96108	1.04810
$k=7$	147.36367	0.61831	0.00287	0.95356	0.70219
$k=6$	151.08969	0.75366	0.00355	0.95240	0.83174
$k=5$	142.02988	1.29472	0.00409	0.95895	1.38609
$k=4$	138.96187	0.99391	0.00168	0.99245	0.97659
$k=3$	**189.95941**	**0.41215**	**0.00002**	**0.99990**	**0.56737**
$k=2$	67.04706	2.93795	0.05495	0.81996	4.24084
$k=1$	—	2.74934	—	—	2.23490

Table 13.3 SCLUST and NBCLUST.

SCLUST	Calinski and Harabasz	C-index	Γ-index
$k=5$	156.35205	0.00574	0.95018
$k=4$	165.37531	0.00527	0.96158
$k=3$	**189.95941**	**0.00002**	**0.99990**
$k=2$	67.04706	0.05495	0.81996
$k=1$	—	—	—

Table 13.4 Selection of the most discriminant variable.

Variable	cor	ctr
1	96.98	99.91
2	2.29	0.09

The hypervolumes test and the gap test are applied to the hierarchies of partitions generated by SHICLUST in the two-dimensional space associated with the first variable. The analysis of the indices leads to the conclusion that there are three clusters in the data.

For example, Table 13.5 presents the values of the statistics S associated with the hypervolumes test for the centroid clustering method. From Table 13.5, the first value of k that can be considered as close to 1 is $k_0=4$. So we conclude that the best partition contains $4-1=3$ clusters.

Finally, when the hypervolumes test and the gap test are applied to the partitions generated by SCLUST, we conclude also that the natural partition contains three clusters. The indices associated with the test statistic S for the hypervolumes test are presented in Table 13.6.

Table 13.5 SHICLUST and the hypervolumes test.

Centroid	Var.1
$k=1$	—
$k=2$	0.578
$k=3$	0.526
$k=4$	0.874
$k=5$	0.947
$k=6$	0.955
$k=7$	0.891
$k=8$	0.617

Table 13.6 SCLUST and hypervolumes.

SCLUST	Var.1
k=1	—
k=2	0.578
k=3	0.526
k=4	0.954
k=5	0.961
k=6	0.815
k=7	0.847

Table 13.7 Table of oil and fats.

Sample	Specific gravity	Freezing point	Iodine value	Saponification value
Linseed oil	[0.930 : 0.935]	[−27 : −18]	[170 : 204]	[118 : 196]
Perilla oil	[0.930 : 0.937]	[−5 : −4]	[192 : 208]	[188 : 197]
Cottonseed oil	[0.916 : 0.918]	[−6 : −1]	[99 : 113]	[189 : 198]
Sesame oil	[0.92 : 0.926]	[−6 : −4]	[104 : 116]	[187 : 193]
Camellia oil	[0.916 : 0.917]	[−21 : −15]	[80 : 82]	[189 : 193]
Olive oil	[0.914 : 0.919]	[0 : 6]	[79 : 90]	[187 : 196]
Beef tallow	[0.86 : 0.87]	[30 : 38]	[40 : 48]	[190 : 199]
Hog fat	[0.858 : 0.864]	[22 : 32]	[53 : 77]	[190 : 202]

13.10.2 Real interval-valued data: oils and fats

Ichino's (1988) data are displayed in Table 13.7. This data set is made up of eight oils, described in terms of four interval-valued variables: specific gravity (y_1), freezing point (y_2), iodine value (y_3) and saponification value (y_4).

13.10.2.1 SCLUST

SCLUST and NBCLUST are applied to the oils and fats data, with the Hausdorff distance. The indices of the three number-of-clusters criteria are given in Table 13.8. The maximum value of the Calinski and Harabasz index is obtained for $k=3$. The C-index is close to 0 when $k=3$. The maximum value of the Γ-index is 0.63636; it corresponds to three clusters. SCLUST and NBCLUST agree on the following three-cluster partition: C_1 = {beef, hog}, C_2 = {cottonseed, sesame, camellia, olive}, C_3 = {linseed, perilla}.

13.10.2.2 SHICLUST

We consider SHICLUST, NBCLUST (with the Hausdorff distance) and the hypervolumes test. The results are given in Table 13.9. The other distances (L_1 and L_2) give similar results. The best four-cluster partition given by the hypervolumes test is C_1={beef, hog}, C_2={camellia, olive, cottonseed, sesame}, C_3={linseed}, C_4={perilla}. The best three-cluster partition is the same as the one given by SCLUST and NBCLUST.

Table 13.8 SCLUST and NBCLUST.

SCLUST	Calinski and Harabasz	C-index	Γ-index
7	0.34849	0.08378	0.57333
6	2.22292	0.15840	0.51304
5	2.54987	0.19063	0.50877
4	3.61575	0.19063	0.50877
3	**13.46313**	**0.03633**	**0.63636**
2	3.16596	0.42138	0.25000
1	—	—	—

Table 13.9 SHICLUST, NBCLUST and the hypervolumes test: Ichino's data.

SHICLUST	Hypervolumes test	Calinski and Harabasz	J-index	C-index	Γ-index	Beale test
Single linkage	4	4	2	4	4	2
Complete linkage	4	3	4	2	3	3
Centroid	4	3	3	4	3	3
Ward	4	3	4	2	3	3

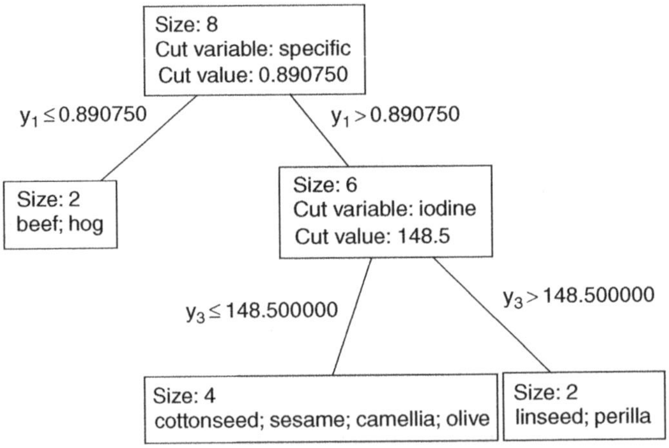

Figure 13.2 SCLASS decision tree.

13.10.2.3 SCLASS

The output of SCLASS is a decision tree (Figure 13.2). This has advantage of allowing the clusters to be interpreted in terms of the original variables. The three-cluster partition given by SCLASS is $C_1 = \{\text{beef, hog}\}$, $C_2 = \{\text{cottonseed, sesame, camellia, olive}\}$, $C_3 = \{\text{linseed, perilla}\}$. A symbolic object of the assertion type (Bock and Diday, 2000) is associated with each of the clusters:

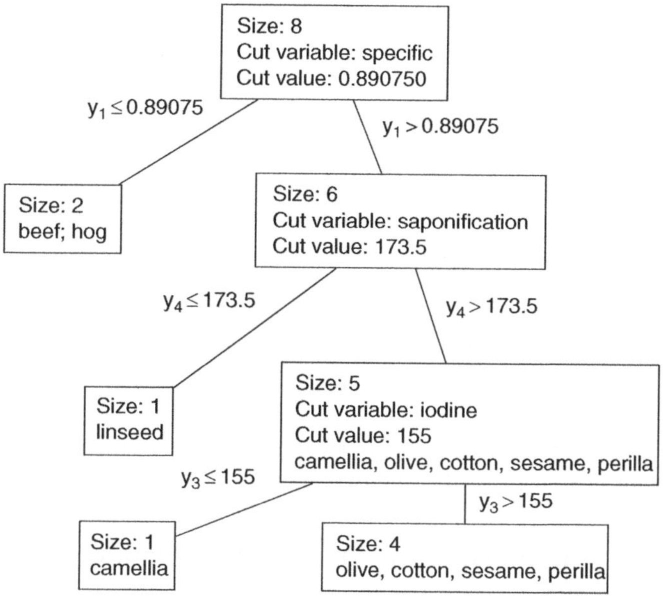

Figure 13.3 DIV decision tree.

$$C_1 : a_1 = ([y_1 \leq 0.890750]),$$
$$C_2 : a_2 = ([y_1 > 0.890750] \wedge [y_3 \leq 148.5]),$$
$$C_3 : a_3 = ([y_1 > 0.890750] \wedge [y_3 > 148.5]).$$

The extensions of the symbolic objects in E correspond to the three clusters of the partition:

$$\text{Ext}_E(a_1) = \{\text{beef, hog}\},$$
$$\text{Ext}_E(a_2) = \{\text{cottonseed, sesame, camellia, olive}\},$$
$$\text{Ext}_E(a_3) = \{\text{linseed, perilla}\}.$$

For example, the object 'linseed' is classified in C_3 because

$$y_1(\text{linseed}) = [0.930; 0.935] \text{ and } (0.930 + 0.935)/2 > 0.89075$$
$$y_3(\text{linseed}) = [170; 204] \text{ and } (170 + 204)/2 > 148.5.$$

13.10.2.4 DIV

The application of DIV to Ichino's data leads to the decision tree in Figure 13.3. The best partition into three clusters is different from the one given by the other clustering methods. The best partition into four clusters is the same for all methods. We remark that the decision

Table 13.10 Merovingian buckles: six categorical multi-valued variables.

Variables	Categories
Fixation	iron nail; bronze bump; none
Damascening	bichromate; predominant veneer; dominant inlaid; silver monochrome
Contours	undulations; repeating motives; geometric frieze
Background	silver plate; hatching; geometric frame
Inlaying	filiform; hatching banner; dotted banner; wide ribbon
Plate	arabesque; large size; squared back; animal pictures; plait; circular

trees associated with the two monothetic divisive methods DIV and SCLASS are different. This can be explained by the fact that the two methods use different cutting criteria.

13.10.3 Real multi-valued data: Merovingian buckles

This set of symbolic data is constituted by 58 Merovingian buckles described by six multi-valued symbolic variables. These variables and the corresponding categories are presented in Table 13.10 (the complete data set is available at http://www-rocq.inria.fr/sodas/WP6/data/data.html). The 58 buckles were examined by archaeologists, who identified two natural clusters. SCLUST and SHICLUST have been applied to the data set, with the three previously defined distances (L_1, L_2 and de Carvalho).

13.10.3.1 SCLUST

Table 13.11 presents the values of three stopping rules obtained with SCLUST and the de Carvalho dissimilarity measure. The table shows the number of clusters in each partition, and the indices given by the three methods for the determination of the number of clusters available for non-hierarchical procedures.

The 'optimal' number of clusters is the value of k corresponding to the maximum value of the indices for the Calinski and Harabasz method and the C-index, and the minimum

Table 13.11 SCLUST and NBCLUST.

SCLUST	Calinski and Harabasz	C-index	Γ-index
8	28.52599	0.03049	0.95382
7	30.37194	0.04221	0.95325
6	23.36924	0.05127	0.92168
5	31.03050	0.04869	0.93004
4	32.66536	0.05707	0.95822
3	41.24019	0.03788	0.96124
2	**51.13007**	**0.01201**	**0.99917**
1	—	—	—

Table 13.12 SHICLUST and NBCLUST.

Complete linkage	Calinski and Harabasz	J-index	C-index	Γ-index	Beale test
8	31.07487	2.39966	0.00790	0.99238	3.72846
7	33.83576	2.35439	0.00860	0.99117	0.00000
6	34.66279	2.28551	0.01338	0.98565	**2.34051**
5	28.87829	3.56668	0.03270	0.94615	3.59244
4	34.53483	1.48947	0.03986	0.93230	1.20779
3	29.45371	2.91271	0.02761	0.95592	1.88482
2	**51.47556**	**1.53622**	**0.00932**	**0.99010**	**1.24133**
1	—	5.27061	—	—	3.20213

value for the Γ-index. The three methods agree on two clusters of buckles. Furthermore, these two clusters are exactly the ones identified by the archaeologists.

SHICLUST

The five best rules from the Milligan and Cooper (1985) study are applied at each level of the four hierarchies. The results for the complete linkage clustering method associated with the L_2 dissimilarity measure are presented in Table 13.12.

The Calinski and Harabasz method, the C-index and the Γ-index indicate that there are two clusters in the data set. For the Duda and Hart method and the Beale test, standard scores have to be fixed. Adopting the recommended values (Milligan and Cooper, 1985), these methods detect respectively two clusters for the J-index, and two or six clusters for the Beale test. The partition into six clusters given by the Beale test is close to another classification of the buckles into seven clusters given by the same archeologists.

Equivalent results are found with the complete linkage algorithm associated with the two other distances (L_1 and de Carvalho), and by the other clustering procedures (single linkage, centroid and Ward) with the L_1, L_2 and de Carvalho distances.

13.10.4 Real modal data: e-Fashion stores

This data set (available at http://www.ceremade.dauphine.fr/~touati/exemples.htm) describes the sales for 1999, 2000 and 2001 in a group of stores (items of clothing and accessories) located in six different countries. The 13 objects are the stores (Paris 6, Lyon, Rome, Barcelona, Toulouse, Aix-Marseille, Madrid, Berlin, Milan, Brussels, Paris 15, Paris 8, London). Eight modal variables (Table 13.13) are recorded on each of the 13 objects, describing the items sold in these stores. For example, the variable 'family product' has 13 categories (dress, sweater, T-shirt, ...). The proportion of sales in each store is associated with all these categories. On the other hand, the variable 'month' describes the proportion of sales for each month of the year.

We apply SCLUST and NBCLUST with the three distances (L_1, L_2, de Carvalho). The resulting partitions are analysed by the methods for the determination of the number of clusters. The results are given in Tables 13.14 and 13.15 (de Carvalho distance), 13.16 and 13.17 (L_1 distance) and 13.18 and 13.19 (L_2 distance).

Table 13.13 The eight modal variables in the e-Fashion data set.

	Modal variables	Number of categories
1	item tag	153
2	category	31
3	family product	12
4	colour tag	160
5	range of colours	17
6	month	12
7	level of sale	5
8	number	5

Table 13.14 SCLUST and NBCLUST – de Carvalho distance.

	Calinski and Harabasz	C-index	Γ-index
SCLUST	2	4	1

Table 13.15 SHICLUST and NBCLUST – de Carvalho distance.

	Calinski and Harabasz	J-index	C-index	Γ-index	Beale test
Single linkage	2	2	2	2	2
Complete linkage	2	2	2	2	2
Centroid	2	2	2	2	2
Ward	2	2	2	2	2

Table 13.16 SCLUST and NBCLUST – L_1 distance.

	Calinski and Harabasz	C-index	Γ-index
SCLUST	4	2	2

Table 13.17 SHICLUST and NBCLUST – L_1 distance.

	Calinski and Harabasz	J-index	C-index	Γ-index	Beale test
Single linkage	3	3	2	2	3
Complete linkage	4	3	2	2	3
Centroid	4	3	2	2	3
Ward	4	3	2	2	3

Table 13.18 SCLUST and NBCLUST – L_2 distance.

	Calinski and Harabasz	C-index	Γ-index
SCLUST	2	2	2

Table 13.19 SHICLUST and NBCLUST – L_2 distance.

	Calinski and Harabasz	J-index	C-index	Γ-index	Beale test
Single linkage	2	2	2	2	2
Complete linkage	2	2	2	2	2
Centroid	2	2	2	2	2
Ward	2	2	2	2	2

The analysis of these tables and of the corresponding partitions leads to the following comments:

- The partition into two clusters is interesting. In almost all cases (except for SCLUST associated with the de Carvalho distance) the partition into two clusters is the same. The first cluster contains the London store, while all the other stores belong to the second cluster. If we analyse the results and examine the (zoom star) graphical representations (Noirhomme-Fraiture and Rouard, 2000; see also Chapter 7 of the present volume) associated with each of the symbolic variables (Figures 13.4–13.7), London distinguishes itself from the other stores by two variables: the type of 'sales promotion' (number) and the number of 'accessories' sold (family product).
- When a classification into more than two clusters is considered, in all cases a class is obtained with only two objects: Madrid and Rome. This can mainly be explained by the fact that these two stores sell mainly accessories (family product), and that these sales are equally distributed during the year (month).

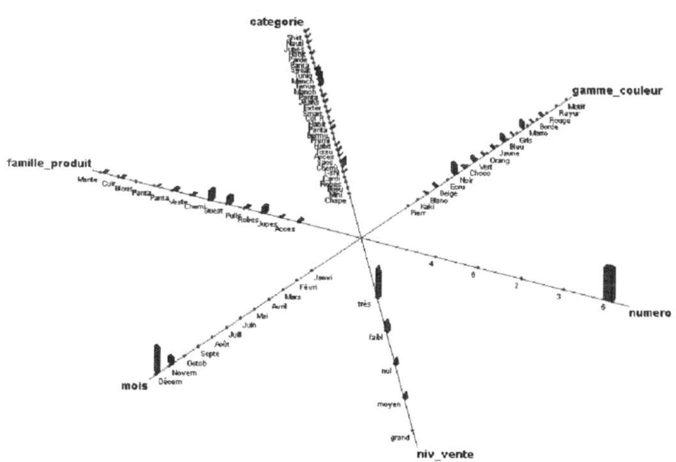

Figure 13.4 Zoom star for London.

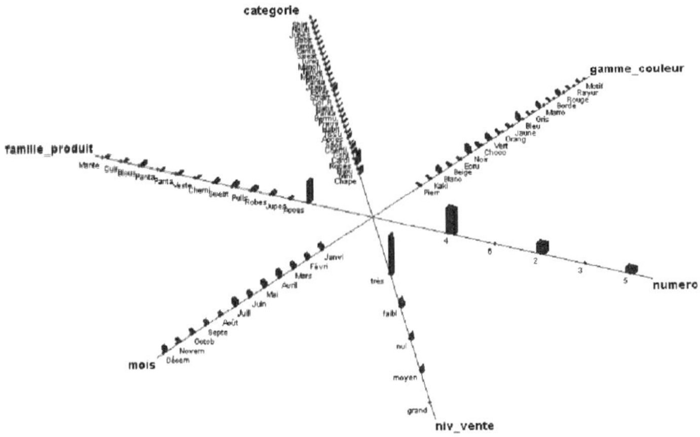

Figure 13.5 Zoom star for Madrid.

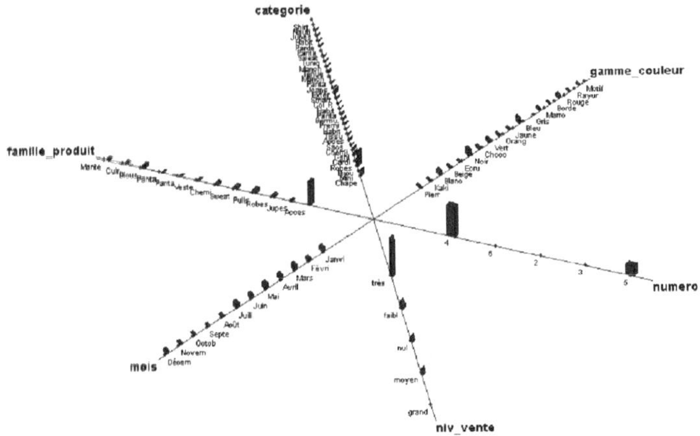

Figure 13.6 Zoom star for Rome.

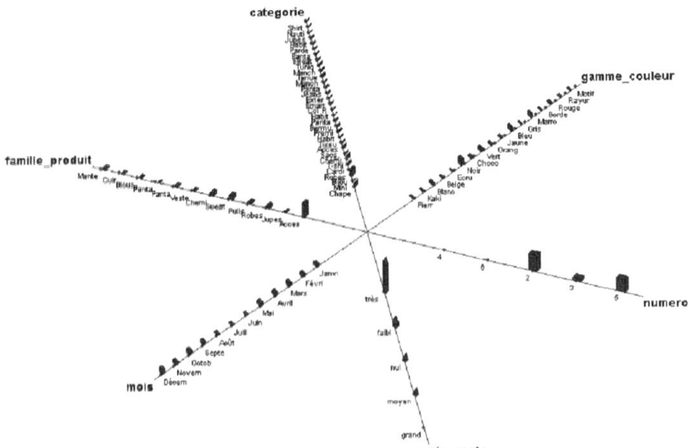

Figure 13.7 Zoom star for Paris.

13.11 Conclusion

In this chapter we have been concerned with the determination of the number of clusters for symbolic objects described by interval-valued, multi-valued and modal variables. In order to generate partitions we have chosen four symbolic clustering modules: SHICLUST, SCLUST, DIV and SCLASS. The determination of the best number of natural clusters has been undertaken, by proposing and using a symbolic extension of two hypothesis tests based on the homogeneous Poisson process and five classic number-of-clusters criteria well known in the scientific literature. These symbolic clustering procedures and methods for the determination of the number of clusters were applied to artificial and real data sets.

The methods for the determination of the number of clusters do not always detect the same structure. We therefore recommend using several methods or tests, carefully comparing and analysing the results obtained, in order to reach a conclusion as to the true structure of the data.

References

Baker, F.B. and Hubert, L.J. (1975) Measuring the power of hierarchical cluster analysis. *Journal of the American Statistical Association*, 70, 31–38.

Beale, E.M.L. (1969) Euclidean cluster analysis. *Bulletin of the International Statistical Institute*, 43, 92–94.

Bock, H.-H. and Diday, E. (2000) *Analysis of Symbolic Data.* Berlin: Springer-Verlag.

Brito, P. (2000) Hierarchical and pyramidal clustering with complete symbolic objects. In H.-H. Bock and E. Diday (eds), *Analysis of Symbolic Data Analysis*, pp. 312–323. Berlin: Springer-Verlag.

Brito, P. (2002) Hierarchical and pyramidal clustering for symbolic data. *Journal of the Japanese Society of Computational Statistics*, 231–244.

Calinski, T. and Harabasz, J. (1974) A dendrite method for cluster analysis. *Communications in Statistics*, 3, 1–27.

Celeux, G., Diday, E., Govaert, G., Lechevallier, Y. and Ralambondrainy, H. (1989) *Classification automatique des données.* Paris: Bordas.

Chavent, M. (1998) A monothetic clustering method. *Pattern Recognition Letters*, 19, 989–996.

Cox, D.R. and Isham, V. (1980) *Point Processes.* London: Chapman & Hall.

Diday, E. (1972) *Nouveaux concepts et nouvelles méthodes en classification* automatique. Thesis, Université Paris VI.

Diday, E. and J. Simon. (1976) Clustering analysis. In K.S. Fu (ed), *Digital Pattern Recognition*, pp. 47–94. Springer-Verlag, Berlin.

Duda, R.O. and Hart, P.E. (1973) *Pattern Classification and Scene Analysis.* Wiley, New York.

Gordon, A.D. (1998) How many clusters? An investigation of five procedures for detecting nested cluster structure. In C. Hayashi, N. Ohsumi, K. Yajima, Y. Tanaka, H.-H. Bock, and Y. Baba (eds), *Data Science, Classification, and Related Methods*, pp. 109–116. Springer-Verlag, Tokyo.

Hardy, A. (1983) Statistique et classification automatique – un modèle. un nouveau critère, des algorithmes, des applications. Doctoral thesis, University of Namur, Belgium.

Hardy, A. (1996) On the number of clusters. *Computational Statistics and Data Analysis*, 83–96.

Hardy, A. (2004) Les méthodes de classification et de détermination du nombre de classes: du classique au symbolique. In M. Chavent, O. Dordan, C. Lacomblez, M. Langlais, and B. Patouille (eds), *Comptes rendus des Onzièmes Rencontres de la Société Francophone de Classification*, pp. 48–55.

Hardy, A. and André, P. (1998) An investigation of nine procedures for detecting the structure in a data set. In A. Rizzi, M. Vichi and H.-H. Bock (eds), *Advances in Data Science and Classification*, pp. 29–36. Springer-Verlag, Berlin.

Hardy, A. and Lallemand, P. (2004) Clustering of symbolic objects described by multi-valued and modal variables. In D. Banks, L. House, F.R. McMorris, P. Arabie and W. Gaul (eds), *Classification, Clustering, and Data Mining Applications*, pp. 325–332. Springer-Verlag, Berlin.

Hardy, A. and Rasson, J.P. (1982) Une nouvelle approche des problèmes de classification automatique. *Statistique et Analyse des Données*, 7(2), 41–56.

Hardy, A., Lallemand, P. and Lechevallier, Y. (2002) La détermination du nombre de classes pour la méthode de classification symbolique SCLUST. In *Actes des Huitièmes Rencontres de la Société Francophone de Classification*, pp. 27–31.

Hubert, L.J. and Levin, J.R. (1976) A general statistical framework for assessing categorical clustering in free recall. *Psychological Bulletin*, 83, 1073–1080.

Ichino, M. (1988) General metrics for mixed features – the cartesian space theory for pattern recognition. In *Proceedings of the 1988 IEEE International Conference on Systems, Man and Cybernetics*, Volume 1, pp. 494–497. International Academic Publishers, Beijing.

Karr, A.F. (1991) *Point Processes and Their Statistical Inference*. Marcel Dekker, New York.

Kubushishi, T. (1996) On some applications of the point process theory in cluster analysis and pattern recognition. Doctoral thesis, University of Namur, Belgium.

Milligan, G.W. and Cooper, M.C. (1985) An examination of procedures for determining the number of clusters in a data set. *Psychometrika*, 50, 159–179.

Moore, M. (1984) On the estimation of a convex set. *Annals of Statistics*, 12(3), 1090–1099.

Noirhomme-Fraiture, M. and Rouard, M. (2000) Visualising and Editing Symbolic Objects. In H.-H. Bock, and E. Diday, (eds), *Analysis of Symbolic Data Analysis*, pp. 125–138. Springer-Verlag, Berlin.

Pirçon, J.Y. (2004) Le clustering et les processus de Poisson pour de nouvelles méthodes monothétiques. Doctoral thesis, University of Namur, Belgium.

Rasson, J.P. and Granville, V. (1996) Geometrical tools in classification. *Computational Statistics and Data Analysis*, 23, 105–123.

Rasson, J.P. and Kubushishi, T. (1994) The gap test: an optimal method for determining the number of natural classes in cluster analysis. In E. Diday, Y. Lechevallier, M. Schader, P. Bertrand, and B. Butschy (eds), *New Approaches in Classification and Data Analysis*, pp. 186–193. Springer-Verlag, Berlin.

Ripley, B.D. and Rasson, J.P. (1977) Finding the edge of a Poisson forest. *Journal of Applied Probability*, 14, 483–491.

Silverman, B.W. (1981) Using kernel density estimates to investigate multimodality. *Journal of the Royal Statistical Society B*, 43, 97–99.

Silverman, B.W. (1986) *Density Estimation for Statistics and Data Analysis*. Chapman & Hall, London.

Verde, R., de Carvalho, F. and Lechevallier, Y. (2000) A dynamical clustering algorithm for multi-nominal data. In H. Kiers, J.P. Rasson, P. Groenen and M. Schader, (eds), *Data Analysis, Classification, and Related Methods*, pp. 387–393. Springer-Verlag, Berlin.

14

Stability measures for assessing a partition and its clusters: application to symbolic data sets

Patrice Bertrand and Ghazi Bel Mufti

14.1 Introduction

Cluster validation can be carried out on the basis of the degree of accordance between the data set and the collection of clusters examined. This type of approach leads to the (sometimes implicit) assumption of a hypothesis on the characteristics of the clusters being examined. For example, the validation indices of Calinski and Harabasz (1974) and Krzanowski and Lai (1985) are both based on the adequacy criterion of minimal cluster inertia, so that they both tend to favour spherical clusters. For an overview of the validation methods that are based on the accordance between the data set and its partitioning, we refer the reader to Dubes and Jain (1979), Jain and Dubes (1988) and the introductory section in Milligan (1996). A different approach validates clusters on the basis of their stability after the impact of removing a few objects from the data set: see, for example, Levine and Domany (2001), Ben-Hur *et al.*, (2002), Tibshirani and Walther (2005) and Bertrand and Bel Mufti (2006). Stability-based methods of cluster validation that come within this approach, generally make it unnecessary to resort to any criteria of accordance between the data and the set of clusters. With the aim of assessing clusters and partitions obtained on symbolic data sets, we propose in this text an improvement of the stability-based method of validation that was introduced by Bertrand and Bel Mufti (2006).

In Section 14.2 we recall the definitions and some properties of the stability measures that were introduced by Bertrand and Bel Mufti (2006) for assessing partitions of classical numeric data sets. Then, in Section 14.3, we consider the discrimination between the case of

Symbolic Data Analysis and the SODAS Software Edited by E. Diday and M. Noirhomme-Fraiture

a data set that is structured into clusters that are well separated and whose cohesions are high, and the case of an homogeneous data set. One approach is to perform Monte Carlo tests that assess how likely the values taken by the stability indices are under a null model that specifies the absence of cluster structure. However, Monte Carlo simulations are frequently computationally expensive, particularly in the case of symbolic data sets. For this reason, we will introduce a different approach that aims to identify the objects that are intermediate between two or more clusters. In Section 14.4 several recommendations are given on how to interpret the numerical values taken by the stability measures. Finally, in Section 14.5, we illustrate our approach on two symbolic data sets introduced in previous chapters of this book: the first includes 59 Merovingian buckles described by multi-valued variables (see Chapter 13), and the second includes 60 meteorological stations in China that are described by interval-valued variables (see Chapter 11).

14.2 Stability measures

In this section we recall the definitions of the stability measures that were introduced and investigated in detail by Bertrand and Bel Mufti (2006) for the case of classical numerical data sets.[1] These stability measures assess either a cluster or a partition, with respect to either or both of the following criteria: cluster isolation and cluster cohesion. Informally speaking, the aim of each of these stability measures is to estimate the reproducibility of clustering results after removal of a few objects from the collection of (symbolic or numeric) objects to be partitioned. This motivation is justified because cluster stability is generally intended to hold if and only if membership of the clusters is not affected by small perturbations of the data set.

We also recall some structural properties, such as the property that each stability measure of a partition is a linear combination of the stability measures of its clusters. The proofs of these properties were established in Bertrand and Bel Mufti (2006).

We denote by E the collection of objects to be partitioned and by $\mathcal{P}$ a partition of E into k clusters that was obtained by applying some partitioning method $\mathcal{M}$ to E. We do not make any assumption either about the type of partitioning method $\mathcal{M}$ or about the type of description available for the objects in E, which therefore may be either numerical or symbolic.

14.2.1 Sampling the data set and partitioning the obtained samples

First, a large number N of samples, denoted by $E^{(1)}, \ldots, E^{(N)}$, are randomly selected from E, the value of N being selected a posteriori. More precisely, N is defined as the smallest value such that each stability measure is computed with a desired precision (the detailed description of the determination of N is delayed to the end of Section 14.2.2 in the case of cluster stability defined with the isolation criterion).

Since our aim is to measure cluster stability, the cluster structure of the samples $E^{(j)}$ $(j = 1, \ldots, N)$ should not depart significantly from the cluster structure $\mathcal{P}$ of E. This constraint leads us to draw each sample $E^{(j)}$ of E according to the so-called *proportionate* sampling scheme (see, for example, Hansen *et al.*, 1993):

---

[1] See also Bel Mufti (1998) for a preliminary version of some of these definitions.

1. Choose a sampling ratio, say p. Here the value of p must be large, since our aim is to estimate the stability of clusters after removal of *only a few* objects from the whole data set.

2. Select randomly, and without replacement, $[p \mid C_i \mid]$ objects[2] from each cluster C_i of $\mathcal{P}$, for $i = 1, \ldots, k$. Therefore, the sizes of the samples $E^{(1)}, \ldots, E^{(N)}$ are identical. Denoting this common size by n', it is clear that $n' = \sum_{i=1}^{k} [p \mid C_i \mid]$, and thus n' is approximately equal to $[pn]$.

Each sample $E^{(j)}$ $(j = 1, \ldots, N)$ is then partitioned into k clusters by applying the method $\mathcal{M}$. We will denote by $\mathcal{P}^{(1)}, \ldots, \mathcal{P}^{(N)}$ the k-partitions on samples $E^{(1)}, \ldots, E^{(N)}$ thus obtained. These partitions $\mathcal{P}^{(1)}, \ldots, \mathcal{P}^{(N)}$ are referred to as *reference partitions*.

14.2.2 Measuring the stability of a cluster with respect to the criterion of isolation

Consider an arbitrary cluster of partition $\mathcal{P}$, say C, and any reference partition, say $\mathcal{P}^{(j)}$. Let us first observe that $\mathcal{P}^{(j)}$ must contain some information about the stability of C, since $\mathcal{P}^{(j)}$ is obtained by applying $\mathcal{M}$ on the subset $E^{(j)}$ that is a type of perturbation of the initial data set E as previously noticed. Based on this remark, the following logical rule, denoted by (α), expresses the fact that cluster C is isolated from the rest of the data.

Rule (α): *isolation of cluster C.* Consider a sample $E^{(j)}$ together with its associated reference partition $\mathcal{P}^{(j)}$. If one, and only one, of two sampled objects belongs to C, then these two objects are not clustered together by the reference partition.

The reader may refer to Vaillant *et al.* (2004) for an exhaustive comparison of most of the indices proposed in the literature for assessing the quality of a (logical) rule. Based on this comparison, Loevinger's index H (see Loevinger, 1947) is a suitable choice for measuring the extent to which logical rule (α) is satisfied. Given any logical rule, say $a \to b$, where a and b denote two events, Loevinger's index H for assessing rule $a \to b$ is defined by $H(a \to b) = 1 - \Pr(a \cap \neg b)/\Pr(a)\Pr(\neg b)$. Loevinger's index for rule (α) is denoted hereafter by $t^{\text{is}}(C, j)$. It is easily proved that

$$t^{\text{is}}(C, j) = 1 - \frac{n'(n'-1) m_{(j;C,\overline{C})}}{2n'_C(n' - n'_C)\, m_{(j)}}, \tag{14.1}$$

where n' is the common size of the samples $E^{(j)}$ with $j = 1, \ldots, N$ (see Section 14.2.1), $m_{(j)}$ is the number of pairs of objects that are clustered together by $\mathcal{P}^{(j)}$, and $m_{(j;C,\overline{C})}$ is the number of pairs of sampled objects that are in the same cluster of $\mathcal{P}^{(j)}$ and for which exactly one of the two objects belongs to C. The extent to which rule (α) is satisfied is finally measured by the average value of all values $t^{\text{is}}(C, j)$ for $j = 1, \ldots, N$, and is denoted by $\overline{t}^{\text{is}}_N(C)$.

Let us define the random variable $T^{\text{is}}(C)$ as Loevinger's index of rule (α) when this rule is computed for a random sample, say $E^{(R)}$, instead of the (observed) sample $E^{(j)}$ that

---

[2] We denote by $[x]$ the integer part of any real number x.

was considered when defining $t^{\text{is}}(C, j)$. Using the central limit theorem, $\overline{t}_N^{\text{is}}(C)$ provides an (unbiased) estimation of the expected value of variable $T^{\text{is}}(C)$, and a standard 95% confidence interval for $E(T^{\text{is}}(C))$. The parameter N is then chosen large enough so that the length of this confidence interval is less than some desired threshold value. In any case, the value of N must be chosen greater than or equal to 100 in order to obtain a faithful estimation of the standard deviation of the statistic $T^{\text{is}}(C)$ that is computed.

14.2.3 Measuring the stability of a cluster with respect to the criterion of cohesion

The stability measure of cluster C, with respect to the criterion of cohesion, is defined using the same approach as for the previous case concerning the criterion of cluster isolation. Based on the information contained in an arbitrary reference partition, we then consider the following logical rule, denoted by (β), that asserts that cluster cohesion of C is high.

Rule ($\boldsymbol{\beta}$): *cohesion of cluster C.* Consider a sample $E^{(j)}$ together with its associated reference partition $\mathcal{P}^{(j)}$. If two (sampled) objects belong to cluster C, then these two objects are clustered together in partition $\mathcal{P}^{(j)}$.

Loevinger's index of rule (β), denoted by $t^{\text{co}}(C, j)$, can be expressed as

$$t^{\text{co}}(C, j) = 1 - \frac{n'(n'-1)m_{(\bar{j};\, C, C)}}{n'_C(n'_C - 1)m_{(\bar{j})}}, \tag{14.2}$$

where $m_{(\bar{j})}$ is the number of pairs of objects that are not clustered together by $\mathcal{P}^{(j)}$, and $m_{(\bar{j};\, C, C)}$ is the number of pairs of sampled objects that are not clustered together by $\mathcal{P}^{(j)}$ and belong to C. The extent to which rule (β) is satisfied is finally measured by the average value of all values $t^{\text{co}}(C, j)$ for $j = 1, \ldots, N$, and is denoted hereafter by $\overline{t}_N^{\text{co}}(C)$.

The choice of N is made following the procedure of determination of N that was described in the case of the stability measure based on the criterion of isolation. Let us define the random variable $T^{\text{co}}(C)$ as Loevinger's index of rule (β) when this rule is computed for a random sample, say $E^{(R)}$, instead of the (observed) sample $E^{(j)}$ that was considered when defining $t^{\text{co}}(C, j)$. Based on the central limit theorem, the unbiaised estimation $\overline{t}_N^{\text{co}}(C)$ of $E(T^{\text{co}}(C))$ provides a standard 95% confidence interval of $E(T^{\text{co}}(C))$. As for the case of cluster stability based on the criterion of isolation, the value of N is again chosen large enough so that this confidence interval has a length smaller than some desired threshold. If stability measures based on the criteria of isolation and cohesion are both computed, then N is defined as the smallest number of samples such that the length of each confidence interval of the two (expected) stability measures is less than the given desired threshold.

14.2.4 Measuring the stability of a cluster with respect to the criterion of global validity

When assessing the global validity of an individual cluster C, it is required that the stability measure evaluates the joint quality of the two rules (α) and (β), in order to assess the effects of sampling on both the isolation and the cohesion of cluster C. With this aim in mind, an extension of Loevinger's index was introduced in order to measure the overall degree of validity of the two rules (α) and (β); see Bertrand and Bel Mufti (2006) for more details.

The extension of Loevinger's index for these two rules leads to the following definition of the overall stability measure of cluster C, denoted hereafter by $t^{\text{va}}(C, j)$:

$$t^{\text{va}}(C,j) = \frac{2\,(n' - n'_C)\, m_{(j)}\, t^{\text{is}}(C,j) + (n'_C - 1)\, m_{(\bar{j})}\, t^{\text{co}}(C;\, j)}{2\,(n' - n'_C)\, m_{(j)} + (n'_C - 1)\, m_{(\bar{j})}}. \tag{14.3}$$

The measure of cluster validity of C is then defined as the average of values $t^{\text{va}}(C, j)$, $j = 1, \ldots, N$. This average is denoted by $\bar{t}_N^{\text{va}}(C)$. Unfortunately, the overall stability measure of a cluster cannot be expressed as some weighted mean of its two stability measures according to the criteria of isolation and cohesion. This is because of the well-known fact that the properties of isolation and cohesion are mutually dependent.

14.2.5 Stability measures of a partition

Two stability measures of a partition are defined using the same approach as in the previous definitions of the stability measures of an individual cluster, that is, by assessing a logical rule that expresses some property concerning either cluster isolation or cluster cohesion. For example, the stability measure of partition $\mathcal{P}$, with respect to the criterion of isolation, is defined by assessing the following rule:

Rule (γ): *isolation of partition* $\mathcal{P}$. Given any arbitrary reference partition $\mathcal{P}^{(j)}$, if two sampled objects are not clustered together by $\mathcal{P}$, then they are not clustered together in partition $\mathcal{P}^{(j)}$.

We denote by $t^{\text{is}}(\mathcal{P}, j)$ Loevinger's index as computed for assessing rule (γ) on the basis of the information provided only by the reference partition $\mathcal{P}^{(j)}$. It can be proved that

$$t^{\text{is}}(\mathcal{P}, j) = \sum_{C\in\mathcal{P}} \frac{n'_C(n' - n'_C)}{2m_{(\bar{\mathcal{P}})}}\, t^{\text{is}}(C, j). \tag{14.4}$$

Therefore, the estimation of the cluster isolation of partition $\mathcal{P}$ is defined as the average of the $t^{\text{is}}(\mathcal{P}, j)$, for $j = 1, \ldots, N$. This average, denoted by $\bar{t}_N^{\text{is}}(\mathcal{P})$, can be expressed as

$$\bar{t}_N^{\text{is}}(\mathcal{P}) = \sum_{C\in\mathcal{P}} \frac{n'_C(n' - n'_C)}{2m_{(\bar{\mathcal{P}})}}\, \bar{t}_N^{\text{is}}(C). \tag{14.5}$$

Concerning the criterion of cohesion, we follow the same type of approach. Denoting by $\bar{t}_N^{\text{co}}(\mathcal{P})$ the estimation of the overall cluster cohesion of partition $\mathcal{P}$, it has been proved that

$$\bar{t}_N^{\text{co}}(\mathcal{P}) = \sum_{C\in\mathcal{P}} \frac{n'_C(n'_C - 1)}{2\, m_{(\mathcal{P})}}\, \bar{t}_N^{\text{co}}(C). \tag{14.6}$$

Therefore, $\bar{t}_N^{\text{is}}(\mathcal{P})$ and $\bar{t}_N^{\text{co}}(\mathcal{P})$ can be seen as linear combinations of all the estimations of (single) cluster stability measures based on the criteria of isolation and cohesion, respectively.

Denoting by $\overline{t}_N^{\text{va}}(\mathcal{P}, j)$ the estimation of the cluster validity of partition $\mathcal{P}$, estimated on the basis of the information provided only by the partition $\mathcal{P}^{(j)}$, it has been proved that

$$t^{\text{va}}(\mathcal{P}, j) = \frac{m_{(j)}\, m_{(\overline{\mathcal{P}})}\, t^{\text{is}}(\mathcal{P}, j) + m_{(\bar{j})}\, m_{(\mathcal{P})}\, t^{\text{co}}(\mathcal{P}, j)}{m_{(j)}\, m_{(\overline{\mathcal{P}})} + m_{(\bar{j})}\, m_{(\mathcal{P})}}. \tag{14.7}$$

Finally, the overall cluster validity of partition $\mathcal{P}$ is estimated by the average, denoted by $\overline{t}_N^{\text{va}}(\mathcal{P})$, of all the $t^{\text{va}}(\mathcal{P}, j)$, $j = 1, \ldots, N$.

14.3 The case of a homogeneous data set

In the case of homogeneous data sets, the values taken by cluster stability measures can be arbitrarily high or low. For example, assume that 4-partitions are obtained by running the K-means algorithm on the following two types of data set:

(i) the objects of the data set are drawn uniformly inside a square in the Euclidean plane;

(ii) the objects of the data set are drawn uniformly inside a circle in the Euclidean plane.

It is clear that the stability measures obtained for the 4-partitions are much more stable in case (i) than they are in case (ii); see Bertrand and Bel Mufti (2006), where stability measures for case (i) were computed. In fact, for data sets in case (ii), each diameter of the circle defines a symmetry axis, which induces instability, whereas in case (i) two perpendicular axes define four square regions, each containing a corner of the square, so that the four square regions form 'natural' clusters that are stable when running the K-means algorithm.

In order to identify the case of a homogeneous data set, one can perform Monte Carlo tests that assess how likely the values taken by the stability indices are under a null model that specifies the absence of cluster structure. The level of statistical significance (with respect to the null hypothesis specified by the null model) of each previously defined stability measure can be determined, both for any cluster C and for the partition $\mathcal{P}$, by following the general three-step procedure:

1. Randomly generate data sets according to a data null model specifying that the data set is homogeneous.
2. Compute the distribution of the values of the stability index for clusters that are obtained by applying the partitioning method $\mathcal{M}$ to simulated data sets generated in step 1.
3. Compute the level of statistical significance of the observed value of the stability index, on the basis of the previous empirical distribution. If the level of significance is low, say less than 5%, then the hypothesis of a homogeneous data set is rejected.

It should be noted that the above procedure, for the determination of the levels of significance, is very similar to the cluster validation test proposed by Gordon (1994). In particular, the choice of an appropriate null model for data requires careful attention (see Gordon, 1996). The symbolic data null model used by our approach involves randomly and uniformly selecting symbolic objects within the convex hull of the examined symbolic data set E. The computation of the convex hull of a symbolic data set is not straightforward: it requires coding the symbolic objects examined as points in a Euclidean space. When the variables are classical quantitative variables, such a coding is clearly not needed. When the variables are interval-valued, a simple coding involves representing each interval as a point (x, y) where x is the centre of the interval and y its length. The other types of symbolic variables are not taken into account by our approach.

When the number of variables is too large (here, 'large' means greater than 10), an alternative of interest in order to avoid the exponential increase in computation due to the determination of convex hulls is to use the ellipsoidal model. The ellipsoidal null model involves randomly selecting the symbolic objects from a multivariate normal distribution whose variance–covariance matrix is the data set variance–covariance matrix. The model still requires the symbolic objects to be coded as points in a Euclidean space.

A rather different alternative for identifying the homogeneity of a data set is to determine the objects that are intermediate between two or several clusters. Compared with the two previous approaches, this alternative can be very useful particularly in explaining the degrees of stability of the clusters. The identification of homogeneous data sets should be less precise, though it should facilitate the detection of homogeneous subsets of the data set. Let us define more precisely this alternative approach. Denoting by w and w' two objects in E, and by $\mathcal{Q}$ a partition of E, we introduce the following notation:

$$J(w) = \{j \in \{1, \ldots, N\} : w \in E^{(j)}\},$$

$$J(w, w') = \{j \in \{1, \ldots, N\} : w, w' \in E^{(j)}\},$$

$$\mathcal{Q}^\star(w) = \{w'' \in E \setminus \{w\} : w \text{ and } w'' \text{ belong to the same cluster of } \mathcal{Q}\}.$$

We now borrow the notion of partial membership from the theory of soft classification (fuzzy clustering), though we endow it with a different meaning. For the object w, we define *partial membership* (or just *membership*) of the cluster C as the probability that, in a random sample of E, an object distinct from w and clustered with w belongs to $C \setminus \{w\}$; we denote this probability by $M(w, C)$. Intuitively, this probability measures the membership score of cluster C for the object w as the theoretical frequency with which w is clustered with an object of C, when considering the partition of some random sample of E.

Following Qiu and Joe (2005), we will say that an object is *vague* if none of its membership scores in a cluster is 1 (and 0 in the other clusters). A vague point will be said to be *intermediate* between clusters, say $C_1, \ldots C_r$, if its membership score in each of these r clusters is different from 0 and is 0 in each other cluster.

Each membership score $M(w, C)$ can be estimated by its corresponding empirical frequency, $\widehat{M}(w, C)$, defined by

$$\widehat{M}(w, C) = \frac{1}{|J(w)|} \sum_{j \in J(w)} \frac{|\mathcal{P}_j^\star(w) \cap C|}{|\mathcal{P}_j^\star(w)|}. \tag{14.8}$$

Clearly, for any arbitrary object in E, the sum of its (theoretical or estimated) membership scores in all the clusters of $\mathcal{P}$ is 1.

We also introduce the membership resemblance R defined as follows. The *membership resemblance* between two objects w and w', denoted by $R(w, w')$, is the probability that w

and w' are clustered together by the partition obtained on a random sample of E, assuming this random sample contains w and w'. $R(w, w')$ is estimated by the empirical probability $\widehat{R}(w, w')$, in other words, by the observed frequency of random samples for which w and w' are clustered together:

$$\widehat{R}(w, w') = \frac{|\{j \in J(w, w') : \mathcal{P}_j(w) = \mathcal{P}_j(w')\}|}{|J(w, w')|}. \tag{14.9}$$

Both the estimated score $\widehat{M}(w, C)$ and the estimated resemblance $\widehat{R}(w, w')$ should be computed for values of $|J(w)|$ and $|J(w, w')|$ sufficiently large to ensure that the lengths of the standard 95% confidence intervals for the expected values, $M(w, C)$ and $R(w, w')$, are less than some desired threshold.

Once the membership scores of each cluster are computed for each object, the vague points are identified, together with the pairs of clusters for which there exist intermediate objects. The membership resemblance index may then be computed for any two objects that are intermediate between two given clusters, say C_1 and C_2. The hypothesis that the subset $C_1 \cup C_2$ of the data set is homogeneous can be dismissed when there are only a few intermediate points between the two clusters C_1 and C_2, and/or when the membership resemblance of the intermediate points is high: see Section 14.5 for an illustration of this case.

14.4 The interpretation of the stability measures

In this section we will consider different ways of interpreting the stability measures of a partition. The stability measures estimate the inherent stability in each of the characteristics of isolation, cohesion and validity, for both the examined partition $\mathcal{P}$ and all its clusters.

Recall first that each stability measure evaluates some logical rule, say $A \Rightarrow B$, using Loevinger's index. This means that each stability measure can attain a maximum value of 1, and that it is not bounded below so that it can take, in theory, an arbitrary negative value. Otherwise, if $A \Rightarrow B$ denotes the logical rule that is assessed by some stability measure, then this stability measure takes the value 0 if and only if there is a perfect independence between the premise A and the conclusion B. However, this independence seldom occurs because the clusters are not chosen randomly, but obtained by using the same partitioning method $\mathcal{M}$ either for the partition examined or for the reference partitions generated on the samples. Since Loevinger's index increases with the degree to which the assessed rule is satisfied, the stability measures are very rarely negative.

It should be noted that the values of stability measures are *relative* measures. In particular, any value taken by a stability measure (different from 0 and 1) may indicate quite different degrees of stability depending on the values taken by different parameters concerning the data set or the partitioning method $\mathcal{M}$, such as the number n of objects, the dimension of the space that includes the data set, or the number k of clusters determined by the method $\mathcal{M}$. In other words, the numerical values taken by a stability measure only allow comparison of the degrees of stability of clusters having approximatively the same size and which were obtained by the same partitioning method applied to the same data set. Otherwise these values do not indicate which of the two clusters (or partitions) can be viewed as more stable: in this case, a real difference between two stability measures could occur without the presence of a real difference in the stability of the two clusters (or partitions) examined.

A second way to interpret the stability measures takes into account the levels of statistical significance (p-values) of the stability measures. The p-values are a by-product of the test of the null hypothesis of a homogeneous data set, and their definition was briefly given in the previous section. The reader may refer to Section 3.2 in Bertrand and Bel Mufti (2006) for a detailed description of the procedure of the test, in particular for the choice of the number of data sets to be simulated under the null model. The p-values are more intrinsic stability measures than the (raw) stability measures. They provide a kind of normalized measure of the departure from the case of a set of objects uniformly distributed in the convex hull of the data set examined. It may occur that the order between two values of a (raw) stability measure is reversed when considering the order between the two associated p-values. This case might happen when one stability measure is high only because of the specific shape of the data set: for example, if the shape of a homogeneous data set is a square, then a partition into four clusters will tend to be highly stable when running the method of K-means, but the associated p-value will not be significant for all that.

Note that the (raw) stability measures do not take account of the form of the input data required by the symbolic partitioning method $\mathcal{M}$, whereas computing the p-values requires, in practice, that the input data be a pattern matrix where each variable is either an interval-valued variable or a classical numerical variable. This restriction on the type of data results from the fact that computing the levels of significance requires use of a null data model based on an approximation of the minimal convex hull (or, failing that, the minimal ellipsoid) containing the data set, in order to provide a faithful and 'data-influenced' null data model.

Alternatively, we propose here a third way that is based on the computation of the membership scores between clusters. This approach does not require that the input data be of any specific type, and by analysing the impact of intermediate objects on the stability of clusters, the membership scores enable us to detect the case where two or more clusters are not well separated. This way of interpreting the values taken by stability measures will be discussed and illustrated in the rest of this chapter.

14.5 Application to real data sets

In this section we apply our approach in order to assess two partitions that are considered elsewhere in this book. First, we examine the 2-partition of the data set including 59 Merovingian buckles that are described by six multi-valued symbolic variables (see Chapter 13). Then we will assess the 5-partition of the meteorological stations data set characterized by 12 interval-valued variables (Chapter 11). Each partition was obtained by running the clustering method SCLUST introduced in Chapter 11. In accordance with our previous experience and previous reports on stability-based methods of cluster validation (Levine and Domany, 2001; Ben-Hur *et al.*, 2002), the value of the sampling ratio p should belong in the interval $[0.7, 0.9]$. In what follows, the stability measures for the partitions will be computed using the sampling ratio $p = 0.8$.

14.5.1 The Merovingian buckles data set

The symbolic data set which is investigated here includes 59 Merovingian buckles, that are described by six multi-valued symbolic variables (Table 14.1; see also Chapter 13). For example, the symbolic description of buckle no. 55 is:

$$s55 = [\text{Fixation} = \text{bronze bump}]$$
$$\wedge\ [\text{Damascening} = \text{dominant inlaid; silver monochrome}]$$
$$\wedge\ [\text{Contours} = \text{NULL}] \wedge [\text{Background} = \text{geometric frame}]$$
$$\wedge\ [\text{Inlaying} = \text{NULL}] \wedge [\text{Plate} = \text{squared back; plait}]$$

According to the NBCLUST validation method presented in Chapter 13, the number of clusters in a partition obtained by running the SCLUST method on these data is estimated to be 2. Here, our aim is then to determine the stability of such a 2-partition after sampling the data set. Table 14.2 indicates the number of sampled buckles that are drawn from each cluster, according to the procedure of proportionate sampling (with $p = 0.8$).

Recall that the number of sampled objects that are drawn from each cluster does not depend on the sample being considered: here, each sample contains 16 buckles from cluster 1 and 32 buckles from cluster 2.

Table 14.3 presents the stability measures for the 2-partition: each was computed by running SCLUST on 500 samples drawn according to the proportionate sampling procedure. When the criterion examined is either the isolation or the cohesion, the stability measure of the 2-partition can be viewed as a weighted average of the stability measures of the two individual clusters (see Section 14.2.5). The weights of this weighted average are indicated in parentheses in the corresponding two columns of Table 14.3. All the stability measures were computed with a precision[3] of at least 0.5%.

The 2-partition is clearly seen to be stable: cluster 2 is characterized by high stability values both in its isolation (0.990) and its cohesion (0.978), and consequently has a high global stability measure (0.985), while cluster 1 is even more stable than cluster 2, since its

Table 14.1 Symbolic descriptors of the Merovingian buckles.

Variables	Descriptors
Fixation	iron nail; bronze bump; none
Damascening	bichromate; predominant veneer; dominant inlaid; silver monochrome
Contours	undulations; repeating motives; geometric frieze
Background	silver plate; hatching; geometric frame
Inlaying	filiform; hatching banner; dotted banner; wide ribbon
Plate	arabesque; large size; squared back; animal pictures; plait; circular

Table 14.2 Number of sampled buckles in each assessed cluster.

Cluster	Size	Number of sampled buckles
1	20	16
2	39	32

[3] The precision of a stability measure (or any statistical estimation) is half of the length of the (standard) approximate 95% confidence interval computed for the expected value of the stability measure.

Table 14.3 Stability measures for the 2-partition.

		Isolation	Cohesion	Validity
Cluster	1	0.990 (0.500)	1 (0.195)	0.992
	2	0.990 (0.500)	0.978 (0.805)	0.985
Partition		0.990	0.982	0.986

Table 14.4 Membership scores for buckle no. 20.

Buckle (w)	Cluster	$\|J(w)\|$	Membership scores in clusters	
			1	2
no. 20	2	420	0.22	0.78

global stability is estimated at level 0.992, and its stability measure with respect to cohesion is 1. In addition to the fact that cluster 1 is characterized by a maximal stability measure with respect to cohesion, these stability values suggest that for some samples, a few elements of cluster 2 should be clustered together with all elements of cluster 1. Since the input space is not naturally embedded in some Euclidean space, the p-values of stability measures cannot be computed. However, membership scores enable us to gain more insight into the lack of cluster isolation. Indeed, it appears that there is only one buckle that is vague, namely buckle no. 20 which belongs to cluster 2. Table 14.4 presents the membership scores of buckle no. 20 in the two clusters (with a precision of at least 5%).

These scores confirm the suggestion mentioned above in order to explain the slight lack of isolation between the two clusters. Additionally, we removed buckle no. 20 and redid both the 2-partitioning of the data set and its assessment on the basis of stability measures. All stability values are then equal to 1. Therefore the only reason for the slight lack of isolation between the two clusters is the existence of the intermediate buckle no. 20.

14.5.2 The meteorological stations data set

The meteorological stations data set consists of the monthly temperatures observed in 60 meteorological stations in China during the year 1988. Each observation is coded in a table as the interval of the minima and maxima for each month. These observations were taken from the Long-Term Instrumental Climatic Data Base of the People's Republic of China. Here is the description of one of these 60 meteorological stations:

$$
\begin{aligned}
\text{ChangSha\_1988} = {} & [\text{January} = [2.7, 7.4]] \wedge [\text{February} = [3.1, 7.7]] \\
& \wedge [\text{March} = [6.5, 12.6]] \wedge [\text{April} = [12.9, 22.9]] \\
& \wedge [\text{May} = [19.2, 26.8]] \wedge [\text{June} = [21.9, 31]] \\
& \wedge [\text{July} = [25.7, 34.8]] \wedge [\text{August} = [24.4, 32]]
\end{aligned}
$$

$$\wedge\ [\text{September} = [20, 27]] \wedge [\text{October} = [15.3, 22.8]]$$
$$\wedge\ [\text{November} = [7.6, 19.6]] \wedge [\text{December} = [4.1, 13.3]]$$

In Chapter 11, this data set was clustered into five clusters in order to illustrate the SCLUST partitioning method. This 5-partition is in fact not the most valid according to both NBCLUST (see Chapter 13) and the stability measures indicated in Table 14.5. However, we will assess the 5-partition and each of its clusters, since this partition is quite appropriate for illustrating various possibilities of interpretation provided by stability measures. Stability measures will be computed on the basis of co-membership defined by running SCLUST on 500 samples of the data set. The common structure of these 500 samples is described by Table 14.6, which gives the numbers of sampled stations that are drawn in each (initial) cluster.

The stability measures for the 5-partition are given in Table 14.7. With the exception of cluster 1, for which the stability measures were computed with a precision of at least 1.7%, all the stability measures concerning this partition and its clusters were computed with a precision of at least 1%.

It turns out that cluster 5 is characterized by high stability values both in its isolation and its cohesion (0.999). The stability measure of the validity of cluster 2 is also high (0.977),

Table 14.5 Stability measures of the k-partitions when $k \in [2, 10]$.

k	2	3	4	5	6	7	8	9	10
Stability measure of partition validity	**0.99**	0.91	0.90	0.96	0.95	**0.99**	0.78	0.97	0.82

Table 14.6 Number of sampled stations in all (assessed) clusters.

Cluster	Size	Number of sampled stations
1	7	6
2	10	8
3	13	11
4	17	14
5	13	11

Table 14.7 Stability measures for the 5-partition.

		Isolation	Cohesion	Validity
Cluster	1	0.899 (0.135)	0.859 (0.061)	0.892
	2	0.969 (0.171)	1.000 (0.115)	0.977
	3	0.937 (0.219)	0.939 (0.225)	0.937
	4	0.979 (0.257)	0.955 (0.373)	0.969
	5	0.999 (0.219)	0.999 (0.225)	0.999
Partition		0.962	0.961	0.961

and this cluster is characterized by a perfect stability in its cohesion. Even if clusters 3 and 4 are less stable, since their stability values for the criterion of validity are 0.937 and 0.969, respectively, these two clusters can be deemed as stable. Cluster 1 is assessed by a quite low stability value, namely 0.892. The stability measures are higher than 0.96 both for the isolation and the cohesion of the 5-partition. Moreover, when considering the criterion of validity, the stability measure is equal to 0.961. Thus, the partition can be deemed to be stable.

Since the stations are described by 12 interval-valued variables, the computation of the p-values of the stability measures would require us to compute a convex hull in a 24-dimensional space in order to use the data-influenced null model. To avoid computations of such complexity, we will decide whether two or several clusters form a homogeneous subset of the data on the basis of the number of intermediate stations between these clusters. According to the membership scores of all the clusters computed for each station, it appears that there are 45 stations that are vague. Table 14.8 shows the distribution of the vague points by cluster. However, only 5 of these 45 vague stations have at least one membership score for which the (standard) 95% confidence interval has lower bound less than 0.95. In what follows, we will consider that these five stations are the only significant vague stations. They are stations no. 5, no. 13, no. 26, no. 44 and no. 48. Table 14.9 presents their membership scores in each cluster. The scores in this table were computed with a precision of at least 1% for stations no. 13 and no. 26, and with a precision of at least 8% for the other three stations no. 5, no. 44 and no. 48.

Notice that the memberships of cluster 1 for stations no. 5 and 44, are 0.53 and 0.33 respectively, while the memberships of cluster 3 for the same two stations are 0.46 and

Table 14.8 Number of vague stations in each cluster.

Cluster	Number of vague stations
1	7
2	10
3	13
4	2
5	13

Table 14.9 Membership scores for significant vague stations.

			Membership scores in clusters				
Station (w)	Cluster	$\|J(w)\|$	1	2	3	4	5
5	1	405	0.53	0	0.47	0	0
13	1	425	0.95	0	0.05	0	0
26	1	442	0.95	0	0.05	0	0
44	3	436	0.33	0	0.67	0	0
48	4	450	0	0.28	0	0.72	0

0.67, respectively. Thus both stations are intermediate between the two clusters 1 and 3. Moreover, the membership resemblance between these two stations is 0.77, which means that they were clustered together in the partitions obtained for 77% of the samples containing both station no. 5 and station no. 44. Stations no. 13 and no. 26 are also intermediate between cluster 1 and cluster 3, but they are of less interest because their membership scores in cluster 1 are both equal to 0.95 with a precision of at least 1%. These results confirm the lack of isolation and cohesion of cluster no. 1. It might also be emphasized that the presence of two intermediate stations between the two clusters 1 and 3, especially station no. 5, damages cluster 1 more than it does cluster 3. Nevertheless this does not mean that the union of clusters 1 and 3 forms a homogeneous subset of the data set, since the number of intermediate stations, namely 2, is not large enough when compared with the sizes of the two clusters, namely 7 and 13.

Station no. 48's membership scores in clusters 2 and 4 are 0.28 and 0.72, respectively. Therefore, station no. 48 can be considered as intermediate between clusters 2 and 4. The lack of isolation of cluster 2 can thus be explained by the presence of this intermediate station, taking into account that the size of cluster 2 is rather small. We may also notice that there is no (significant) intermediate station belonging to cluster 5, which confirms the high stability of this cluster.

Finally, the above detailed examination of the stability measures, together with the membership scores for all the stations, leads to the following conclusions:

- Cluster 5 is clearly stable.
- Cluster 2 has perfect cohesion, but its isolation is not perfect probably because of the intermediate station no. 48 that belongs to cluster 4.
- Cluster 4 is rather stable but both its isolation and its cohesion are perturbed by one of its stations (48) that is significantly intermediate between clusters 4 and 2.
- Clusters 3 and 1, especially cluster 1, are not very stable. This is mainly due to the presence of the two stations, no. 5 and no. 44, that are intermediate between the two clusters. In particular, station no. 5, which belongs to cluster 1, has membership scores close to 0.5 in the two clusters and this is likely to seriously damage the stability of cluster 1 both in its isolation and its cohesion.
- The 5-partition as a whole can be deemed to be rather stable. This assertion is confirmed by the values of the weights of the clusters in the decomposition of the partition stability, in particular by the two small weights of cluster 1, which is the least stable cluster.
- Due to the fact that there are only three stations that are really vague, it can be asserted that the data set is not homogeneous.

In order to reinforce the above conclusions, we redid both the partitioning of the data set into five clusters and its assessment by computing the stability measures, after removing the three intermediate stations no. 5, no. 44 and no. 48. Table 14.10 contains the stability measures related to the 5-partition obtained on the reduced data set.

The stability measures were computed after applying SCLUST to 500 samples of the reduced data set. With the exception of cluster 5 whose global validity is assessed 0.976, instead of 0.999 for the whole data set, all the stability measures are markedly larger than

Table 14.10 Stability measures for the 5-partition of the reduced data set.

		Isolation	Cohesion	Validity
Cluster	1	1.000 (0.121)	1.000 (0.047)	1.000
	2	1.000 (0.180)	0.999 (0.131)	1.000
	3	0.985 (0.228)	0.976 (0.257)	0.982
	4	0.985 (0.242)	0.992 (0.308)	0.987
	5	0.970 (0.228)	0.986 (0.257)	0.976
Partition		0.986	0.9898	0.987

those obtained on the whole data set. Moreover, it can be observed that the precision on each stability measure is improved – at least 0.6% instead of 1% for the whole data set. This confirms the intuition that the more stable the partition, the higher the precision of the stability measures is. It can be observed here that the new clusters (obtained for the reduced data set) are not necessarily the initial clusters without the removed intermediate stations. For example, new cluster 1 is exactly the initial cluster 1 without station no. 5, which is intermediate between clusters 1 and 3, but cluster 5 does not contain any intermediate stations, while new cluster 5 is enlarged with station no. 21 (initially assigned to cluster 4), and loses station no. 40 (now assigned to new cluster 3). The lack of stability of cluster 5 in the reduced data set is probably due to these unexpected changes: this is confirmed by the membership scores for the reduced data set that identify station no. 21 (respectively no. 40) as an intermediate station between new clusters 5 and 4 (respectively 5 and 3).

References

Bel Mufti, G. (1998) Validation d'une classe par estimation de sa stabilité. Doctoral thesis, Université Paris-Dauphine, Paris.

Ben-Hur, A., Elisseeff, A. and Guyon I. (2002) A stability based method for discovering structure in clustered data. In R.B. Altman (eds), *Pacific Symposium on Biocomputing, 2002*, pp. 6–17. Singapore: World Scientific.

Bertrand, P. and Bel Mufti, G. (2006) Loevinger's measures for assessing cluster stability. *Computational Statistics and Data Analysis*, 50(4): 992–1015.

Calinski, R.B. and Harabasz, J. (1974) A dendrite method for cluster analysis. *Communications in Statistics*, 3: 1–27.

Dubes, R. and Jain, A.K. (1979) Validity studies in clustering methodologies. *Pattern Recognition*, 11: 235–254.

Gordon, A.D. (1994) Identifying genuine clusters in a classification. *Computational Statistics and Data Analysis*, **18**: pp. 561–581.

Gordon, A.D. (1996) Null models in cluster validation. In W. Gaul and D. Pfeiffer (eds), *From Data to Knowledge: Theoretical and Practical Aspects of Classification, Data Analysis, and Knowledge Organization*, pp. 32–44. Berlin: Springer-Verlag.

Gordon, A.D. (1999) *Classification* (2nd edition). Boca Raton, FL: Chapman & Hall.

Hansen, M.H., Hurwitz, W.N. and Madow, W.G. (1993) *Sample Survey Methods and Theory*, Vol. 1: *Methods and Applications*. New York: John Wiley & Sons, Inc.

Jain, A.K. and Dubes, R. (1988) *Algorithms for Clustering Data*. Englewood Cliffs, NJ: Prentice Hall.

Krzanowski, W.J. and Lai, Y.T. (1985) A criterion for determing the number of groups in data set using sum of squares clustering. *Biometrics*, 44: 23–34.

Levine, E. and Domany, E. (2001) Resampling method for unsupervised estimation of cluster validity. *Neural Computation*, 13(11): 2573–2593.

Loevinger, J. (1947) A systemic approach to the construction and evaluation of tests of ability. *Psychological Monographs*, 61(4).

Milligan, G.W. (1996) Clustering validation: results and implications for applied analyses. In P. Arabie, L.J. Hubert, and G. De Soete (eds.), *Clustering and Classification*, pp. 341–375 River Edge, NJ: World Scientific.

Qiu, W. and Joe, H. (2006) Separation index and partial membership for clustering. *Computational Statistics and Data Analysis*, 50(3): 585–603.

Tibshirani, R. and Walther, G. (2005) Cluster validation by prediction strength. *Journal of Computational & Graphical Statistics*, 14(3): 511–528.

Vaillant, B., Lenca, P. and Lallich S. (2004) A clustering of interestingness measures. In E. Suzuki, and S. Arikawa, (eds), *Discovery Science*, pp. 290–297. Berlin: Springer-Verlag.

15

Principal component analysis of symbolic data described by intervals

N. Carlo Lauro, Rosanna Verde and Antonio Irpino

15.1 Introduction

Principal component analysis (PCA) aims to visualize, synthesize and compare units on factor spaces with the minimum loss of information (e.g. minimum distortion of the distance between original data). When such units are represented by points, all that is of interest is their position in space. However, when they are represented by *boxes* in a multidimensional space, as is the case with symbolic objects described by interval-valued variables, it is necessary to take into account not only their *location* but also their *size* and *shape*. That is to say, two points can only be differentiated by their location in space, but two boxes can also be differentiated by their size (the volume of the box) and shape (one box can be narrow or wide in one or more dimensions compared to another). Several approaches have been developed to take these three aspects into consideration when defining the similarity between boxes. These approaches start with different theoretical and methodological assumptions.

According to the symbolic data analysis (SDA) paradigm, considering the input, the technique of analysis and the output, we may have two families of analysis: symbolic (input)–classical (treatment)–symbolic (output) and symbolic–symbolic–symbolic. The former

Symbolic Data Analysis and the SODAS Software Edited by E. Diday and M. Noirhomme-Fraiture

family came first historically and is based on a symbolic input table, a suitable numerical coding of data, a treatment with some classical data analysis technique, and a suitable transformation of classical results into a symbolic description. To this family belong PCA on vertices (VPCA), center (CPCA) and symbolic objects (SPCA).

VPCA and CPCA (Cazes *et al.*, 1997; Chouakria *et al.*, 1998) involve a two-step analysis based first on the numerical coding of the vertices of a box or their centre and then performing a classical PCA on these coded data.

SPCA (Lauro and Palumbo, 2000), implemented in the SODAS software, stresses the fact that a box is a cohesive set of vertices that depends also on its size and shape, introducing a more consistent way to treat units as a complex data representation by introducing suitable constraints for the vertices belonging to the same object. This approach overcomes the drawback of the previous approach, where vertices are treated as single independent units described by points. Both approaches represent boxes on factor planes as rectangles of minimum area enclosing the projections of vertices for each box. Such rectangles are also interpreted as symbolic objects.

In order to avoid loss of information due to the data transformation, more recently, the interval algebra introduced by Moore (1966) is considered for a different approach to the PCA on boxes.

Among the interval algebra theorems for the computation of interval data functions, one has been emphasized for the treatment of such kinds of data: 'If a system of equations has a symmetric coefficient matrix, then the midpoints of the interval solutions are equal to the solution of the interval midpoints' (Lawson and Hanson, 1974). This theorem permitted the development of new analysis methods not based merely on the representation of intervals by means of their extreme points (the vertices of the boxes), but based on coding the intervals by their centres or *midpoints* and *radii*. In this direction, an intermediate family of analyses have been developed. These take a symbolic table as input, classical techniques are extended to take into account some interval algebra theorems or definitions, and the output is reconstructed symbolic data according to the same theorems of interval algebra. This family can be considered as a *hybrid* approach as the treatement step is neither fully classical nor fully based on interval algebra.

We refer in particular to the methods called midpoints radii principal component analysis (MRPCA: Palumbo and Lauro, 2003), where classic linear algebra techniques are used to treat intervals coded as a pair (midpoint, radius). This is a hybrid approach in the sense that it takes into consideration some theorems of interval algebra but uses a classic linear algebra algorithm to analyse data by simply reconstructing boxes ex post on the factor planes. Within this approach is a new method called 'spaghetti' PCA (Irpino, 2006). This technique starts from a global representation of a box by means of its main diagonal, expressed as a function, and then performs a PCA of the associated correlation matrix.

In order to accomplish the symbolic–symbolic–symbolic paradigm of analysis Gioia and Lauro proposed an approach developed using interval linear algebra called IPCA (Gioia and Lauro, 2006). This approach is fully consistent with the interval nature of the descriptors of boxes, and performs a PCA of an interval correlation matrix allowing interval eigenvalues and eigenvectors with interval components.

The aim of this chapter is to review these different results, with particular reference to those methods implemented in SODAS.

15.2 Principal component analysis of interval data matrices: the input

In the SDA framework, various authors have proposed approaches designed to extend factorial data analysis techniques so as to study the relationships between symbolic objects (SOs) in a reduced subspace.

Statistical units described by interval variables can be considered as special cases of symbolic data, in which only numerical (interval and single-value) variables are considered. Moreover, the SDA approach to interval data treatment offers many useful tools that can be helpful in interpreting the results. For these reasons our approach to interval data representation is presented by adopting the notation and definitions of SDA. Let Ω be a set of SOs ω_i $(1 \leq i \leq n)$ described by p variables or descriptors $Y = \{y_1, y_2, \ldots, y_p\}$ with domain in $(D_1, D_2, \ldots, D_p)$.

Nowadays, SDA is based either on numerical treatments of suitably coded SOs followed by symbolic interpretations of results, or on symbolic methods that directly process the symbolic descriptors. In the following we use the first framework to analyse SOs described only by quantitative interval variables. Henceforth, an observed value of a variable y_j no longer represents a single measure, as in the classic data analysis, but refers to the lower $\underline{y}_j$ and the upper $\overline{y}_j$ bounds of an interval.

To process, by numerical methods, SOs described by interval variables, they are recoded by combining min and max values into the so-called vertices data matrix. In the simple case of $p = 2$, the description of the generic SO ω_i is associated with the ith row of the interval data matrix $\mathbf{Y}$:

$$\mathbf{Y} = \begin{bmatrix} \vdots & \vdots \\ y_{i1} & y_{i2} \\ \vdots & \vdots \end{bmatrix} = \begin{bmatrix} \vdots & \vdots \\ [\underline{y}_{i1}, \overline{y}_{i1}] & [\underline{y}_{i2}, \overline{y}_{i2}] \\ \vdots & \vdots \end{bmatrix}. \tag{15.1}$$

The related vertices coordinates with respect to the new variables z_1 and z_2 – having the same domains as y_1 and y_2, respectively – correspond to the rows of the matrix Z_i:

$$Z_i = \begin{bmatrix} \underline{y}_{i1} & \underline{y}_{i2} \\ \underline{y}_{i1} & \overline{y}_{i2} \\ \overline{y}_{i1} & \underline{y}_{i2} \\ \overline{y}_{i1} & \overline{y}_{i2} \end{bmatrix} \begin{matrix} \\ \\ \\ \end{matrix} \quad (\text{columns } z_1, z_2). \tag{15.2}$$

In a geometric view (Figure 15.1) the generic SO ω_i is represented by a rectangle, having $2^2 = 4$ vertices corresponding to all possible (min, max) combinations.

In the general case of p variables, each coding matrix Z_i is constituted by 2^p rows and p columns. The coding matrix $\mathbf{Z}$ is obtained by stacking the n coding matrices Z_i (with $1 \leq i \leq n$). $\mathbf{Z}$ has $N = n \times 2^p$ rows and p columns, and represents the numerical coding of the n SOs. Without loss of generality, we can assume that the z_j are standardized.

Other methods start from a different codification of the interval data matrix. The interval data matrix is transformed into two matrices containing the centres and the radii of intervals.

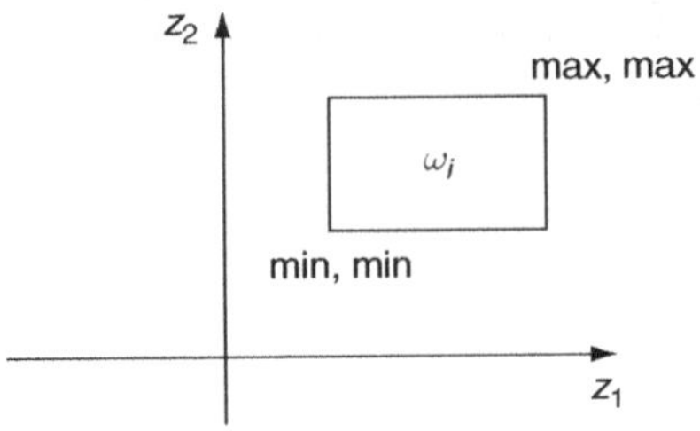

Figure 15.1 SO graphical representation in two dimensions.

Considering a generic interval $y_{ij} \equiv \left[\underline{y}_{ij}, \overline{y}_{ij}\right]$, it is possible to rewrite it as an ordered couple $y_{ij} \equiv \left(m_{ij}, r_{ij}\right)$ where $m_{ij} = (\underline{y}_{ij} + \overline{y}_{ij})/2$ represents its midpoint and $r_{ij} = (\overline{y}_{ij} - \underline{y}_{ij})/2$ represents its half-length or radius:

$$\mathbf{Y} \equiv (\mathbf{M}, \mathbf{R}) \equiv \left(\left[m_{ij}\right], \left[r_{ij}\right]\right) \qquad \text{for } i = 1, \ldots, n \text{ and } j = 1, \ldots, p. \tag{15.3}$$

In the following we list the most interesting approaches considering also the decomposition criterion underlying the single PCA.

15.3 Symbolic–numerical–symbolic PCAs

15.3.1 The first approaches to the PCA of interval data: VPCA and CPCA

Cazes *et al.* (1997) and Chouakria *et al.* (1998) developed an extension of PCA to SOs described by interval variables. They named their method vertices principal component analysis (VPCA). Specifically, this approach, based on a geometric representation of the SOs as hypercubes, is a classical PCA on a suitable numerical coding of their vertices considered as statistical units. VPCA looks for the best representation of the SOs on a factor plane, by optimizing the total variance of all the vertices of the hypercubes.

The technique works by decomposing the correlation matrix associated with the vertices of hyper-rectangles expressed in terms of lower and upper bounds. Let p be the interval variables and n the number of observations. The analysis is performed on the matrix coding the vertices that has $2pn$ rows and p columns.

In the original proposal, VPCA consists of performing a classic PCA on the standardized $\mathbf{Z}$ matrix. In this way, vertices are elements of the subspace $\mathbb{R}^p$, whereas the p quantitative descriptors are elements of $\mathbb{R}^N$. VPCA looks for a suitable subspace to represent SOs and, from a dual point of view, to represent the p variables. As in classical PCA the optimal subspace is here spanned by the axes $\mathbf{v}_m$ (with $1 \leq m \leq p$), maximizing the sum of squares of projected vertex coordinates $\boldsymbol{\psi}_m = \mathbf{Z}\mathbf{v}_m$:

$$\boldsymbol{\psi}'_m \boldsymbol{\psi}_m = \mathbf{v}'_m \mathbf{Z}'\mathbf{Z}\mathbf{v}_m, \tag{15.4}$$

where $\mathbf{v}'_m \mathbf{v}_{m'} = 0$ for $m \neq m'$ and $\mathbf{v}'_m \mathbf{v}_{m'} = 1$ for $m = m'$. Therefore, the characteristic equation of VPCA in $\mathbb{R}^N$ is given by:

$$\frac{1}{N} \mathbf{Z}'\mathbf{Z}\mathbf{v}_m = \lambda_m \mathbf{v}_m, \qquad 1 \leq m \leq p, \tag{15.5}$$

where $\mathbf{v}_m$ and λ_m are the generic eigenvector and the generic eigenvalue, respectively, associated with the matrix $N^{-1}\mathbf{Z}'\mathbf{Z}$.

Performing the analysis in $\mathbb{R}^p$, we define the equation

$$\frac{1}{N} \mathbf{Z}\mathbf{Z}'\mathbf{w}_m = \lambda_m \mathbf{w}_m, \qquad 1 \leq m \leq p, \tag{15.6}$$

which has the same non-zero eigenvalues as (15.2), but different eigenvectors, obeying the relation $\mathbf{v}_m = \lambda_m^{-1/2} \mathbf{Z}'\mathbf{w}_m$. In VPCA, the principal axes can be interpreted in terms of the variables z_j having maximal contributions (Lebart *et al.*, 1995).

CPCA decomposes the correlation matrix of the centres of intervals and then projects on the factor axes the vertices as supplementary points. Let p be the number of interval variables and n the number of observations. The analysis is performed on the matrix coding the centres that has n rows and p columns.

The characteristic equation of CPCA in $\mathbb{R}^n$ is given by

$$\frac{1}{n} \tilde{\mathbf{M}}'\tilde{\mathbf{M}}\mathbf{v}_m = \lambda_m \mathbf{v}_m, \qquad 1 \leq m \leq p, \tag{15.7}$$

where $\mathbf{v}_m$ and λ_m are the generic eigenvector and the generic eigenvalue, respectively, associated with the matrix $\tilde{\mathbf{M}}'\tilde{\mathbf{M}}$ where $\tilde{\mathbf{M}}$ is the matrix $\mathbf{M}$ standardized by the standard deviation of the centres.

15.3.2 SPCA

In VPCA vertices are treated as independent statistical units. In this case several pieces of information about boxes, such as their size and shape, are lost. In order to overcome this drawback, Lauro and Palumbo (2000) proposed a new version of PCA on vertices. They first introduced a cohesion constraint in the codification of boxes by their vertices, emphasizing the membership of vertices belonging to the boxes. The technique they developed allows the maximization of the inter-object variance being, in this sense, similar to CPCA.

The method is based on the maximization of the *between* (SOs) variance matrix:

$$\frac{1}{N} \mathbf{Z}'\mathbf{A}\,(\mathbf{A}'\mathbf{A})^{-1}\,\mathbf{A}'\mathbf{Z}, \tag{15.8}$$

where $\mathbf{A}$ is a Boolean matrix ($N \times n$) describing the membership of the N vertices in the n SO representations. If all SOs have the same number of vertices, for example N/n, $\mathbf{A}'\mathbf{A} = (N/n)\mathbf{I}_n$. If SO descriptions are constrained by hierarchical or logical dependencies, or when some interval descriptions are thin, $\mathbf{A}'\mathbf{A}$ is a diagonal matrix with the ith generic element of the main diagonal equal to the number of vertices of the ith SO.

The axes of maximum inertia are obtained in $\mathbb{R}^N$ as solutions to the following characteristic equation:

$$\frac{1}{N} [\mathbf{Z}'\mathbf{A}(\mathbf{A}'\mathbf{A})^{-1}\mathbf{A}'\mathbf{Z}]\tilde{\mathbf{v}}_m = \frac{1}{N} \mathbf{Z}'\mathbf{P}_{\mathbf{A}}\mathbf{Z} = \tilde{\lambda}_m \tilde{\mathbf{v}}_m \tag{15.9}$$

where $\tilde{\lambda}_m$ are the eigenvalues of the matrix $\mathbf{Z}'\mathbf{P}_A\mathbf{Z}$ and $\mathbf{V}_m$ are the associated eigenvectors (for $m = 1, \ldots, M$; with M the maximum number of non-null eigenvalues), defined under the orthonormality constraints already expressed in (15.1). Since

$$\mathbf{P}_A = \mathbf{A}(\mathbf{A}'\mathbf{A})^{-1}\mathbf{A}' \tag{15.10}$$

is an orthogonal projector matrix, this approach can be seen as a particular case of the so-called PCA with respect to a reference subspace (D'Ambra and Lauro, 1982). We recall that this consists of a PCA of variables projected onto a subspace, here spanned by the columns of **A**.

The vertex coordinates, of the generic hypercube associated with an SO ω_i, on the axis m, are given by the vector $\tilde{\psi}_{i,m} = Z_i\tilde{\mathbf{v}}_m$.

The analysis in $\mathbb{R}^p$ consists of solving the characteristic equation

$$(\mathbf{A}'\mathbf{A})^{-1/2}(\mathbf{A}'\mathbf{Z}\mathbf{Z}'\mathbf{A})(\mathbf{A}'\mathbf{A})^{-1/2}\tilde{\mathbf{w}}_m = \tilde{\lambda}_m\,\tilde{\mathbf{w}}_m, \tag{15.11}$$

where $\tilde{w}_m = (A'A)^{-1/2}A'Z\tilde{v}_m$ except for a constant $1/N$.

The contributions of the variables are defined as in VPCA, whereas it has been proposed to evaluate the representation of SOs considering all the vertices of each SO and representing only SOs having high values of relative contribution (computed as the mean of the relative contributions of the vertices belonging to the SO).

The SO images – as in VPCA – are made by *maximum covering area rectangles*. Nevertheless, in the proposed approach, the constraint on the vertices allows, according to the optimized criterion, the area of the SOs representation to be reduced.

The previous treatment of interval data is an extension of the so-called multiple PCA (Escofier and Pagès, 1988) where the data in the input are points (the vertices or centres).

Alternatively, the partial PCA can be used to better stress the differences among SOs. Later, a partial PCA where hypercube vertices are centred with respect to their min value is shown.

Comparing the SPCA with the CPCA proposed by Cazes *et al.* (1997), where a PCA is performed only on the centres of the interval descriptions standardized by the correlation matrix of the centres of variables, several minor improvements have been introduced.[1]

In order to take into consideration only the *sizes* and the *shapes* of the boxes associated with a multi-valued interval data description, Lauro and Palumbo (2000) proposed a PCA on the range transformation of data (RTPCA).

In the classical PCA, each statistical unit is represented by a point, while in SDA we have to cope with SOs' shape and size. In order to consider the latter structural elements, we use the *range transformation* $V_j^i = \left[\overline{y}_{ij} - \underline{y}_{ij}\right]$, which reveals useful information for studying *size* and *shape*. In SDA an important role is played by the *description potential* (DP) measure, which is the *hypervolume* associated with an SO, computed as the Cartesian

[1] First of all, data are standardized by the standard deviation of the vertices which are all considered to be active units for the analysis, while in PCA on centres they are considered as supplementary units. More generally, it can be considered to be an improvement of CPCA as it can also be performed when data descriptions are constrained by logical or hierarchical rules.

product $V_1^i \times \cdots \times V_j^i \times \cdots \times V_p^i$ of the p descriptors associated with the SO ω_i, $1 \leq i \leq N$ (de Carvalho, 1992, 1997).

When interval descriptors need to be normalized, the DP of a SO a_i is formally defined as:

$$\pi(a_i) = \prod_{j=1}^{p} \mu(\overline{V}_j^i), \tag{15.12}$$

where $\mu(\overline{V}_j^i)$ is the normalized range with respect to the domain $D_j = \bar{V}_j^i$ of the interval descriptor y_j: O_j. As the DP measure tends to zero if at least one $\bar{V}_j$ is close to zero, we prefer to use the following alternative measure, called the *linear description potential* (LDP). The LDP of ω_i is defined by de Carvalho (1997) as

$$\sigma(a_i) = \sum_{j=1}^{p} (\bar{V}_j^i). \tag{15.13}$$

Let $\omega_1, \ldots, \omega_n$ be a set of SOs described by p interval descriptors, and $\mathbf{X}$ the $(n \times p)$ matrix with generic term $x_{ij} = \sqrt{\overline{V}_j^i}$.

The method proposed in this section performs a factor decomposition of the quantity $\sum_{i=1}^{n} \sigma(a_i)$ allowing a graphical representation which, differently from VPCA, shows the descriptors' influence on the total LDP.

From a geometric point of view, the *range transformation* implies an affine translation of each object, so that the vertices min $= \{\underline{y}_{i1}, \ldots, \underline{y}_{ij}, \ldots, \underline{y}_{ip}\}$ are reported in the origin (see Figure 15.2). It is easy to see that, given the orthogonality relationship between pairs of sides of each hypercube, the search for a suitable subspace in order to visualize the size and shape of each SO can be simply realized on the n max vertices as a PCA with respect to the origin (non-centred PCA): $\overline{y}_{i1}, \ldots, \overline{y}_{ij}, \ldots, \overline{y}_{ip}$.[2] We refer to this approach as *principal component analysis on the range transformation* (RTPCA) of interval variables. The cohesion of vertices is respected, and the hypercube can be easily visualized by projecting all the other vertices, that have not concurred in the analysis, as supplementary points.

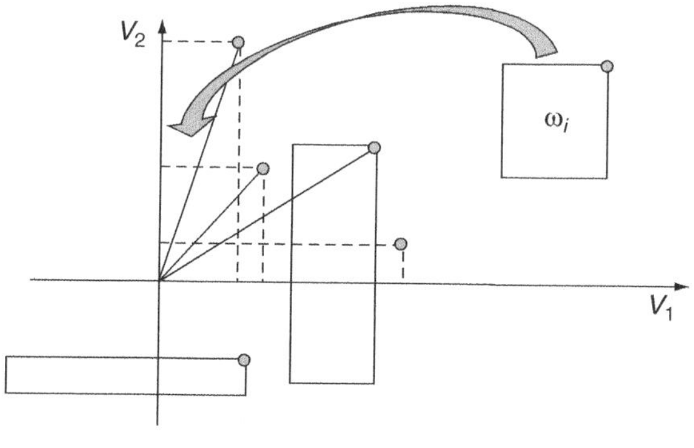

Figure 15.2 SO transposition to the origin.

[2] Notice that, in this way the curse of dimensionality, which affects both VPCA and SPCA, can be overcome. The total number of points is reduced from $n \times 2^p$ to n.

The PCA performed on the matrix **X** decomposes the LDP criterion,

$$\left(\mathrm{tr}(\mathbf{X}\mathbf{X}') = \mathrm{tr}(\mathbf{X}'\mathbf{X}) = \sum_i \sigma(a_i)\right), \tag{15.14}$$

according to the following characteristic equation:

$$\mathbf{X}'\mathbf{X}\mathbf{t}_m = \mu_m \mathbf{t}_m \qquad 1 \leq m \leq p, \tag{15.15}$$

or equivalently

$$\mathbf{X}\mathbf{X}'\mathbf{u}_m = \mu_m \mathbf{u}_m, \qquad 1 \leq m \leq p, \tag{15.16}$$

where μ_m is the mth eigenvalue ($\sum_m \mu_m = \sum_i \sigma(a_i)$) and $\mathbf{u}_m$ and $\mathbf{t}_m$ are the associated eigenvectors in $\mathbb{R}^p$ and $\mathbb{R}^n$ respectively. Both analyses are defined under the usual orthonormality constraints.

The SO ω_i representation in the optimal subspace $m^* < p$ can be obtained by the matrix $\mathbf{\Phi}$, whose elements are obtained as the juxtaposition of the first m^* ($1 \leq m^* \leq p$) axes:

$$\mathbf{\Phi} = [\phi_1 \cdots \phi_m \cdots \phi_{m^*}] \tag{15.17}$$

with $\phi_m = \mathbf{X}\mathbf{t}_m$.

In RTPCA the amount of *information* contribution associated with each axis is given by the corresponding eigenvalue μ_m. The ratio between the squared coordinate and the eigenvector, both with respect to the axis m,

$$\mathrm{CTA}_{im} = \frac{\phi_{im}^2}{\mu_m}, \tag{15.18}$$

is a measure of the contribution of the SO ω_i to principal axis m. The relative contribution, giving the quality of the SO representation, is measured as

$$\mathrm{CTR}_{im} = \frac{\sum_i \phi_{im}^2}{\sum_j x_{ij}^2}, \qquad \text{with } m = \{1, \ldots, m^*\}.$$

Because the matrix **X** has all positive entries, the eigenvector $\mathbf{u}_1$ and the factor $\mathbf{t}_1$ also have all positive values. Thus the first axis is easily interpreted in terms of SO *size factors*, while the following ones discriminate SOs according to their *shape* features. Their interpretation depends on the contribution (squared coordinates) of the original variables to the axis: $\mathrm{CTA}_{jm} = \mathbf{t}_{jm}^2$.

It is worth noting that in RTPCA, SOs can be more simply represented by single points (max vertex coordinates). Therefore, close points refer to SOs whose LDPs are mainly influenced by the same variables. In other words, the closeness means that hypercubes, which represent SOs, are quite similar for shape and size.[3]

[3] A similar approach to the analysis of the length of intervals was proposed by Cazes (2002).

VPCA (Lauro and Palumbo, 2000) allows the evaluation of SOs with respect to their positioning in the space of the recoded descriptors. On the other hand, RTPCA analyses interval data emphasizing SOs *size* and *shape*. Lauro and Palumbo (2000) presented a mixed strategy aiming to combine VPCA and RTPCA to improve SOs representation when their differences in terms of scale and structural (*size* and *shape*) are taken into account, defining the so-called *principal components analysis on symbolic objects (SPCA)*. They outlined the following three-step approach:

1. Perform the RTPCA of $\mathbf{Y}$ in order to extract the principal axes that better represent the size and shape of SOs.
2. Transform $\mathbf{Z}$ into $\hat{\mathbf{Z}} = \mathbf{P}_\mathbf{A}\mathbf{Z}$, allowing SO vertex cohesion to be taken into account.
3. Perform a PCA of the projections of the rows of $\hat{\mathbf{Z}}$ on $\mathbf{\Phi}$ by the projection matrix $\mathbf{P}_\Phi = \mathbf{\Phi}(\mathbf{\Phi}'\mathbf{\Phi})^{-1}\mathbf{\Phi}'$, in order to stress the *size* and *shape* information.

This approach is based on the solution of the characteristic equation

$$\hat{\mathbf{Z}}'\mathbf{P}_\Phi\hat{\mathbf{Z}} = \mathbf{Z}'\mathbf{A}(\mathbf{A}'\mathbf{A})^{-1/2}\mathbf{P}_\Phi(\mathbf{A}'\mathbf{A})^{-1/2}\mathbf{A}'\mathbf{Z}\mathbf{s}_m = \rho_m\mathbf{s}_m, \tag{15.19}$$

where the diagonal matrix $(\mathbf{A}'\mathbf{A})^{-1}$ has been decomposed into $(\mathbf{A}'\mathbf{A})^{-1/2}(\mathbf{A}'\mathbf{A})^{-1/2}$ in order to ensure symmetry, and $\mathbf{s}_m$ and ρ_m are the mth eigenvector and eigenvalue. The $m = 1, \ldots, M$ eigenvectors are calculated under the orthonormality constraints $\mathbf{s}'_m\mathbf{s}_{m'} = 0$ $(m \neq m')$ and $\mathbf{s}'_m\mathbf{s}_m = 1$.

The interpretation of results depends on the choice of $\mathbf{P}_\Phi$. In fact, considering $\mathbf{P}_\Phi$ as a weighting system related to the *size* and the *shape* of the SOs, the projection matrix allows different aspects of the set of SOs to be emphasized. The generic diagonal term of $\mathbf{P}_\Phi$ is equal to the quantity in (15.16):

$$\phi_i(\phi'_i\phi_i)^{-1}\phi'_i = \frac{\sum_m \phi^2_{i,m}}{\mu_m}. \tag{15.20}$$

In order to show the size of SOs we shall include the first principal component ϕ_1 in the definition of $\mathbf{P}_\Phi$; alternatively, the shape aspects could be emphasized by dropping the first principal component.

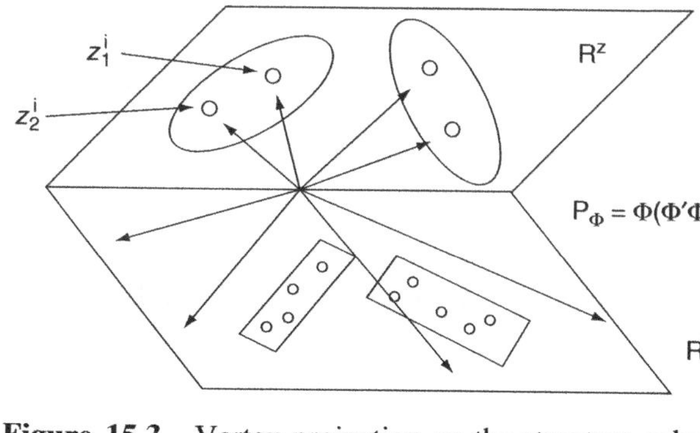

Figure 15.3 Vertex projection on the structure subspace.

Figure 15.3 shows an intuitive schematic of the procedure. The top of the figure represents vertices in $\mathbb{R}^{\mathbf{Z}}$ space, and the two ellipses refer to two different SOs. By means of $\mathbf{P}_{\Phi}$, these vertices are projected into $\mathbb{R}^{\mathbf{T}}$, and then VPCA is performed.

15.4 Interval algebra based methods

15.4.1 The hybrid approach

Moore's (1966) interval algebra, based on the theory of *hybrid numbers*, defines a set of operations that can be performed on intervals (sum, difference, multiplication, etc.) on the basis of the following theorem (Moore, 1966): if *f(x)* is a continuous function where each variable $x_i, i = 1, \ldots, n$, is present no more than once, then $f([x]_1, \ldots, [x]_n) = \{f(x_1, \ldots, x_n) \,|x_i \in [x]_i\}$.

This theorem allowed several new techniques to treat interval-valued data especially in numerical computing fields. The use of this theorem and of the interval algebra in a statistical perspective is currently the subject of great debate in the scientific community. While much work has already been done towards the solution of linear systems of equation of interval data, a lot of work has to be done in order to achieve significant results in the treatment of non-linear problems such as the treatment of quadratic forms. In order to introduce some concepts related to the algebra of intervals into the PCA treatment, the definition of a suitable codification of the boxes must be clarified.

Recall the result due to Lawson and Hanson (1974) quoted in Section 15.1. This statement has sparked a new way of seeing the analysis of interval data. It means that coding intervals as pairs of *midpoints* and *radii* may be more fruitful than coding them as pairs of *minima* and *maxima*.

The first way to deal with PCA in an interval algebra framework has been deepened by Gioia and Lauro (2006). This approach, called IPCA (PCA based on interval algebra), will be discussed in Section 15.4.2.

15.4.1.1 MRPCA

Let us consider a quadratic extension of the distance proposed by Neumaier (1990):

$$d\left(y_{ij}, y_{i'j}\right) = \left|m_{ij} - m_{i'j}\right| + \left|r_{ij} - r_{i'j}\right| \tag{15.21}$$

(the distance is better known as the Hausdorff distance in $\mathbb{R}$ between two intervals). If $[\mu]_I$ represents the mean interval of $[a]_I$, $[b]_I$ and $[c]_I$, the following is true:

$$([a]_I - [\mu]_I) + ([b]_I - [\mu]_I) + ([c]_I - [\mu]_I) = 0. \tag{15.22}$$

$[\mu]_I$ can also be expressed in terms of radii and midpoints:

$$[\mu]_I = (\overline{m}, \overline{r}). \tag{15.23}$$

Taking into account the definition of distance between intervals, we have the following measure of variability:

$$\overline{\text{var}}(j) = \frac{\sum_i \left(m_{ij} - \overline{m}_j\right)^2}{n} + \frac{\sum_i \left(r_{ij} - \overline{r}_j\right)^2}{n} + \frac{2\sum_i \left|m_{ij} - \overline{m}_j\right| \cdot \left|r_{ij} - \overline{r}_j\right|}{n}. \tag{15.24}$$

Given a generic interval variable Y_j, the 'variance' is defined as the sum of three components: the variance between midpoints; the variance between radii; and twice a measure of congruence between midpoints and radii.

PCA is generalized to interval variables, described by midpoints and radii, by maximizing the projections of distances between any interval and the mean interval.

The generic term of the data matrix is given by the standardized interval

$$\tilde{d}_{ij} = \frac{d([y]_I\, ij, [\mu]_I\, j)}{\sqrt{\overline{\text{var}}(j) \cdot n}} = \frac{|m_{ij} - \overline{m}_j| + |r_{ij} - \overline{r}_j|}{\sqrt{\overline{\text{var}}(j) \cdot n}}. \tag{15.25}$$

The full matrix of the correlations is

$$\left[\left(\mathbf{M}'\boldsymbol{\Sigma}^{-1}\mathbf{M}\right) + \left(\mathbf{R}'\boldsymbol{\Sigma}^{-1}\mathbf{R}\right) + \left(\mathbf{M}'\boldsymbol{\Sigma}^{-1}\mathbf{R} + \mathbf{R}'\boldsymbol{\Sigma}^{-1}\mathbf{M}\right)\right]. \tag{15.26}$$

The distance formulation allows calculation of principal components (PCs) in three steps:

(i) a PCA of the matrix of midpoints,

$$\mathbf{M}\boldsymbol{\Sigma}^{-1}\mathbf{u}_k^m = \lambda_k^m \mathbf{u}_k^m, \tag{15.27}$$

where $\mathbf{u}_k^m$ and λ_k^m $(1 \leq k \leq p)$ are defined under the usual orthonormality constraints;

(ii) a PCA of the matrix of radii,

$$\mathbf{R}\boldsymbol{\Sigma}^{-1}\mathbf{u}_k^r = \lambda_k^r \mathbf{u}_k^r \tag{15.28}$$

with the same orthonormality constraints on $\mathbf{u}_k^r$ and λ_k^r $(1 \leq k \leq p)$;

(iii) and an interval reconstruction step by the projection of suitably rotated ranges into the midpoints space spanned by the eigenvectors associated with the PCA of the midpoints matrix.

A radius rotation matrix is obtained by maximizing the Tucker congruence coefficient between midpoints and radii:

$$f(\mathbf{T}) = \sum_l \frac{\mathbf{t}_l'\mathbf{M}_l' r}{(\mathbf{t}_l'\mathbf{M}'\mathbf{M}\mathbf{t})^{1/2}\,(\mathbf{r}_l'\mathbf{r}_l)^{1/2}} \tag{15.29}$$

under the constraint $\mathbf{T}'\mathbf{T} = \mathbf{I}$.

An alternative approach to MRPCA has been proposed by D'Urso and Giordani (2004). The technique is based on the assumption that the midpoints and the radii are modelled by means of the same components. It follows that the MRPCA model can be seen as a special case of simultaneous component analysis with invariant pattern (SCAP). SCA is a generalization of PCA proposed by Kiers and ten Berge (1989) when observations on the same variables have been observed in more than one population. Instead of analysing the observations separately, MRPCA finds components that explain as much variance as possible in all populations simultaneously, based on a matrix that is the vertical juxtaposition of the matrix of midpoints and radii. It is based on the decomposition of a distance between multidimensional boxes computed as the sum of the Euclidean distances between all the vertices of a box and all the vertices.

15.4.1.2 Spaghetti PCA

A further *hybrid* approach to the PCA of multidimensional interval data was proposed by Irpino (2006).[4] The so-called *spaghetti PCA* decomposes the correlation matrix of the main diagonals of the hyper-rectangles representing multidimensional interval data. The multidimensional boxes are represented by their main diagonals (md_i), considered as segments of uniform points described in the following way:

$$md_i(t) = \overbrace{\left\{ \begin{array}{c} \cdots \\ y_{ij} + t(\bar{y}_{ij} - y_{ij}) \\ \cdots \end{array} \right.}^{\text{Vertex notation}} = \overbrace{\left\{ \begin{array}{c} \cdots \\ m_{ij} + r_{ij}(2t-1), \\ \cdots \end{array} \right.}^{\text{Midpoint–radius notation}} \qquad j = 1, \ldots, p, 0 \le t \le 1. \tag{15.30}$$

The technique decomposes the correlation matrix of the descriptors of such segments according to the following formulae for the computation of the mean, standard deviation, covariance and correlation, respectively:

$$\mu_j = \frac{1}{n}\sum_i \int_0^1 [m_{ij} + r_{ij}(2t-1)]\,\mathrm{d}t = \frac{1}{n}\sum_i m_{ij}, \tag{15.31}$$

$$\sigma_j = \sqrt{\frac{1}{n}\sum_i \int_0^1 [m_{ij} + r_{ij}(2t-1)]^2\,\mathrm{d}t - \mu_j^2} = \sqrt{\frac{1}{n}\sum_i (c_{ij}^2 + 1/3 r_{ij}^2) - \mu_j^2}, \tag{15.32}$$

$$\begin{aligned} \mathrm{cov}(Y_j, Y_k) &= \frac{1}{n}\sum_i \int_0^1 [m_{ij} + r_{ij}(2t-1)][m_{ik} + r_{ik}(2t-1)]\,\mathrm{d}t - \mu_j\mu_k \\ &= \frac{1}{n}\sum_i (m_{ij}m_{ik} + 1/3 r_{ij}r_{ik}) - \mu_j\mu_k, \end{aligned} \tag{15.33}$$

$$\mathrm{corr}(Y_j, Y_k) = \frac{\mathrm{cov}(Y_j, Y_k)}{\sigma_j\sigma_k}. \tag{15.34}$$

Considering the standardized main diagonals as

$$\tilde{y}_i(t) = \frac{md_i(t) - \boldsymbol{\mu}}{\boldsymbol{\sigma}} = \overbrace{\left\{ \begin{array}{c} \cdots \\ \dfrac{[y_{ij} + t(\bar{y}_{ij} - \bar{y}_{ij})] - \mu_j}{\sigma_j} \\ \cdots \end{array} \right.}^{\text{Vertex notation}} = \overbrace{\left\{ \begin{array}{c} \cdots \\ \dfrac{[m_{ij} + r_{ij}(2t-1)] - \mu_j}{\sigma_j}, \\ \cdots \\ j = 1, \ldots, p, 0 \le t \le 1, \end{array} \right.}^{\text{Midpoint–radius notation}} \tag{15.35}$$

the method decomposes the correlation matrix of the main diagonals:

$$\frac{1}{n}\tilde{\mathbf{Y}}'\tilde{\mathbf{Y}}\mathbf{u}_m = \lambda_m \mathbf{u}_m, \qquad 1 \le m \le p, \tag{15.36}$$

[4] Originally the method was developed to treat oriented interval data but can be used without loss of generality on classical interval data.

under the usual orthonomality constraint on the $\mathbf{u}_m$.

The principal innovation introduced by spaghetti PCA is that the main diagonals represent, at the same time, the information related to the *position* (by means of centres), the *size* (by means of the lengths) and the *shape* (by means of the slopes) of a box in a multidimensional space. It allows the discovery of how much of the total inertia of data is related to the inertia of centres and how much to the internal variability of the boxes.

15.4.2 IPCA

Starting from Moore's algebra of intervals, Gioia and Lauro (2006) propose a PCA of interval data according to the linear algebra developed for this kind of data. Indeed, starting from the novel approach for treating interval-valued variables (Gioia and Lauro, 2005), the methodology consists of using both the interval algebra and the optimization theory for the decomposition of the interval correlation matrix of a set of multidimensional interval data. This is the first extension of PCA to interval data consistent with interval computation theory.

An $n \times n$ interval matrix is the set

$$\mathbf{Y}^{\mathbf{I}} = [\underline{\mathbf{Y}}, \overline{\mathbf{Y}}] = \{ \mathbf{Y} : \underline{\mathbf{Y}} \leq \mathbf{Y} \leq \overline{\mathbf{Y}} \} \tag{15.37}$$

where $\underline{\mathbf{Y}}$ and $\overline{\mathbf{Y}}$ are $n \times n$ matrices which satisfy

$$\underline{\mathbf{Y}} \leq \overline{\mathbf{Y}}, \tag{15.38}$$

and whose elements are standardized according to the interval standard deviation computed using formulae presented in the Appendix to this chapter. The aim is to use, if possible, the interval algebra instruments to adapt the mathematical models, on the basis of the classical PCA, to the case in which an interval data matrix is given. Let us suppose that the interval-valued variables have been previously standardized according to the method proposed by Gioia and Lauro (2005).

It is known that the classical PCA on a real matrix $\mathbf{Y}$, in the space spanned by the variables, solves the problem of determining $m \leq p$ axes $\mathbf{u}_\alpha$, $\alpha = 1, \ldots, m$, such that the sum of the squared projections of the point units on $\mathbf{u}_\alpha$ is maximum:

$$\mathbf{u}'_\alpha \mathbf{Y}'\mathbf{Y}\mathbf{u}_\alpha = \max, \qquad 1 \leq \alpha \leq m, \tag{15.39}$$

under the constraints

$$\mathbf{u}'_\alpha \mathbf{u}_\beta = \begin{cases} 0, & \text{for } \alpha \neq \beta, \\ 1, & \text{for } \alpha = \beta. \end{cases} \tag{15.40}$$

The above optimization problem may be reduced to the eigenvalue problem:

$$\mathbf{Y}'\mathbf{Y}\mathbf{u}_\alpha = \lambda \mathbf{u}_\alpha, \qquad 1 \leq \alpha \leq m. \tag{15.41}$$

When the data are of interval type, $\mathbf{Y}^{\mathbf{I}}$ may be substituted into (39) and the interval algebra may be used for the products; equation (15.39) becomes an *interval eigenvalue problem* of the form

$$(\mathbf{Y}^{\mathbf{I}})^{\mathrm{T}} \mathbf{Y}^{\mathbf{I}} \mathbf{u}^I_\alpha = \lambda^I \mathbf{u}^I_\alpha, \tag{15.42}$$

which has the interval solutions

$$\left[\lambda_\alpha(\mathbf{Z}):\mathbf{Z}\in(\mathbf{Y^I})^\mathrm{T}\mathbf{Y^I}\right],\quad \left[\mathbf{u}_\alpha(\mathbf{Z}):\mathbf{Z}\in(\mathbf{Y^I})^\mathrm{T}\mathbf{Y^I}\right]\qquad \alpha=1,\ldots,p, \tag{15.43}$$

that is, the set of αth eigenvalues of any matrix $\mathbf{Z}$ contained in the interval product $\left(\mathbf{Y^I}\right)^\mathrm{T}\mathbf{Y^I}$, and the set of the corresponding eigenvectors respectively.

Using the interval algebra for solving problem (42), the *interval solutions* will be computed but, generally, these intervals are *oversized* with respect to the *intervals of solutions* that we are searching for. This will be discussed in more detail below.

For the sake of simplicity, let us consider the case $p=2$, where two interval-valued variables

$$\mathbf{Y}_1^\mathbf{I}=\left(Y_{i1}=\left[\underline{y}_{i1},\overline{y}_{i1}\right]\right),\quad i=1,\ldots,n,\qquad \mathbf{Y}_2^\mathbf{I}=\left(Y_{i2}=\left[\underline{y}_{i2},\overline{y}_{i2}\right]\right),\quad i=1,\ldots,n, \tag{15.44}$$

have been observed on the n units considered. $\mathbf{Y}_1^\mathbf{I}$ and $\mathbf{Y}_2^\mathbf{I}$ assume an *interval of values* on each statistical unit: we do not know the exact value of the components y_{i1} or y_{i2} for $i=1,\ldots,n$, only the *range* in which this value falls. In the proposed approach the task is to contemplate *all possible values* of the components x_{i1}, x_{i2}, each of which in its own interval of values $Y_{i1}=\left[\underline{y}_{i1},\overline{y}_{i1}\right]$, $Y_{i2}=\left[\underline{y}_{i2},\overline{y}_{i2}\right]$ for $i=1,\ldots,n$. Furthermore, for each different set of values $y_{11},y_{21},\ldots,y_{n1}$ and $y_{12},y_{22},\ldots,y_{n2}$, where $y_{ij}\in\left[\underline{y}_{ij},\overline{y}_{ij}\right]$, $i=1,\ldots,n$, $j=1,2$, a different cloud of points in the plane is uniquely by determined and the PCA on that set of points must be computed. Thus, with *interval PCA (IPCA)* we mean to determine the *set* of solutions of the classical *PCA* on each set of point units, set which is univocally determined for any different choice of the point units each of which is in its own box of variation.

Therefore, the *interval of solutions* which we are looking for are the set of the *αth axes*, each of which maximizes the sum of square projections of a set of points in the plane, and the set of the *variances* of those sets of points respectively. This is equivalent to solving the optimization problem (15.41), and thus the eigenvalue problem (15.42) for each matrix $\mathbf{Y}\in\mathbf{Y^I}$.

In light of the above considerations, the background to approaching directly the interval eigenvalue problem (15.38), comes out by observing that the following inclusion holds:

$$\left(\mathbf{Y^I}\right)^\mathrm{T}\mathbf{Y^I}=\left\{\mathbf{YX}\ \mid\ \mathbf{Y}\in(\mathbf{Y^T})^\mathbf{I},\ \mathbf{X}\in\mathbf{Y^I}\right\}\supset\left\{\mathbf{Y^TY}\ \mid\ \mathbf{Y}\in\mathbf{Y^I}\right\}. \tag{15.45}$$

This means that the interval matrix $\left(\mathbf{Y^I}\right)^\mathrm{T}\mathbf{Y^I}$ also contains matrices which *are not* of the form $\mathbf{Y^TY}$. Thus the *interval eigenvalues* and the *interval eigenvectors* of (15.42) will be *oversized* and, in particular, will *include* the set of all eigenvalues and the set of the corresponding eigenvectors of any matrix of the form $\mathbf{Y^TY}$ contained in $\left(\mathbf{Y^I}\right)^\mathrm{T}\mathbf{Y^I}$.

This drawback may be solved by computing an interval eigenvalue problem considering, in place of the product

$$(\mathbf{Y}')^\mathbf{I}\mathbf{Y^I}=\left\{\mathbf{YX}\ \mid\ \mathbf{Y}\in(\mathbf{Y}')^\mathbf{I},\ \mathbf{X}\in\mathbf{Y^I}\right\}, \tag{15.46}$$

the set of matrices

$$\mathbf{\Theta^I}=\left\{\mathbf{Y'Y}\ \mid\ \mathbf{Y}\in\mathbf{Y^I}\right\}, \tag{15.47}$$

i.e., the set of all matrices given by the product of a matrix multiplied by its transpose. For computing the *αth* eigenvalue and the corresponding eigenvector of $\boldsymbol{\Theta}$, still denoted by λ^I_α $\mathbf{u}^I_\alpha$, some results of the interval linear algebra presented by Gioia and Lauro (2006) may be used.

The *correlation interval matrix* will be indicated by $\boldsymbol{\Gamma}^{\mathbf{I}} = (\mathbf{corr}^{\mathbf{I}}_{ij})$ where $\mathbf{corr}^{\mathbf{I}}_{ij}$ is the *interval of correlations* between $\mathbf{Y}^{\mathbf{I}}_i$, $\mathbf{Y}^{\mathbf{I}}_j$ (Gioia and Lauro, 2005), computed using the formula in the Appendix. Notice that while the ijth component of $\boldsymbol{\Gamma}^I$ is the *interval of correlations* between $\mathbf{Y}^{\mathbf{I}}_i$, $\mathbf{Y}^{\mathbf{I}}_j$, the ijth component of $(\mathbf{Y}')^{\mathbf{I}}\mathbf{Y}^{\mathbf{I}}$ is an interval which includes that interval of correlations and also contains redundant elements.

The interval eigen-equation decomposed by IPCA is then

$$\boldsymbol{\Gamma}^{\mathbf{I}}\mathbf{u}^I_\alpha = \lambda^I\mathbf{u}^I_\alpha. \tag{15.48}$$

It is important to remark that $\boldsymbol{\Theta}^{\mathbf{I}} \subset \boldsymbol{\Gamma}^{\mathbf{I}}$; then the eigenvalues/eigenvectors of $\boldsymbol{\Gamma}^{\mathbf{I}}$ will also be oversized with respect to those of $\boldsymbol{\Theta}^{\mathbf{I}}$.

The *αth interval axis* or *interval factor* will be the αth interval eigenvector associated with the *αth* interval eigenvalue in decreasing order.[5]

The *orthonormality* between couples of interval axes must be interpreted according to:

$$\forall \boldsymbol{u}_\alpha \in \boldsymbol{u}^I_\alpha \text{ such that } \boldsymbol{u}'_\alpha\boldsymbol{u}_\alpha = 1, \exists\, \boldsymbol{u}_\beta \in \boldsymbol{u}^I_\beta \text{ with } \alpha \neq \beta \text{ such that } \boldsymbol{u}'_\beta\boldsymbol{u}_\beta = 1, \quad \boldsymbol{u}'_\alpha\boldsymbol{u}_\beta = 0.$$

Thus two interval axes are orthonormal to one another if, given a unitary vector in the first interval axis, there exists a unitary vector in the second one so that their scalar product is zero.

15.5 Visualizing PCA results on factor planes

PCAs developed for interval-valued data allow the representation of projected boxes in three different ways: enclosing the projected vertices by means of minimum covering area rectangles (MCARs) for VPCA, CPCA and SPCA; using reconstruction formulae using midpoint and radius rotation operators for MRPCA; or using the inner product interval operator for IPCA. On the other hand, in PCA it is also possible to represent the projection of the original variables on the so-called circle of correlation. Indeed, for each type of PCA the projection of results (boxes and variables) on factor spaces is done according to the following formulae.

- *VPCA*. The matrix of vertices $\mathbf{Z}_{(N\times p)}$ is multiplied by the mth eigenvector in order to calculate the coordinates of the vertices on the mth factorial axis:

$$\boldsymbol{\psi}_m = \mathbf{Z}\mathbf{v}_m. \tag{15.49}$$

[5] Considering that the αth eigenvalue of $\boldsymbol{\Sigma}$ is computed by perturbing the αth eigenvalue of $(\mathbf{Y}^c)'\mathbf{Y}^c$, the ordering on the interval eigenvalues is given by the natural ordering of the corresponding scalar eigenvalues of $(\mathbf{Y}^c)'\mathbf{Y}^c$.

Considering the ith SO, having in general 2^p vertices, $\psi_{im} = [\min(\boldsymbol{\psi}_{jm}), \max(\boldsymbol{\psi}_{jm})]$, if the jth vertex belongs to the ith SO, is the interval coordinate of the ith SO for the mth axis. The correlations plot is plotted according to (15.6):

$$\varphi = \mathbf{Z}'\mathbf{w}_m. \qquad (15.50)$$

- *CPCA*. The matrix of vertices $\mathbf{Z}^*_{(N\times p)}$ (considered as supplementary units) is multiplied by the mth eigenvector of the correlation matrix of centres in order to calculate the coordinates of the vertices on the mth factorial axis:

$$\boldsymbol{\psi}_m = \mathbf{Z}^*\mathbf{v}_m. \qquad (15.51)$$

Considering the ith SO, having in general 2^p vertices, $\psi_{im} = [\min(\boldsymbol{\psi}_{jm}), \max(\boldsymbol{\psi}_{jm}),$ where the jth vertex belongs to the ith SO, is the interval coordinate of the ith SO for the mth axis. The correlations plot is plotted according to the calculation of correlations between the variables and the new PCs. Given the generic jth original variable and the mth PC, the correlation is equal to

$$\varphi_{jm} = \mathrm{corr}(\mathbf{M}_{.j}, \tilde{\mathbf{M}}_{.j}\mathbf{v}_m). \qquad (15.52)$$

- *SPCA*. The matrix of cohesive vertices, $\hat{\mathbf{Z}} = \mathbf{P}_\mathbf{A}\mathbf{Z}$, is premultiplied by the $\mathbf{P}_\Phi$ projection operator on the subspace spanned by the decomposition of radii and then they are finally multiplied by the mth eigenvector in order to calculate the coordinates of the vertices on the mth factorial axis:

$$\boldsymbol{\psi}_m = \mathbf{P}_\Phi\hat{\mathbf{Z}}s_m. \qquad (15.53)$$

Considering the ith SO, having in general 2^p vertices, $\psi_{im} = [\min(\boldsymbol{\psi}_{jm}), \max(\boldsymbol{\psi}_{jm})]$, where the jth vertex belongs to the ith SO, is the interval coordinate of the ith SO for the mth axis. The correlations plot is plotted according to the calculation of correlations between the variables and the new PCs. Given the generic jth original variable and the mth PC the correlation is equal to

$$\varphi_{jm} = \mathrm{corr}(\hat{\mathbf{Z}}_{.j}, \boldsymbol{\psi}_m). \qquad (15.54)$$

The visualization of SOs on factor planes by MCARs, proposed in symbolic factorial data analysis, even if it is the easiest way to represent and describe projected SOs, seems not to be the most satisfactory shape for representing the projections of the boxes associated with them. MCARs allow simple descriptions of the SOs by means of symbolic interval descriptions given by the interval coordinates on the factor variables, but also provide an overgeneralization in the descriptions of the projected SOs (Lauro *et al.*, 2000). The same problem has been faced by the interval computation research group that referred to the MCAR as an *interval hull* and the overgeneralization problem as a *wrapping effect*.

An alternative visualization can be furnished by convex hulls (Verde and De Angelis, 1997; Porzio *et al.*, 1998), constructed on the projected box vertices. Nevertheless, even if it seems, from a geometrical point of view, the most capable visualization of the boxes images, the symbolic description of SOs is hard to achieve. Indeed, a description of the SOs with respect to the factor axes cannot be expressed in terms of symbolic interval values. The description of the convex hull is given by the simplex generated by edges obtained as linear bounded combinations of the new factor variables.

Another important aspect of the representation by means of convex hulls is that they require considerable computing resources. Irpino *et al.* (2003) have developed a fast algorithm for the generation of a 2D convex hull. Given p interval variables, the hypercube which describes an SO has $n = 2^p$ vertices. From a computational point of view the number of projected points (hypercube vertices), on which the convex hull has been computed, increases exponentially when the number of variables increases (for example, 10 intervals generate 1024 vertices). If no structure on data is defined, the best algorithm for the generation of a 2D convex hull needs at least $n \log n = 2^p \log 2^p$ comparisons. Irpino *et al.* (2003) introduced a faster algorithm that has a lower number of comparison operations due to the geometrical properties of the boxes and to the linearity of the projection operators.

The algorithm (2DPP-CH, 2D Projected Parallelogram Convex Hull) is based on the following principles:

1. In the original space, a box is a geometrical shape symmetric with respect the centre of gravity.
2. It is possible to prove that the edges of the 2D convex hull drawn on the factorial plane correspond to the projections of the sides of the box in the original space.
3. As the projection operator is a linear combination of the original variables, the vertices projected onto a factor plane are symmetric to the projection of the hypercube barycentre. Thus, given p interval variables the maximum number of extreme points of the 2D convex hull is $2p$. Once we have calculated p consecutive extreme points, the other p are given by the property of the symmetry to the centre of gravity of the shape. It is easy to prove that from a point (a projected vertex of the original box) it is possible to go only in p different directions.

Given p interval variables the number of comparisons is equal to

$$\sum_{i=1}^{p}(p-i) = \frac{p^2+p}{2} \ll 2^p \log 2^p.$$

The convex hull (Figure 15.4) allows better visualization of the projection of the hypercube describing an SO. However, its description in terms of symbolic assertion is not so easy. In fact, the convex hull is a polygon that can be described by the simplex of the half planes starting from its edges.

In the same paper, Irpino *et al.* (2003) propose a novel graphical representation of the projected box called a PECS (Parallel Edges Connected Shape) which is a compromise between the great interpretability of MCARs and the non-overgeneralized description of convex hulls.

- *MRPCA*. In order to represent the boxes on the kth factor axes MRPCA considers a reconstruction formula based on the following quantities: $\boldsymbol{\psi}_k^m = \tilde{\mathbf{M}}\mathbf{u}_k^m$ as the coordinates

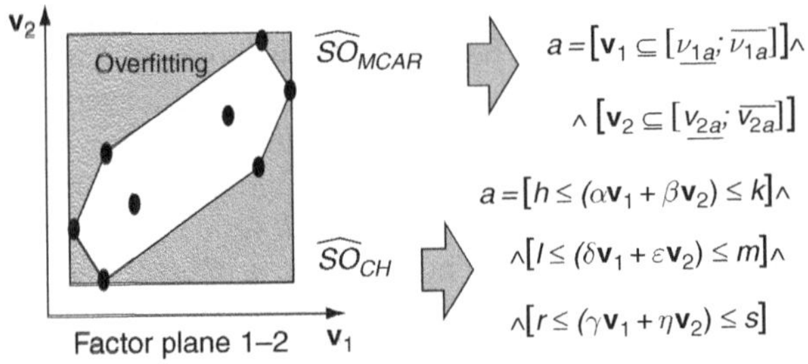

Figure 15.4 Representing and describing SO using MCAR ($\widehat{SO}_{\mathrm{MCAR}}$) and CH ($\widehat{SO}_{\mathrm{CH}}$).

of the projected standardized midpoints $\tilde{\mathbf{M}}$ on the kth factor axis; $\boldsymbol{\psi}_k^r = \tilde{\mathbf{R}}\mathbf{u}_k^r$ as the coordinates of the projected standardized radii $\tilde{\mathbf{R}}$ on the kth factor axis; $\boldsymbol{\psi}_k^{*r} = T(\tilde{\mathbf{R}}\mathbf{u})_k^r$ as the coordinates of the projected standardized rotated radii $T(\tilde{\mathbf{R}})$ on the kth factor axis. Then the coordinate interval of the ith SO for the kth axis is calculated as:

$$\psi_{ik} = \left[\psi_{ik}^m - \psi_{ik}^{*r}, \psi_{ik}^m + \psi_{ik}^{*r}\right]. \tag{15.55}$$

Two circles of correlations are plotted: one for the components of centres and one for the radius components. Given the generic jth original variable and the kth PC, the two correlations are computed:

$$\varphi_{jk}^m = \mathrm{corr}(\tilde{\mathbf{M}}_{.j}, \boldsymbol{\psi}_k^m), \tag{15.56}$$

$$\varphi_{jk}^r = \mathrm{corr}(\tilde{\mathbf{R}}_{.j}, \boldsymbol{\psi}_k^r). \tag{15.57}$$

- *Spaghetti PCA*. The representation of the main diagonal of the ith box is the projection of all the points belonging to the diagonal on the factor axes:

$$\psi(t)_{ik} = \sum_{j=1}^{p} \left[\tilde{m}_{ij} + \tilde{r}_{ij}(2t-1)\right] u_{jk}, \qquad 0 \leq t \leq 1. \tag{15.58}$$

The circle of correlations is plotted according to the calculation of correlations between the variables and the new PCs. Given the generic jth original variable and the kth PC, the correlation is computed using (15.34):

$$\varphi_{jk} = \mathrm{corr}(\tilde{\mathbf{Y}}_{.j}, \boldsymbol{\psi}(t)_{.k}). \tag{15.59}$$

- *IPCA*. This technique uses the dot product developed for intervals. In this case the coordinate interval of the ith box on the kth factor axis is obtained as for classical PCA, but in this case the elements of the vectors and matrices involved in the calculation are intervals:

$$\boldsymbol{\psi}_k^{\mathbf{I}} = \mathbf{Y}^{\mathbf{I}}\mathbf{u}_k. \tag{15.60}$$

The circle of correlations is plotted according to the calculation of correlations between the variables and the new PCs. Given the generic jth original variable and the kth PC the correlation interval is computed as

$$\varphi^I_{jk} = \text{corr}(\tilde{\mathbf{Y}}^I_{.j}, \boldsymbol{\psi}^I_k). \tag{15.61}$$

15.6 A comparative example

In this section, we present the main results of the application of the PCAs described above to the Ichino (1988) oil data set. The data set (Table 15.1) describes eight different classes of oils characterized by four quantitative interval-valued variables.

We start by showing the main results of the VPCA where a PCA on the standardized vertices is performed. The standardization of vertices is done using the standard deviation of each variable computed by taking into consideration all the vertices of the boxes. The correlation matrix to decompose is described in Table 15.2. The total inertia of vertices is decomposed and provides the eigenvalues in Table 15.3. The main graphical results are

Table 15.1 Symbolic data table: Ichino oils data set.

Name	Specific gravity	Freezing point	Iodine value	Saponification
Linseed	[0.93, 0.94]	[−27, −18]	[170, 204]	[118, 196]
Perilla	[0.93, 0.94]	[−5, −4]	[192, 208]	[188, 197]
Cottonseed	[0.92, 0.92]	[−6, −1]	[99, 113]	[189, 198]
Sesame	[0.92, 0.93]	[−6, −4]	[104, 116]	[187, 193]
Camellia	[0.92, 0.92]	[−21, −15]	[80, 82]	[189, 193]
Olive	[0.91, 0.92]	[0, 6]	[79, 90]	[187, 196]
Beef tallow	[0.86, 0.87]	[30, 38]	[40, 48]	[190, 199]
Hog fat	[0.86, 0.86]	[22, 32]	[53, 77]	[190, 202]

Table 15.2 VPCA: correlation matrix of vertices.

	GRA	FRE	IOD	SAP
GRA	1.0000			
FRE	−0.8982	1.0000		
IOD	0.7470	−0.6399	1.0000	
SAP	−0.2900	0.3679	−0.3741	1.0000

Table 15.3 VPCA: eigenvalues and explained inertia.

Eigenvalues	Explained inertia	% of total inertia	Cumulative
E1	2.732	68.29%	68.29%
E2	0.809	20.23%	88.52%
E3	0.380	9.50%	98.02%
E4	0.079	1.98%	100.00%

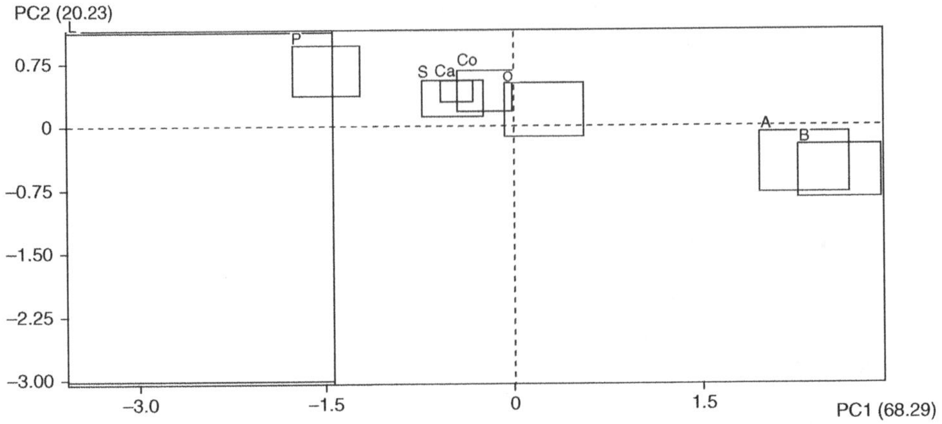

Figure 15.5 VPCA: SOs representation on the first factor plane.

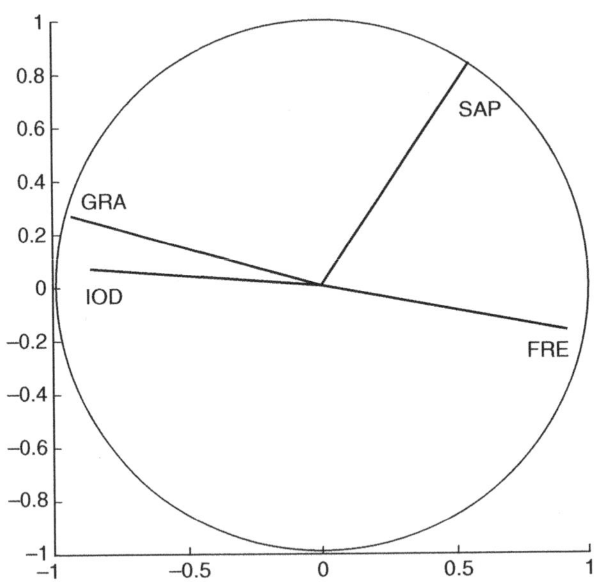

Figure 15.6 VPCA: Circle of correlation on the first factor plane.

shown in Figures 15.5 and 15.6 for the first factor plane, which explains the 88.52% of the total inertia. The representation of the boxes (Figure 15.5) has been performed using the MCARs of the vertices projected onto the factor plane for each box. The correlation circle (Figure 15.6) represents the correlation between original variables and the first two factor axes (Table 15.4) as in classical PCA.

CPCA performs a principal component analysis only of the centres of the boxes. Each box is then represented only by its centre. The matrix to decompose (Table 15.5) is then the correlation matrix of the standardized centres. The standardization is performed according

Table 15.4 VPCA: correlations between original variables and principal components.

	PC1	PC2	PC3	PC4
GRA	−0.934	0.265	−0.113	0.210
FRE	0.915	−0.163	0.325	0.174
IOD	−0.857	0.062	0.508	−0.061
SAP	0.536	0.842	0.061	−0.028

Table 15.5 CPCA: correlation matrix of the centres of boxes.

	GRA	FRE	IOD	SAP
GRA	1.0000			
FRE	−0.9213	1.0000		
IOD	0.7652	−0.6591	1.0000	
SAP	−0.4581	0.5842	−0.5933	1.0000

Table 15.6 CPCA: eigenvalues and explained inertia.

Eigenvalues	Explained inertia	% of total inertia	Cumulative
E1	3.0094	75.24%	75.24%
E2	0.6037	15.09%	90.33%
E3	0.3483	8.71%	99.04%
E4	0.0386	0.96%	100.00%

to the standard deviation of centres. The eigendecomposition allows the eigenvalues listed in Table 15.6.

It is worth noting that even if the first factor plane explains 90.33% of the total inertia, this inertia is related only to the centres, losing, in this case, the inertia internal to the boxes. Indeed, in order to obtain the projection of boxes on the first factor plane (Figure 15.7) by means of the MCARs, the vertices are projected as supplementary units according after their standardization using the standard deviation of centres. Figure 15.8 reports the correlation between the original variables and the first two factorial axes (Table 15.7).

SPCA allows the inertia of vertices to be taken into consideration according to a cohesion constraint. The vertices are then projected in the space of transformed radii and a PCA is performed. The covariance matrix of the standardized and projected (in the transformed radii space) vertices is described in Table 15.8. The decomposition produces the first two eigenvalues as described in Table 15.9. Axes–variables correlations are shown in Table 15.10.

The first factor plane is capable of synthesizing 98.40% of the inertia. The MCARs associated with the vertices of each box are shown in Figure 15.9. In this plot, objects are more separated than in VPCA and or CPCA. Figure 15.10 shows the correlation circle that represents the correlation between the original variables and the first two factor axes.

Turning to MRPCA, we first consider partial analysis based on the matrix of centre (or midpoint) values standardized according to (15.25). In Tables 15.11, 15.12 and 15.13

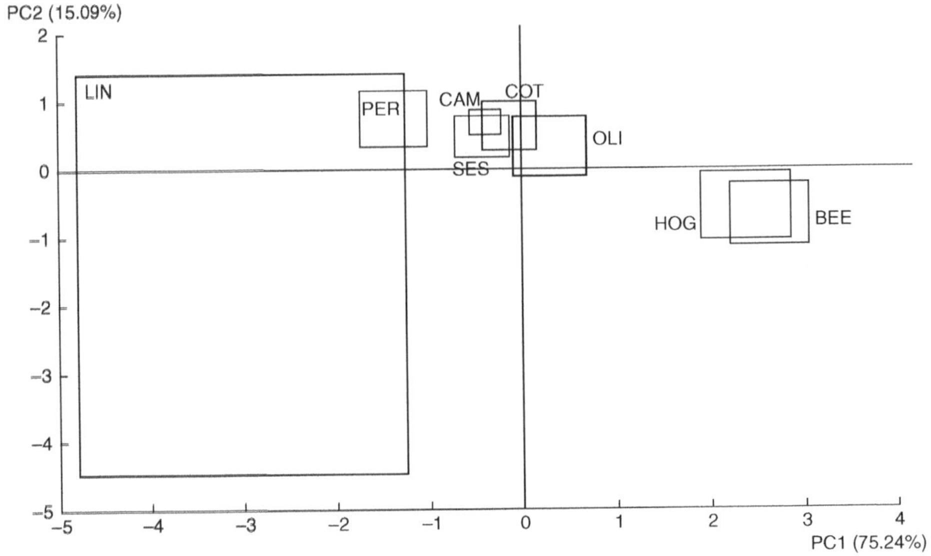

Figure 15.7 CPCA: SOs representation on the first factor plane.

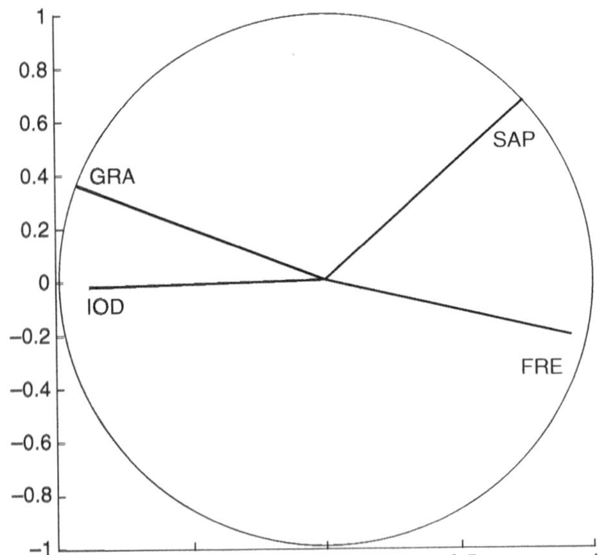

Figure 15.8 CPCA: circle of correlation on the first factor plane.

Table 15.7 VPCA: correlations between original variables and principal components.

	PC1	PC2	PC3	PC4
GRA	−0.9236	0.3544	0.0454	−0.1393
FRE	0.9237	−0.2020	−0.3017	−0.1223
IOD	−0.8722	−0.0312	−0.4855	0.0516
SAP	0.7366	0.6606	−0.1395	0.0398

Table 15.8 SPCA: covariance matrix of the projected vertices onto the subspace spanned by RTPCs.

	GRA	FRE	IOD	SAP
GRA	0.39660			
FRE	−0.32074	0.36770		
IOD	0.54456	−0.44184	0.81726	
SAP	−0.30035	0.36602	−0.40215	0.39581

Trace = 1.99737

Table 15.9 SPCA: eigenvalues and explained inertia.

Eigenvalues	Explained inertia	% of total inertia	Cumulative
E1	1.74248	88.12%	88.12%
E2	0.20337	10.28%	98.40%

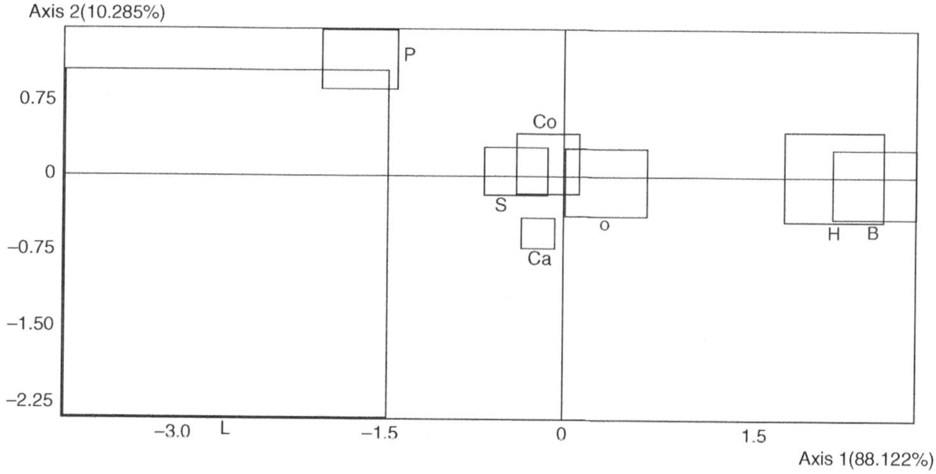

Figure 15.9 SPCA: SOs representation on the first factor plane.

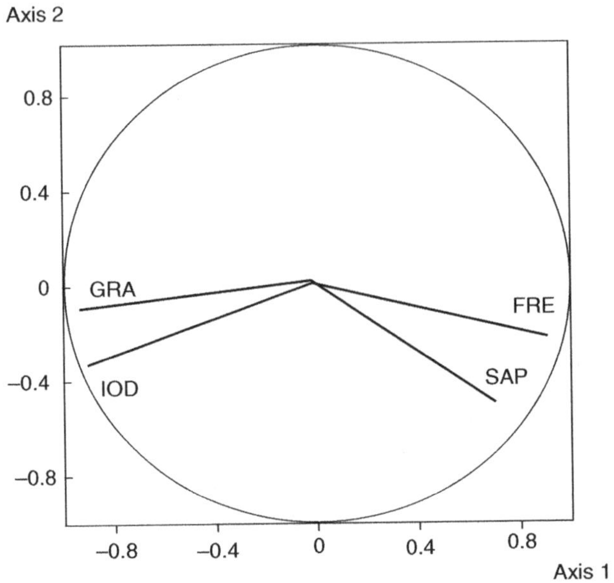

Figure 15.10 SPCA: Circle of correlation on the first factor plane.

Table 15.10 SPCA: correlations between original variables and principal components.

	PC1	PC2
GRA	−0.92679	−0.09624
FRE	0.90282	−0.22163
IOD	−0.90781	−0.3432
SAP	0.70057	−0.49518

Table 15.11 MRPCA: covariance matrix of the standardized midpoints ($\mathbf{M}'\boldsymbol{\Sigma}^{-1}\mathbf{M}$).

	GRA	FRE	IOD	SAP
GRA	0.8704			
FRE	−0.8128	0.8943		
IOD	0.6688	−0.5839	0.8775	
SAP	−0.2186	0.2826	−0.2842	0.2616

Trace = 2.9038 (72.59%)

are respectively reported the standardized covariance matrices of midpoints, radii and the midpoints vs. radii. The sum of the three matrices is reported in Table 15.14. This is a classical PCA on the interval midpoints whose solutions are given by the eigensystem in (15.27).

Table 15.12 MRPCA: covariance matrix of the standardized radii ($\mathbf{R}'\mathbf{\Sigma}^{-1}\mathbf{R}$).

	GRA	FRE	IOD	SAP
GRA	0.0066			
FRE	−0.0018	0.0057		
IOD	0.0010	0.0025	0.0069	
SAP	0.0116	0.0163	0.0336	0.2457

Trace= 0.2649 (6.62%)

Table 15.13 MRPCA: covariance matrix of the standardized midpoints and radii ($|\mathbf{M}'|\mathbf{\Sigma}^{-1}|\mathbf{R}|$).

	GRA	FRE	IOD	SAP
GRA	0.0615	0.0611	0.0555	0.2515
FRE	0.0611	0.0500	0.0683	0.3075
IOD	0.0543	0.0578	0.0578	0.3253
SAP	0.0245	0.0239	0.0368	0.2464

Trace= 0.4157 (10.39%)

Table 15.14 MRPCA: sum of the covariance matrices ($\mathbf{M}'\mathbf{\Sigma}^{-1}\mathbf{M} + \mathbf{R}'\mathbf{\Sigma}^{-1}\mathbf{R} + |\mathbf{M}'|\mathbf{\Sigma}^{-1}|\mathbf{R}| + |\mathbf{R}'|\mathbf{\Sigma}^{-1}|\mathbf{M}|$).

	GRA	FRE	IOD	SAP
GRA	1.0000			
FRE	−0.6924	1.0000		
IOD	0.7796	−0.4553	1.0000	
SAP	0.0690	0.6303	0.1115	1.0000

Trace = 4 (100%)

Similarly to the PCA on midpoints, a PCA on radii (also standardized according to (15.25)) is performed solving (15.28); see Table 15.15. Both midpoints and radii PCAs admit an independent representation. The quantity $\sum_k (\lambda_k^m + \lambda_k^r) \leq p$ but it does not include the entire variability because the residual inertia, given by the midpoints–radii interconnection, has not yet been taken into account.

The interval bounds over the PCs are derived from the midpoints and radii coordinates, if PCs of radii are superimposed on the PCs of midpoints. This can be achieved if radii are rotated proportionally to their connections with midpoints.

Palumbo and Lauro (2003) proposed, as orthogonal congruence rotation criterion, to maximize the congruence coefficient proposed by Tucker between midpoints and radii.

As in single-valued PCA, it is also possible in interval-valued variables PCA to define some indicators that are related to interval contribution.

Measures of explanatory power are defined with respect to the partial analyses (midpoints and radii) as well as with respect to the global analysis. The authors observed that the variability associated with each dimension is expressed by its eigenvalue. For example, the proportion of variability associated with the first dimension is given by

$$\frac{\lambda_1^m + \lambda_1^r}{\mathrm{tr}(\mathbf{M}'\boldsymbol{\Sigma}^{-1}\mathbf{M} + \mathbf{R}'\boldsymbol{\Sigma}^{-1}\mathbf{R})},$$

Table 15.15 MRPCA: eigenvalues of the matrices of the covariance of the midpoints and of the covariance of the radii.

Eigenvalues	Midpoints inertia	%	Radii inertia	%	Midipoints plus radii inertia	%	% with respect the the trace of the matrix in Table 15.4
E1	2.3593	81.25%	0.252	95.17%	2.6113	82.41%	65.28%
E2	0.3316	11.42%	0.008	3.02%	0.3396	10.72%	8.49%
E3	0.1823	6.28%	0.0027	1.02%	0.185	5.84%	4.63%
E4	0.0306	1.05%	0.0021	0.79%	0.0327	1.03%	0.82%
Total	2.9038		0.2648		3.1686		79.22%

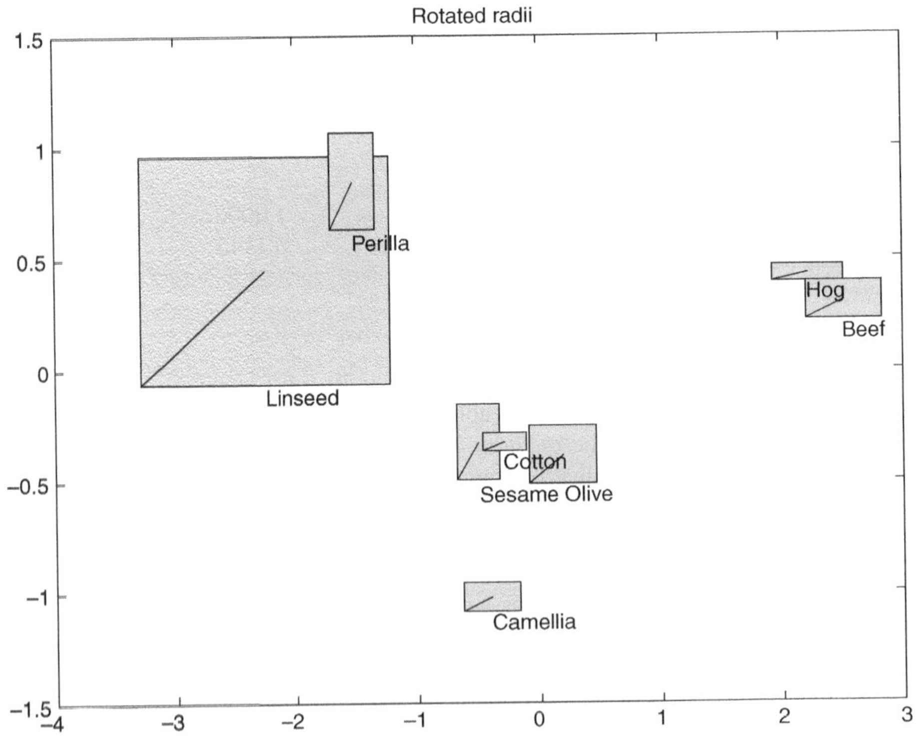

Figure 15.11 MRPCA: representation of SOs on the first factor plane after the rotation of radii.

where λ_1^m and λ_1^r represent the first eigenvalues related to the midpoints and radii, respectively. They express a partial information; indeed, there is a residual variability that depends on the midpoints and radii connection that cannot be explicitly taken into account.

The MCARs associated with each box are represented in Figure 15.11. In this plot, objects are more separated than in VPCA, CPCA or SPCA. In different cases, if the correlation of midpoints with radii has a stronger impact on the total inertia, MRCA achieves a better separation of the projected boxes. Figure 15.12 shows the correlation circle that represents the correlation between the original variables and the first two factorial axes for the midpoints (Table 15.16) and for the radii (Table 15.17).

'Spaghetti' PCA performs a PCA of the main diagonals of the boxes associated with a multidimensional interval description of SOs. The interval bounds and their midpoints

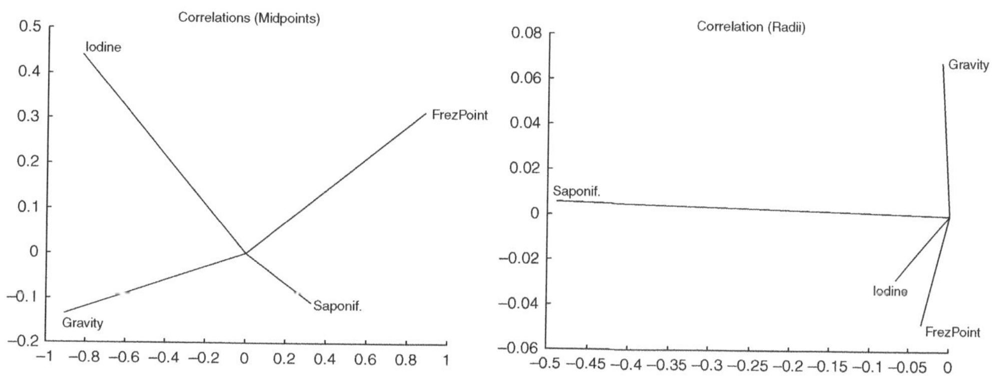

Figure 15.12 MRPCA: correlation plot of midpoints and of radii on the first factor plane.

Table 15.16 MRPCA: correlations between original variables and principal components for the midpoints.

	PC1	PC2	PC3	PC4
GRA	−0.9618	−0.1724	−0.1727	−0.1241
FRE	0.9363	0.3138	−0.1157	−0.1073
IOD	−0.8710	0.4799	−0.0940	0.0480
SAP	0.6321	−0.2442	−0.7226	0.1365

Table 15.17 MRPCA: correlations between original variables and principal components for the radii.

	PC1	PC2	PC3	PC4
GRA	−0.2906	0.8762	0.3539	−0.1502
FRE	−0.4391	−0.7143	0.5336	−0.1107
IOD	−0.8248	−0.1121	−0.1964	−0.5182
SAP	−0.9999	0.0031	−0.0037	0.0143

Table 15.18 Spaghetti PCA: matrix of correlation.

	GRA	FRE	IOD	SAP
GRA	1.0000			
FRE	−0.9083	1.0000		
IOD	0.7648	−0.6441	1.0000	
SAP	−0.3466	0.5172	−0.4379	1.0000

Table 15.19 Spaghetti PCA: eigenvalues and decomposition of inertia.

Eigen values	Explained inertia	% of total inertia	Cumulative	Expl. inertia by midpoints	% of inertia due to midpoints	Expl. inertia by radii	% of inertia due to radii
E1	2.8532	71.33%	71.33%	2.81782	70.45%	0.03538	0.88%
E2	0.7235	18.09%	89.42%	0.42954	10.74%	0.29396	7.35%
E3	0.3772	9.43%	98.85%	0.36207	9.05%	0.01513	0.38%
E4	0.0461	1.15%	100.00%	0.03923	0.98%	0.00687	0.17%
Total	4.0000			3.64867	91.22%	0.35133	9.78%

and radii are standardized according to (15.35). The elements of the correlation matrix (Table 15.18) to be decomposed are computed according to (15.36).

The PCA presented is capable of treating simultaneously the inertia due to the midpoints and radii. Irpino (2006) shows their additive properties allowing the computation of the inertia related to the radii and to the midpoints. The main results of the eigendecomposition are shown in Table 15.19. The inertia due to the internal variability of the boxes (the radii component) is 9.78% of the total inertia.

Figure 15.3 shows the main diagonals projected on the first factor plane that explains 89.42% of the total inertia. The MCAR built according the main diagonals allows a comparison with the other methods. In this case, 'Spaghetti' PCA seems to discriminate the projected boxes better than the other PCAs. Figure 15.14 shows the circle of the correlations between the original variables and the first two factorial axes (Tab. 15.20).[6]

IPCA firstly uses interval algebra and optimization theory to solve the PCA problem when data are described by intervals. IPCA extends to the interval numeric data the classic PCA. Using an extension of the classic basic statistics to interval data, Gioia and Lauro (2006) propose to apply an eigendecomposition of the interval correlation matrix (Table 15.21) of interval data standardized according the interval standard deviation developed by the same authors (Gioia and Lauro, 2005). The eigendecomposition, according to suitable optimization techniques, produces the interval eigenvalues described in Table 15.22.

[6] The greyed parts on the graph represent the intersection between the unit circle and the rectangle defined by the interval correlation. Indeed, all the values external to the circle are not consistent with the measure of the correlation (that belongs to the interval $[-1, 1]$).

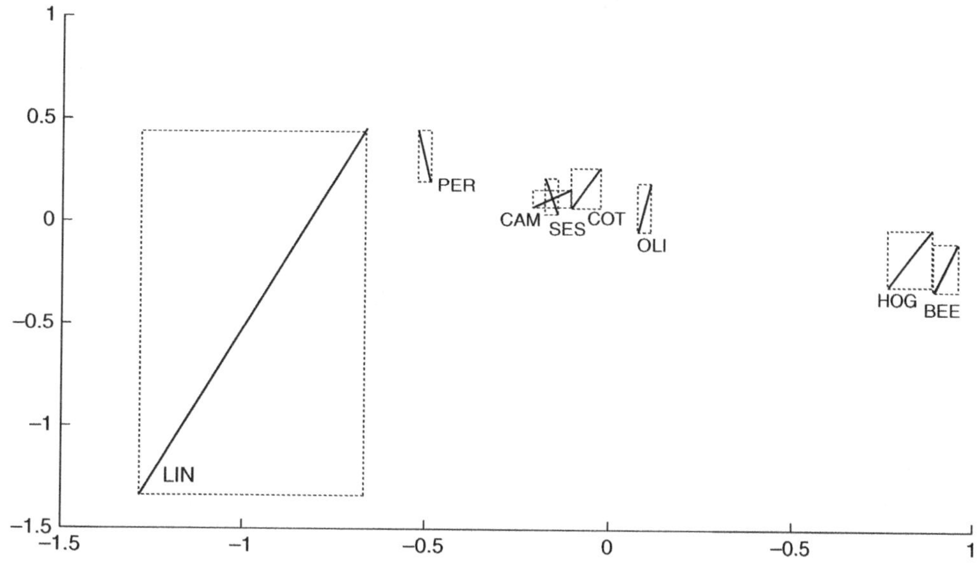

Figure 15.13 Spaghetti PCA: representation of SOs on the first factor plane.

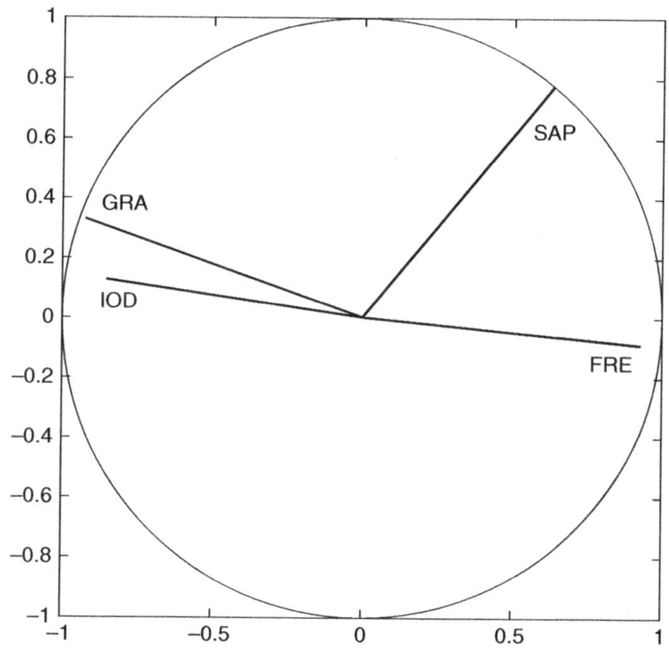

Figure 15.14 Spaghetti PCA: circle of correlation on the first factor plane.

Table 15.20 Spaghetti PCA: correlations between original variables and principal components.

	PC1	PC2	PC3	PC4
GRA	−0.9264	−0.3211	0.1213	−0.1545
FRE	0.9283	0.0881	−0.3356	−0.1338
IOD	−0.8549	−0.1292	−0.4998	0.0520
SAP	0.6344	−0.7720	−0.0052	0.0403

Table 15.21 IPCA: matrix of correlation intervals.

	GRA	FRE	IOD	SAP
GRA	[1.00,1.00]			
FRE	[−0.97, −0.80]	[1.00,1.00]		
IOD	[0.62,0.88]	[−0.77, −0.52]	[1.00,1.00]	
SAP	[−0.64, −0.16]	[0.30,0.75]	[−0.77, −0.34]	[1.00,1.00]

Table 15.22 IPCA: eigenvalue intervals and decomposition of the inertia.

Eigenvalues	Explained inertia	% of total inertia
E1	[2.45,3.40]	[61%,86%]
E2	[0.68,1.11]	[15%,32%]
E3	[0.22,0.33]	[4%,9%]
E4	[0.00,0.08]	[0%,2%]

Table 15.23 IPCA: correlation intervals between original variables and the first two principal caomponents.

	PC1	PC2
GRA	[−0.99, −0.34]	[−0.99,0.99]
FRE	[0.37,0.99]	[−0.99,0.99]
IOD	[−0.99, −0.20]	[−0.99,0.99]
SAP	[−0.25,0.99]	[−0.99,0.99]

The main graphical results are reported in Figure 15.5 which shows the box representation on the first factor plane and circle of interval correlations between the original variable and the first two principal components (Table 15.23). In this case data are less separated than in the other methods. This is due to the well-known *wrapping* effect of interval operations, especially when linear transformations of data are taken into consideration or when computing non-linear functions of interval data.

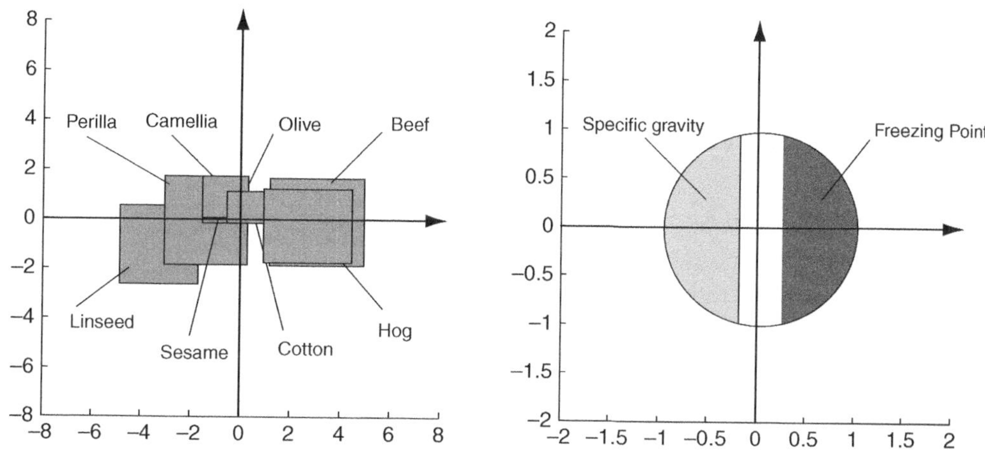

Figure 15.15 IPCA: representation of SOs on the first factor plane and the correlation circle.

15.7 Conclusion and perspectives

The present chapter has described the history of the PCAs developed for the analysis of SOs described by interval data, observing how, from a symbolic–numerical–symbolic treatment of the former PCAs, first a symbolic–hybrid–symbolic treatment was developed that allowed new PCAs to be defined, emphasizing the role of the *size* and of the *shape* of the boxes into the analysis, leading eventually to a symbolic–symbolic–symbolic PCA where data are intervals and the computations are based on interval algebra methods. In all other methods, intervals are coded into bounds, midpoints, radii, etc. and then treated with the usual data analysis techniques based on classical algebra (this means 'by construction'). In the case of ICPA data are 'naturally' treated using 'interval algebra'.

Whereas VPCA and CPCA work on some boxes representative points considered as independent of each other, IPCA was the first attempt to treat boxes, with the great advantage of using well established methods (like classical PCA) which are easy to interpret.

SPCA allowed the cohesiveness of vertices and the size and shape to come into play, but continued to work separately on some boxes' representative points.

MRPCA introduced some interval algebra results in the treatment of interval data, but suffers, for example, from the subjective choice of the rotation operator and continues to treat separately the space spanned by the midpoints and the space spanned by the radii of boxes.

'Spaghetti' PCA, even if it unifies the treatment of midpoints and radii, continued to be a method based on classical numerical treatment.

IPCA, despite being the first fully consistent method for the PCA of interval data (data as well as methods are well defined by the interval algebra theory), suffers from drawbacks deriving from the wrapping effect of the interval operations. In this direction, it is useful to refine the IPCA in order to reduce the *wrapping* effect using also some *paving* techniques for the representation of results and for the computation of more consistent eigenpairs.

Appendix.

Correlation matrix computation and the standardization of data for IPCA
Given two single-valued variables: $X_r = (x_{ir})$, $X_s = (x_{is})$, $i = 1, \ldots, n$, it is known that the correlation between X_r and X_s is given by

$$\mathrm{corr}(X_r, X_s) = h(x_{1,r}, \ldots x_{n,r}; x_{1,s}, \ldots x_{n,s}) = \frac{\mathrm{cov}(X_r, X_s)}{\sqrt{\mathrm{var}(X_r)}\sqrt{\mathrm{var}(X_s)}}. \tag{15.62}$$

Let us consider now the following interval-valued variables:

$$X_r^I = (X_{ir} = [\underline{x}_{ir}, \overline{x}_{ir}]) \quad , \quad X_s^I = (X_{is} = [\underline{x}_{is}, \overline{x}_{is}])_i \qquad i = 1, \ldots, n.$$

The *interval correlation* is computed as follows (Gioia and Lauro 2005):

$$\mathrm{corr}(X_r^I, X_s^I) = \left[\min h(x_{1,r}, \ldots, x_{n,r}; x_{1,s}, \ldots, x_{n,s}), \max h(x_{1,r}, \ldots, x_{n,r}; x_{1,s}, \ldots, x_{n,s})\right] \tag{15.63}$$

where $h(x_{1,r}, \ldots, x_{n,r}; x_{1,s}, \ldots, x_{n,s})$ is the function in (15.62).

Analogously, given the single-valued variable X_r, the *standardized* $S_j = (s_{ir})_i$, of X_r is given by

$$s_{ir} = \frac{x_{ir} - \bar{x}_r}{\sqrt{n \cdot \sigma_r^2}}, \qquad i = 1, \ldots, n, \tag{15.64}$$

where $\overline{x}_r$ and σ_r^2 are the mean and the variance of X_r, respectively.

Given an interval-valued variable X_r^I, following the approach of Gioia and Lauro (2005), the component s_{ir} in (15.64) for each $i = 1, \ldots, n$, transforms into the following function:

$$s_{ir}(x_{ir}, \ldots x_{nr}) = \frac{x_{ir} - \bar{x}_r}{\sqrt{n \cdot \sigma_r^2}} \tag{15.65}$$

as x_{ir} varies in $[\underline{x}_{ir}, \overline{x}_{ir}]$, $i = 1, \ldots, n$. The *standardized interval* component s_{ir}^I of X_r^I may be computed by minimizing/maximizing function (15.64), i.e. calculating the following set:

$$s_{ir}^I = \left[\min s_{ir}(x_{ir}, \ldots x_{nr}), \max s_{ir}(x_{ir}, \ldots x_{nr})\right]. \tag{15.66}$$

This is the interval of the standardized component s_{ir} that may be computed when each component x_{ir} ranges in its interval of values. For computing the interval standardized matrix S^I of an $n \times p$ matrix $\mathbf{X}^\mathbf{I}$, interval (15.66) may be computed for each $i = 1, \ldots, n$ and each $r = 1, \ldots, p$. Given a real matrix $\mathbf{X}$ and denoting by $\mathbf{S}$ the standardized version of $\mathbf{X}$, we define the product matrix $\mathbf{SS}' = (ss'_{ij})$. Given an interval matrix $\mathbf{X}^\mathbf{I}$, the product of $\mathbf{S}^\mathbf{I}$ by its transpose will not be computed by the interval matrix product $(\mathbf{S}')^I\mathbf{S}^\mathbf{I}$ but by minimizing/maximizing each component of **SS'** when x_{ij} varies in its interval of values. The interval matrix is

$$(ss'_{ij})^I = \left[\min ss'_{ij}(x_{ij}, \ldots x_{nj}), \max ss'_{ij}(x_{ij}, \ldots x_{nj})\right]. \tag{15.67}$$

References

Cazes, P. (2002) Analyse factorielle d'un tableau de lois de probabilité. *Revue de Statistique Appliquée*, L (3), 5–24.

Cazes, P., Chouakria, A., Diday, E. and Schektman, Y. (1997) Extension de l'analyse en composantes principales à des données de type intervalle. *Revue de Statistique Appliquée*, XIV(3), 5–24.

Chouakria, A., Diday, E. and Cazes, P. (1998) An improved factorial representation of symbolic objects. In *Studies and Research, Proceedings of the Conference on Knowledge Extraction and Symbolic Data Analysis (KESDA'98)*, pp. 276–289. Luxembourg: Office for Official Publications of the European Communities.

D'Ambra, L. and Lauro, C. N. (1982) Analisi in componenti principali in rapporto a un sottospazio di riferimento. *Rivista di Statistica Applicata* 15(1), 51–67.

de Carvalho, F.A.T. (1992) *Méthodes descriptives en analyse de données symboliques*. Doctoral thesis Université Paris IX Dauphine, Paris.

de Carvalho, F.A.T. (1997) Clustering of constrained symbolic objects based on dissimilarity functions. Indo-French Workshop on Symbolic Data Analysis and its Applications, University of Paris IX.

D'Urso, P. and Giordani, P. (2004) A least squares approach to principal component analysis for interval valued data. *Chemometrics and Intelligent Laboratory Systems*, 70(2), 179–192.

Escofier, B. and Pagès, J. (1988), *Analyse factorielles multiples*. Dunod, Paris.

Gioia, F. and Lauro, N.C. (2005) Basic statistical methods for interval data. *Statistica applicata*, 1.

Gioia, F. and Lauro, N.C. (2006) Principal component analysis on interval data. *Computational Statistics*, 21(2), 343–363.

Ichino, M. (1988) General metrics for mixed features – the cartesian space theory for pattern recognition. In *Proceedings of the 1988 IEEE International Conference on Systems, Man and Cybernetics*, Vol. 1, pp. 494–497. International Academic Publishers, Beijing.

Irpino, A. (2006) 'Spaghetti' PCA analysis: An extension of principal components analysis to time dependent interval data. *Pattern Recognition Letters*, 27(5), 504–513.

Irpino, A., Verde, R. and Lauro N. C. (2003) Visualizing symbolic data by closed shapes. In M. Shader, W. Gaul and M. Vichi (eds), *Between Data Science and Applied Data Analysis*, pp. 244–251. Springer-Verlag, Berlin.

Kiers, H.A.L. and ten Berge J.M.F. (1989) Alternating least squares algorithms for simultaneous components analysis with equal component weight matrices for all populations. *Psychometrika*, **54**, 467–473

Lauro, N.C. and Palumbo, F. (2000) Principal components analysis of interval data: a symbolic data analysis approach. *Computational Statistics*, 15(1), 73–87.

Lauro, N.C., Verde, R. and Palumbo, F. (2000) Factorial data analysis on symbolic objects under cohesion constrains. In H.A.L Kiers, J.-P. Rasson, P.J.F. Groenen and M. Schader (eds), *Data Analysis, Classification, and Related Methods*, Springer-Verlag, Berlin.

Lawson, C. L. and Hanson, R. J. (1974) *Solving Least Squares Problems*. Prentice Hall, Englewood Cliffs, NJ.

Lebart L. Morineau, A. and Piron, M. (1995) *Statistique exploratorie multidimensionelle*. Dunod, Paris.

Moore, R.E. (1966) *Interval Analysis*. Prentice Hall, Englewood Cliffs, NJ.

Neumaier, A. (1990) *Interval Methods for Systems of Equations,* Cambridge University Press, Cambridge.

Palumbo, F. and Lauro, N.C. (2003) A PCA for interval valued data based on midpoints and radii. In H. Yanai, A. Okada, K. Shigemasu, Y. Kano, and J.J. Meulman (eds), *New Developments in Psychometrics*. Springer-Verlag, Tokyo.

Porzio, G., Ragozini, G. and Verde, R. (1998) Generalization of symbolic objects by convex hulls. In *Actes des XXXèmes Journées de Stabistique*, Rennes, pp. 458–460.

Verde, R. and De Angelis, P. (1997) Symbolic objects recognition on a factorial plan. In *NGUS'97*, Bilbao, Spain.

16

Generalized canonical analysis

N. Carlo Lauro, Rosanna Verde and Antonio Irpino

16.1 Introduction

The aim of generalized canonical analysis of symbolic objects (SGCA) is to analyse the relationships between symbolic object (SO) descriptors and SOs on a factor plane. In the symbolic data analysis (SDA) framework this technique allows the treatment of SOs whose descriptors are of different types (interval, categorical multi-valued, modal).

The SGCA of symbolic data (Verde 1997, 1999) is based on an extension of classic generalized canonical analysis (Volle, 1985). The decomposition criterion in the analysis is the squared multiple correlation index computed on coded symbolic variables. The analysis is performed on a coding matrix of the SO descriptions where each row identifies the coordinates of the vertices of the hypercubes associated with each SO in the representation space.

Similarly to the other factor methods (principal component analysis, factor discriminant analysis) employed in SDA, SGCA is a multi-step symbolic–numerical–symbolic procedure. This means it is based on an intermediate coding phase in order to homogenize different kinds of descriptors using, for instance, a *fuzzy* or a *crisp* (*complete disjunctive*) coding system. Therefore, the *symbolic* nature of the input data is retrieved in a *symbolic* interpretation of the output.

As with SPCA, in the analysis we consider a cohesion constraint in order to preserve the unity of the SO in the analysis. The optimized criterion is additive and each component is an expression of the power of each descriptor to discriminate between SOs. According to the classic SO representation, the SOs are geometrically visualized by hypercubes, in the original coding or factor space. The coordinates of the vertices of the hypercubes, after the data coding process, are the values of the row vectors of the table obtained from the numerical transformation of the symbolic data.

Symbolic Data Analysis and the SODAS Software Edited by E. Diday and M. Noirhomme-Fraiture

SGCA looks for factor axes, as a linear combination of the coded symbolic descriptors, in order to analyse, on the reduced subspace (factor plane), both relationships among SOs and associations between the categories of the symbolic variables. Such geometrical results allow SOs to be compared with respect to shape, size and location of their images on the factor plane, and the associations between descriptors to be analysed graphically.

SGCA can also be performed when logical rules and taxonomies are considered in the SO descriptions, which reduce the representation space and introduce a hierarchy on the categories of categorical multi-valued variables.

In order to consider such supplementary information, the original SO description space has to be decomposed into a consistent subspace according to the spatial constraints deriving from conditions defined by the logical rules. A taxonomy on the categories of a categorical multi-valued variable allows the data to be represented with respect to different levels of the hierarchy by going up or down in the *tree structure*.

As SGCA is usually considered as a general factor analysis procedure, it can be implemented to homogenize and quantify the symbolic predictors that in factor discriminant analysis on SOs can be of different type (see Chapter 18). Moreover, when all SOs are described by categorical multi-valued variables, SGCA leads to multiple correspondence analysis on symbolic data (SMCA).

16.2 SGCA input data

Let E be a set of SOs described by p symbolic variables $Y_1, \ldots, Y_p$ of different types (categorical multi-valued, interval, modal) with domain in D_j, $j=1, \ldots, p$ (Bock and Diday, Chapter 3, 2000). Symbolic data are collected in a symbolic data table $\mathbf{X}$. The rows contain the symbolic descriptions of the SOs and the columns the symbolic descriptors. Each cell of the table contains different types of data: a set of categories for the categorical multi-valued variable, an interval of values for the interval variables and a frequency, probability or belief distribution for modal variables. SGCA can take into account logical dependence rules, taxonomies and missing values present in the SOs description.

16.3 Strategy

Like most factor techniques, SGCA is a *symbolic–numerical–symbolic* technique (Diday, 1987) and consists of the following steps:

(i) symbolic data coding process;

(ii) extension of the classical GCA to the transformed symbolic data;

(iii) Symbolic interpretation of the results according to the nature of the original symbolic data.

16.3.1 Coding of the symbolic descriptors

GCA is based on the optimization of a numerical criterion. In order to extend this technique to symbolic data analysis, both numerical and categorical descriptors have to be homogenized. Therefore, the first phase of the proposed approach involves the numerical coding of the descriptors $Y_j (j=1, \ldots, p)$ of each SO in a table $\mathbf{Z}_{ij}$.

For instance, the symbolic description

$$s(i) = [\text{income}(i) = [1.0, 3.5]] \wedge [\text{profession}(i) = \{\text{employer}, \ \text{worker}\}] \wedge [\text{sex}(i) = \{\text{M}(0.55), \text{F}(0.45)\}]$$

contains the interval description of income, the categorical multi-valued description of profession and the modal description of sex.

The system of coding used differs according to the type of descriptor. In particular, if Y_j is a categorical multi-valued variable, the coding table $\mathbf{Z}_{ij}$, associated with the ith SO and the jth variable, is a binary matrix of 0/1 values. The ith SO is coded with respect to k_{ij} rows of $\mathbf{Z}_{ij}$, k_{ij} being the categories that it presents for the descriptor Y_j; in the above example, taking the domain of the profession variable as the set of categories {employer, worker, manager}, the coded description of the profession is done in two rows as follows:

$$\begin{bmatrix} \text{employer} & \text{worker} & \text{manager} \\ \hline 1 & 0 & 0 \\ 0 & 1 & 0 \end{bmatrix}.$$

If Y_j is a modal variable, the ith SO is coded in the table $\mathbf{Z}_{ij}$ according to the frequency or probability values that it assumes for the categories of Y_j. In the above example, taking the domain of the sex variable as the set of couples (category, weight) $\{\text{M}(\pi_\text{M}), \text{F}(\pi_\text{F})\}$, where the weights are relative frequencies, the coded description of the profession is done in one row as follows:

$$\begin{bmatrix} \text{M} & \text{F} \\ \hline 0.55 & 0.45 \end{bmatrix}.$$

If Y_j is an interval variable, it is categorized and codified according to a *fuzzy* coding system (e.g. using basic spline functions which allow the numerical information of the original numerical variables to be preserved). Usually, linear functions are used to code a numerical variable, with values in [0,1], with respect to a certain number of categories (e.g. low, medium and high). Thus, the two bounds of the interval of values, assumed by the ith SO, are coded in two different rows of the coding matrix $\mathbf{Z}_{ij}$. However, we have observed that a fuzzy coding of the only interval bounds with respect to three categories (low, medium, high) does not allow us to retrieve all the points of the interval according to the usual reconstruction formulae of the B-splines (De Boor, 1978). In order to overcome this numerical drawback, the coding of each interval must be performed not only for the two bounds but also for an intermediate value coincident with the medium point of the coding. In this way, the interval is split into two consecutive intervals [lower bound, medium value], [medium value, upper bound] and they are coded according to the fuzzy system (e.g. semilinear) in four rows.

In order to code the interval description of income for the description presented above, we need to categorize the interval variable and then, using a semilinear B-spline, to code the interval in a fuzzy way. Let us assume that 0 represents the low level of income, 2 the medium level and 4 the high level. Figure 16.1 shows a scheme for the fuzzy coding of the interval description.

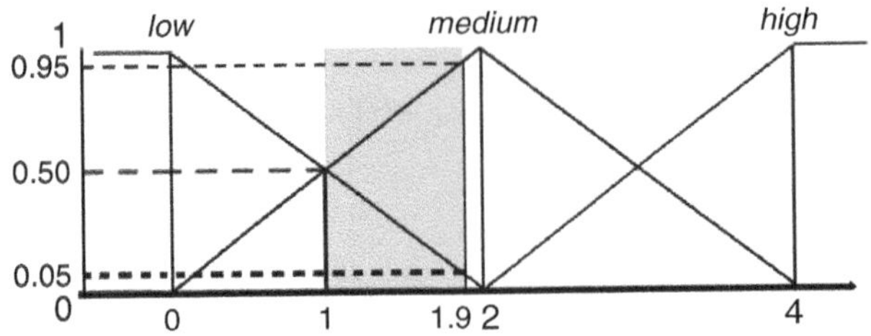

Figure 16.1 Scheme for the B-spline fuzzy coding of an interval description.

To be consistent with the coding system and the topological spaces involved, the fuzzy coding of the interval description is done as follows:

$$[1.0, 1.9] \Rightarrow \begin{matrix} 1.0 \Rightarrow \\ 1.9 \Rightarrow \end{matrix} \left[\begin{array}{ccc} & \text{Income} & \\ \text{low} & \text{medium} & \text{high} \\ \hline 0.5 & 0.5 & 0 \\ 0.05 & 0.95 & 0 \end{array} \right].$$

The complete coding of the symbolic description associated with the ith SO, as reported in the above example, is then the following:

$$\mathbf{Z}_i = [\mathbf{Z}_{i1} | \mathbf{Z}_{i2} | \mathbf{Z}_{i3}] = \left[\begin{array}{ccc|ccc|cc} & \text{Income} & & & \text{Profession} & & \text{Sex} & \\ \text{low} & \text{medium} & \text{high} & \text{employer} & \text{worker} & \text{manager} & \text{M} & \text{F} \\ \hline 0.5 & 0.5 & 0 & 1 & 0 & 0 & 0.55 & 0.45 \\ 0.05 & 0.95 & 0 & 1 & 0 & 0 & 0.55 & 0.45 \\ 0.5 & 0.5 & 0 & 0 & 1 & 0 & 0.55 & 0.45 \\ 0.05 & 0.95 & 0 & 0 & 1 & 0 & 0.55 & 0.45 \end{array} \right].$$

This represents a large increase in the number of the rows of the matrix $\mathbf{Z}_{ij}$ with consequent effect on the computational cost of the procedure.

In the space of the coding descriptors, the geometrical representation of the generic ith SO, in terms of hypercubes, is obtained by the Cartesian product of the $\mathbf{Z}_{ij}$ rows ($j = 1, \ldots, p$) and is denoted $\mathbf{Z}_{ij}$. By a vertical concatenation of all the matrices $\mathbf{Z}_{ij}$ (for all the SOs: $i = 1, \ldots, n$) we obtain p matrices $\mathbf{Z}_j (j = 1, \ldots, p)$, of dimension $N \times k_j$, where k_j is the number of categories of Y_j when it is a categorical multi-valued or modal variable. If Y_j is numerical (quantitative single-valued or interval) then k_j is the number of categories (usually $k_j = 2$ or 3) into which the variable has been categorized and coded according to a fuzzy system.

Finally, the number N of rows depends on the number of variables of different type which influence, in several ways, the number of coding rows of each $\mathbf{Z}_{ij}$.

Let q be the number of categorical multi-valued variables and g the number of interval descriptors. Then the number of rows of each $\mathbf{Z}_{ij}$ is $N_i = 2^{g+f_i} \times \prod_{j=1}^{q} k_{ij}$ (where f_i is the number of intervals in the ith SO description that contain the medium values of the interval variable domains with respect to which the fuzzy coding is done). The number of rows of the matrices $\mathbf{Z}_{ij} (j = 1, \ldots, p)$ is $N = \sum_{i=1}^{n} N_i$.

Finally, the global coding matrix, denoted by $\mathbf{Z}_{N\times K}$, containing all the coordinates of the hypercube vertices representing the n SOs, can be also seen as formed by the juxtaposition of the p binary or fuzzy coding matrices:

$$\mathbf{Z}_{N\times K} = \left[\mathbf{Z}_1 | \ldots | \mathbf{Z}_j | \ldots | \mathbf{Z}_p\right] \tag{16.1}$$

where N is the number of rows of the upgraded coding matrices $\mathbf{Z}_j$ (for $j=1,\ldots,p$). The total number K of categories of all the Y_j transformed variables is equal to $K = h \times g + \sum_{j=1}^{q} k_j$, where h is the number of categories associated with interval or numerical variables coded in a fuzzy way (usually $h=3$) and k_j is the number of categories for each of the q categorical or modal variables Y_j.

Moreover, in the coding phase, logical relations among the descriptors, as well as taxonomies, can be taken into account as will be shown in Sections 16.3.2 and 16.3.3.

16.3.2 Logical rules in the SO description space

Logical conditions reduce the SO description space, according to logical coherence constraints expressed by relationships between symbolic variables, as

$$\text{if } [Y_k \in S_k] \text{ then } [Y_j = \text{NA}] \quad \text{or} \quad \text{if } [Y_k \in S_k] \text{ then } [Y_j = S_j].$$

In the first case, we have a *hierarchical dependence* (HD) rule which restricts the domains of a descriptor Y_k if another descriptor Y_j assumes a subset of values S_j of its domain D_j. This means that if Y_k assumes values in $S_k \subseteq D_k$ then Y_j becomes *non-applicable*. In the second case, we have a *logical dependence* (LD) rule which provides a restriction of the domain of a descriptor Y_j to a subset $S_j \subseteq D_j$ if another descriptor Y_k assumes a subset of values $S_k \subseteq D_k$.

We describe a redefinition of SOs in the presence of a sequence of logical inductions between their descriptors. SOs are geometrically represented by hypercubes, where each dimension corresponds to the values taken by a descriptor. Of course, the original SO description space is reduced whenever dependence rules are introduced. Thus, the real dimension of the hypercubes associated with the SOs goes down. In fact, *induction* allows us to identify a subset of the space of the description that is not admissible.

Therefore, we aim to look for a subset of disjoint symbolic sub-objects consistently with restrictions given by the rules. A possible solution was proposed by Csernel and de Carvalho (1998) using the *normal symbolic form* algorithm when the dependence rules are structured into a tree of dependence graphs. Alternatively, we can redefine the logical rules in terms of *implication* (LD) and *equivalence* (HD) relations between variables, typical of first-order logic, and consider their properties in order to search for a coherent decomposition of the SOs, independently of the order of the multiple rules (for more details, see Irpino and Verde, 2004).

Briefly, the SO decomposition scheme, here proposed, can be synthesized as follows. Let $\left[Y_j \subseteq S_j\right] \rightarrow [Y_k \subseteq S_k]$ be a dependence rule between two descriptors Y_j and $Y_k (j \neq k)$, and $P_{jk} = D_j \times D_k$ the Cartesian product of the description space of the two descriptors. Denoting the premise as $a = \left[Y_j \subseteq S_j\right] \times D_k$ and the consequence as $b = [Y_k \subseteq S_k] \times D_j$, we may write the rule r as $a \rightarrow b$, and in logic form, $\bar{a} \vee b$, where $\bar{a}$ is the complement of a. A dependence rule allows us to identify a subset of P_{jk} that is not admissible in the description

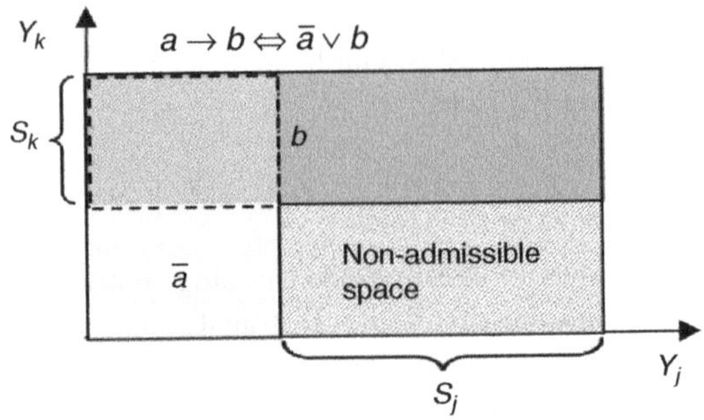

Figure 16.2 Decomposition of the P_{jk} space.

space. In particular, this subset is given by the expression $a \wedge \overline{b}$ that is the complement of $\bar{a} \vee b$.

For instance, if $[Y_j \geq l_j] \to [Y_k \geq l_k]$ the space of admissibility (and the complementary space of non-admissibility) is graphically represented in Figure 16.2.

Given the ith SO, an element of the set E, which takes values d_{ij} and d_{ik} for Y_j and Y_k, the assertion $s_i = [Y_j = d_{ij}] \wedge [Y_k = d_{ik}]$ is decomposed into two sub-objects with descriptions that are consistent with the condition expressed by the logical rule:

$$(s_i \wedge \bar{a}) : s_{i1} = [Y_j = (d_{ij} \cap \bar{S}_j)] \wedge [Y_k = d_{ik}]$$
$$(s_i \wedge b) : s_{i2} = [Y_j = d_{ij}] \wedge [Y_k = (d_{ik} \cap S_k)].$$

These descriptions give the graphical representations of the SOs shown in Figure 16.3. As we can observe, the two descriptions overlap. This means that a redundancy in the SO representation space has been generated by this decomposition. To solve this problem, we introduce the third excluded rule: $\bar{a} \vee a$. Applied to the rule, it allows a partition of the admissible space instead of an overlapping of the admissible space (Figure 16.4). In such a

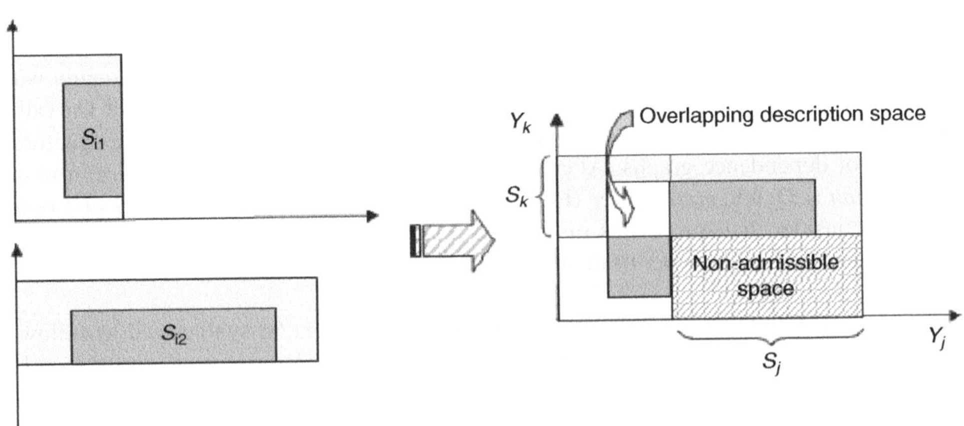

Figure 16.3 Representations of the two symbolic sub-objects.

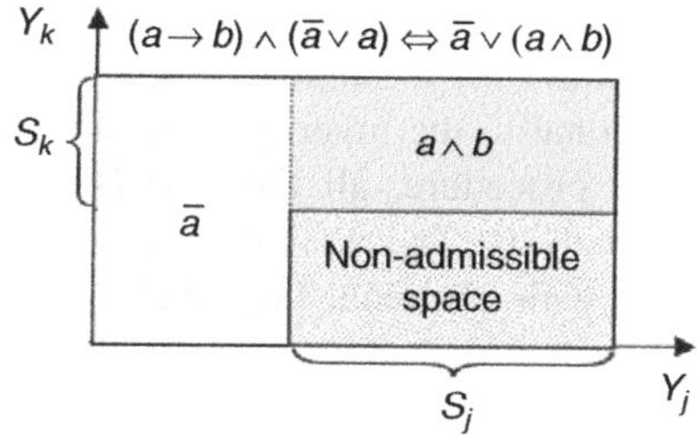

Figure 16.4 Decomposition of the P_{jk} space into non-overlapping.

way, since $d_{ik} \cap S_k = \emptyset$, the object s_i can be now decomposed into the two sub-objects s_{i1} and s_{i2} having the following 'no-overlap' descriptions:

$$(s_i \wedge \bar{a}) : s_{i1} = \left[Y_j = (d_{i,j} \cap \bar{S}_j)\right] \wedge \left[Y_k = d_{i,k}\right]$$
$$(s_i \wedge (a \wedge b)) : s_{i2} = \left[Y_j = (d_{i,j} \cap S_j)\right] \wedge \left[Y_k = (d_{i,k} \cap S_k)\right].$$

The graphical representation of the two descriptions of the symbolic sub-objects is shown in Figure 16.5.

Because of the splitting of the initial SOs due to the rules into sub-objects, the number of SOs to be considered in the analysis increases. Indeed, the analysis in the presence of rules is performed using the sub-objects instead of the initial SOs. For simplicity, we leave denoted by n the total number of SOs, considering the initial ones, when no split is performed, otherwise n denotes the number of the whole set of sub-objects.

The sub-objects are coded according to the same system of coding of the original SOs. An indicator variable G is also introduced in order to keep the information about the membership of the coded sub-objects in the same original SO. The binary matrix associated with the variable G is then denoted **G**. Its dimensions are N and n.

N represents the number of rows in which the vertices of the objects or sub-objects have been coded (N increases considerably depending on to the number of splits of the SOs into sub-objects) and n is the number of elements of the set E of SOs.

Further, on the basis of the rules, some objects can be split into several sub-objects, while others remain undivided. In order to avoid the split SOs playing a greater role in the

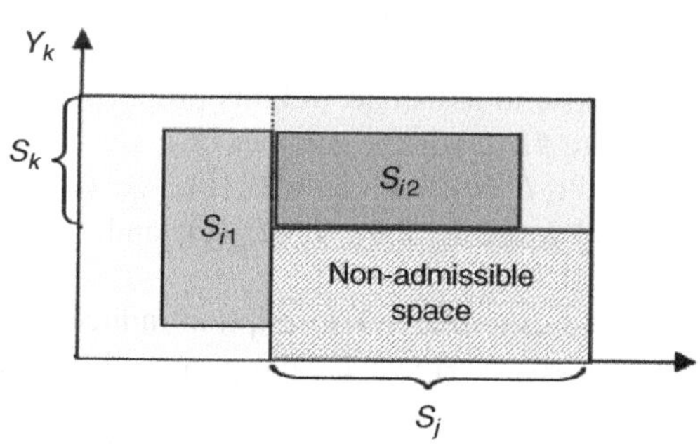

Figure 16.5 Representations of the two symbolic sub-object descriptions with no overlap.

analysis than undivided ones, because of an increase in the number of rows N_i associated with their descriptions in the matrix **Z**, a suitable weight system has to be considered. Thus, assigning weights proportional to the inverse of the number of symbolic sub-objects generated by the decomposition procedure, all the new (split and undivided) SOs of E will play the same role in the analysis. The weights can be easily gathered in a diagonal matrix $\mathbf{\Pi}$, of dimension $n \times n$, with elements the inverse of the number of sub-objects into which each SO has been decomposed. For example, if there are only two SOs, if the first SO has been split into four sub-objects and the last SO is undivided, we have $n = 5$. The first four sub-objects have a 0.25 weight (1/4) while for the last one (the undivided) the weight is equal to 1. For simplicity, we can also consider the matrix $\tilde{\mathbf{G}} = \mathbf{G}\Pi$ where the ith column is equal to zero except for those elements indicating the membership of the vertices to the ith SO (undivided object or sub-object) that are equal to the weights in $\mathbf{\Pi}$.

16.3.3 Taxonomical structure on the categories of the symbolic categorical multi-valued descriptors

The SGCA is the only factorial technique in SDA which allows categorical multi-valued variables to be analysed. In particular, when all the SO descriptors are categorical multi-valued, SGCA is to be considered as a multiple correspondence analysis on SOs.

Taxonomies related to hierarchical relations between the categories of categorical single- and multi-valued variables can cause different effects in the analysis of symbolic data related to the different degrees of generality of the categories (e.g. we have different results if SOs are described by the variable 'eye colour' categorized as {black, brown, green, blue} instead of {dark, pale}). Therefore, to consider relationships among categories in SO descriptions, we suggest relating a suitable coding of symbolic data to each level of the hierarchical structure.

A sequence of levels of a hierarchy can be defined for a categorical multi-valued variable Y_j as follows. Starting with the root node r (level 0), the first level is constituted by the direct descendant nodes (both terminals or internals) of r, e.g. $L_1(r)$; the second level is formed by the direct descendant nodes (both terminals or internals) of the nodes in $L_1(r)$, e.g. $L_2(a), L_2(b), \ldots$, with $a, b, \ldots \in L_1(r); \ldots$; the hth level contains as elements all the nodes descending from all the nodes f (terminals or internals) belonging to the level $h-1$; the last level contains all the terminal nodes of the tree structure.

In order to take into account the different level of generalization on the categories, we propose to weight the categories assumed by the SOs associated with the nodes at each level of the hierarchical structure of the categorical multi-valued variable Y_j. Let $a, b \in L(r)$; at level 1, we propose to associate weights proportionally to the path distance: $\theta_a = \theta_b = 1/\#L_1(r) = 0.5$ (where $\#L_1(r)$ is the number of nodes at first level). The weights taken by the elements $x, y, \ldots \in L_h(g), s, t, \ldots \in L_h(k), \ldots$ (with $g, k \ldots \in L_{h-1}(f)$, for all f at level h–1) at the level h are $\theta_x = \theta_g \cdot 1/\#L_h(g)$, and so on. Thus, the system of coding results is normalized to 1 at each level.

In SDA techniques, taxonomies on the SO description induce a reduction in the number of categories in the description space. However, the proposed system of coding, based on *fuzzy* transformation, allows us to keep the structure information on the SOs.

In accordance with the SGCA aim of representing the relationships between the categories of multi-valued variables, we analyse the data, considering as categories all the terminal leaves. In order to display the intermediate nodes (more general categories) of the hierarchical structure, we project the centre of gravity of the coordinates of their represented sons' leaves. For example, we analyse data considering the variable 'eye colour' categorized as {black, brown, green, blue}. In order to display the categories {dark, pale} we respectively compute the centre of gravity of {black, brown} and of {green, blue} and then we project them.

16.3.4 GCA on the matrix Z under cohesion constraint

The procedure is carried out on the global coding matrix $\mathbf{Z}_{N\times K}$ that can also be considered as constituted by the juxtaposition of the p fuzzy and/or disjunctive coding matrices, $\mathbf{Z} = [\mathbf{Z}_1|\ldots|\mathbf{Z}_j|\ldots|\mathbf{Z}_p]$. If a previous decomposition of the SOs into sub-objects by logical rules has been provided, the matrix $\mathbf{Z}$ is replaced by the matrix $\tilde{\mathbf{Z}} = \tilde{\mathbf{G}}'Z$. This takes into account the membership of the sub-objects in the same SOs and the appropriate system of weights to grant the same importance to all the SOs in the analysis. For simplicity, we continue to denote by $\mathbf{Z}$ the matrix of coded data.

In order to keep the cohesion among the vertices of the hypercube associated with each SO, we introduce a cohesion constraint on the rows of the matrix $\mathbf{Z}$. Therefore, $\mathbf{A}_{N\times n}$ is an indicator matrix that identifies the membership of the N vertices in the different n SOs. Moreover, to preserve the unity of the decomposed SO, consistently with the conditions expressed by logical rules, they will be assigned with the same code to their original SO in the indicator matrix $\mathbf{A}$.

The maximized criterion expresses the explanatory power of the coded descriptors with respect to the set of hypercube vertices that geometrically represent each SO. The factor axes are obtained as the solution of the following characteristic equation:

$$\frac{1}{N}\mathbf{A}'\mathbf{Z}\Sigma^{-1}\mathbf{Z}'\mathbf{A}\xi_m = \tilde{\boldsymbol{\mu}}_m\tilde{\xi}_m \tag{16.2}$$

under the orthonormality constraints $\tilde{\xi}_m'\tilde{\xi}_m = 1$ and $\tilde{\xi}_m'\tilde{\xi}_{m'} = 0$ (for $m \neq m'$), where $\boldsymbol{\Sigma}$ is a block diagonal matrix of dimension $K \times K$ as follows:

$$\boldsymbol{\Sigma} = \begin{bmatrix} \mathbf{Z}_1'\mathbf{Z}_1 & \mathbf{0} & \cdots & \mathbf{0} \\ \mathbf{0} & \mathbf{Z}_2'\mathbf{Z}_2 & \cdots & \mathbf{0} \\ \vdots & \vdots & \ddots & \vdots \\ \mathbf{0} & \mathbf{0} & \cdots & \mathbf{Z}_p'\mathbf{Z}_p \end{bmatrix}.$$

The maximum number of non-trivial eigenvalues is $K-p+1$ (provided that the total number of all categories $K = \sum_{j=1}^{p} k_j$ is less than N).

If the matrix $\mathbf{A}$ is assumed centred, the trace of the matrix decomposed in the analysis corresponds, up to a constant, to the sum of the Goodman–Kruskal τ indices (Goodman and Kruskal, 1954), computed for each categorized descriptor (binary or fuzzy coded):

$$\mathrm{tr}\left(\mathbf{A}'\mathbf{Z}\Sigma^{-1}\mathbf{Z}'\mathbf{A}\right) = \sum_{j=1}^{p}\tau_{Z_j}. \tag{16.3}$$

The eigenvectors of dual space are given by

$$\tilde{\beta}_m = \frac{1}{\sqrt{\tilde{\mu}_m}} \boldsymbol{\Sigma}^{-1}\mathbf{Z}'\mathbf{A}\xi_m.$$

The vector of the coordinates of the vertices belonging to the ith SO for the mth factor axis is:

$$\tilde{\varphi}_{im} = \mathbf{Z}_i\boldsymbol{\Sigma}^{1/2}\tilde{\beta}_m \qquad m = 1, \ldots, K - p + 1.$$

In the GCA the amount of *information* contribution associated with each axis is given by the corresponding eigenvalue $\tilde{\mu}_m$. The ratio between the squared coordinate and the eigenvector norm, both with respect to the axis m,

$$\mathrm{CTA}_{im} = \frac{1}{\tilde{\mu}_m}\widetilde{\varphi}'_{im}\widetilde{\varphi}_{im},$$

is a global measure of the contribution of the ith SO to mth axis. It is expressed as the sum of the CTA values of all the vertices belonging to it. The relative contribution, giving the quality of the SO representation, is given by

$$\mathrm{CTR}_{im} = \left(\sum\nolimits_{j=1,\ldots,M} \widetilde{\varphi}'_{ij}\widetilde{\varphi}_{ij}\right)^{-1} \widetilde{\varphi}'_{im}\widetilde{\varphi}_{im},$$

where $M \leq K - p + 1$ is the number of the non-trivial eigenvectors. The contributions of the categories are computed according to their correlation with respect to the new axes.

16.4 Example

Example 16.1. In the following example, vertices SGCA is performed on the file car.xml, which describes 11 characteristics of 33 car models. Once SGCA is executed, textual and graphical outputs are available. The input symbolic data table is shown in Table 16.1. The inertia explained by the first six factors is 95%, as shown in Table 16.2.

Considering the first three factor axes, the coordinate intervals of the 33 car models are presented in Table 16.3. The positions of the 33 car models on the first factor plane are shown in Figure 16.6.

It is also possible to consider the coordinates of the categories on the first three factor axes (Table 16.4) and their graphical representation (Figure 16.7).

In order to interpret the axes in terms of the original categories it is possible to obtain the axis description table (Table 16.5), based on the absolute contribution of each of the categories to the axes (the column sum is equal to 1). For each of the first three axes, the categories are ordered by their absolute contribution and by the sign of the coordinates of the categories; not all the categories are displayed, only those one that have a contribution greater than the mean contribution of all categories to the axis. The sign (− or +) represents the axis orientation, and when a category contributes with a + (−) sign it is used to explain the positive (negative) half of the axis.

In this example, the first factor axis separates those cars having high values for Length, Height, Width, and Axes distance on the right (luxury cars: Audi A8, Audi A7, Lancia K, BMW 7, Mercedes S), from those having low Width, medium Height, high Acceleration

Table 16.1 Symbolic data table.

Model	Price	Cubic capacity	Fuel P = Petrol, D = diesel	Traction F = front, R = rear I = 4WD	Max speed	Acceleration	Axes distance	Length	Width	Height
Alfa 145	[27.8,33.5]	[1370,1910]	P , D	F	[185,211]	[8.3,11.2]	[254,254]	[406,406]	[171,171]	[143,143]
Alfa 156	[41.5,62.2]	[1598,2492]	P	F	[200,227]	[8.5,10.5]	[260,260]	[443,443]	[175,175]	[142,142]
Alfa 166	[64.4,88.7]	[1970,2959]	P	F	[204,211]	[9.8,9.9]	[270,270]	[472,472]	[182,182]	[142,142]
Aston Martin	[260.5,460]	[5935,5935]	P	R	[298,306]	[4.7,5]	[259,269]	[465,467]	[183,192]	[124,132]
Audi A3	[40.2,68.8]	[1595,1781]	P	F, I	[189,238]	[6.8,10.9]	[250,251]	[415,415]	[174,174]	[142,142]
Audi A6	[68.2,140.2]	[1781,4172]	P	F, I	[216,250]	[6.7,9.7]	[276,276]	[480,480]	[181,181]	[145,145]
Audi A8	[123.8,171.4]	[2771,4172]	P	I	[232,250]	[5.4,10.1]	[289,289]	[503,503]	[188,188]	[144,144]
BMW 3 series	[45.4,76.3]	[1796,2979]	P	R , I	[201,247]	[6.6,10.9]	[273,273]	[447,447]	[174,174]	[142,142]
BMW 5 series	[70.2,198.7]	[2171,4398]	P	R	[226,250]	[6.7,9.1]	[283,283]	[478,478]	[180,180]	[144,144]
BMW 7 series	[104.8,276.7]	[2793,5397]	P	R	[228,240]	[7,8.6]	[293,307]	[498,512]	[186,186]	[143,143]
Ferrari	[240.2,391.6]	[3586,5474]	P	R	[295,298]	[4.5,5.2]	[260,260]	[476,476]	[192,192]	[130,130]
Punto	[19.2,30.8]	[1242,1910]	P , D	F	[155,170]	[12.2,14.3]	[246,246]	[380,384]	[166,166]	[148,148]
Fiesta	[19.2,24.7]	[1242,1753]	P , D	F	[167,167]	[13.1,13.9]	[245,245]	[383,383]	[163,163]	[132,132]
Focus	[27.4,34]	[1596,1753]	P , D	F	[185,193]	[10.8,11]	[262,262]	[415,415]	[170,170]	[143,143]
Honda NSK	[205.2,215.2]	[2977,3179]	P	R	[260,270]	[5.7,6.5]	[253,253]	[414,414]	[175,175]	[129,129]
Lamborghini	[413,423]	[5992,5992]	P	I	[335,335]	[3.9,3.9]	[265,265]	[447,447]	[204,204]	[111,111]
Lancia Y	[19.8,29]	[1242,1242]	P	F	[158,174]	[11.2,14.1]	[238,238]	[372,372]	[169,169]	[144,144]
Lancia K	[58.8,81.3]	[1998,2959]	P	F	[212,220]	[8.9,9.2]	[270,270]	[469,469]	[183,183]	[146,146]

Table 16.1 *(continued)*.

Model	Price	Cubic capacity	Fuel P = Petrol, D = diesel	Traction F = front, R = rear I = 4WD	Max speed	Acceleration	Axes distance	Length	Width	Height
Maserati GT	[155,159.5]	[3217,3217]	P	R	[280,290]	[5.1,5.7]	[266,266]	[451,451]	[182,182]	[131,131]
Mercedes SL	[132.8,262.5]	[2799,5987]	P	R	[232,250]	[6.1,9.7]	[252,252]	[447,447]	[181,181]	[129,129]
Mercedes C	[55.9,115.2]	[1998,3199]	P	R	[210,250]	[5.2,11]	[272,272]	[453,453]	[173,173]	[143,143]
Mercedes E	[69.2,389.4]	[1998,5439]	P	R	[222,250]	[5.7,9.7]	[283,283]	[482,482]	[180,180]	[144,144]
Mercedes S	[128.2,394.3]	[3199,5786]	P	R	[210,240]	[7.2,8.4]	[297,309]	[504,516]	[186,186]	[144,144]
Nissan Micra	[18.4,24.1]	[998,1348]	P	F	[150,164]	[12.5,15.5]	[236,236]	[375,375]	[160,160]	[144,144]
Corsa	[19.2,30.6]	[973,1796]	P	F	[155,202]	[9,17]	[249,249]	[382,382]	[165,165]	[144,144]
Vectra	[36.4,49.0]	[1598,2171]	P , D	F	[193,207]	[10.5,12.5]	[264,264]	[450,450]	[171,171]	[143,143]
Porsche	[147.7,246.4]	[3387,3600]	P	R , I	[280,305]	[4.2,5.2]	[235,235]	[443,444]	[177,183]	[130,131]
Twingo	[16.9,23.4]	[1149,1149]	P	F	[151,168]	[11.7,13.4]	[235,235]	[343,343]	[163,163]	[142,142]
Rover 25	[21.4,33.0]	[1119,1994]	P , D	F	[160,185]	[10.7,15]	[251,251]	[399,399]	[169,169]	[142,142]
Rover 75	[50.4,65.3]	[1796,2497]	P	F	[195,210]	[10.2,11.6]	[275,275]	[475,475]	[178,178]	[143,143]
Skoda Fabia	[19.5,32.6]	[1397,1896]	P , D	F	[157,183]	[11.5,16.5]	[246,246]	[396,396]	[165,165]	[145,145]
Skoda Octavia	[27.4,48.6]	[1585,1896]	P , D	F	[190,191]	[11.1,11.8]	[251,251]	[452,452]	[173,173]	[143,143]
Passat	[39.6,63.4]	[1595,2496]	P , D	F , I	[192,220]	[9.6,12.7]	[270,270]	[470,470]	[175,175]	[146,146]

Table 16.2 SGCA eigenvalues.

Eigenvalues	Inertia	Percentage of explained inertia	Cumulative % of inertia	Histogram 0– – –50%– – –100%
1	0.00088	48.483	48.483	|************
2	0.00041	22.669	71.152	|*****
3	0.00021	11.725	82.877	|**
4	0.00010	5.240	88.117	|*
5	0.00007	3.944	92.061	|*
6	0.00005	2.937	94.998	|

Table 16.3 Coordinate intervals of car models.

Model	Factor 1	Factor 2	Factor 3
Alfa 145	[−0.0102, −0.0009]	[−0.0080, 0.0009]	[−0.0124, −0.0090]
Alfa 156	[−0.0015, 0.0061]	[0.0027, 0.0086]	[−0.0109, −0.0088]
Alfa 166	[0.0085, 0.0146]	[0.0036, 0.0080]	[−0.0084, −0.0068]
Aston Martin	[−0.0057, 0.0008]	[0.0124, 0.0167]	[0.0104, 0.0123]
Audi A3	[−0.0048, 0.0063]	[−0.0010; 0.0105]	[−0.0108, −0.0067]
Audi A6	[0.0125, 0.0265]	[0.0015, 0.0150]	[−0.0093, −0.0029]
Audi A8	[0.0247, 0.0287]	[0.0096, 0.0127]	[−0.0049, −0.0008]
BMW 3 series	[−0.0083, 0.0125]	[0.0103, 0.0238]	[−0.0108, 0.0034]
BMW 5 series	[0.0047, 0.0139]	[0.0173, 0.0224]	[0.0016, 0.0071]
BMW 7 series	[0.0103, 0.0185]	[0.0141, 0.0206]	[0.0053, 0.0102]
Ferrari	[−0.0019, 0.0024]	[0.0140, 0.0181]	[0.0094, 0.0117]
Punto	[−0.0127, −0.0029]	[−0.0174, −0.0080]	[−0.0114, −0.0082]
Fiesta	[−0.0205, −0.0140]	[−0.0139, −0.0070]	[−0.0092, −0.0069]
Focus	[−0.0090, −0.0030]	[−0.0055, 0.0004]	[−0.0123, −0.0100]
Honda NSK	[−0.0019, −0.0010]	[0.0204, 0.0211]	[0.0061, 0.0070]
Lamborghini	[0.0058, 0.0060]	[0.0042, 0.0043]	[0.0039, 0.0039]
Lancia Y	[−0.0126, −0.0084]	[−0.0121, −0.0098]	[−0.0087, −0.0075]
Lancia K	[0.0124, 0.0183]	[0.0024, 0.0066]	[−0.0094, −0.0077]
Maserati GT	[0.0009, 0.0016]	[0.0223, 0.0226]	[0.0071, 0.0077]
Mercedes SL	[−0.0025, 0.0036]	[0.0172, 0.0234]	[0.0035, 0.0087]
Mercedes C	[−0.0047, 0.0080]	[0.0181, 0.0246]	[−0.0002, 0.0059]
Mercedes E	[0.0041, 0.0148]	[0.0150, 0.0223]	[0.0013, 0.0089]
Mercedes S	[0.0097, 0.0199]	[0.0116, 0.0192]	[0.0060, 0.0106]
Nissan Micra	[−0.0181, −0.0135]	[−0.0154, −0.0120]	[−0.0077, −0.0065]
Corsa	[−0.0153, −0.0033]	[−0.0124, −0.0035]	[−0.0102, −0.0072]
Vectra	[−0.0065, 0.0029]	[−0.0041, 0.0045]	[−0.0122, −0.0091]
Porsche	[−0.0067, 0.0081]	[0.0095, 0.0192]	[−0.0014, 0.0106]
Twingo	[−0.0173, −0.0141]	[−0.0149, −0.0129]	[−0.0074, −0.0065]

Table 16.3 (*continued*).

Model	Factor 1	Factor 2	Factor 3
Rover 25	[−0.0179, −0.0055]	[−0.0127, −0.0014]	[−0.0115, −0.0076]
Rover 75	[0.0061, 0.0122]	[0.0009, 0.0057]	[−0.0090, −0.0076]
Skoda Fabia	[−0.0170, −0.0051]	[−0.0152, −0.0050]	[−0.0111, −0.0073]
Skoda Octavia	[−0.0083, −0.0002]	[−0.0062, 0.0007]	[−0.0116, −0.0088]
Passat	[0.0012, 0.0178]	[−0.0062, 0.0124]	[−0.0132, −0.0084]

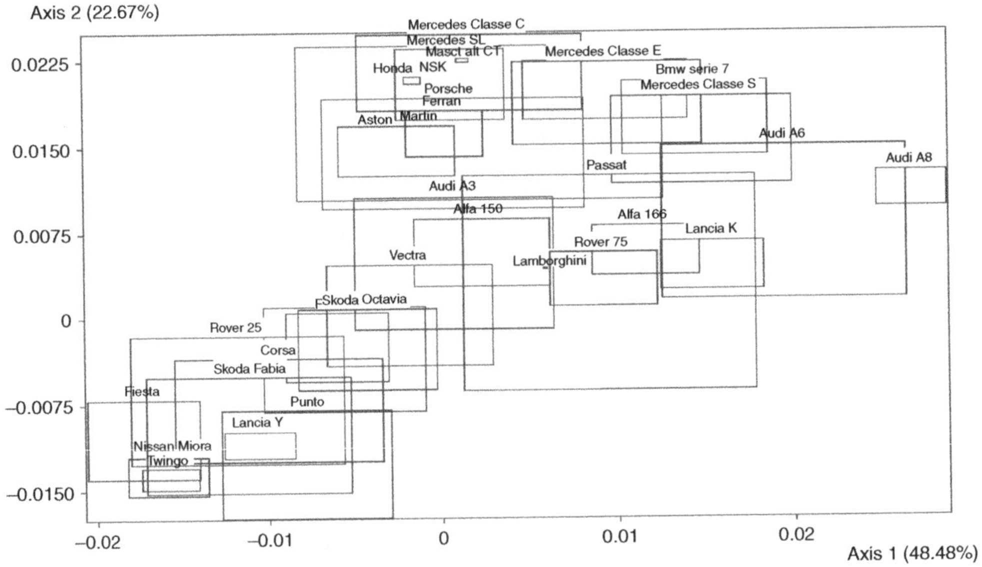

Figure 16.6 Representation of objects on the first factor plane.

Table 16.4 Coordinates of categories.

Variable–Category	Factor 1	Factor 2	Factor 3
Price–LOWetc	−0.00105	−0.00024	−0.00017
Price–MED	0.00075	0.00020	0.00004
Price–HI	−0.00078	−0.00051	0.00067
Cub.Cap.–LOW	−0.00075	−0.00057	−0.00012
Cub.Cap.–MED	0.00024	0.00027	−0.00002
Cub.Cap.–HI	0.00019	−0.00036	0.00042
Fuel–Pet	0.00006	0.00005	0.00001
Fuel–Die	−0.00062	−0.00050	−0.00014
Traction–Fro	0.00013	−0.00035	−0.00010
Traction–Rea	−0.00057	0.00042	0.00028
Traction–4wd	0.00055	0.00000	−0.00021
Top_Speed–LOW	−0.00133	−0.00071	0.00004

Top_Speed–MED	0.00028	0.00015	−0.00008
Top_Speed–HI	−0.00056	−0.00029	0.00066
Accelerati–LOW	−0.00027	0.00004	0.00032
Accelerati–MED	0.00025	0.00008	−0.00006
Accelerati–HI	−0.00166	−0.00076	−0.00001
Axes_dist.–LOW	−0.00109	−0.00084	0.00022
Axes_dist.–MED	−0.00026	0.00017	−0.00021
Axes_dist.–HI	0.00147	−0.00005	0.00056
Length–LOW	−0.00053	−0.00152	0.00011
Length–MED	−0.00103	0.00054	−0.00011
Length–HI	0.00234	−0.00072	0.00020
Width–LOW	−0.00193	−0.00054	0.00020
Width–MED	0.00013	0.00028	−0.00014
Width–HI	0.00180	−0.00145	0.00083
Height–LOW	0.00071	−0.00139	0.00070
Height–MED	−0.00183	0.00068	0.00006
Height–HI	0.00207	−0.00068	−0.00012

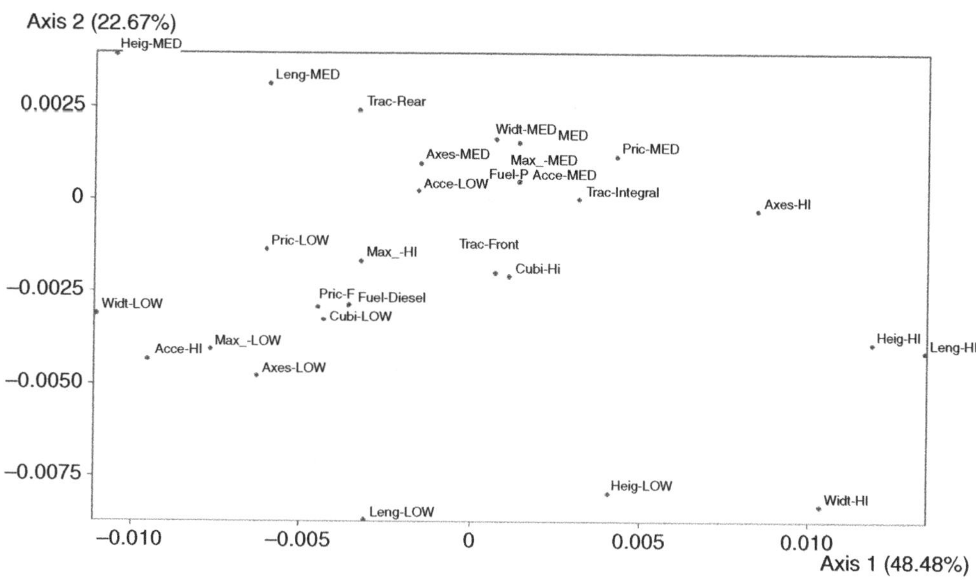

Figure 16.7 Representation of descriptor categories on the first factor plane.

time, low Max speed, and Axes distance on the left (city cars: Fiesta, Twingo, Nissan Micra, Lancia Y, Punto, Skoda Fabia, Corsa).

The second factor axis separates those cars having low values for Length, Height, Axes distance, Max speed, high Width, Acceleration time, Length, and Height on the bottom, and medium Height on the top.

The quality of representation of the car models and of the categories is presented in Tables 16.6 and 16.7 (the row sum, considering all the factors, is equal to 1).

Table 16.5 Description of first three axes.

Axis 1			Axis 2			Axis 3		
Category	Contribution	Direction	Category	Contribution	Direction	Category	Contribution	Direction
Width–LOWetc	0.1079	−	Length–LOW	0.1949	−		Central zone	
Height–MED	0.0975	−	Width–HI	0.1774	−	Cub.Cap.–HI	0.0581	+
Accelerati–HI	0.0798	−	Height–LOW	0.1633	−	Axes_dist.–HI	0.1008	+
Top_Speed–LOW	0.0515	−	Axes_dist.–LOW	0.0595	−	Top_Speed–HI	0.1412	+
Axes_dist.–LOW	0.0347	−	Accelerati–HI	0.0488	−	Price–HI	0.1448	+
	Central zone		Length–HI	0.0441	−	Height–LOW	0.1568	+
Axes_dist.–HI	0.0629	+	Top_Speed–LOW	0.0427	−	Width–HI	0.2229	+
Width–HI	0.0947	+	Height–HI	0.0395	−			
Height–HI	0.1248	+		Central zone				
Length–HI	0.1598	+	Height–MED	0.0389	+			

Table 16.6 Quality of representation of categories.

Variable–Category	Factor 1	Factor 2	Factor 3
Price–LOWetc	0.71944	0.08015	0.07777
Price–MED	0.71352	0.11282	0.00976
Price–HI	0.15122	0.13977	0.46465
Cub.Cap.–LOW	0.38572	0.47638	0.04181
Cub.Cap.–MED	0.24418	0.61856	0.00817
Cub.Cap.–HI	0.02879	0.21594	0.56622
Fuel–Pet	0.22441	0.31252	0.04685
Fuel–Die	0.22441	0.31252	0.04685
Traction–Fro	0.04413	0.68502	0.11846
Traction–Rea	0.27653	0.32394	0.27766
Traction–4WD	0.37901	0.00000	0.21978
Top_Speed–LOW	0.57612	0.35095	0.00257
Top_Speed–MED	0.47484	0.28717	0.17189
Top_Speed–HI	0.10131	0.05979	0.58390
Accelerati–LOW	0.11247	0.00492	0.64491
Accelerati–MED	0.54873	0.12673	0.14287
Accelerati–HI	0.53317	0.23920	0.00002
Axes_dist.–LOW	0.29710	0.37424	0.05071
Axes_dist.–MED	0.10503	0.09433	0.29486
Axes_dist.–HI	0.38561	0.00108	0.23025
Length–LOW	0.04466	0.77098	0.00770
Length–MED	0.57489	0.34042	0.02628
Length–HI	0.78253	0.15866	0.02343
Width–LOW	0.65856	0.11123	0.02886
Width–MED	0.04444	0.46997	0.22792
Width–HI	0.26017	0.35773	0.22807
Height–LOW	0.04743	0.38445	0.18735
Height–MED	0.67289	0.19727	0.00253
Height–HI	0.70631	0.16405	0.00922

Table 16.7 Quality of representation of car models.

Model of cars	Factor 1	Factor 2	Factor 3
Alfa 145	0.72332	0.17920	0.09055
Alfa 156	0.50905	0.16755	0.26814
Alfa 166	0.33844	0.15235	0.22672
Aston Martin	0.04958	0.03065	0.72973
Audi A3	0.77424	0.00332	0.13272
Audi A6	0.99525	0.00053	0.00383
Audi A8	0.88858	0.01859	0.06114
BMW series 3	0.26553	0.66887	0.04796
BMW series 5	0.56972	0.20938	0.18998

Table 16.7 (*continued*).

Model of cars	Factor 1	Factor 2	Factor 3
BMW series 7	0.62047	0.00001	0.31757
Ferrari	0.01150	0.03381	0.78588
Punto	0.28041	0.66108	0.01245
Fiesta	0.69028	0.29979	0.00000
Focus	0.76081	0.10337	0.11268
Honda NSK	0.48561	0.07372	0.32892
Lamborghini	0.01307	0.26489	0.46007
Lancia Y	0.48736	0.47477	0.00237
Lancia K	0.88462	0.01309	0.05806
Maserati GT	0.13512	0.25314	0.43605
Mercedes SL	0.23857	0.12816	0.45581
Mercedes C	0.05818	0.90112	0.00556
Mercedes E	0.58373	0.11597	0.25734
Mercedes S	0.57947	0.01873	0.34119
Nissan Micra	0.53961	0.42488	0.00012
Corsa	0.56231	0.39840	0.01030
Vectra	0.73794	0.02634	0.18221
Porsche	0.32859	0.00008	0.43790
Twingo	0.52722	0.42221	0.00044
Rover 25	0.75608	0.21591	0.02007
Rover 75	0.08267	0.02772	0.67780
Skoda Fabia	0.54782	0.41526	0.00928
Skoda Octavia	0.76716	0.11807	0.10323
Passat	0.16776	0.13973	0.49262

References

Bock, H.-H. (2000) Symbolic data. In H.-H. Book and E. Diday (eds), *Analysis of Symbolic Data*, Chapter 3. Springer-Verlag, Berlin.

Csernel, M. and de Carvalho, F.A.T. (1998) Towards a normal symbolic form. In C. Hayashi, K. Yajima, H.-H. Bock, N. Ohsumi, Y. Tanaka and Y. Baba (eds), *Data Science, Classification and Related Methods*, pp. 379–386. Springer-Verlag, Tokyo,

De Boor, C. (1978) *A Practical Guide to Splines*. Springer-Verlag, New York.

Diday, E. (1987) Introduction à l'approche symbolique en analyse des données. *Journées Symbolique-Numérique*, Université Paris Dauphine.

Goodman, L.A. and Kruskal, W.H. (1954) Measures of association for cross-classifications. *Journal of the American Statistical Association*, 49, 732–764.

Irpino A. and Verde R. (2004) Analysis of symbolic data under constraints. *Atti della XLII Riunione Scientifica della SIS*, Bari, Italy.

Verde, R. (1997) Symbolic object decomposition by factorial techniques. Indo-French Meeting, LISE-CEREMADE, Université Paris IX Dauphine.

Verde, R. (1999) Generalised canonical analysis on symbolic objects. In M. Vichi and O. Opitz (eds), *Classification and Data Analysis: Theory and Application*, pp. 195–202. Springer-Verlag, Berlin.

Volle, M. (1985) *L'Analyse des données*. Economica, Paris.

Part III
SUPERVISED METHODS

17

Bayesian decision trees

Jean-Paul Rasson, Pascale Lallemand and Séverine Adans

17.1 Introduction

The Bayesian decision tree provides a discriminant decision procedure that solves the following discriminant analysis problem (Bardos, 2001). Let $\Omega = \{x_1, \ldots, x_n\}$ be a set of n individuals described by p explanatory variables y_i $(i = 1, \ldots, p)$ and a class variable C, with values in $\{1, \ldots, m\}$ where m is the number of classes. Discriminant analysis tries to predict the unknown C for an individual $\tilde{x}$ on the basis of its p characteristics $y_i(\tilde{x})$ $(i = 1, \ldots, p)$ and a training set. The steps of this symbolic discrimination procedure are: to represent the given partition in the form of a Bayesian decision tree; to create a rule that is able to assign a new individual to one class of a prior partition.

17.2 The underlying population and variables

We consider a population $\Omega = \{x_1, \ldots, x_n\}$ where each individual is described by two categories of variables:

- p predictors $y_1, \ldots, y_p$ (interval-valued variables);
- the class variable C (a categorical single-valued variable).

The variable y_j is an interval-valued variable if, for all $x_k \in \Omega$, $y_j(x_k)$ is a closed and bounded interval in $I\!R$: $y_j(x_k) = [\underline{x}_{kj}, \bar{x}_{kj}]$. The class variable C specifies the class of an individual x_k in the training set in the form of a unique value $C(x_k) = c_k \in \{1, \ldots, m\}$ and should be determined for each individual $\tilde{x}$ to be classified. The data table in Table 17.1 summarizes the above specifications.

Symbolic Data Analysis and the SODAS Software Edited by E. Diday and M. Noirhomme-Fraiture

Table 17.1 A symbolic data table and new data for $\tilde{x}$.

	y_1	$\ldots$	y_j	$\ldots$	y_p	C
x_1	$y_1(x_1)$	$\ldots$	$y_j(x_1)$	$\ldots$	$y_p(x_1)$	c_1
$\vdots$		$\vdots$		$\vdots$		
x_j	$y_1(x_j)$	$\ldots$	$y_j(x_j)$	$\ldots$	$y_p(x_j)$	c_j
$\vdots$		$\vdots$		$\vdots$		
x_n	$y_1(x_n)$	$\ldots$	$y_j(x_n)$	$\ldots$	$y_p(x_n)$	c_n
$\tilde{x}$	$y_1(\tilde{x})$	$\ldots$	$y_j(\tilde{x})$	$\ldots$	$y_p(\tilde{x})$	?

The case of interval data is explained in detail in Chapter 9 – see Section 9.3.7.

Briefly, let us consider an interval $[a, b]$. Instead of using the lower and upper bounds (a and b) of this interval, we will use its centre, $M = a + b/2$, and its half-length, $L = b - a/2$ (Bock and Diday, 2000).

17.3 Creation of the decision tree

We have chosen to create a *binary* tree following our experience with the system used by credit companies. These companies work in a monothetic fashion, that is, if they have a set of individuals to be classified:

- they take the variables one by one;
- they choose a 'cut value';
- they decide which individual has a value of the variable greater than (or less than) this cut value.

In this section we explain this procedure.

17.3.1 What is a binary tree?

The binary tree construction process is as follows. We begin with a first node, called the *root*. By using a 'cut rule' (explained later), the individuals of this node are separated into a left node and a right node. If the node can no longer be split, it is called *terminal*. The process continues until we have only terminal nodes (see Figure 17.1).

The entire construction of the tree, then, has to follow three steps:

- the selection of the splits;
- the decision to declare a node terminal or to continue splitting;
- the assignment of each terminal node to a class.

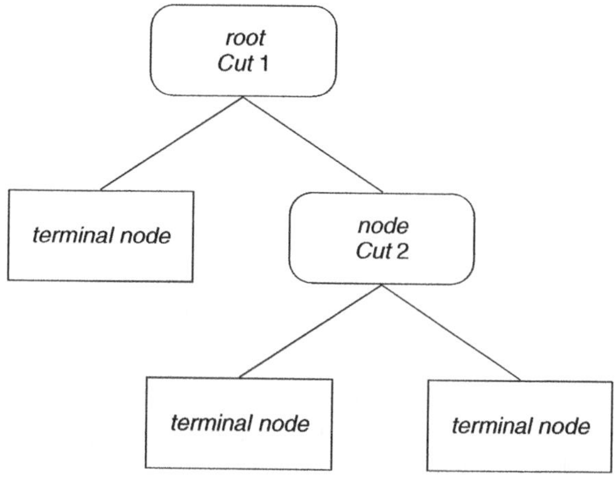

Figure 17.1 Example of a binary tree.

17.3.2 Set of binary questions

In the framework of the recursive algorithm, each node division step is performed on the basis of a single variable (suitably chosen) and yes/no answers to specific binary questions. The set of binary questions that will be useful for the recursive partition procedure is based on the Bayesian rule (Bertholet, 2003; Breiman *et al.*, 1984).

17.3.2.1 Bayesian rule

Suppose that for the population Ω, the data vector x is distributed with density functions $f_k(x)(k=1,\ldots,m)$. Suppose also that the prior probabilities p_k $(k=1,\ldots,m)$ are known. By definition, the Bayesian rule assigns x to the population with the largest value of $p_k f_k(x)$. Thus, the acceptance region for the class k is

$$P_i=\{x|p_i f_i(x)\geq p_j f_j(x)\forall j=1\ldots m\},\qquad i=1,\ldots,k. \tag{17.1}$$

17.3.2.2 Determining the prior probabilities

The maximum likelihood rule or the Bayesian rule is the splitting criterion if the density function and the prior probabilities of belonging to the classes are known.

If the prior probabilities are unknown, two choices are available to determine them. We can use:

- the uniform prior probabilities based on the number of classes (then (17.1) is the maximum likelihood rule),

$$\hat{p}_k=\frac{1}{m},\qquad \text{for } k=1,\ldots,m;$$

- the prior probabilities based on the proportions observed in the training set,

$$\hat{p}_k=\frac{n_k}{n},\qquad \text{for } k=1,\ldots,m,$$

where n_k is the number of data in class k and n the total number of data in the training set.

17.3.2.3 Kernel method

To estimate the intensity of a non-homogeneous Poisson process, we use a non-parametric method, the *kernel method* (Silverman, 1981; 1986). The kernel estimator, in fact a sum of 'bumps' placed around the observations, is defined by

$$\hat{f}_l(x) = \frac{1}{n_l} \sum_{i=1}^{n_l} \frac{1}{h} K\left(\frac{x - X_i}{h}\right), \qquad x \in I\!R, l = 1, \ldots, m,$$

where K is called the *kernel* (which is symmetric, continuous, and is such that $\int K(x)dx = 1$, $K \geq 0$) and determines the shape of the bumps, and $h > 0$ is a parameter determining the width of the bumps, called the *smoothing parameter*.

The most important point is the determination of a suitable value h; this is further detailed in Section 9.3.4.

17.3.2.4 Construction of questions

To a binary question, an individual answers 'yes' or 'no' according to a binary function. In the case of the Bayesian decision tree,

- the binary question is 'Is $p_1 f_1(x) > p_2 f_2(x)$?'.
- the binary function is defined by

$$q_B^{(x)} = \begin{cases} \text{true}, & \text{if } p_1 f_1(x) > p_2 f_2(x), \\ \text{false}, & \text{if } p_1 f_1(x) \leq p_2 f_2(x). \end{cases}$$

- the bipartition (B_1, B_2) of B induced by the binary question is

$$B_1 = \{x \in B \mid q_B(x) = \text{true}\},$$
$$B_2 = \{x \in B \mid q_B(x) = \text{false}\}.$$

By convention, when building a dichotomous hierarchy of classes with a 'left' and 'right' subclass (node) for any class, we will assume that membership of the left node results from the property $[p_1 f_1(x) > p_2 f_2(x)]$, while membership of the right node is based on the property $[p_1 f_1(x) \leq p_2 f_2(x)]$.

17.3.3 Cut variable

As already stated, our method works with a single variable at each split. This variable must therefore be the most discriminant one, that is, the one which leads to the 'purest' nodes. The purity of a node is actually the number of individuals which are correctly classified in it (by leave-one-out and/or bootstrap methods), where the classification is verified in relation to the class variable C. This number can be obtained by the Bayesian rule computed for each variable. The choice of the most discriminant variable, at one step, will result in selecting the variable which minimizes some impurity measure, the number of misclassified individuals, at this step.

17.3.4 Splitting criteria

For each variable, by a dichotomic process, we find the largest value of the parameter h, giving a specified number of nodes of the associated intensities, strictly larger than one. Once this h determined, we divide the set of individuals in the node into two groups, for which the impurity measure is minimum. Proceeding variable by variable, we will be able to select the one which generates the purest nodes. This procedure is recursively performed until some stopping rules are fulfilled:

- the number of points in a node is less then a predefined value;
- there is only one class in a node.

17.3.5 Pruning method

At the end of the splitting process, we obtain a huge tree. The best subtree is selected. Indeed, we have developed, under the hypothesis of a non-homogeneous Poisson process, a tree-pruning method that takes the form of a classical hypothesis test, the *gap test* (Kubushishi, 1996; Rasson and Kubushishi, 1994). This method is detailed in Section 9.3.6.

17.4 Decision rule and discriminant analysis

To each terminal node an individual is allocated. This individual, described by the path in the tree that provides it, defines a decision rule which is able to classify new individuals into one of the two classes.

The cut variable and its associated cut value will form the *cut rule*, used later for discriminant analysis. Subsequently, using the cut value given by the Bayesian rule, we can compute the value of the variable concerned, in order to properly classify the new individual.

17.5 Example

We have applied our discriminant analysis method to an artificial data set (Table 17.2). This data set consists of 20 individuals, described by two interval variables and a categorical single-valued variable. The latter represents the class of the individual, and takes value 1 or 2, or is set to 99 for new individuals to be classified. The data set is shown graphically in Figure 17.2, in which the dark rectangles represent the individuals to be classified.

As already described, the first step is the determination of the cut rule. The graphical representation in Figure 17.3 illustrates this.

The binary question used for the cut rule is

$$q = [\text{Variable } 1 \leq 8.31].$$

Then we have:

$$\begin{aligned} \text{class } 1: & \ [\text{Variable } 1 \leq 8.31], \\ \text{class } 2: & \ [\text{Variable } 1 > 8.31]. \end{aligned}$$

Table 17.2 Artificial data set.

Sample	Variable 1	Variable 2	Class
artif 1	[2, 4]	[4, 5.5]	1
artif 2	[5, 7]	[6.5, 8]	1
artif 3	[1.5, 2.5]	[6, 7.5]	1
artif 4	[1, 3]	[2.5, 3.5]	1
artif 5	[5, 6]	[3.5, 4.5]	1
artif 6	[3.5, 6]	[1.5, 3]	1
artif 7	[1, 2.5]	[0.5, 2]	1
artif 8	[9.5, 11.5]	[5, 6]	2
artif 9	[11, 13]	[6.5, 8]	2
artif 10	[12, 13]	[4, 6]	2
artif 11	[11.5, 13.5]	[2, 3.5]	2
artif 12	[9.5, 13.5]	[0.5, 1.5]	2
artif 13	[8.5, 10.5]	[2.5, 4]	2
artif 14	[12.5, 13.5]	[8.5, 9.5]	2
artif 15	[9.5, 10.5]	[6.5, 8.5]	2
artif 16	[9.5, 12]	[9, 10]	2
artif 17	[4, 5.5]	[9, 10.5]	99
artif 18	[8.5, 9]	[0.5, 1.5]	99
artif 19	[3, 4]	[7.5, 8.5]	99
artif 20	[12, 14]	[10.5, 11]	99

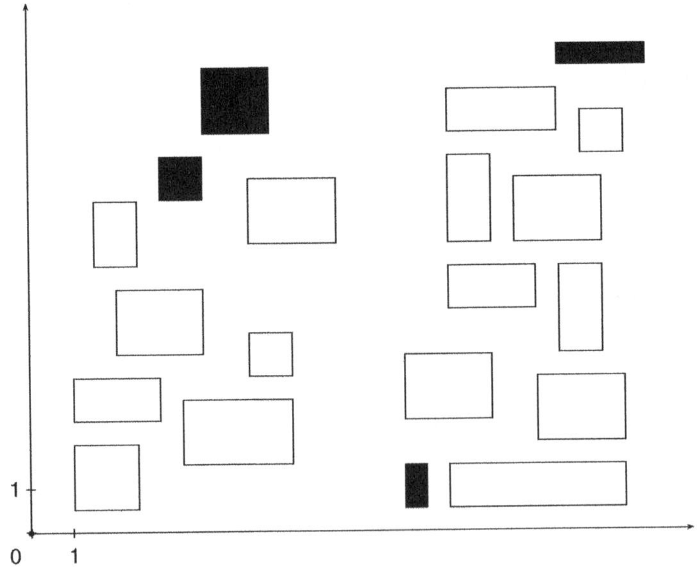

Figure 17.2 Graphical representation of the artificial data set.

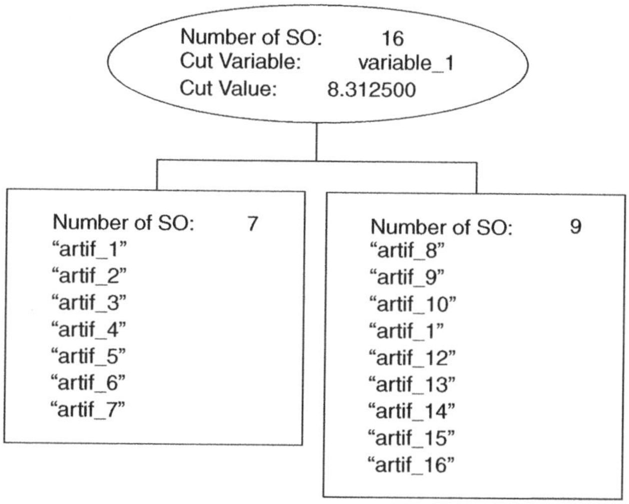

Figure 17.3 Tree representing the classification of the first 16 objects.

This leads to the following result:

```
CLASSIFICATION:
----------
INDIVIDUAL  CLASS 1  CLASS 2
    16         1        0
    17         0        1
    18         1        0
    19         0        1
```

It is easy to see that the results are correct. For example, if we consider individual no. 17, the centre of the interval for the first variable is equal to 4.75 (< 8.31), so it is assigned to the first class.

References

Bardos, M. (2001) *Analyse discriminante: Application au risque et scoring financier*. Paris: Dunod.

Bertholet, V. (2003) Apports de l'estimation de densité aux arbres de discrimination: Applications réelles. Doctoral thesis. Namur: Facultés Universitaires Notre-Dame de la Paix.

Bock, H.-H. and Diday E. (2000) *Analysis of Symbolic Data: Exploratory Methods for Extracting Statistical Information from Complex Data*. Berlin: Springer-Verlag.

Breiman, L., Friedman J., Olshen, R. and Stone, C.(1984) *Classification and Regression Trees*. Belmont, CA: Wadsworth.

Kubushishi T. (1996) On some applications of the point process theory in cluster analysis and pattern recognition. Doctoral thesis. Namur: Facultés Universitaires Notre-Dame de la Paix, Namur, Belgium.

Rasson, J.P. and Kubushishi T. (1994) The gap test: an optimal method for determining the number of natural classes in cluster analysis. In E. Diday, Y. Lechevallier, M. Schader, P. Bertrand and

B. Burtschy, (eds), *New Approaches in Classification and Data Analysis*, pp. 186–193. Berlin: Springer-Verlag.
Silverman, B. (1981) Using kernel density estimates to investigate multimodality. *Journal of the Royal Statistical Society, B*, 43: 97–99.
Silverman, B. (1986) *Density Estimation for Statistics and Data Analysis*. London: Chapman & Hall.

18

Factor discriminant analysis

N. Carlo Lauro, Rosanna Verde and Antonio Irpino

18.1 Introduction

The aim of discrimination techniques is to predict the membership of statistical units in r a priori classes based on the values that they take for p predictive variables (predictors), usually assumed to be numerical. In general, discrimination techniques look for a classification rule, as a function of the predictors. This may be a linear, quadratic or logistic function or an expression for an underlying probability density function.

Factor discriminant analysis on symbolic objects (SFDA) (Lauro *et al.*, 2000) is concerned with defining suitable geometrical classification rules to classify SOs into a priori classes. It looks for the best discriminant subspace as a linear combination of the predictors. It begins by selecting the predictors according to their discriminant power. Then, because the predictors can be both quantitative and categorical variables, a quantification step is performed in order homogenize such variables. Generalized canonical analysis on symbolic data (SGCA; see Chapter 16) has been proposed as a quantification procedure for coding the descriptors in the SFDA. Further, the additivity of the optimized criterion in the SGCA can enable a suitable selection of the best discriminant predictors. Finally, a classical FDA is carried out on the new factors, obtained by the quantification process.

Different geometrical classification rules are proposed on the images of SOs and classes in the factor plane. Different assignment criteria have been proposed based on the concepts of classes of SOs (i.e., the class is the set of member SOs) and SO classes (i.e., the class is a new SO summarizing the set of member SOs). In the first case the assignment of a new SO to a class is governed by the distance from the nearest, furthest or by the centre of gravity of all the elements of the class (simple, average or complete linkage). We observe that the distance is a suitable measure, defined between intervals in terms of the coordinates of the SOs on the factor axes.

Symbolic Data Analysis and the SODAS Software Edited by E. Diday and M. Noirhomme-Fraiture

In the second case, an SO is classified using as criterion the minimum generalization of the characteristics of a class in order to include the new SO.

18.2 Principles of the method

SFDA is performed on a set E of SOs, belonging to r a priori classes, and described by p predictors, which can be of any of the five types described in Chapter 1. SOs are represented by hypercubes, with each side corresponding to the value taken by an interval descriptor; while the values assumed for categorical variables identify more hypercubes, one for each category, in a non-metric space (see SGCA).

SFDA can also take into account some other external information on the structure of the predictors in the form of logical or hierarchical rules and taxonomies defined on categorical descriptors. Like the other factor methods, SFDA is based on a *symbolic* (input)–*numerical* (treatment)–*symbolic* (output) approach (Diday, 1987).

SFDA is performed in the following main steps:

(i) quantification of the SO descriptors;

(ii) FDA on the transformed predictors;

(iii) definition of the classification geometrical rules;

(iv) symbolic interpretation of the results (factor discriminant axes and classification rules) according to the original data.

18.2.1 Quantification of class predictors

The *first phase* of the SFDA procedure involves the quantification of predictors, which corresponds to quantification of the descriptive variables of the SOs belonging to the a priori classes.

As SO descriptors are of several types, SGCA (see Chapter 16) is used to code and select them as well as to transform these variables into a new set of numerical predictors.

We briefly recall the process of homogenization of the SO descriptors (Section 16.3.1). Depending on the nature of the variables, the SOs are coded in p tables $\mathbf{Z}_j$ (for $j=1, \ldots, p$), where $\mathbf{Z}_j$ is:

- a *binary* coding matrix of 0/1 values if y_j is a categorical multi-valued variable with k_j different categories;
- a *fuzzy* coding matrix of values in [0,1] if y_j is a modal variable and the values are the relative frequencies or probabilities that each SO assumes for the categories of the variable;
- a *fuzzy* coding matrix if y_j is a quantitative single-valued or interval variable, categorized and codified according to a *fuzzy system of coding* (e.g. using basic spline functions), to preserve the numerical information of the original numerical variables. In this context, linear functions are usually used to code the numerical variables with respect to three categories (e.g. low, medium and high). The two bounds of each interval are coded in two different rows of the coding matrix $\mathbf{Z}_j$.

The global coding matrix **Z** is obtained by the Cartesian product of the rows, corresponding to the same SO, of the various $\mathbf{Z}_j$.

From the solution of the characteristic equation of SGCA, we obtain the eigenvalues and eigenvectors denoted by $\tilde{\mu}_m$, $\tilde{\xi}_m$ and, in the dual space, $\tilde{\beta}_m$. Using $\tilde{\beta}_m$, it is possible to compute the coordinates of the vertices of the hypercubes associated with the description of the SOs, denoted by $\tilde{\varphi}_{i,m}$ (Section 16.3.4). These vectors of coordinates are then assumed in SFDA as the quantified predictors, according to the classical FDA on qualitative variables (Celeux and Nakache, 1994)

18.2.2 Selection of variables

A critical point of SFDA is the selection of the symbolic predictors to better discriminate the a priori classes, especially when the SFDA is performed with a confirmative aim.

There are several reasons why the selection of variables is important in discriminant analysis:

- The reduction of the number of variables involves a remarkable reduction of computation time and cost caused by the reduction in the amount of data that need to be dealt with.
- By reducing the number of variables, redundancy is reduced. It is worth considering variables which cannot improve the accuracy of the analysis.
- The effect of the previous point is the so-called *curse of dimensionality*: there is a point beyond which each variable introduced in the analysis involves a decrease in the correct classification ratio (CCR). In fact, a higher correct classification rate can be often achieved by using fewer variables in the analysis.

Excluding a variable with a lower discriminant power can provide an improvement in the performance of the analysis in terms of higher CCR. In the same way, it can involve a remarkable reduction of the dimensions of the matrix **X** and a consequent decrease in the computational time and cost.

The criterion defined for selecting symbolic variables is based on a generalization of the Goodman–Kruskal capability index $\tau_{Y|X_j}$ (Goodman and Kruskal, 1954), for binary and *fuzzy* coded variables, which is consistent with the criterion optimized in the quantification phase for symbolic categorical predictors.

The trace of the matrix in the SFDA quantification step is decomposed, up to a constant, in terms of the generalized Goodman–Kruskal index $\tau_{Y|X_j}$, computed for each categorized variable $\mathbf{X}_j$:

$$\mathrm{tr}\left(\frac{1}{N}\left[\mathbf{G}'\mathbf{X}\boldsymbol{\Delta}_x^{-1}\mathbf{X}'\mathbf{G}-\frac{p}{N}(\mathbf{G}'\mathbf{1}\mathbf{1}'\mathbf{G})\right]\right)=\sum_{\alpha}\lambda_{\alpha}=\sum_{j=1}^{p}\tau_{Y|X_j}.$$

The quantification of categorical variables is based on an additive approach that allows the capability index to be used to select predictors according to their discriminant power. Therefore, the selection procedure involves the following steps:

1. Compute the $\tau_{Y|X_j}$ indices for the response variable **Y** and each explicative variable $\mathbf{X}_j$.
2. Sort the $\tau_{Y|X_j}$ indices in ascending order and compute the cumulative percentage $\tau_{Y|X_j}$.
3. Evaluate the decrease in the multiple *tau* index $\tau_{Y|X_1,\ldots,X_j}$ by dropping the variables with the lower $\tau_{Y|X_j}$ with respect to the aggregate multiple $\tau_{Y|X_1,\ldots,X_p}$ with all p variables:

$$\frac{\tau_{Y|X_1,\ldots,X_j}-\tau_{Y|X_1,\ldots,X_{j-1}}}{\tau_{Y|X_1,X_2,\ldots,X_p}}=\frac{\left(\tau_{Y|X_1}+\cdots+\tau_{Y|X_j}\right)-\left(\tau_{Y|X_1}+\cdots+\tau_{Y|X_{j-1}}\right)}{\tau_{Y|X_1,X_2,\ldots,X_p}}$$
$$=\frac{\tau_{Y|X_j}}{\tau_{Y|X_1,X_2,\ldots,X_p}}.$$

4. Remove as many variables $\mathbf{X}_j$ as necessary to get a cumulate percentage of the sum of the $\tau_{Y|X_j}$ indices larger than a predefined percentage α (in general, greater than 75%).

18.3 The analysis on the transformed predictors

The *second phase* assumes as new predictors the SO coordinate variables on the SGCA factor axes $\tilde{\phi}_j, \forall j=1,\ldots,M$, collected in a matrix $\tilde{\mathbf{\Phi}}=\left[\tilde{\phi}_1,\ldots,\tilde{\phi}_M\right]$. Then, denoting by $\mathbf{C}_{N\times r}$ the indicator matrix of the membership of the object vertices in r a prori classes, the factor discriminant axes are obtained as solutions of the following characteristic equation:

$$\left[(\tilde{\mathbf{\Phi}}'\mathbf{H}\tilde{\mathbf{\Phi}})^{-1}(\tilde{\mathbf{\Phi}}'\mathbf{HC})(\mathbf{C}'\mathbf{HC})^{-1}(\mathbf{C}'\mathbf{H}\tilde{\mathbf{\Phi}})\right]\nu_m=\kappa_m\nu_m, \qquad (18.1)$$

where the column vectors of $\tilde{\mathbf{\Phi}}$ are centred, while the columns of the matrix **C** are not centred; **H** is the diagonal matrix of the weights of the hypercube vertices, with generic term h_{ii} equal to $|N_s|/N$ ($i=1,\ldots,N$, $s=1,\ldots,n$, and N_s is the set of vertices associated with the sth object), according to the number of vertices of the SOs in each class. Moreover, κ_m and ν_m are the mth eigenvalue and eigenvector of the matrix in (18.1).

The number of axes to be retained in the FDA analysis is chosen in the usual way – the percentage of variance explained by the first q axes (with $q\le\min(N, M-1)$).

The coordinates of the bounds of the mth ($1\le m\le q$) interval associated with the sth SO are given by $\tilde{s}_{sm}\equiv\left[\underline{\psi_{im}},\overline{\psi_{im}}\right]$, where $\underline{\psi_{im}}=\min_{i\in N_s}(\psi_{im})$, $\overline{\psi_{im}}=\max_{i\in N_s}(\psi_{im})$, and $\psi_{im}=\tilde{\phi}_{im}\nu_m$ is the coordinate of the generic ith vertex belonging to the sth SO and N_s is its number of vertices. The projected object in the reduced subspace is then described as:

$$\tilde{s}_s=\left[\left[\underline{\psi_{i1}},\overline{\psi_{i1}}\right];\ldots;\left[\underline{\psi_{im}},\overline{\psi_{im}}\right];\ldots;\left[\underline{\psi_{iq}},\overline{\psi_{iq}}\right]\right].$$

In particular, each SO and class is described by an assertion containing the interval coordinates for each factor axis.

18.3.1 Symbolic interpretations of the factor analysis results

Further aspects, related to the factor analyses on SOs, concern the quality, the interpretation of the results on factor planes, and the stability of the factorial configuration.

As in classic factor discriminant analysis, the quality of the representation of the set of SOs on factor planes is directly related to the percentage of total variability explained by the factor axes used for the definition of the factor plane.

The most immediate interpretation of factor axes is done with respect to the contributions of the original variables. The measures of descriptors' contribution to the axes are expressed, as in the classical PCA, as the squared correlation variable/factor (see Lebart *et al.*, 1995). In particular, it is possible to compute the contribution of each predictor using its contribution to the GCA quantification phase for each new predictor $\tilde{\phi}_j, \forall j = 1, \ldots, M$. In SFDA it is possible to compute the contribution of each $\tilde{\phi}_j, \forall j = 1, \ldots, M$, to the factor discriminant axes. We may then compute the contribution of each original predictor to the FDA factor axes by multiplying the contribution to GCA factor axes (or to the new predictors $\tilde{\phi}_j, \forall j = 1, \ldots, M$) and the contribution of each $\tilde{\phi}_j, \forall j = 1, \ldots, M$, to the factor discriminant axes.

18.3.2 Definition of the classification rule

The last phase is represented by the definition of alternative geometrical classification rules. We suppose that both SOs and classes are represented on factor planes by rectangles. The classification of a generic SO in a class C_i is defined according to two events:

(i) If an SO image is included in the representation of C_i (of the training set) it is assigned to this class C_i.

(ii) If an SO is partially or completely outside all the represented classes or it is in an overlapping area between two or more classes, we need to consider a suitable measure of similarities to assign it to a single class C_i.

To define a geometric classification rule, we propose three approaches, two based on the description potential $\pi(.)$, defined by de Carvalho (1992) as the volume obtained by the Cartesian product of the SO descriptor values, and one based on the Hausdorff distance between polygons:

1. *Ichino–de Carvalho dissimilarity index.* This is based on the following measure between projected objects represented as hypercubes:

$$d(\tilde{s}_j, \tilde{s}_s) = \sqrt[m]{\sum_{\alpha} \left(p_\alpha \psi_\alpha(\tilde{s}_j, \tilde{s}_s)\right)^m}$$

where $\tilde{s}_j$ and $\tilde{s}_s$ are the factor representation of two SOs, j and s, p_α is the αth eigenvalue, m is the metric, and

$$\psi_\alpha(\tilde{s}_j, \tilde{s}_s) = \frac{\mu(\tilde{s}_{s\alpha} \oplus \tilde{s}_{j\alpha}) - \mu(\tilde{s}_{s\alpha} \cap \tilde{s}_{j\alpha}) + \gamma\left(2\mu(\tilde{s}_{s\alpha} \cap \tilde{s}_{j\alpha}) - \mu(\tilde{s}_{s\alpha}) - \mu(\tilde{s}_{j\alpha})\right)}{\mu(\tilde{s}_{s\alpha} \oplus \tilde{s}_{j\alpha})},$$

where $\mu(\tilde{s}_{j\alpha})$ is the length of the coordinate interval of object j on the αth axis, $\mu(\tilde{s}_{s\alpha} \oplus \tilde{s}_{j\alpha})$ is the length of the coordinate interval of the conjunction of the two coordinate intervals of the objects j and s on the αth axis, and $\mu(\tilde{s}_{s\alpha} \cap \tilde{s}_{j\alpha})$ is the length of the intersection of the coordinate intervals of the two objects on the αth axis. Computation of the dissimilarity between two objects requires the choice of a suitable metric (1 = city block, 2 = Euclidean, ...), and a bound $\gamma \in]0,1[$ that allows the intersection between objects to be emphasized.

2. *Descriptor potential increase* (dpi) index. This measure is based on the concept of descriptor potential increase. It involves computing how much a class image has to be enlarged in order to contain the image of the object to be affected. The dpi is given by the following formula:

$$\text{dpi}\left(\tilde{s}_j, \tilde{s}_s\right) = \frac{\prod_{\alpha=1}^{m} \mu\left(\tilde{s}_{s\alpha} \oplus \tilde{s}_{j\alpha}\right) - \prod_{\alpha=1}^{m} \mu\left(\tilde{s}_{j\alpha}\right)}{\prod_{\alpha=1}^{m} \mu\left(\tilde{s}_{j\alpha}\right)}$$

where $\tilde{s}_j$ represents the jth class, $\tilde{s}_s$ the sth object to be affected, $\mu\left(\tilde{s}_{j\alpha}\right)$ is the interval coordinate length of the jth 'object class' on the αth axis, and $\mu\left(\tilde{s}_{s\alpha} \oplus \tilde{s}_{j\alpha}\right)$ of the conjunction of the two intervals.

3. *Hausdorff distance*. This is based on the Hausdorff distance between two objects computed on the interval coordinates on the factorial axes. This measure requires only the choice of the metric (1 = city block, 2 = Euclidean, ...). The Hausdorff distance is the maximum distance from one set to the nearest element of another set. More formally, the Hausdorff distance from set A to set B is a maximin function, defined as

$$h(A, B) = \max_{a \in A} \left\{ \min_{b \in B} \{d(a, b)\} \right\}, \tag{18.2}$$

where a and b are the points of the sets A and B respectively, and $d(a, b)$ is a metric between points.

Generally, the Hausdorff distance is oriented (we could also say *asymmetric*), which means that most of the time $h(A, B) \neq h(B, A)$. This asymmetry is a property of maximin functions, while a minimin function is symmetric.

A more general definition of Hausdorff distance is given by

$$H(A, B) = \max\left(h(A, B); h(B, A)\right). \tag{18.3}$$

The two distances $h(A, B)$ and $h(B, A)$ are sometimes termed the *forward* and *backward* Hausdorff distances from A to B. Although the terminology is not universally accepted among authors, (18.3) is what is usually meant when talking about Hausdorff distance. Henceforth, we will refer to (18.3) as *Hausdorff distance*.

First of all, let us show how to compute a Hausdorff distance between two symbolic data described by interval variables. This type of data is a particular case of symbolic data, collected in a symbolic table, where each cell contains an interval of real values.

Given two symbolic data **A** and **B**, described by p interval variables, they can be represented by the following vectors of intervals:

$$\mathbf{A} = \left(y_1^A, y_2^A, \ldots, y_i^A, \ldots, y_p^A\right) = \left(\left[\underline{y_1^A}, \overline{y_1^A}\right], \left[\underline{y_2^A}, \overline{y_2^A}\right], \ldots, \left[\underline{y_i^A}, \overline{y_i^A}\right], \ldots, \left[\underline{y_p^A}, \overline{y_p^A}\right]\right)$$

$$\mathbf{B} = \left(y_1^B, y_2^B, \ldots, y_i^B, \ldots, y_p^B\right) = \left(\left[\underline{y_1^B}, \overline{y_1^B}\right], \left[\underline{y_2^B}, \overline{y_2^B}\right], \ldots, \left[\underline{y_i^B}, \overline{y_i^B}\right], \ldots, \left[\underline{y_p^B}, \overline{y_p^B}\right]\right).$$

A and **B** are visualized by two parallelepipeds in R^p. The Hausdorff distance between the observed intervals on **A** and **B** with respect to the variable i can be computed as follows:

$$h_i(\mathbf{A},\mathbf{B}) = \begin{cases} 0 & \text{if } \left[\underline{y_i^A}, \overline{y_i^A}\right] \subseteq \left[\underline{y_i^B}, \overline{y_i^B}\right], \\ \max \begin{bmatrix} \min\left(\left|\underline{y_i^A} - \underline{y_i^B}\right|, \left|\underline{y_i^A} - \overline{y_i^B}\right|\right), \\ \min\left(\left|\overline{y_i^A} - \underline{y_i^B}\right|, \left|\overline{y_i^A} - \overline{y_i^B}\right|\right) \end{bmatrix} & \text{otherwise,} \end{cases} \tag{18.4}$$

$$H_i(\mathbf{A},\mathbf{B}) = \max\left[h_i(\mathbf{A},\mathbf{B}), h_i(\mathbf{B},\mathbf{A})\right]. \tag{18.5}$$

Choosing a suitable L_r norm, the Hausdorff distance between the two vectors of intervals is equal to

$$h(\mathbf{A},\mathbf{B}) = \left|\sum_{i=1}^{p} (h_i(\mathbf{A},\mathbf{B}))^r\right|^{1/r} \quad h(\mathbf{B},\mathbf{A}) = \left|\sum_{i=1}^{p} (h_i(\mathbf{B},\mathbf{A}))^r\right|^{1/r}, \tag{18.6}$$
$$H(\mathbf{A},\mathbf{B}) = \max\left(h(\mathbf{A},\mathbf{B}); h(\mathbf{B},\mathbf{A})\right).$$

If the distance between a new SO s and a class is evaluated for all the SOs belonging to such class (in (1) and (3)), the classification criterion can be the single, average or complete linkage, as in classical clustering techniques.

In SFDA, we compute the Hausdorff distance $H\left(\tilde{s}_j, \tilde{s}_s\right)$, where $\tilde{s}_j$ is the representation of the jth class and $\tilde{s}_s$ is the representation of the sth object to be affected, once a metric and a linkage criterion are chosen.

18.4 Example

The following example involves a set of 12 concepts (sex–age typology of people in Portugal) described by 37 interval variables and one categorical single-valued variable that identifies the membership of the SOs in three classes (partition into three clusters; see Table 18.1). The 37 predictors describe the interval of proportion observed for each of them in the 12 typologies of people observed. If all the predictors have been chosen, they are quantified via GCA and we have the eigenvalues for the FDA phase shown in Table 18.2. The analysis on the whole set of predictor variables allows the representation of objects and classes of objects on the first factor plane shown in Figure 18.1.

It is easy to see that the three classes are overlapping and, in particular, that class 2 is included in class 3. Table 18.3 shows what predictor has contributed to defining the first two factor axes, ordered on the basis of their contribution, from the negative to the positive half of the axis (the category on the top has the highest contribution and a negative sign for its coordinate, while the category on the bottom has the highest contribution and a positive sign for its coordinate). We recall that each numerical variable is fuzzy-coded (see GCA) into a categorical variable with three levels (high, medium and low). In our case we have the contribution of each category to the factor axes. The number of categories (see GCA) in our case is equal to 111. The categories represented are those having a contribution greater than $1/111 = 0.009 = 0.9\%$.

Table 18.1 The symbolic data table.

Typology of people (sex–age)	Full-time	Part-time	Primary studies or less	Secondary studies – professional	University studies	Without studies	Employee	Family worker	Other professional status	Self-employed with employees	Self-employed without employees	Agriculture, cattle, hunting, forestry and fishing	Construction
Men 15–24	[94.5,96.5]	[3.5,5.5]	[54.4,59]	[36.5,41]	[1.4,2.7]	[1.8,3.2]	[86.8,89.8]	[4,6.1]	[0.2,0.8]	[0.9,2]	[3.8,5.8]	[4.1,6.2]	[23.9,28]
Men 25–34	[96.3,97.6]	[2.4,3.7]	[57.5,61.1]	[27.6,31]	[6.8,8.8]	[2.9,4.3]	[80.2,83.1]	[0.9,1.8]	[0.3,0.9]	[4.3,6]	[10.1,12.4]	[4.4,6]	[21.7,24.8]
Men 35–44	[97.9,98.8]	[1.2,2.1]	[60.7,64.2]	[23.9,27]	[6.7,8.6]	[3.7,5.2]	[71.3,74.5]	[0.1,0.4]	[0.3,0.9]	[9.2,11.3]	[14.7,17.3]	[5.4,7.1]	[19.4,22.3]
Men 45–54	[96.6,97.9]	[2.1,3.4]	[67.1,70.7]	[16.5,19.5]	[6.8,8.9]	[4.3,6.1]	[65.5,69.2]	[0.1,0.6]	[0.2,0.8]	[11.3,13.8]	[17.7,20.7]	[8.4,10.6]	[12.8,15.5]
Men 55–64	[87.2,90.1]	[9.9,12.8]	[63.3,67.7]	[7.1,9.7]	[4.4,6.5]	[18.8,22.6]	[47.8,52.5]	[0.7,1.7]	[0.4,1.3]	[11.3,14.4]	[32.7,37.1]	[24,28.1]	[8.9,11.8]
Men 65 +	[53.2,59.8]	[40.2,46.8]	[45.1,51.7]	[4.2,7.3]	[2.9,5.6]	[38.3,44.8]	[12.6,17.3]	[3.6,6.4]	[1.3,3.2]	[6.1,9.7]	[66.9,72.9]	[65.6,71.7]	[1.1,3]
Women 15–24	[89.1,92.3]	[7.7,10.9]	[46.4,51.9]	[39.5,44.9]	[5.5,8.3]	[1,2.4]	[88.9,92.1]	[3.4,5.7]	[0.1,0.8]	[0.1,0.9]	[2.9,5.1]	[3.1,5.4]	[0.5,1.7]
Women 25–34	[88.2,90.7]	[9.3,11.8]	[49.1,53.2]	[29.3,33]	[14.6,17.6]	[1.1,2.1]	[82.9,85.9]	[1.5,2.7]	[0.8,1.7]	[1.8,3.1]	[8.6,11]	[5.7,7.7]	[0.7,1.6]
Women 35–44	[87.4,89.9]	[10.1,12.6]	[56.9,60.8]	[22.1,25.4]	[12.1,14.8]	[3.1,4.7]	[74.6,77.9]	[1.2,2.2]	[0.7,1.5]	[3.8,5.4]	[14.9,17.8]	[8.9,11.2]	[0.6,1.4]
Women 45–54	[80,83.4]	[16.6,20]	[61.6,65.8]	[12.1,15.1]	[9,11.7]	[10.9,13.8]	[65.3,69.4]	[2.6,4.1]	[0.5,1.3]	[4.3,6.3]	[21.3,25]	[14.6,17.9]	[0.4,1.2]
Women 55–64	[61.6,67]	[33,38.4]	[43.6,49.1]	[2.9,5.1]	[4.6,7.2]	[41,46.5]	[41.1,46.6]	[3.1,5.3]	[1,2.4]	[3.5,5.9]	[42.9,48.4]	[34.4,39.8]	[0,0.6]
Women 65 +	[35.3,43.3]	[56.7,64.7]	[22.2,29.3]	[0.8,3]	[1.3,3.8]	[66.1,73.6]	[12.7,18.6]	[4,7.9]	[1.7,4.4]	[1.5,4.2]	[68.9,76.1]	[66.7,74.1]	[0,0.6]

Table 18.1 *(continued)*.

Typology	Electricity, gas and water	Financial intermediation	Hotels and restaurants	Manufacturing	Mining and quarrying	Other services	Public administration	Real state, renting and business activities	Transport, storage and communication	Wholesale and retail trade, repairs	Armed forces	Clerks	Craft and related trades workers	Elementary occupations
M.15–24	[0.2,0.8]	[0.8,1.9]	[4.2,6.3]	[29.2,33.5]	[0,0.5]	[2.2,3.8]	[3.5,5.4]	[2.8,4.5]	[2.5,4.1]	[14.1,17.5]	[1.5,2.8]	[5.7,8]	[39.7,44.3]	[12.3,15.5]
M.25-34	[0.6,1.3]	[2,3.2]	[3.3,4.8]	[24.8,28.1]	[0.3,0.9]	[5.6,7.4]	[5,6.8]	[3.3,4.8]	[4.1,5.7]	[14.1,16.8]	[1.3,2.2]	[6.4,8.4]	[34.1,37.6]	[7.2,9.2]
M.35–44	[1.1,2]	[1.6,2.6]	[3.2,4.6]	[21.8,24.8]	[0.3,0.9]	[6.3,8.2]	[8.2,10.3]	[3.1,4.4]	[5.3,7.1]	[13.6,16.1]	[0.8,1.6]	[6.1,7.9]	[31.5,34.9]	[6.1,7.9]
M.45–54	[0.8,1.6]	[1.9,3.1]	[3.4,5]	[24.3,27.7]	[0.4,1.1]	[7.2,9.4]	[7.5,9.7]	[2.3,3.6]	[6.7,8.8]	[12.8,15.5]	[0.5,1.2]	[6.7,8.8]	[26.2,29.7]	[5.9,7.8]
M.55–64	[0.6,1.6]	[1.1,2.3]	[2.5,4.2]	[18.6,22.4]	[0.3,1.1]	[5.2,7.5]	[4.8,7.1]	[1.7,3.1]	[4.7,6.9]	[14,17.4]	[0,0.4]	[3.7,5.7]	[19.6,23.5]	[7.9,10.6]
M.65+	[0,0.7]	[0,0.6]	[0.6,2.2]	[3.7,6.6]	[0,0]	[3.5,6.4]	[1.5,3.5]	[1.4,3.4]	[1,2.8]	[8.4,12.4]	[0,0]	[0.7,2.3]	[4.4,7.6]	[6.1,9.7]
W.15–24	[0,0]	[0.5,1.7]	[8.1,11.3]	[35.7,41]	[0,0.1]	[15.7,19.9]	[1.3,2.8]	[4.8,7.4]	[0.7,2]	[16.1,20.4]	[0,0.6]	[11.9,15.8]	[20.9,25.6]	[9.5,13]
W.25–34	[0,0.3]	[1.1,2.2]	[5.6,7.6]	[27.7,31.3]	[0,0.3]	[28,31.7]	[3.2,4.7]	[4.4,6.2]	[1.6,2.7]	[11.4,14.2]	[0,0.1]	[12.9,15.7]	[15.6,18.7]	[13.2,16.1]
W.35–44	[0.1,0.6]	[1.6,2.7]	[5.6,7.5]	[21.8,25.1]	[0,0.2]	[31.8,35.5]	[5.7,7.7]	[2.8,4.3]	[1.1,2.1]	[9.6,12.1]	[0,0.1]	[11.2,13.8]	[12,14.7]	[17.7,20.8]
W.45–54	[0.1,0.5]	[1,2]	[6.9,9.3]	[15,18.2]	[0,0.2]	[31.2,35.4]	[5.8,8.1]	[2.4,4]	[1.1,2.3]	[9.8,12.6]	[0,0]	[10.3,13.1]	[7.4,9.8]	[20.5,24.2]
W.55–64	[0,0.2]	[0.4,1.4]	[4.1,6.6]	[8.1,11.4]	[0,0]	[25.1,30.1]	[2.5,4.6]	[0.9,2.3]	[0.3,1.3]	[11.1,14.8]	[0,0]	[2.2,4.1]	[5.4,8.2]	[25.3,30.3]
W.65+	[0,0]	[0,0.9]	[0.3,2]	[3.2,6.7]	[0,0]	[9.2,14.5]	[0.3,2]	[0,1.2]	[0,0.6]	[6.8,11.5]	[0,0.2]	[0.3,2.1]	[1.8,4.6]	[11,16.6]

Table 18.1 *(continued)*.

Typology	Legislators, senior officials and managers	Plant and machine operators and assemblers	Professionals	Service workers and shop and market sales workers	Skilled agricultural and fishery workers	Technicians and associate professionals	Divorced or separated	Married	Single	Widow(er)	Class
M.15–24	[1,2.2]	[10.2,13.2]	[0.9,2]	[9.3,12.2]	[3.3,5.1]	[4.4,6.5]	[0,0.6]	[12.7,15.9]	[83.8,87]	[0,0]	Class 1
M.25–34	[4.1,5.8]	[12.1,14.6]	[5.3,7.1]	[9.5,11.8]	[3.1,4.5]	[6.8,8.8]	[1.3,2.2]	[62.4,66]	[32.3,35.8]	[0,0]	Class 1
M.35–44	[9.4,11.6]	[10.8,13.1]	[5.3,7]	[9,11.1]	[4.1,5.6]	[7.1,9.1]	[1.3,2.3]	[88.7,90.8]	[7.2,9.2]	[0.1,0.4]	Class 3
M.45–54	[12.2,14.8]	[12.8,15.5]	[4.8,6.6]	[6.6,8.7]	[6,8]	[7.5,9.7]	[1.7,2.8]	[93,94.9]	[2.4,3.7]	[0.4,1.1]	Class 3
M.55–64	[12.5,15.8]	[10,13]	[2.8,4.6]	[5.5,7.9]	[21.1,25.1]	[4,6.1]	[0.7,1.7]	[93.5,95.6]	[1.4,2.7]	[1.5,2.9]	Class 3
M.65+	[7,10.7]	[0.9,2.7]	[2.1,4.4]	[2.9,5.6]	[61.5,67.8]	[1,2.8]	[0.2,1.3]	[84.1,88.7]	[1.4,3.4]	[8.4,12.5]	Class 2
W.15–24	[0.8,2.2]	[7.4,10.5]	[2.6,4.6]	[25.4,30.3]	[2,3.8]	[5.1,7.8]	[0.3,1.4]	[25.7,30.7]	[68.4,73.4]	[0,0.1]	Class 1
W.25–34	[2.3,3.6]	[5.2,7.1]	[10.2,12.8]	[18.1,21.3]	[4.2,6]	[7.4,9.7]	[2.2,3.5]	[71,74.7]	[21.9,25.4]	[0.3,1]	Class 1
W.35–44	[5.4,7.3]	[3.4,5]	[7.6,9.8]	[15.7,18.7]	[7.3,9.5]	[8.8,11.2]	[5.1,7]	[82.9,85.8]	[7.1,9.2]	[1,1.9]	Class 1
W.45–54	[6.8,9.2]	[2.2,3.7]	[5.2,7.3]	[15.3,18.7]	[12.7,15.8]	[7.7,10.2]	[5.3,7.5]	[80.3,83.7]	[4.7,6.8]	[4.8,6.9]	Class 3
W.55–64	[6.1,9.1]	[0.3,1.3]	[2.1,4.1]	[11.1,14.9]	[30.2,35.4]	[3.8,6.2]	[3.3,5.6]	[70,75]	[5.1,7.8]	[14.5,18.7]	Class 2
W.65+	[3.9,7.7]	[0,0]	[1,3.4]	[3.4,7]	[63,70.7]	[0.6,2.7]	[0.9,3.1]	[51.4,59.5]	[7.2,12]	[29.1,36.7]	Class 2

Table 18.2 SFDA eigenvalues.

Eigenvalues	Inertia	Percentage of explained inertia	Cumulative % of inertia	Histogram 0– – –– 50%– – – 100%
1	0.00583	81.247	81.247	|***************
2	0.00105	14.669	95.916	|**
3	0.00029	4.084	100.000	|*

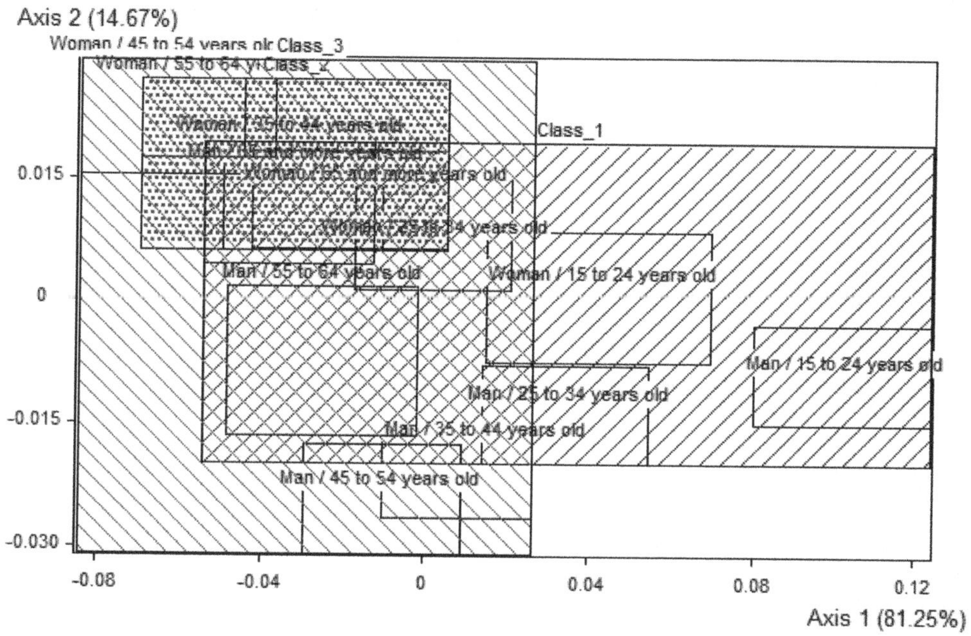

Figure 18.1 Classes and objects represented on the first factorial plane, without the automatic selection of the predictors.

The first factor, which explains 81.25% of the total inertia, differentiates those objects (on the negative half) with a higher percentage (with respect to the average percentage) of workers in service sectors and as clerks from those (on the positive half) with a higher percentage (with respect to the average percentage) of workers in manufacturing or as family workers.

The second factor, which explains 14.67% of the total inertia, differentiates those objects (on the negative half) with a higher percentage of workers in industrial sectors (electricity, transport) and with secondary and professional studies from those (on the positive half) with a higher percentage of people with university studies, low level of secondary and primary studies and manufacturers and craft and related trades workers.

Table 18.3 Contribution of predictors to axes.

Axis 1			Axis 2		
Variable–Category	Contribution (%)	Half	Variable–Category	Contribution (%)	Half
Service_wo–HI	8.080	–	Electricit–HI	6.244	–
Wholesale_–LOW	5.024	–	Legislator–HI	5.579	–
Primary_st–LOW	4.900	–	Other_serv–MED	5.464	–
Elementary–HI	3.872	–	Self-emplo–HI	5.448	–
Clerks–HI	3.145	–	Transport–HI	4.629	–
Transport–LOW	3.089	–	Constructi–MED	4.306	–
Without_st–MED	3.075	–	Family_wor–LOW	3.986	–
Hotels_and–HI	3.060	–	Elementary–LOW	3.794	–
Plant_and_–LOW	3.007	–	Financial_–HI	3.146	–
Agricultur–MED	2.919	–	Service_wo–LOW	2.932	–
Skilled_ag–MED	2.860	–	Mining_and–HI	2.581	–
Technician–HI	2.795	–	Craft_and_–MED	2.469	–
Public_adm–HI	2.558	–	Part-time–HI	2.124	–
Self-emplo–MED	1.869	–	Full-time–LOW	2.124	–
Manufactur–LOW	1.719	–	Widow/Wido–HI	2.115	–
Widow/Wido–MED	1.616	–	Married–HI	1.845	–
Primary_st–HI	1.579	–	Without_st–HI	1.599	–
Employee–MED	1.505	–	Divorced_o–MED	1.598	–
Divorced_o–HI	1.464	–	Skilled_ag–HI	1.417	–
Other_serv–MED	1.414	–	Public_adm–HI	1.285	–
Craft_and_–LOW	1.409	–	Agricultur–HI	1.154	–
Other_prof–HI	1.262	–			
Full-time–MED	1.178	–			
Part-time–MED	1.178	–			
University–HI	1.159	–			
Self-emplo–MED	1.134	–			
Secondary_–LOW	1.109	–			
Legislator–MED	1.104	–			
Married–MED	1.098	–			
Other_serv–HI	1.079	–			
Armed_forc–LOW	1.026	–			
Constructi–LOW	1.012	–			
Real_state–LOW	0.979	–			
Central zone			Central zone		
Family_wor–HI	1.118	+	Other_serv–HI	0.980	+
Manufactur–HI	1.700	+	Divorced_o–HI	1.133	+
			Hotels_and–HI	1.162	+
			Married_–MED	1.430	+
			Secondary_–LOW	1.589	+
			University–HI	1.900	+
			Wholesale_–LOW	1.947	+
			Manufactur–LOW	2.318	+
			Primary_st–LOW	2.491	+
			Craft_and_–LOW	2.965	+

HI = High level for the variable, MED = medium level and LOW = low level.

If we use the automatic selection procedure based on the capability index $\tau_{Y|X_j}$, Table 18.4 shows the ranking for the predictors and which were selected for the FDA phase. In this case only the following 22 out of 37 predictors are chosen as active in the analysis:

Table 18.4 Ranking of the descriptors on the basis of their predictivity power.

Selection	Predictor	$\tau_{Y\|X_j}$	Cum.	% of $\tau_{Y\|X_j}$	Cum% $\tau_{Y\|X_j}$
Refused	Real_estate	0.087	0.09	0.42	0.42
Refused	Financial_intermedia	0.120	0.21	0.57	0.99
Refused	Mining_and_quarrying	0.128	0.34	0.62	1.61
Refused	Electricity	0.155	0.49	0.74	2.35
Refused	Other_professional_s	0.161	0.65	0.78	3.13
Refused	Service_workers_and_	0.191	0.84	0.92	4.04
Refused	Hotels_and_restauran	0.254	1.10	1.22	5.26
Refused	Clerks	0.278	1.37	1.33	6.60
Refused	Wholesale_and_retail	0.402	1.78	1.93	8.53
Refused	Family_worker	0.426	2.20	2.04	10.60
Refused	Plant_and_machine_op	0.460	2.66	2.21	12.80
Refused	Full-time	0.473	3.14	2.27	15.10
Refused	Part-time	0.473	3.61	2.27	17.30
Refused	Technicians_and_asso	0.525	4.13	2.52	19.80
Refused	Transport	0.536	4.67	2.57	22.40
Accepted	Armed_forces	0.552	5.22	2.65	25.10
Accepted	Other_services	0.566	5.79	2.72	27.80
Accepted	Public_administratio	0.571	6.36	2.74	30.50
Accepted	Craft_and_related_tr	0.664	7.02	3.19	33.70
Accepted	Widow/Widower	0.667	7.69	3.20	36.90
Accepted	Skilled_agricultural	0.676	8.36	3.25	40.20
Accepted	Manufacturing	0.677	9.04	3.25	43.40
Accepted	Construction	0.679	9.72	3.26	46.70
Accepted	Divorced_or_Separated	0.708	10.40	3.40	50.10
Accepted	Elementary_occupatio	0.713	11.10	3.42	53.50
Accepted	University_studies	0.722	11.90	3.47	57.00
Accepted	Agriculture	0.733	12.60	3.52	60.50
Accepted	Professionals	0.738	13.30	3.54	64.00
Accepted	Without_studies	0.740	14.10	3.55	67.60
Accepted	Primary_studies_or_l	0.746	14.80	3.58	71.20
Accepted	Self-employed_with_e	0.787	15.60	3.78	74.90
Accepted	Legislators	0.831	16.40	3.99	78.90
Accepted	Married	0.832	17.30	3.99	82.90
Accepted	Secondary_studies_-_	0.855	18.10	4.11	87.00
Accepted	Single	0.870	19.00	4.17	91.20
Accepted	Self-employed_withou	0.910	19.90	4.37	95.60
Accepted	Employee	0.924	20.80	4.44	100.00

1. Primary studies or less
2. Secondary studies and professional
3. University studies
4. Without studies
5. Employee
6. Self-employed with employees
7. Self-employed without employees
8. Agriculture, cattle, hunting, forestry and fishing
9. Construction
10. Manufacturing
11. Other services
12. Public administration
13. Armed forces
14. Craft and related trades workers
15. Elementary occupations
16. Legislators, senior officials and managers
17. Professionals
18. Skilled agricultural and fishery workers
19. Divorced or separated
20. Married
21. Single
22. Widow/widower

The SFDA eigenvalues associated with the selection of 22 predictors are shown in Table 18.5.

The main result is that the inertia explained is greater in FDA (the total inertia is 0.011 23) with the variable selection based on the index $\tau_{Y|X_j}$ than on the full selection of

Table 18.5 SFDA eigenvalues.

Eigenvalues	Inertia	Percentage of explained inertia	Cumulative % of inertia	Histogram 0– – – – – 50%– – –100%
1	0.00901	80.227	80.227	|*************
2	0.00171	15.259	95.487	|**
3	0.00051	4.513	100.000	|*

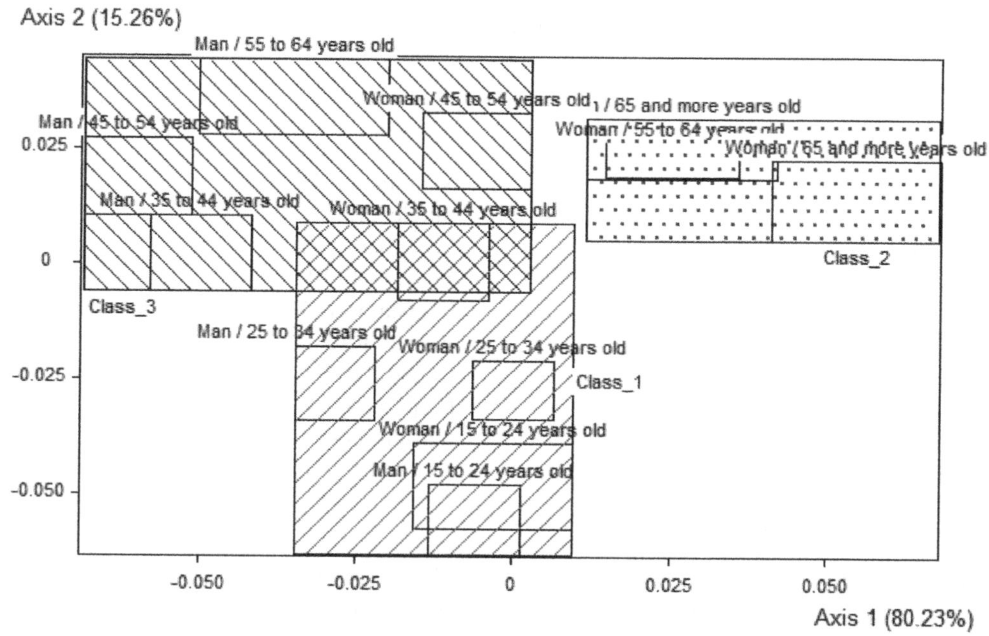

Figure 18.2 Classes and objects represented on the first factorial plane, using the automatic selection of the predictors.

predictors (the total inertia is 0.007 17). This is quite evident from the first factor plane (Figure 18.2).

In this case we observe that classes are more separated and in particular the first axis separates class 2 from classes 3 and 1, while the second axis separates class 1 and class 3. Table 18.6 shows what predictor has best contributed to the definition of the first two factor axes ordered on the basis of their contribution from the negative to the positive half of the axis. Recall that each numerical variable is fuzzy-coded (see GCA) into a categorical variable with three levels (high, medium and low). In our case we have the contribution of each category to the factor axes. The number of categories (see GCA) in our case is equal to 66. The categories represented are those having a contribution greater than $1/66 = 0.015 = 1.5\%$.

The first factor, which explains 80.23% of the total inertia, differentiates those objects which are more characterized, with a higher percentage (with respect to the average percentage), by workers in positions of responsibility such as self-employed with employees and managers, from those (on the positive half) more characterized by older people who have a higher percentage (with respect to the average percentage) of people with secondary and professional studies, a higher percentage (with respect to the average percentage) of workers that are self-employed without employees, but also a higher percentage (with respect to the average percentage) of people without studies and of widowers.

The second factor, which explains 15.26% of the total inertia, differentiates those objects (on the negative half) with a higher percentage (with respect to the average percentage) of single people, working as manufacturers and with secondary and professional studies from those (on the positive one) with a higher percentage (with respect to the average percentage) of widowers, without studies and low level of primary studies completed.

Table 18.6 Contribution of predictors to axes.

Axis 1			Axis 2		
Variable–category	%contr	Half	Variable–category	%contr	Half
Constructi–MED	3.274	–	Manufactur–HI	3.664	–
Public_adm–HI	3.048	–	Single–HI	3.337	–
Legislator–HI	2.966	–	Secondary_–HI	3.042	–
Primary_st–HI	2.543	–	Married–LOW	2.909	–
Self-emplo–HI	2.535	–	Legislator–LOW	2.518	–
Self-emplo–MED	2.075	–	Armed_forc–HI	2.442	–
Employee–MED	1.984	–	Constructi–HI	2.331	–
Secondary_–MED	1.951	–	Employee–HI	2.176	–
Married–HI	1.692	–	Self-emplo–LOW	2.164	–
Elementary–LOW	1.576	–	Craft_and_–HI	2.051	–
Central zone			Central zone		
Self-emplo–LOW	1.568	+	Skilled_ag–MED	1.545	+
Legislator–LOW	1.606	+	Agricultur–MED	1.611	+
Armed_forc–HI	1.738	+	Elementary–HI	2.436	+
University–LOW	1.771	+	Secondary_–LOW	2.768	+
Single–HI	1.811	+	Craft_and_–LOW	2.943	+
Secondary_–LOW	1.854	+	Agricultur–HI	2.979	+
Employee–HI	1.893	+	Skilled_ag–HI	3.016	+
Profession–LOW	1.894	+	Employee–LOW	3.144	+
Married–LOW	1.931	+	Self-emplo–HI	3.190	+
Public_adm–LOW	2.086	+	Manufactur–LOW	3.374	+
Craft_and_–LOW	2.197	+	Without_st–HI	3.556	+
Manufactur–HI	2.488	+	Widow/Wido–HI	3.957	+
Manufactur–LOW	2.996	+	Primary_st–LOW	4.879	+
Skilled_ag–HI	3.039	+			
Agricultur–HI	3.077	+			
Employee–LOW	3.111	+			
Primary_st–LOW	3.366	+			
Secondary_–HI	3.420	+			
Self-emplo–HI	3.447	+			
Without_st–HI	3.705	+			
Widow/Wido–HI	3.990	+			

Axes description contribution, direction (+ positive, – negative)
HI = High level for the variable, MED = medium level and LOW = Low level.

18.4.1 Comparison of dissimilarities used for classification step

To compare and choose the best dissimilarity (Hausdorff distance, Ichino–de Carvalho, pdi) and criteria of classification (single, average and complete linkage) to use, we present a comparison of the possible rules that can be chosen, both when all predictors are chosen and

Table 18.7 Misclassification rate using different dissimilarity measure and linkage criteria.

Linkage	Misclassification rate		
	Hausdorff distance Euclidean metric	Ichino–de Carvalho dissimilarity $\gamma = 0.5$ and Euclidean metric	Potential description increase index
Single	50.00%	75.00%	
Average	33.34%	59.33%	50.00%
Complete	50.00%	50.00%	

Table 18.8 Misclassification rate using different dissimilarity measure and linkage criteria.

Linkage	Misclassification rate		
	Hausdorff distance Euclidean metric	Ichino–de Carvalho dissimilarity $\gamma = 0.5$ and Euclidean metric	Potential description increase index
Single	16.66%	50.00%	
Average	16.66%	16.66%	33.34%
Complete	8.33%	8.33%	

when automatic selection based on the Goodman–Kruskal τ has been used. To calculate the misclassification index we have performed the following procedure: for each object, we omit it from the data set, we identify classes and we use the dissimilarity and the classification criteria in order to reassign it.

Table 18.7 compares the misclassification rates using the three dissimilarity or distance measures and linkages, for complete selection of predictors. On the basis of our example, we note that the Hausdorff distance is generally the best choice of assignment rule, with misclassification rates not greater than the other measures using the three linkages. In particular, the best rule to use is Hausdorff distance and average linkage.

Table 18.8 compares the misclassification rates using the three dissimilarity or distance measures and linkages, in the case of automatic selection based on the τ Goodman–Kruskal predictivity index (22 variables on 37). On the basis of our example, we may observe that the Hausdorff distance is generally the best choice of assignment rule, with misclassification rates not greater than the other measures using the three linkages. In particular, the best rule to use is the Hausdorff distance or Ichino–de Carvalho dissimilarity and complete linkage.

References

Celeux, G. and Nakache, J.P. (1994) *Analyse discriminante sur variables qualitatives*, Polytechnica.

de Carvalho, F.A.T. (1992) *Méthodes descriptives en analyse de données symboliques*. Doctoral thesis, Université Paris IX Dauphine.

Diday, E. (1987) Introduction à l'approche symbolique en analyse des données, *Journées Symbolique-Numerique*, Université Paris IX Dauphine.

Goodman, L.A. and Kruskal, W.H. (1954). Measures of association for cross-classifications. *Journal of the American Statistical Association*, 49, 732–764.

Lauro, C., Verde, R. and Palumbo, F. (2000) Factorial discriminant analysis on symbolic objects. In H.-H. Bock and E. Diday (eds), *Analysis of Symbolic Data*. Springer-Verlag, Berlin.

Lebart, L., Morineau, A. and Piron, M. (1995) *Statistique exploratorie multidimensionelle*. Dunod, Paris.

19

Symbolic linear regression methodology

Filipe Afonso, Lynne Billard, Edwin Diday and Mehdi Limam

19.1 Introduction

Symbolic data can occur naturally in their own right on any sized data set. They can occur as a result of aggregation of a (usually, but not necessarily) large or extremely large data set into one of a smaller size so as to make it more manageable for subsequent analysis, with specific aggregation designed to answer some particular scientific question. A symbolic data set can also be the result after organizing a (classical or symbolic) data set, via the assertion mechanism, to enable appropriate statistical analyses to assist in answering fundamental scientific question(s) behind the assertion(s). A review of what constitutes symbolic data and how such data contrast with classical data can be found in Billard and Diday (2003) and Bock and Diday (2000). A key distinction is that a classical observation is a single point in the space $\mathbb{R}^p$, whereas a symbolic observation is a hypercube in $\mathbb{R}^p$. A review of how an initial data set is transformed into a symbolic data set is provided in Chapter 5 and also in Bock and Diday (2000). The purpose of the present chapter is to propose ways to adapt classical ideas in regression analysis to enable regression analysis on quantitative symbolic data. Billard and Diday introduced regression for interval-valued symbolic data (2000) and histogram-valued symbolic data (2002). Other approaches have been considered in, for example, de Carvalho *et al.* (2004) and Lima Neto *et al.* (2005). This chapter considers how to implement regression analysis in the presence of taxonomical and hierarchical dependent variables. The method is described and tested on a data set simulated from real statistics. Finally, we present an example of symbolic linear regression in order to study concepts.

Symbolic Data Analysis and the SODAS Software Edited by E. Diday and M. Noirhomme-Fraiture

19.2 State of the art

19.2.1 Linear regression

In a linear regression model on classical data, we wish to explain a dependent variable Y in terms of a linear function of explanatory variables X_j, $j=1,\ldots,k$:

$$Y=\beta_0+\beta_1 X_1+\cdots+\beta_k X_k+\epsilon=\boldsymbol{\beta}\,\mathbf{X}+\boldsymbol{\epsilon}, \tag{19.1}$$

where $\mathbf{X}=(1, X_1,\ldots, X_k)$ is the vector of regressor or predictor variables, $\boldsymbol{\beta}=(\beta_0,\ldots,\beta_k)$ is the vector of regression parameters, and $\boldsymbol{\epsilon}$ is the residual error vector with zero mean and standard deviation $\sigma\mathbf{I}$. In order to find the best vector $\boldsymbol{\beta}$, we minimize the sum of the squared residual errors. We obtain

$$\boldsymbol{\beta}^*=(\mathbf{X}'\mathbf{X})^{-1}\mathbf{X}'\mathbf{Y}, \tag{19.2}$$

where $\boldsymbol{\beta}^*=(\beta_0^*,\ldots,\beta_k^*)$ is the vector of estimated regression parameters. In the special case where $p=1$, (19.2) reduces to

$$\beta_1^*=\sum_{i=1}^{n}(X_i-\bar{X})(Y_i-\bar{Y})/\sum_{i=1}^{n}(X_i-\bar{X})^2 \tag{19.3}$$

and

$$\beta_0^*=\bar{Y}-\beta_1^*\bar{X}, \tag{19.4}$$

where $\bar{X}$ and $\bar{Y}$ are the sample means of the observed X and Y values.

19.2.2 Linear regression with interval-valued data[1]

[1]For interval-valued data, denote

$$Y_i=[a_i, b_i] \quad \text{and} \quad \mathbf{X}_i=([c_{i1}, d_{i1}],\ldots,[c_{ik}, d_{ik}]).$$

Then the elements of the variance–covariance matrix $\mathbf{V}=(\text{cov}(X_{j_1}, X_{j_2}))$, $j_1, j_2=1,\ldots,k$, are, for $j_1\neq j_2$,

$$\begin{aligned}\text{cov}(X_{j_1}, X_{j_2})=&\frac{1}{4n}\sum_{i=1}^{n}(c_{ij_1}+d_{ij_1})(c_{ij_2}+d_{ij_2})\\&-\frac{1}{4n^2}\left[\sum_{i=1}^{n}(c_{ij_1}+d_{ij_1})\right]\left[\sum_{i=1}^{n}(c_{ij_2}+d_{ij_2})\right],\end{aligned} \tag{19.5}$$

and, for $j_1=j_2$ (Betrand and Goupil, 2000),

$$\text{var}(X_{j_1})=\frac{1}{4n}\sum_{i=1}^{n}(c_{ij_1}+d_{ij_1})^2-\frac{1}{4n^2}\left[\sum_{i=1}^{n}(c_{ij_1}+d_{ij_1})\right]^2. \tag{19.6}$$

[1] This section is based on Billard and Diday (2000, 2002).

Likewise (Bertrand and Goupil, 2000) the variance of Y, σ_Y^2, is found from (19.6) replacing c_{ij} and d_{ij} everywhere by a_i and b_i, respectively. Also, the symbolic means of the interval-valued X and Y are

$$\bar{Y} = \frac{1}{2n}\sum_{i=1}^{n}(a_i + b_i), \quad \bar{X}_j = \frac{1}{2n}\sum_{i=1}^{n}(c_{ij} + d_{ij}), \quad j = 1, \ldots, k. \tag{19.7}$$

The $(j+1)$th element of the $(k+1) \times 1$ vector $\mathbf{X}'\mathbf{Y}$ becomes

$$(x_j' y_j) = \frac{1}{2}\sum_{i=1}^{m}(c_{ij} + d_{ij})(a_i + b_i), \quad j = 1, \ldots, k. \tag{19.8}$$

The symbolic regression model fitted to these interval-valued observations is found from (19.2) using the symbolic statistics of (19.5)–(19.8).

19.2.3 Linear regression with histograms[2]

For histogram-valued observations, let us write the dependent observations as

$$Y_i = \{p_{i1}[a_{i1}, b_{i1}], \ldots, p_{is_i}[a_{is_i}, b_{is_i}]\}$$

where p_{is_i} is the relative frequency for the subinterval $[a_{is_q}, b_{is_q}]$, $s_q = 1, \ldots, s_i$, $i = 1, \ldots, n$; that is, for observation i, the observed histogram takes values on s_i subintervals. Note that in general $s_{i_1} \neq s_{i_2}$, $i_1, i_2 = 1, \ldots, n$. Similarly, the predictor variables X_j, when histogram-valued, are denoted by, for observation i,

$$X_{ij} = \{p_{ij1}[c_{ij1}, d_{ij1}], \ldots, p_{ijs_{ij}}[c_{ijs_{ij}}, d_{ijs_{ij}}]\},$$

where $p_{ijs_{q_iq_j}}$ is the relative frequency for the subinterval $[c_{ijs_{q_iq_j}}, d_{ijs_{q_iq_j}}]$, $s_{q_iq_j} = 1, \ldots, s_{ij}$, $i = 1, \ldots, n$, $j = 1, \ldots, k$; that is, for the observation i, the observed histogram on the X_j variable takes values on s_{ij} subintervals. In general, $s_{i_1j_1} \neq s_{i_2j_2}$ for $i_1 \neq i_2$ and/or $j_1 \neq j_2$. However, for the special case where a given histogram-valued observation Y_i (say) is interval-valued, then $s_i = 1$; likewise for any X_{ij} that is interval-valued, $s_{ij} = 1$.

The histogram-valued counterparts of the interval-valued terms in (19.5)–(19.8) used in (19.2) to fit a regression model are defined as follows. From Billard and Diday (2002),

$$\begin{aligned}\mathrm{cov}(X_{j_1}, X_{j_2}) = {} & \frac{1}{4n}\sum_{i=1}^{n}\left\{\sum_{s_{q_iq_{j_1}}=1}^{s_{ij_1}}\sum_{s_{q_iq_{j_2}}=1}^{s_{ij_2}} p_{ij_1s_{q_iq_{j_1}}} p_{ij_2s_{q_iq_{j_2}}} (c_{ij_1s_{ij_1}} + d_{ij_1s_{ij_1}})(c_{ij_2s_{ij_2}} + d_{ij_2s_{ij_2}})\right\} \\ & - \frac{1}{4n}\left[\sum_{i=1}^{n}\sum_{s_{q_iq_{j_1}}=1}^{s_{ij_1}} p_{ij_1s_{ij_1}}(c_{ij_1s_{ij_2}} + d_{ij_1s_{ij_1}})\right]\left[\sum_{i=1}^{n}\sum_{s_{q_iq_{j_2}}=1}^{s_{ij_2}} p_{ij_2s_{ij_2}}(c_{ij_2s_{ij_2}} + d_{ij_2s_{ij_2}})\right].\end{aligned} \tag{19.9}$$

[2]This section is based on Billard and Diday (2002).

Again from Billard and Diday (2002),

$$\text{var}(X_j) = \frac{1}{4n} \sum_{i=1}^{n} \sum_{s_{q_iq_j}=1}^{s_{ij}} p_{ijs_{q_iq_j}} (c_{ijs_{q_iq_j}} + d_{ijs_{q_iq_j}})^2 - \frac{1}{4n^2} \left[\sum_{i=1}^{n} \sum_{s_{q_iq_j}=1}^{s_{ij}} p_{ijs_{q_iq_j}} (c_{ijs_{q_iq_j}} + d_{ijs_{q_iq_j}}) \right]^2, \tag{19.10}$$

$$\bar{Y} = \frac{1}{n} \sum_{i=1}^{n} \sum_{s_q=1}^{s_i} p_{is_q} (a_{is_q} + b_{is_q}), \tag{19.11}$$

$$\bar{X}_j = \frac{1}{n} \sum_{i=1}^{n} \sum_{s_{q_i}j=1}^{s_{ij}} p_{ijs_{q_i}j} (c_{ijs_{q_i}j} + d_{ijs_{q_i}j}), \quad j = 1, \ldots, k. \tag{19.12}$$

Finally, we have the $(j+1)$th element of $\mathbf{X}'\mathbf{Y}$, for $j = 1, \ldots, k$, given by

$$(x_j'y_j) = \frac{1}{2} \sum_{i=1}^{n} \left[\sum_{s_{q_iq_j}=1}^{s_{ij}} p_{ijs_{q_iq_j}} (c_{ijs_{q_iq_j}} + d_{ijs_{q_iq_j}}) \right] \left[\sum_{s_q=1}^{s_i} p_{is_q} (a_{is_q} + b_{is_q}) \right]. \tag{19.13}$$

19.3 Taxonomic and hierarchical dependent variables

Taxonomic variables are variables organized in a tree in order to express several levels of generality. Taxonomies are, for example, useful for classifying flora and fauna, or geographical zones (towns, regions, countries). We will study an example with two taxonomic variables, 'zone' and 'age' (Figure 19.1).

Hierarchical dependent variables (or mother–daughter variables) are those organized in a hierarchy. Some variables, the daughters, exist only when another variable, the mother, takes a restricted set of values. There can be several levels (as when X has a daughter Y who in turn has a daughter Z, and so on). For example, in Figure 19.2, 'radius' ('side') is only meaningful if the mother variable, 'form', takes the value {circle} ({square}).

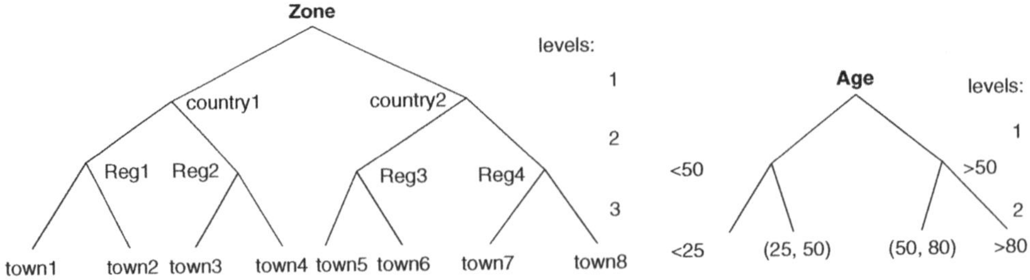

Figure 19.1 Two taxonomic variables.

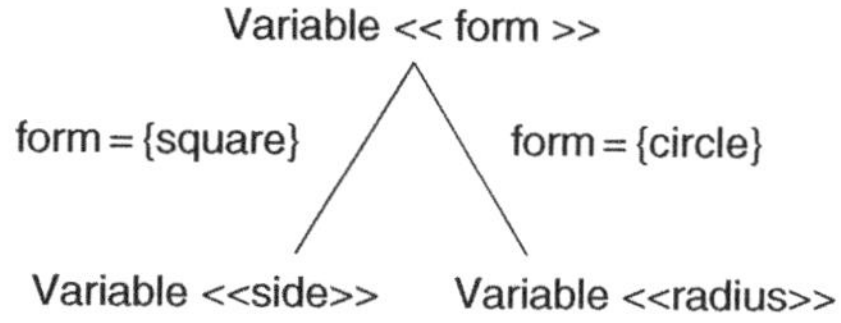

Figure 19.2 Hierarchically dependent variables.

19.4 Linear regression for taxonomic variables

19.4.1 Input data

In general, suppose there are t taxonomical variables, with the jth such variable having t_j levels, $j = 1, \ldots, t$. To illustrate the methodology, suppose there are two taxonomic variables, 'zone' and 'age' (as in Figure 19.1). Suppose the taxonomic variable 'age' divides into young (as in age < 50) and old (age ≥ 50) age ranges (at level $v = 1$, $X_1 =$ age range, say) with these in turn dividing into ages ≤ 25 and $(25, 50)$, and $[50, 80)$ and ≥ 80 years, respectively (at level $v = 2$, $X_2 =$ age, say). Here, $t_1 = 2$. Suppose that 'zone' consists of two possible countries each of which has two possible regions with the regions in turn comprising two towns. There are three levels, at level $v = 3$, $Z_3 =$ town, at level $v = 2$, $Z_2 =$ region, and at level $v = 1$, $Z_1 =$ country, with $t_2 = 3$.

Then, assume the data are as in Table 19.1 where 'zone' and 'age' are the explanatory variables and $Y =$ income is the quantitative dependent variable. The problem is that we are dealing with values at different levels of generality. Indeed, for individuals 5, 9, 11 and 15 we know the age range X_2 but not the age X_1. For individuals 4, 5, 6, 8 and 18, we know the country or the region but not the town. In this example, 'age' and 'zone' are the taxonomical explanatory variables and 'income' is the quantitative dependent variable.

19.4.2 Method of regression

In this method, $t^* = \max_j(t_j)$ regressions are performed, one at each level. Thus, for each level, we increase the generality of the values at lower levels and decrease the generality of

Table 19.1 Data matrix with taxonomies.

Individual	Age	Zone	Income	Individual	Age	Zone	Income
1	$[25, 50)$	town1	3000	11	< 50	town8	3000
2	< 25	town3	2100	12	$[50, 80)$	town7	2800
3	≥ 80	town3	2400	13	$[50, 80)$	town7	3800
4	< 25	region2	2000	14	$[25, 50)$	town6	2300
5	< 50	region1	1900	15	< 50	town4	2800
6	$[25, 50)$	country1	2500	16	$[25, 50)$	town1	3100
7	≥ 80	town5	3000	17	≥ 80	town1	3300
8	$[50, 80)$	region3	2500	18	$[50, 80)$	country1	3100
9	≥ 50	town5	3200	19	$[50, 80)$	town7	3400
10	$[25, 50)$	town7	2600	20	< 25	town4	1500

the values at higher levels to the relevant level. For example, at the level $t=3$, we perform a regression at the level of the towns Z_3 and of the age X_2. Thus, if the zone variable is known only to be of region $Z_2=$ region1, with no information about the value of the town Z_3 variable, then a weight of (1/2, 1/2) is given to the possible values (town1, town2). Moreover, if the zone variable is known only to be of country $Z_1=$ country1 (say) with no information about the value of the town Z_3 variable, then a weight of (1/4, 1/4, 1/4, 1/4) is given to the possible values (town1, town2, town3, town4) in country1. At the level $t=2$, a regression is fitted across the regions Z_2 and the age range X_1. Thus, the towns are aggregated up to their regions (e.g., town3 is replaced by region2) and if the zone variable is known only to be of country $Z_1=$ country1 (country2) with no information about the value of the region Z_2 variable, then a weight of (1/2, 1/2) is given to the possible values, here region1 and region2 (region3 and region4). At the level $t=1$, we perform a regression at the level of the countries Z_1 and of the age range X_1. Therefore, the towns and regions are aggregated up to their countries (town1 is replaced by country1, region3 is replaced by country2, . . .). Thus, the data of the first ten observations in Table 19.1 under this method assume the form of Table 19.2. In this weighting scheme, it is assumed all possible values of X_j, Z_j are equally likely.

Thus the regression equations at the levels $t=3$, 2, 1, respectively, are:

$$Y = 3432.0 - 822.9U_1^{(1)} - 41.0U_2^{(1)} + 394.9U_3^{(1)} - 271.3V_1^{(1)} - 1819.9V_2^{(1)} - 738.9V_3^{(1)} - 622.9V_4^{(1)} - 533.6V_5^{(1)} - 1296.8V_6^{(1)} - 567.9V_7^{(1)},$$

where $U_i^{(1)}$, $i=1,\ldots,4$, represent the binary variables associated with the four ages, and $V_j^{(1)}$, $j=1,\ldots,8$, are the corresponding variables for the eight towns;

$$Y = 3337.6 - 544.0U_1^{(2)} - 72.4V_1^{(2)} - 716.6V_2^{(2)} - 451.6V_3^{(2)},$$

Table 19.2 Weight matrix at (a) level $t=3=\max(3,2)$; (b) level 2; (c) level 1.

(a)

	Age				Zone							
i	<25	[25, 50)	[50, 80)	≥ 80	town1	town2	town3	town4	town5	town6	town7	town8
1	0	1	0	0	1	0	0	0	0	0	0	0
2	1	0	0	0	0	0	1	0	0	0	0	0
3	0	0	0	1	0	0	1	0	0	0	0	0
4	1	0	0	0	0	0	0.5	0.5	0	0	0	0
5	0.5	0.5	0	0	0.5	0.5	0	0	0	0	0	0
6	0	1	0	0	0.25	0.25	0.25	0.25	0	0	0	0
7	0	0	0	1	0	0	0	0	1	0	0	0
8	0	0	1	0	0	0	0	0	0.5	0.5	0	0
9	0	0	0.5	0.5	0	0	0	0	1	0	0	0
10	0	1	0	0	0	0	0	0	0	0	1	0

Table 19.2 *(Continued)*

(b)

	Age		Zone			
i	<50	≥ 50	region1	region2	region3	region4
1	1	0	1	0	0	0
2	1	0	0	1	0	0
3	0	1	0	1	0	0
4	1	0	0	1	0	0
5	1	0	1	0	0	0
6	1	0	0.5	0.5	0	0
7	0	1	0	0	1	0
8	0	1	0	0	1	0
9	0	1	0	0	1	0
10	1	0	0	0	0	1

(c)

	Age		Zone	
i	<50	≥ 50	country1	country2
1	1	0	1	0
2	1	0	1	0
3	0	1	1	0
4	1	0	1	0
5	1	0	1	0
6	1	0	1	0
7	0	1	0	1
8	0	1	0	1
9	0	1	0	1
10	1	0	0	1

where $U_i^{(2)}$, $i=1, 2$, are the indicator variables for the two age ranges and $V_j^{(2)}$, $j=1, \ldots, 4$, are the indicator variables for the four regions; and

$$Y = 3131.9 - 529.0U_1^{(3)} - 229.0V_1^{(3)},$$

where $U_i^{(3)}$, $i=1, 2$, and $V_j^{(3)}$, $j=1, 2$, are the indicator variables for age range and country, respectively.

Note that when using the predictor variables in an indicator variable format, we have to delete one of the possible values (at level $t=2$, ≥ 80 for age, and at level $t=3$, town8 for zone) in order to be able to invert the $\mathbf{X'X}$ matrix in (19.2).

An advantage of this method is that all observations are used. A disadvantage is that the weights used to apportion missing values at lower levels may not be faithful to the real data set. Other weights can be used. For example, weights corresponding to the proportion or cardinality of those possible values that do occur could be used. Alternatively, these values could be estimated with linear regression using the taxonomic variable as the dependent variable and then the regression can be fitted to the 'completed' data set at level 1.

19.4.3 Example

We illustrate the method with a data set simulated from real statistics. Since there are both taxonomic and non-taxonomic variables, we first demonstrate the methodology when all the explanatory variables are taxonomic (Test 1, say), and then demonstrate it by adding non-taxonomic variables (Test 2).

For Test 1 the data set consists of 10 000 individuals with dependent variable Y as income. The predictor variables are two taxonomical variables, work characteristics X and zone Z. The work variable has two levels, $t_1=2$. Thus, at the first level representing the type of work X_1, individuals either work full-time $X_1=1$, part-time $X_1=2$, or do not work

$X_1 = 3$. At the second level, the variable X_2 is weeks worked and takes four possible values where each of $X_1 = 1$ and $X_1 = 2$ has three branches corresponding to number of weeks worked with values 50 weeks or more ($X_2 = 1$), 27 to 49 weeks ($X_2 = 2$) and 26 weeks or less ($X_2 = 3$). The fourth level value is $X_2 = 4$ which is the only branch pertaining to $X_1 = 3$.

The zone variable Z has four levels, $t_2 = 4$. At the top of the tree, Z_1 represents the division variable with $Z_1 = 1, 2, 3, 4$ being the NorthEast, MidWest, South and West, respectively. Each division has branches at the second level ($v = 2$) corresponding to the region variable Z_2 with two regions in each division, except for the South which has three regions. Each region has two branches at the third $v = 3$ level corresponding to $Z_3 = 1$, 2, if the observation is from a metro or non-metro area, respectively. Finally, at level $v = 4$, Z_4 takes values 1 (2) if the metro area is a central city with fewer (more) than 1 million inhabitants, respectively, $Z_4 = 3$ (4) if the metro area is a non-central city with fewer (more) than 1 million inhabitants and $Z_4 = 5$ if it is a non-metro area. For Test 2, in addition to the taxonomic variables X and Z described in Test 1, the data set contains values for non-taxonomic quantitative variables, namely, W_1 (age), W_2 (glucose), W_3 (cholesterol), W_4 (haemoglobin), W_5 (haematocrit), W_6 (red blood count), W_7 (white blood count) and non-taxonomic qualitative variables W_8 (race = 1, 2 for white, black), W_9 (age group, $W_9 = 1, \ldots, 7$ for age groups 15–24, 25–34, 35–44, 45–54, 55–64, 65–74, and over 74) and W_{10} (diabetes, with $W_{10} = 0$, 1, 2 for no diabetes, mild diabetes and yes, respectively).

Since a data set will typically contain some observations with missing values on (some) lower branches, what happens in these situations is of interest. To study this, three different sets of missing data are generated by random selection. Thus, in the first case all the data are used (see Table 19.3, column (a)), in the second scenario only a few values are missing (12% of the fourth level of the taxonomy, 6% of level 3 and 3% of level 2, column (b)) while in the third scenario there is a large percentage of lower-level values missing (40%, 20%, 10% at levels 4, 3, 2, respectively, column (c)). In order to capture the impact of the missing values on the measure, we use the usual correlation coefficient R. The R values of Test 1 and Test 2 for each of the cases (a)–(c) are shown in Table 19.3.

The quality of the regression persists with the replacement of taxonomic values at the lower levels of the tree by their corresponding values at higher levels. However, if we look at the R values, we observe that level 4 is clearly the best level, as expected, when there are only a few missing values (from column (a), $R = 0.86$ at level 4 and $R = 0.72$ at level 3, for Test 2). However, in the presence of a lot of missing values at the first level,

Table 19.3 R values of Test 1 and Test 2.

	Test 1			Test 2		
Level v	(a)	(b)	(c)	(a)	(b)	(c)
4	0.33	0.31	0.25	0.86	0.84	0.78
3	0.16	0.16	0.14	0.72	0.72	0.70
2	0.05	0.05	0.05	0.59	0.59	0.59
1	0.03	0.03	0.03	0.56	0.56	0.56

the difference between level 4 and level 3 is relatively less important (from column (c), $R = 0.78$ at level 4 and $R = 0.70$ at level 3, for Test 2). As would be expected, fits at level 2 and 1 are not as good. Indeed, for Test 1 with no missing values the correlations are $R = 0.05$ (0.03) at level 2 (1). Consequently, it seems that this method retains the inherent structure of the taxonomy tree, as a taxonomic variable could explain a variable at level $t = k$ but not at level $t = k - 1$. In contrast, when there are many missing values at the kth level, a regression at the $(k - 1)$th level may be interesting.

19.5 Linear regression for hierarchical variables

19.5.1 Input data

Let us consider the regression methodology for mother–daughter variables organized in a hierarchical tree. We describe the methodology through the following example. Suppose there are five predictor variables X_j, $j = 1, \ldots, 5$, organized into the hierarchy as given in Figure 19.3 with a dependent quantitative variable Y. Thus, the variable X_1 takes possible values a, b or c. Then, if for individual i (or symbolic object i) the mother variable $X_1(i) = \{a\}$, the daughter variable X_2 is considered; while if $X_1(i) = \{b\}$, then the daughter variable X_5 is considered. Suppose the data are as given in Table 19.4. Notice that if $X_1(i) = a$, then the variable $X_5(i)$ is non-applicable (written $X_5(i) = \text{NS}$).

19.5.2 Methodology

The regression algorithm proceeds by fitting predictor variables at the top of the hierarchical tree and then moving progressively down the tree, as follows:

1. Fit a (classical or symbolic) regression model to the predictor variable at the top of the tree (here X_1) and the dependent variable Y (along with the other non-mother–daughter variables).
2. Calculate the residuals $Y - Y^*$, where Y^* is the predicted Y value based on the regression model of step 1.
3. Fit a regression model to the daughter predictor variable (of the previous mother predictor variable) on the residuals obtained from that previous regression. The

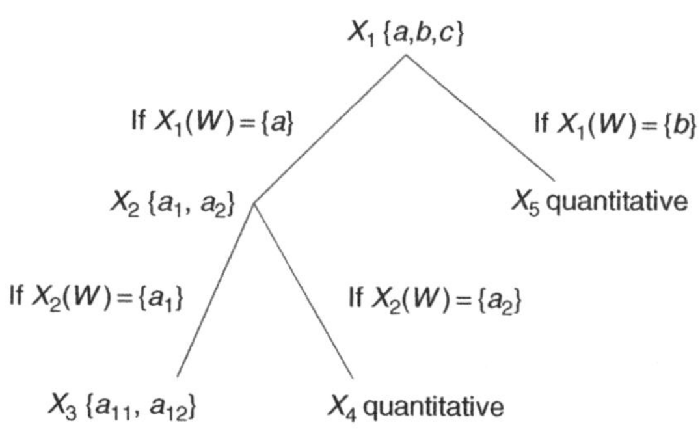

Figure 19.3 Hierarchically dependent variables X_i, $i = 1, \ldots, 5$.

Table 19.4 Data matrix hierarchical variables.

i	Y	X_1	X_2	X_3	X_4	X_5
1	4100	a	a_1	a_{11}	NS	NS
2	3500	a	a_1	a_{11}	NS	NS
3	3200	a	a_1	a_{12}	NS	NS
4	2600	a	a_2	NS	10	NS
5	2500	a	a_2	NS	12	NS
6	2000	a	a_2	NS	20	NS
7	2400	a	a_2	NS	15	NS
8	1800	b	NS	NS	NS	40
9	1700	b	NS	NS	NS	50
10	1500	b	NS	NS	NS	60
11	1400	b	NS	NS	NS	70
12	1200	b	NS	NS	NS	80
13	300	c	NS	NS	NS	NS
14	400	c	NS	NS	NS	NS
15	500	c	NS	NS	NS	NS

daughter variables selected are those with the largest number of relevant individuals (i.e., those individuals with NS values removed).

4. The regression equation from step 3 is 'added' to that from the previous regression fit.

5. Return to step 2, and continue until the base of the hierarchical tree is reached.

19.5.3 Example

Take the data of Table 19.4. At step 1, the fit of Y on X_1, using all observations, gives

$$Y = 400 + 2500a + 1120b.$$

Then, the residuals $(Y - Y^*)$ are as shown in Table 19.5.

The daughter variables of X_1 are X_2 and X_5, of which X_2 has the most observations (seven). Hence, at step 3, we first fit a regression model to X_2 on the residuals $R_1 = Y - Y^*$ using the observations $i = 1, \ldots, 7$. This gives the equation

$$R_1 = Y - Y^* = -525 + 1225a_1,$$

where $X_2 = (a_1, a_2)$.

Then, we repeat step 3 using the daughter variable which has the second most observations, in this case X_5. Thus, we fit a regression model to X_5 on the residuals $R_1 = Y - Y^*$ using the observations $i = 8, \ldots, 12$. This gives the equation

$$R_1 = Y - Y^* = 900 - 15X_5.$$

Table 19.5 Hierarchical variables regression fits.

	Fit of X_1		Fit of X_2, X_5		Fit of X_3, X_4		Final fit	
Y	Y^*	$R_1 = Y - Y^*$	Y^*	$R_2 = R_1 - Y_1^*$	Y^*	$R_3 = R_2 - Y^*$	Y^*	$R_4 = Y - Y^*$
4100	2900	1200	700	500	200	300	3800	300
3500	2900	600	700	−100	200	−300	3800	−300
3200	2900	300	700	−400	−400	0	3200	0
2600	2900	−300	−525	225	252.8	−27.8	2627.5	−27.5
2500	2900	−400	−525	125	133.8	−8.8	2508.5	−8.5
2000	2900	−900	−525	−375	−342.0	−33.0	2032.5	−32.5
2400	2900	−500	−525	25	−44.6	69.6	2330	70
1800	1520	280	300	−20	300	−20	1820	−20
1700	1520	180	150	30	150	30	1670	30
1500	1520	−20	0	−20	0	−20	1520	−20
1400	1520	−120	−150	30	−150	30	1370	30
1200	1520	−320	−300	−20	−300	−20	1220	−20
300	400	−100	400	−100	400	−100	400	−100
400	400	0	400	0	400	0	400	0
500	400	100	400	100	400	100	400	100

There being no other daughter variables of X_1, we proceed to step 4. Thus, we 'add' the two previous equations to the first one to give:

$$Y = 400 + 2500a + 1120b + \delta_{1a}(-525 + 1225a_1) + \delta_{1b}(900 - 15X_5),$$

where $\delta_{ij} = 1$ (0) if $i = j$ $(i \neq j)$. The (new) Y^* can be calculated from the new equation and hence the residuals $R_2 = Y - Y^*$ can be found, as given in Table 19.5.

We now do step 5, that is, we return to step 2 and repeat this process at the next level of the tree. Here there are two variables X_3 and X_4. Since X_4 has more observations (four) than X_3 (three), we fit a regression of X_4 to the residuals R_2 using the observations $i = 4, \ldots, 7$. This gives

$$R_2 = 847.5 - 59.5X_4.$$

Fitting R_2 to the three observations $(i = 1, 2, 3)$ on X_4 gives

$$R_2 = -400 + 600X_3.$$

We proceed to step 4 to 'add' the two previous equations to the first one. The fitted predictions Y^* residuals $R_3 = Y - Y^*$ are displayed in Table 19.5. The resulting regression equation is:

$$\begin{aligned} Y = {} & 400 + 2500a + 1120b + \delta_{1a}(-525 + 1225a_1) \\ & + \delta_{1b}(900 - 15X_5) + \delta_{1a_1}(-400 + 600a_{11}) \\ & + \delta_{1a_2}(847.5 - 59.5X_4). \end{aligned}$$

There being no more variables, the algorithm stops. Fitting this whole regression gives the predicted values Y^* and the corresponding residuals R_4, as shown in Table 19.5.

In effect, the methodology starts with a regression analysis on the data at the top of the hierarchical tree (on the X_1 predictor variable), and then adds regressions calculated on the resulting residuals at each lower branch of the tree corresponding to the daughters. Each daughter variable improves the regression calculated on its mother variable only. This improvement is reflected by the increasing value of the squared correlation coefficient values. In the above example, it can also be shown that performing the regression with the variable at the top of the hierarchy X_1 gives $R = 0.86$ while performing the regression with all the daughter variables gives $R = 0.98$.

This methodology remains efficient even for those hierarchies with several mother–daughter variables since those variables at the bottom of the tree have less importance than those at the top. We note that we cannot apply the standard-type regression analysis of variance to the whole regression. However, it is possible to do this at each daughter regression level. Thus, the F-test values for each of these sub-regressions can guide us as to when the algorithm should be stopped.

19.6 SREG module

These different methods have been implemented in the SODAS2 software. The SREG module provides methods and tests for multiple linear regression on symbolic data that are in the form of intervals, histograms, taxonomic variables, mother–daughter variables, multinominal variables and qualitative histograms.

19.7 Applications of symbolic linear regression

While the methodology was illustrated above on data sets of individuals $i = 1, \ldots, n$, where each variable takes a single (classical) value, the methods developed apply equally to symbolic data. In this case, we wish to study concepts. For example, we may be interested in the regression relation between variables for the concept defined by an age $\times$ race (or age $\times$ race $\times$ gender, or age $\times$ region, etc.) descriptor. Here, the particular choices of concepts will depend on the fundamental questions or interest, and therefore there can be many such choices. In other situations, the data can describe predefined concepts. For example, we may have a data set where each individual is a football team player but we do not wish to study the players but rather the teams composed of players.

To illustrate, in the work–demographic–medical data set studied in Section 19.4.3, suppose we are interested in concepts defined by a cholesterol-group $\times$ work descriptor where cholesterol-group is the qualitative variable constructed from the 28 intervals

$$([57, 99.9], [100, 109.9], [110, 119.9], [120, 129.9], [130, 134.9],$$
$$[135, 139.9], \ldots, [230, 234.9], [235, 244.9], [245, 259.9], [260, 281]),$$

and where work is the taxonomic variable defined in Section 19.4.3. Therefore, we have $28 \times 7 = 196$ concepts. Hence, we may be interested in the relation between cholesterol as the quantitative dependent variable with the variables haematocrit (quantitative), haemogroup (categories are 10 intervals of haemoglobin [10,11], [11.1,11.4], ...,

Table 19.6 Initial classical data matrix.

i	Concepts	Cholesterol	X_2	X_1	$X =$ work	Haematocrit	Haemogroup
1	[150, 154.9] × 11	151	1	1	11	35	[13.9, 14.2]
2	[150, 154.9] × 11	153	1	1	11	32	[13.9, 14.2]
3	[150, 154.9] × 11	154	1	1	11	31	[13.5, 13.8]
4	[180, 184.9] × 12	180	1	2	12	43	[11.0, 11.4]
5	[180, 184.9] × 12	184	1	2	12	46	[11.5, 11.8]

Table 19.7 Concepts built from Table 19.6.

Concepts	Cholesterol	Work	Haematocrit	Haemogroup
[150, 154.9] × 11	[151, 154]	11	[31, 35]	1/3[13.5, 13.8],2/3[13.9, 14.2]
[180, 184.9] × 12	[180, 184]	12	[43, 46]	[11.0, 11.4], [11.5, 11.8]

Table 19.8 Symbolic regression diagnostics dependent variable = cholesterol.

Explanatory variables	Fisher test F	Quantile $f(0.95)$	R
Haematocrit	34.5	4.17	0.42
Haemogroup	32.3	4.17	0.40
Work (level 1)	0.05	2.42	0.01
Multiple regression	17.1	3.31	0.43

[13.9,14.2], [14.3,15.2]), and work (see Table 19.6). After the creation of the concepts, the quantitative variables haematocrit and cholesterol become interval-valued variables (with the DB2SO module of SODAS). Moreover, the haemogroup variable becomes a quantitative histogram-valued variable because the categories are intervals. Finally, as the work variable is a part of the definition of concepts, work remains a taxonomic variable (see Table 19.7).

We present the results (Fisher test F and coefficient of determination R) of the simple symbolic linear regressions and the multiple linear regression with all the explanatory variables in Table 19.8. We can conclude that haematocrit and haemogroup seem to be quite good predictors ($F > f(0.95)$ and $R_{\text{haematocrit}} = 0.42$ and $R_{\text{haemogroup}} = 0.40$, whereas the taxonomic variable work is rejected by the F-test.

19.8 Conclusions and perspectives

This work first developed the methodology to fit regression models to data which internally contain taxonomic and hierarchical dependent variables. Other methods of regression have been discussed and tested in Afonso *et al.* (2003, 2004). This symbolic linear regression is useful when we wish to study concepts instead of individuals. The need to develop mathematical rigour for these new linear regression methods remains an open problem. Future work with regard to non-linear regression is also of interest.

References

Afonso, F., Billard, L. and Diday, E. (2003) Extension des méthodes de régression linéaire aux cas des variables symboliques taxonomiques et hiérarchiques. In *Actes des XXXV Journées de Statistique*, SFDS Lyon 2003, Vol. 1, pp. 89–92.

Afonso, F., Billard, L. and Diday, E. (2004) Symbolic linear regression with taxonomies. In D. Banks, L. House, F.R. McMorris, P. Arabie and W. Gaul (eds), *Classification, Clustering, and Data Mining Applications*, pp. 429–437. Berlin: Springer-Verlag.

Bertrand, P. and Goupil, F. (2000) Descriptive statistics for symbolic data. In H.-H. Bock and E. Diday (eds), *Analysis of Symbolic Data: Exploratory Methods for Extracting Statistical Information from Complex Data*, pp. 103–124. Berlin: Springer-Verlag.

Billard, L. and Diday, E. (2000) Regression analysis for interval-valued data. In H.A.L. Kiers, J.-P. Rasson, P.J.F. Groenen and M. Schader (eds), *Data Analysis, Classification, and Related Methods*, pp. 369–374. Berlin: Springer-Verlag.

Billard, L. and Diday, E. (2002) Symbolic regression analysis. In K. Jajuga, A. Sokolowski and H.-H. Bock (eds), *Classification, Clustering, and Data Analysis: Recent Advances and Applications*, pp. 281–288. Berlin: Springer-Verlag.

Billard, L. and Diday, E. (2003) From the statistics of data to the statistics of knowledge: symbolic data analysis. *Journal of the American Statistical Association*, 98: 470–487.

Bisdorff, R. and Diday, E. (2000) Symbolic data analysis and the SODAS software in official statistics. In H.A.L. Kiers, J.-P. Rasson, P.J.F. Groenen and M. Schader (eds), *Data Analysis, Classification, and Related Methods*, pp. 401–407. Berlin: Springer-Verlag.

Bock, H.H. and Diday, E. (eds) (2000) Similarity and dissimilarity. In H.-H. Bock and E. Diday (eds), *Analysis of Symbolic Data: Exploratory Methods for Extracting Statistical Information from Complex Data*, Chapter 8. Berlin: Springer-Verlag.

de Carvalho, F.A.T., Lima Neto, E. de A. and Tenerio, C.P. (2004) A new method to fit a linear regression model for interval-valued data. In S. Biundo, T. Frühwirth and G. Palm (eds), *KI 2004: Advances in Artificial Intelligence*, pp. 295–306. Berlin: Springer-Verlag.

Lima Neto, E. de A., de Carvalho, F.A.T. and Freire, E.S. (2005) Applying constrained linear regression models to predict interval-valued data. In U. Furbach (ed), *KI 2005: Advances in Artificial Intelligence*, Lecture Notes on Artificial Intelligence 3698, pp. 92–106. Berlin: Springer-Verlag.

20

Multi-layer perceptrons and symbolic data

Fabrice Rossi and Brieuc Conan-Guez

20.1 Introduction

Multi-layer perceptrons (MLPs) are a powerful non-linear regression tool (Bishop, 1995). They are used to model non-linear relationship between quantitative inputs and quantitative outputs. Discrimination is considered as a special case of regression in which the output predicted by the MLP approximates the probability of the input belonging to a given class. Unfortunately, MLPs are restricted to inputs and outputs that belong to a normed vector space such as $\mathbb{R}^n$ or a functional space (Rossi and Conan-Guez, 2005; Rossi *et al.*, 2005). In this chapter, we propose a solution that allows use of MLPs for symbolic data both as inputs and as outputs.

20.2 Background

We briefly recall in this section some basic definitions and facts about MLPs. We refer the reader to Bishop (1995) for a much more detailed presentation of neural networks.

20.2.1 The multi-layer perceptron

The MLP is a flexible and powerful statistical modelling tool based on the combination of simple units called *neurons*. A neuron with n real-valued inputs is a parametric regression model with $n+1$ real parameters given by $E(Y|X) = N(X, \alpha) = T(\alpha_0 + \sum_{j=1}^{n} \alpha_j X_j)$. In this equation, Y is the target variable in $\mathbb{R}$, X is the explanatory variable in $\mathbb{R}^n$ (X_j is the jth coordinate of X), α is the parameter vector in $\mathbb{R}^{n+1}$ and T is a fixed non-linear function

Symbolic Data Analysis and the SODAS Software Edited by E. Diday and M. Noirhomme-Fraiture

called the *activation* function of the neuron. In this very basic regression model, α is the only unknown information.

An MLP is obtained by combining neurons into layers and then by connecting layers. A layer simply consists in using several neurons in parallel, in general with the same activation function for all neurons. In this way we construct a multivariate regression model given by $E(Y|X) = H(X, \alpha_1, \ldots, \alpha_p)$. In this model, Y is now a target variable with values in $\mathbb{R}^p$. Each coordinate of Y is modelled by a simple neuron, i.e. $E(Y_i|X) = T(\alpha_{i,0} + \sum_{j=1}^{n} \alpha_{i,j} X_j)$.

We then combine layers by a simple composition rule. Assume, for instance, that we want to build a multivariate regression model of Y (with values in $\mathbb{R}^p$) on X (with values in $\mathbb{R}^n$). A possible model is obtained with a two-layer MLP using q neurons in its first layer and p neurons in its second layer. In this situation, the model is given by $E(Y_i|X) = T_2(\beta_{i,0} + \sum_{k=1}^{q} \beta_{i,k} Z_k)$, where the q intermediate variables $Z_1, \ldots, Z_q$ are themselves given by $Z_k = T_1(\alpha_{k,0} + \sum_{j=1}^{n} \alpha_{k,j} X_j)$. In fact, the Z_k variables are obtained as outputs from the first layer and used as inputs for the second layer. Obviously, more than two layers can be used.

In this regression model, the activation functions T_1 and T_2 as well as the number of neurons q are known parameters, whereas the vectors α and β have to be estimated using the available data.

20.2.2 Training and model selection

Given a sample of size N, $(Y^i, X^i)_{1 \leq i \leq N}$, distributed as (Y, X), our goal is to construct a model that explains Y given X, i.e., we want to approximate $E(Y|X)$. Theoretically, this can be done with an MLP (see White, 1990).

Let us first consider a fixed architecture, that is, a fixed number of layers, with a fixed number of neurons in each layer and with a given activation function for each layer. In this situation, we just have to estimate the numerical parameters of the model. Let us denote by w the vector of all numerical parameters of the chosen MLP (parameters are also called weights). The regression model is $E(Y|X) = H(X, w)$. As $E(Y|X)$ can be distributed differently from $H(X, w)$, it is possible that for all w, $H(X, w)$ is distinct from $E(Y|X)$; therefore, we have to choose w so as to minimize the differences between $H(X, w)$ and $E(Y|X)$. This is done indirectly by choosing an error measure in the target space ($\mathbb{R}^p$), denoted by d, and by searching for the value of w that minimizes $\mathcal{E}(w) = E(d(Y, H(X, w)))$. In practice, this is done by minimizing the empirical error defined by

$$\widehat{\mathcal{E}}_N(w) = \frac{1}{N} \sum_{i=1}^{N} d(Y^i, H(X^i, w)).$$

This minimization is performed by a gradient descent algorithm, for instance the Broyden–Fletcher–Goldfarb–Shanno method, or a conjugate gradient method (see Press *et al.*, 1992; Bishop, 1995).

In practice, d is taken to be either the quadratic distance or the cross-entropy, depending on the setting: the rationale is to obtain a maximum likelihood estimate in cases where there is actually a w such that $E(Y|X) = H(X, w)$. See Section 20.3.3 for practical examples of the choice of the error distance in the case of symbolic data.

Unfortunately, even if w is optimal and minimizes $\widehat{\mathcal{E}}_N(w)$, the model might be limited because the architecture of the MLP was badly chosen. We therefore have to perform a model selection to choose the number of neurons, and possibly the number of layers and the activation functions. Traditional model selection techniques can be used to perform this task. We can, for instance, compare different models using an estimation of $\mathcal{E}(w)$ constructed thanks to k-fold cross-validation, bootstrap, etc. (see Bishop, 1995).

20.3 A numerical coding approach

As stated in the introduction, MLPs are restricted to real-valued inputs and outputs. Some extensions allow functional inputs (see Rossi and Conan-Guez, 2005; Rossi *et al.*, 2005). Older works have also proposed the use of interval-valued inputs (see Šíma, 1995; Simoff, 1996; Beheshti *et al.*, 1998; see also Rossi and Conan-Guez, 2002, and Section 20.3.4 below) but there is currently no simple way to deal with symbolic data. In this chapter we propose a numerical coding approach that allows MLPs to be used on almost any symbolic data.

20.3.1 Recoding one symbolic variable

In this section, we present a numerical coding scheme for each type of variable. First, single-valued variables are considered (quantitative and categorical variables), then we focus on symbolic variables:

- *Quantitative single-valued.* Obviously, we do not have to do any recoding for a quantitative single-valued variable as this is a standard numerical variable.
- *Categorical single-valued.* The values of a categorical single-valued variable are categories (also called modalities). Let $\{A_1, \ldots, A_m\}$ be the list of those categories. A categorical single-valued variable is recoded thanks to the traditional disjunctive coding, as follows:

$$\begin{array}{lll} A_1 & \text{is recoded as} & (1,0,0,\ldots,0), \\ A_2 & \text{is recoded as} & (0,1,0,\ldots,0), \\ \vdots & \vdots & \vdots \\ A_m & \text{is recoded as} & (0,0,\ldots,0,1). \end{array}$$

  Therefore, we replace the categorical single-valued variable of interest by m numerical variables. It should be noted that if the categories are ordered (if we have $A_1 < A_2 < \cdots < A_m$), then the disjunctive coding scheme is not well adapted to the problem nature. In such a case, a standard numerical coding, where each category is replaced by its rank ($A_i \to i$), should be considered. This gives rise to a numerical variable.

- *Interval.* An interval variable is described by a pair of extreme values, $[a, b]$. We replace one interval variable by two quantitative single-valued variables, $\mu = (a+b)/2$ (the mean of the interval) and $\delta = b - a$, the length of the interval (we refer to this as *mean and length coding*). From a statistical point of view, it is better to use $\log \delta$ rather than δ directly, but some symbolic data include zero-length intervals and it is therefore not always possible to use the logarithmic representation (see Section

20.3.3). Another possibility is to recode $[a, b]$ as a and b, considered as quantitative single-valued variable (*bound-based coding*).

- *Categorical multi-valued.* Categorical multi-valued variables are generalizations of categorical single-valued variables for which the value of the variable is no longer a category but a subset of the set of categories. We use exactly the same coding strategy as above (m numerical variables), but we allow several 1 for a given variable, one in each column corresponding to a category in the subset. For instance, $\{A_1, A_3\}$ is recoded as $(1, 0, 1, 0, \ldots, 0)$.
- *Modal.* Modal variables are generalizations of categorical multi-valued variables for which each category has a weight. We again use the m-variable coding but we use the weights of the categories rather than 0 and 1.

20.3.2 Recoding the inputs

Input data are recoded as explained in the previous section, but additional care is needed. It is indeed well known that MLP inputs must be centred and scaled before being considered for training in order to avoid numerical problems in the training phase (when we minimize $\widehat{\mathcal{E}}_N(w)$). Moreover, this preprocessing must be compatible with the initialization strategy used by the minimizing algorithm. Indeed, all gradient descent algorithms are iterative: we start from a randomly chosen candidate solution w_0 and improve its quality iteratively. In general, w_0 uses small initial values and it is important to be sure that the MLP inputs will belong to the same approximate range of values. It is therefore important to apply centring and scaling to the recoded inputs before the training.

Moreover, it is very common in practice to use regularization in order to improve the quality of the modelling performed by the MLP (and to avoid over-fitting, see Bishop, 1995). This is done by minimizing a new error function $\widehat{\mathcal{R}}_N(w)$ rather than the standard error $\widehat{\mathcal{E}}_N(w)$. The rationale of this new error is to penalize complex models. This can be done, for instance, using the following error function:

$$\widehat{\mathcal{R}}_N(w) = \widehat{\mathcal{E}}_N(w) + \sum_{j=1}^{t} \lambda_j w_j^2. \tag{20.1}$$

In this equation, λ_j is a penalty factor for weight j and is called the *weight decay parameter* for weight j. Its practical effect is to restrict the permitted variation for this weight: when λ_j is small, the actual value of w_j has almost no effect on the penalty term included in $\widehat{\mathcal{R}}_N(w)$ and therefore this weight can take arbitrary values. On the other hand, when λ_j is big, w_j must remain small.

In general we use only one value and $\lambda_j = \lambda$ for all j, but in some situations we use one penalty term per layer (the optimal value of the weight decay parameters is determined by the model selection algorithm). In our situation, the recoding scheme introduces some problems. Indeed, a categorical variable with a lot of categories is translated into a lot of variables. This means that the corresponding numerical parameters will be heavily constrained by the weight decays. Therefore, the recoding method introduces arbitrary differences between variables when we consider them from a regularization point of view. In order to avoid this problem, we normalize the decay parameter.

Let us consider a categorical single-valued variable with five categories translated into five variables $X_1, \ldots, X_5$. For each neuron in the first layer, we have five corresponding weights, $w_{1,1}, \ldots, w_{1,5}$ for the first neuron, $w_{2,1}, \ldots, w_{2,5}$ for the second neuron, etc. The corresponding penalty, $\widehat{\mathcal{R}}_N(w)$, is normally $\lambda \sum_{k=1}^{q} \sum_{i=1}^{5} w_{k,i}^2$ if there are q neurons in the first layer and if we use only weight decay for the whole layer. We propose to replace this penalty by $\frac{1}{5}\lambda \sum_{k=1}^{q} \sum_{i=1}^{5} w_{k,i}^2$, that is, we divide the weight decay parameter used for each variable corresponding to the recoding of a categorical single-valued variable by the category number of this variable.

We use the same approach for extension of the categorical type, that is, the categorical multi-valued and the modal types. We do not modify the weight decay for interval variables as they really are comparable to two variables: for instance, modifying one and not the other is meaningful, which is not the case for categorical and modal variables.

20.3.3 Recoding the outputs

Output data are recoded as explained in Section 20.3.1, but additional care is again needed. First of all, a noise model has to be chosen in order to justify the use of an error measure d. More precisely, while any error measure with derivatives can be used, it is very important to model the way Y behaves around $E(Y|X)$ so as to obtain sensible estimation of w, the weight vector. Given a model of Y (for instance Gaussian with known variance and mean given by $E(Y|X)$, i.e., the output of the MLP), we can choose an error measure that leads to a maximum likelihood estimate for w: in the case of numerical output, using a quadratic error corresponds to assuming that the noise is Gaussian, which is sensible. Second of all, some constraints must be enforced on the outputs of the MLP so as to obtain valid symbolic values: the length of an interval must be positive, the sum of values obtained for a modal variable must be one, etc.

For non-numerical variables, the situation is quite complex. Let us review the different cases.

20.3.3.1 Interval variable

We have here both a noise problem and a consistency problem. Indeed, while the mean of an interval can be arbitrary, this is not the case for its length, which must be positive. Therefore, the output neuron that models the length variable obtained by recoding an interval must use an activation function that produces only positive values. Moreover, while it is sensible to assume that the noise is Gaussian for the mean of the target interval, leading to a quadratic error measure for the corresponding output, this assumption is not valid for the length.

A simple solution can be applied to symbolic data in which there is no zero-length interval. In this situation, rather than recoding $[a, b]$ into two variables $\mu = (a+b)/2$ and $\delta = b - a$, we replace δ by $l = \log \delta$. With this transformation, we can use a regular activation function and model the noise as Gaussian on l. Therefore, the error measure can be the quadratic distance.

Unfortunately, this recoding is not possible for degenerate intervals $[a, a]$ which might be encountered. In this kind of situation, we have to use an adapted activation function and choose a model for the variability of δ. A possibility is to use a gamma distribution (or one of its particular cases such as an exponential distribution or chi-square distribution). This implies the use of a specific error measure.

The case of bound-based recoding is similar. If we recode $[a, b]$ into two variables a and b, we have to ensure that $b \geq a$. There is no direct solution to this problem and we have to rely on specific activation functions. It is therefore simpler to use the mean and length recoding for output variables.

20.3.3.2 Categorical single-valued variable

This case has been already studied by the neural community because it corresponds to a supervised classification problem. Indeed, when we want to classify inputs into classes $A_1, \ldots, A_m$, we construct a prediction function that maps an input into a label chosen in the set $\{A_1, \ldots, A_m\}$. This can be considered similar to the construction of a regression for a categorical single-valued target variable with values in $\{A_1, \ldots, A_m\}$.

In order to train an MLP, we must be able to calculate the gradient of $\widehat{\mathcal{E}}_N(w)$, and therefore the activation functions must be differentiable. As a consequence, an MLP cannot directly output labels. Of course, we will use the disjunctive coding proposed in Section 20.3.1, but the MLP will seldom output exact 0 and 1. Therefore, we will interpret outputs as probabilities.

Specifically, let us assume that the target variable Y is categorical single-valued, with values in $\{A_1, \ldots, A_m\}$. It is therefore translated into m variables $Y_1, \ldots, Y_m$ with values in $\{0, 1\}$ and such that $\sum_{i=1}^m Y_i = 1$. Then the last layer of the MLP must have m neurons. Let us call $T_1, \ldots, T_m$ the outputs of the last layer. Using a softmax activation function (described below and in Bishop, 1995), we can ensure that $T_i \in [0, 1]$ for all i and that $\sum_{i=1}^m T_i = 1$. The natural interpretation for those outputs is probabilistic: T_i approximates $P(Y = A_i|X)$.

The model for the recoded variable is normally $E(Y_i|X) = T_i = T(\beta_{i,0} + \sum_{k=1}^q \beta_{i,k} Z_k)$, where $Z_1, \ldots, Z_q$ are outputs from the previous layer. In order to construct the softmax activation function, we introduce $U_i = \beta_{i,0} + \sum_{k=1}^q \beta_{i,k} Z_k$ and define

$$T_i = \frac{\exp(U_i)}{\sum_{j=1}^m \exp(U_j)}.$$

This activation function implies that $T_i \in [0, 1]$ and that $\sum_{i=1}^m T_i = 1$. Using the probabilistic interpretation, it is easy to construct the likelihood of $(T_1, \ldots, T_m)$ given the observation $(Y_1, \ldots, Y_m)$. It is obviously

$$\prod_{i=1}^m T_i^{Y_i}.$$

The maximum likelihood principle leads to the minimization of the quantity

$$d(Y, T) = -\sum_{i=1}^m Y_i \ln T_i.$$

The corresponding distance is the cross-entropy, which should therefore be used for nominal output.

To summarize, when we have a categorical single-valued output variable with m categories:

- we use disjunctive coding to represent this variable as m numerical variables;
- we use a softmax activation function for the corresponding m output neurons;
- we use the cross-entropy distance to compare the values produced to the desired outputs;
- the actual output of the MLP can be either considered directly as a model variable, or transformed into a categorical single-valued variable by using a probabilistic interpretation of the outputs to translate numerical values into the most likely label.

20.3.3.3 Categorical multi-valued variable

The case of the categorical multi-valued variable is a bit more complex because such a variable does not contain a lot of information about the underlying data it is summarizing. Indeed, if we have, for instance, a value of $\{A_1, A_3\}$, it does not mean that A_1 and A_3 are equally likely. Therefore, we use a basic probabilistic interpretation: we assume that categories are conditionally independent given X.

That said, the practical implementation is very close to that for categorical single-valued variables. Let us consider a categorical multi-valued target variable Y, with values in $\{A_1, \ldots, A_m\}$. It is translated into m variables $Y_1, \ldots, Y_m$ with values in $\{0, 1\}$ (the constraint $\sum_{i=1}^{m} Y_i = 1$ is no longer valid). As for a nominal variable, we denote by $T_1, \ldots, T_m$ the outputs of the last layer of the MLP. We use an activation function such that $T_i \in [0, 1]$, for instance the logistic activation function

$$T(x) = \frac{1}{1 + \exp(-x)}.$$

Then T_i is interpreted as the probability of category A_i appearing in Y. Given that categories are assumed independent, the likelihood of $(T_1, \ldots, T_m)$ given the observation $(Y_1, \ldots, Y_m)$ is again

$$\prod_{i=1}^{m} T_i^{Y_i}.$$

As for a categorical single-valued variable, the maximum likelihood estimation is obtained by using the cross-entropy error distance.

To summarize, when we have a categorical multi-valued output variable with m categories:

- we use the 0/1 coding to represent this variable as m numerical variables;
- we use an activation function with values in $[0, 1]$ for the corresponding m output neurons;
- we use the cross-entropy distance to compare produced values to desired outputs;
- we use a probabilistic interpretation of the outputs – we consider that category A_i belongs to the categorical multi-valued output if and only if $T_i > 0.5$.

20.3.3.4 Modal variable

The modal variable case can be handled in almost exactly the same way as the categorical single-valued variable case. Let us consider a modal variable with support $\mathcal{A} = \{A_1, \ldots, A_m\}$. This is described by a vector $(p_1, \ldots, p_m) \in \mathbb{R}^m$, with the following additional constraints:

- $p_i \in [0, 1]$, for all i;
- $\sum_{i=1}^{m} p_i = 1$.

A modal variable must be interpreted as a probability distribution on $\mathcal{A}$. It is recoded by the vector $(p_1, \ldots, p_m)$. Unfortunately, we do not know exactly how the probability distribution has been constructed. As symbolic descriptions are often summaries, we will assume here that l micro-observations with values in $\mathcal{A}$ were used to construct the estimated probabilities $(p_1, \ldots, p_m)$. This implies that lp_i out of l observations correspond to the category A_i.

Exactly as for a categorical single-valued variable, we use m output neurons with a softmax activation function. We again denote by $T_1, \ldots, T_m$ the corresponding outputs. With the proposed interpretation of the variable, the likelihood of $T_1, \ldots, T_m$ given the observation $(p_1, \ldots, p_m)$ is (by construction lp_i are integers)

$$\frac{l!}{(lp_1)!\ldots(lp_m)!} T_1^{lp_1} T_2^{lp_2} \ldots T_m^{lp_m}.$$

The maximum likelihood principle leads to the minimization of the quantity

$$d(Y, T) = -l \sum_{i=1}^{m} p_i \ln T_i.$$

Of course, l might be removed from this cross-entropy-like distance, but only if each considered value of the modal variable Y comes from l micro-observations. When the number of micro-observations depends on the value of the variable, we must keep this weighting in the error distance. Unfortunately, this value is not always available. When the information is missing, we can use the cross-entropy error distance unweighted.

To summarize, when we have a modal output variable with m categories:

- we use the probabilities associated with the categories to translate the variable into m real-valued variables;
- we use a softmax activation function for the corresponding m output neurons;
- we use the cross-entropy distance to compare the values produced to the desired outputs;
- when the information is available, we use the size of the micro-observations set that has been used to produce the modal description as a weight in the cross-entropy distance;
- thanks to the softmax activation function, the output of the m neurons are probabilities and can therefore be directly translated into a modal variable.

20.3.4 Alternative solutions

Alternative solutions for interval-valued inputs have been proposed by Šíma (1995), Simoff (1996) and Beheshti *et al.* (1998). The basic idea of these works is to use interval arithmetic, an extension of standard arithmetic to interval values (see Moore, 1966). The main advantage of interval arithmetic is that it allows uncertainty to be taken into account: rather than working on numerical values, we work on intervals centred on the considered numerical values.

Unfortunately, these approaches are not really suited to symbolic data. Rossi and Conan-Guez (2002) showed that a recoding approach provides better results than an interval arithmetic approach. The main reason is that extreme values in an interval do not always correspond to uncertainty. In meteorological analysis, for instance, we cannot differentiate broad classes of climate simply by using the mean temperature: extreme values give valuable information, as continental weather is characterized in general by important yearly variation around the mean, while oceanic weather is characterized by smaller yearly variation (see also Section 20.5).

20.4 Open problems

20.4.1 High number of categories

It is unfortunately common to deal with categorical or modal variables with many categories. The proposed recoding method introduces a lot of variables. The practical consequence is a slow training phase for the MLP. In some situations, when the number of categories is really high (100, say), this might even prevent the training from succeeding.

One way around this is to use a lossy encoding in which several categories are merged, for instance based on their frequencies in the data set. Unfortunately, it is very difficult to carry out this kind of simplification while taking target variables into account. Indeed, an optimal unsupervised lossy encoding might lose small details that are needed for a good prediction of target variables.

20.4.2 Multiple outputs

We have shown in Section 20.3.3 how to deal with symbolic outputs. While we can handle almost any type of symbolic data, we have to be extremely careful when mixing symbolic outputs of different types. For instance, it is well known (see Bishop, 1995) that using the quadratic error distance for vector output corresponds to assuming that the noise is Gaussian, with a fixed variance and independent on each output. Departure from this model (for instance, a different variance for each output) is possible but implies the modification of the error measure.

The problem is even more crucial when we mix different types of symbolic data. If we have, for instance, a numerical variable and a categorical single-valued variable, simply using the sum of a quadratic distance and a cross-entropy distance will seldom result in a maximum likelihood estimation of the parameters of the MLP. One at least has to take into account the variance of the noise of the numerical variable. Moreover, the basic solution will be to assume that the outputs are conditionally independent given the input, but this might be a very naive approach.

We do not believe that there is an automatic general solution for this problem and we emphasize the importance of choosing a probabilistic model for the outputs in order to obtain meaningful results.

20.4.3 Other types of symbolic data

Our solution does not deal with additional structure in symbolic data. For instance, we do not take into account taxonomies or rules. One way to deal with taxonomies is to use a hierarchical recoding: the deepest level of the hierarchy is considered as a set of categories for a categorical single-valued variable which leads to a disjunctive coding. Values from higher levels of the hierarchy are coded as categorical multi-valued values, that is, by setting to 1 all categories that are descendants of the considered value.

Another limitation comes from the fact that we cannot deal with missing data: if a symbolic description is not complete for one individual (for instance, one interval variable is missing for this individual), treatment of that individual by the MLP model is not possible. One solution to overcome this limitation is to apply classical imputation methods on recoded variables, that is, to replace missing data by 'guessed' values. Of course, some care has to be taken to respect the semantics of imputed variables. A naive method involves replacing missing values by means of corresponding variables. A more sophisticated method relies on the k-nearest-neighbour algorithm: given a vector in which some coordinates are missing, we calculate its k nearest neighbours among vectors that do not lack these coordinates, and we replace missing values by averages of coordinates of the k nearest neighbours. Usually the meta-parameter k is determined by cross-validation. It is also possible to use this kind of imputation method directly at the symbolic level, that is, before the recoding phase, if some generalized mean operator is available for the data considered (for examples of such operators, see Bock and Diday, 2000).

20.5 Experiments

20.5.1 Introduction

In this section, we use a semi-synthetic example to show how in practice neural net models, and specifically MLPs, can process symbolic objects. Only the specific case of interval variables will be considered in these experiments (categorical multi-valued variables and modal variables will not be addressed here and require additional experiments). Results obtained in this specific example will allow us to tackle three important issues relative to treatment of data involving symbolic objects:

- *Benefits of symbolic objects over standard approaches.* Symbolic approaches allow richer and more complex descriptions of data than standard approaches (for instance, the mean approach where data are simply averaged). We saw at the beginning of this chapter that these complex descriptions imply some specific adaptations of neural net models (adaptation of the activation function, careful use of the weight decay technique, etc.). Moreover, in some cases, the model complexity (which is related to the number of weights) can be higher for symbolic approaches than for standard approaches: indeed, a categorical single-valued variable with a large number of categories implies a high-dimensional input, and therefore a large number of weights. As a direct consequence, estimation of neural net models can be more difficult when dealing with symbolic objects. It is therefore legitimate to wonder if symbolic approaches are worthwhile. In the proposed example, we show that they are indeed very helpful: model performances are improved thanks to symbolic objects.

- *Difficult choice of coding method*. In many real-world problems, the practitioner has at his disposal the raw data, that is, primary data on which no preprocessing stages have been applied. He is therefore free to choose the best coding method to recode such data into symbolic objects: for instance, he can recode the raw data into modal variables, or into intervals in accordance with the nature of the problem. For each of these recoding methods, some meta-parameters have to be set up: for instance, in the case of modal variables, it is up to the practitioner to choose the number of categories. For an interval variable, he has to decide whether bound-based coding or mean and length coding is more appropriate. As we will see in the proposed experiments, such choices have a noticeable impact on model performance.
- *Low-quality data and robustness of models*. Finally, it is not uncommon in many real-world problems to have to deal with low-quality raw data, that is, data with missing values or with noisy measurements. It is therefore tempting to wonder if symbolic approaches can cope successfully with such data. Once again, the proposed experiments will allow us to address this problem: symbolic approaches perform better when dealing with low-quality data than standard approaches.

20.5.2 Data presentation

In order to address the different issues described above, we have chosen a synthetic example based on real-world data: we consider climatic data from China published by the Institute of Atmospheric Physic of Chinese Academy of Sciences, Beijing (see Shiyan *et al.*, 1997). This application concerns mean monthly temperatures, and total monthly precipitations observed in 260 meteorological stations spread over the Chinese territory. In these experiments, we restrict ourselves to the year 1988: each station is therefore described by a single vector (12 temperatures and 12 precipitations). Moreover, we have the coordinates of all the stations (longitude and latitude).

As the goal of this chapter is to show how in practice we can process symbolic objects with neural net models, we represent meteorological information related to each station using symbolic objects (only interval variables are considered in this application). The goal of these experiments is then to infer from the meteorological description of each station (which may be symbolic or not), its location in China (longitude and latitude). This problem is not a real practical application, but it has two interesting characteristics.

First, inference of station location is quite a difficult task, as it is obviously an ill-posed problem. Indeed, we try to use an MLP to model the inverse of the function which maps station location to meteorological description. As this function is not a one-to-one mapping (two stations located far away one from each other in China can have very similar meteorological descriptions), the inverse function is therefore a set-valued function, which is very difficult to model with standard techniques. We will see that MLPs based on symbolic objects perform better in this case than standard approaches.

Secondly, the data are not native symbolic data and we have access to underlying raw data. This allows us to study the effect of the coding on model performance. Moreover, as we are interested in the robustness of symbolic approaches when dealing with low-quality data, we can intentionally degrade the original data, and then study the effects on model performance (we remove from the original data some chosen values). Experiments will show that, using symbolic approaches, models estimated on high-quality data (data with no missing values) perform correctly on low-quality data.

20.5.3 Recoding methods and experimental protocol

We consider four different experiments corresponding to standard and symbolic approaches (the first two can be considered as standard approaches, whereas the last two are symbolic approaches):

1. We simply process the raw data. Therefore, the input of the MLP model is a vector of 24 coordinates (the 12 temperatures and the 12 precipitations associated with a given station over one year). It is worth noticing that in this experiment the model complexity is quite high, as the number of weights is directly linked to the input dimension (24 in this case).

2. We aggregate temperature and precipitation data using a simple average. Therefore, in this case, each station is described by a two-dimensional vector, $(temp_{\text{mean}}, precip_{\text{mean}})$.

3. Temperature data as well as precipitation data are recoded into intervals. In this case, each station is described by its extrema, $([temp_{\text{min}}, temp_{\text{max}}], [precip_{\text{min}}, precip_{\text{max}}])$. The input dimension of the MLP model is 4, as each pair of intervals is submitted as a four-dimensional vector.

4. In order to explore different coding methods, we input into the MLP model the mean and the standard deviation of each variable. This corresponds to a robust interval coding in which the length is estimated using the standard deviation rather than using the actual extreme values. Each station is therefore described by $(temp_{\text{mean}}, temp_{\text{sd}}, precip_{\text{mean}}, precip_{\text{sd}})$. The input dimension of each MLP is 4.

In all the experiments, we use two distinct MLPs: one for the longitude inference, and one for the latitude inference. Of course, it would have been possible to infer the location (longitude and latitude) as a whole with a unique MLP. As we will see in the experiments, such an approach is not well adapted to the nature of the problem: indeed, latitude inference is much easier than longitude inference. Therefore, in order not to penalize one problem over the other, we decided to keep the problems separated.

All the MLPs in this application have a single hidden layer. We test different sizes for the hidden layer: 3, 5, 7, 10, 15, 20, 30 and 40 neurons. In order to estimate models and to compute a good estimate of their real performance, the whole data set is split into three parts. The training set contains 140 stations. This set is used to estimate model parameters using a standard minimization algorithm (a conjugate gradient method). The error criterion is the quadratic error. The second part of the data set is the validation set (60 stations) which is used to avoid over-fitting: the minimization algorithm is stopped when the quadratic error on this validation set is minimized. For each experiment, the minimization is carried out 10 times: the starting point is randomly chosen at each time. The best architecture (the number of hidden neurons and the values of the weights) is chosen according to the quadratic error

Table 20.1 Mean absolute error in degrees and architecture complexity.

Inputs	Longitude	Latitude	Number of weights
Full data (24)	4.07(3)	1.27(30)	860
Mean	7.31(30)	2.51(17)	190
Mean and std. dev.	4.91(20)	1.34(25)	272
Min. and max.	4.73(25)	1.56(40)	392

on the validation set. Finally, as the error on the validation set is not a good estimate of model real performance, we compute the mean absolute error in degrees on the test set (60 stations).

Table 20.1 summarizes the results obtained in the different experiments. These clearly show that longitude inference is a more difficult task than latitude inference: indeed, in the latter, model accuracy is close to 1 degree, whereas in the former the accuracy is at best 4 degrees. On the whole data set, the range for longitude is 56 degrees and the range for latitude is 34.23 degrees. This means that the best relative error for latitude is less than 4%, whereas it is more than 7% for longitude.

This difference can be partly explained by taking a closer look at the characteristics of the Chinese climate: due to its large surface, there is a large diversity of climates in China (from cold and dry to hot and wet). If we focus on temperature and precipitation, some geographical inferences can be made. Mean yearly temperature is of course highly indicative of station latitude (south is hot, north is cooler). However, the coldest region of China is Tibet (south-west), which is not located in the north of China. The total yearly precipitation is informative on the Xinjiang–Guangzhou (NW–SE) axis: the Xinjiang region is very dry, whereas Guangzhou city has a very wet climate (monsoon). Therefore we can see that both variables contribute quite obviously to the inference of the latitude. For the longitude case, accuracy is not as good, as both variables contribute less information on this east–west axis.

If we now study the performance of the different approaches, we can see that the full data approach does best. Symbolic approaches ((min, max) and (mean, sd)) do quite well too, with a performance very close to that of the full data approach. Finally, the mean approach gives the worst results. This leads to the following remarks:

- First, only approaches which are able to model the variability of the meteorological phenomenon over one year can achieve good performance. Indeed, if two distant stations have similar mean temperature and mean precipitation, then the mean approach cannot accurately infer their locations. This is the case, for example, for the cities of Urumqi and Harbin. Urumqi is located in the north-west of China, and Harbin is located in the north-east of China. Urumqi has a continental climate with very hot summer and very cold winter, whereas Harbin has an oceanic climate with cool summer and chilly winter. However, both cities have quite similar mean temperature and mean precipitation over one year. In the case of the other approaches (full data approach and symbolic approaches), the variability of the meteorological phenomenon is preserved in the description (for instance, as explained before, for the temperature description, interval length is greater for continental climates than for oceanic climates), which leads to better performances.

- Even if the full data approach leads to some slight performance improvements over symbolic approaches, the latter should be preferred over the former in practical situations. Indeed, in the case of the full data approach, input dimension is quite high (24), which implies a large number of parameters for the model (860 weights). In the case of symbolic approaches, important information available in the original data, such as variability, has been summarized thanks to a compact description. We can see that this data reduction does not impair performance too much. Moreover, model complexity is lower than for the full data approach (272 for the (mean, sd) coding and 392 for the (min, max) coding), which leads to a faster estimation phase.
- Finally, there is little to choose between the two symbolic representations. The (min, max) coding and (mean, sd) coding give quite similar results. We will see nevertheless in the next section, that both coding are not strictly equivalent.

In order to illustrate the behaviour of the different approaches, Figures 20.1–20.4 show, for each approach, model inferences versus true station locations. Triangles represent the location of the 60 stations which belong to the test set. Line segments represent the location inferred by the model. In each figure, we can see that line segments tend to be horizontal, which corroborates the fact that longitude inference is more difficult than latitude inference.

20.5.4 Low-quality data

We have seen that the full data approach and symbolic approaches perform quite similarly on the Chinese data example. The only advantage at this point of symbolic approaches is

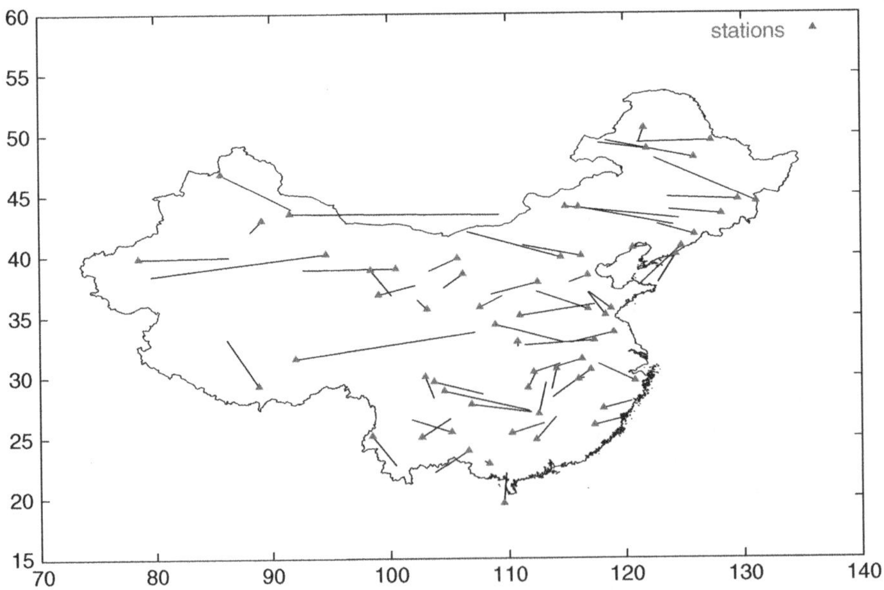

Figure 20.1 Model inferences versus true station locations: full data.

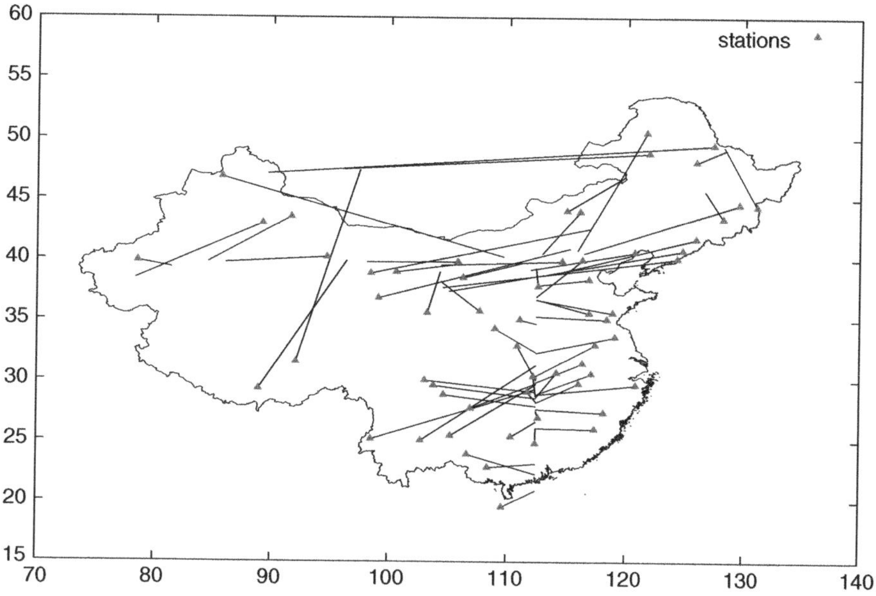

Figure 20.2 Model inferences versus true station locations: mean approach.

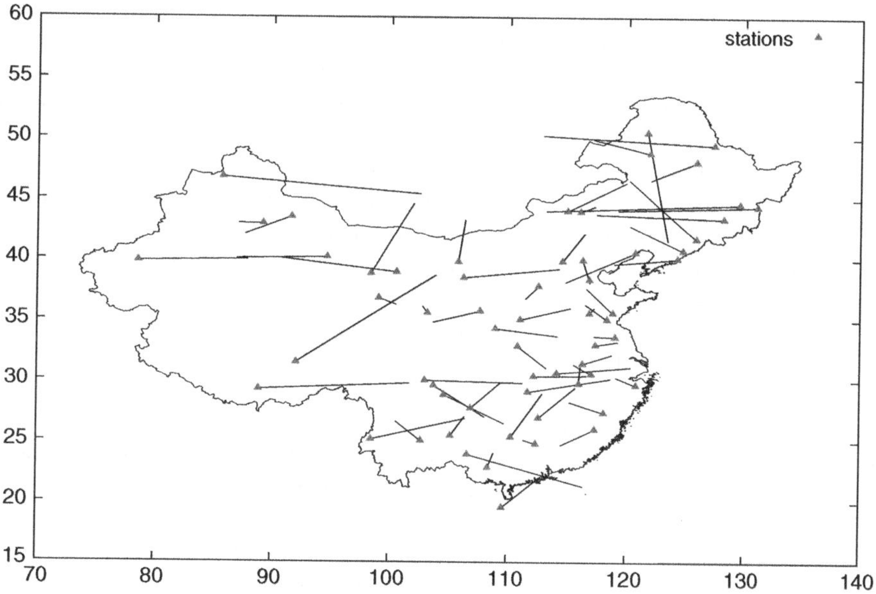

Figure 20.3 Model inferences versus true station locations: minimum and maximum.

that the model complexity (number of weights) is lower. The goal now is to show that models based on symbolic objects are much more robust to low-quality data than those based on the full data approach. Specifically, for each approach (standard and symbolic), we consider the MLP model estimated in the previous section: this estimation was carried

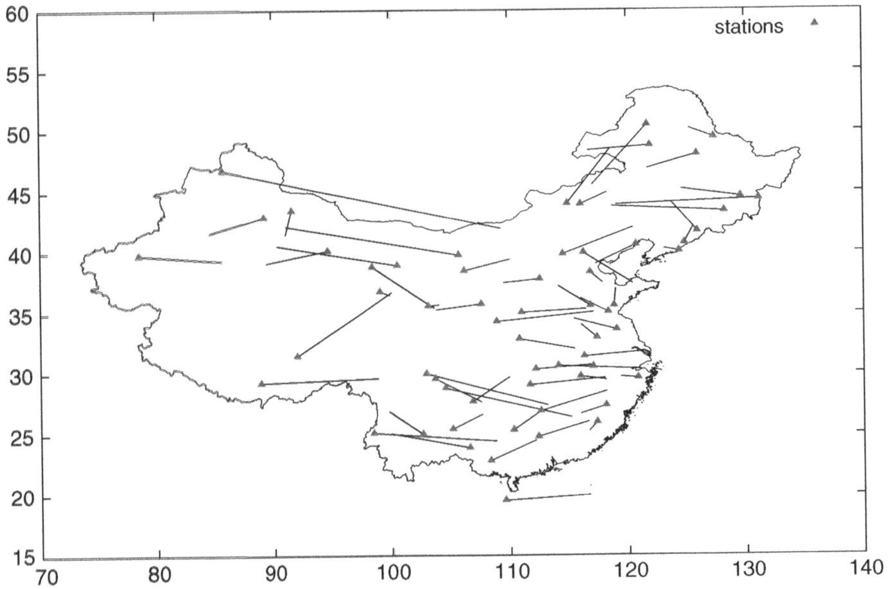

Figure 20.4 Model inferences versus true station locations: mean and standard deviation.

out with high-quality data (no missing values). Our goal is to study the performance of this model when unseen data with missing values (low-quality data) are submitted.

In order to investigate the robustness of the different approaches, we intentionally degrade the Chinese data which belong to the test set (the training set and the validation set are not considered in this section, as we use the models which have been estimated in the previous section). Data degradation is done three times: the first degradation leads to data of medium quality (half the values are missing); the second degradation leads to low-quality data (two-thirds are missing); and the third degradation leads to very low-quality data (three-quarters are missing). Data degradation is done according to the following protocol: we consider for each meteorological station the temperature vector (12 coordinates) and the precipitation vector (12 coordinates). For the first degradation, we remove one coordinate in two for each vector (coordinates 2, 4, 6, 8, 10, 12 are missing values). Therefore the temperature vector is now a six-dimensional vector, just like the precipitation vector. For the second degradation, we remove two coordinates out of three from the original data, and the dimension of each vector is 4 (coordinates 2, 3, 5, 6, 8, 9, 11, 12 are missing values). Finally, in the third experiment we remove three coordinates out of four, and the dimension of each vector is 3 (coordinates 2, 3, 4, 6, 7, 8, 10, 11, 12 are missing values).

Models based on mean, minimum and maximum, or mean and standard deviation calculation can be applied directly to these new data. All we have to do is to recompute each of these quantities with the remaining values. For the full data approach, direct treatment is not so simple, as MLP models have an input dimension of 24, which is incompatible with the data dimension (12 for the first degradation, 8 for the second degradation and 6 for the third degradation). Therefore, in order to submit these data to the full data model, we must replace each missing value by an estimate. Many well-known missing-value techniques can be applied in order to compute these estimates. In order not to penalize the full data approach, we choose to replace missing values by new values computed by linear interpolation. For the sake of clarity, we denote by c_i the ith coordinate. For the first degradation, coordinates c_2, c_4, c_6, c_8, c_{10}, c_{12} are missing. Coordinate c_2 is replaced by $(c_1 + c_3)/2$. We proceed in the same way for coordinates c_4, c_6, c_8, c_{10}. For coordinate c_{12}, we make use of the periodic aspect of the climate: coordinate c_{12} is replaced by $(c_1 + c_{11})/2$ (December is computed from November and January of the same year). For the second degradation, we have $c_2 = (2c_1 + c_4)/3$ and $c_3 = (c_1 + 2c_4)/3$, and so on for coordinates c_5, c_6, c_8, c_9, c_{11}, c_{12}. Finally, for the third degradation, we have $c_2 = (3c_1 + c_5)/4$, $c_3 = (c_1 + c_5)/2$ and $c_4 = (c_1 + 3c_5)/4$, and so on for coordinates c_6, c_7, c_8, c_{10}, c_{11}, c_{12}.

For each of the models estimated in the first experiments, we compute the mean absolute error on the modified test set. Figure 20.5 summarizes the performance of the different approaches for the inference of station latitude with respect to the data degradation level. Figure 20.6 does the same for longitude. The results lead to the following remarks:

- The full data approach is very sensitive to data quality compared to the other approaches. Indeed, data degradation strongly impairs performance (error increases from 4.07 degrees to 12.4 degrees for the longitude when half the values are missing, while increasing from 1.27 degrees to 4.1 degrees for the latitude). Other approaches are not impacted in these proportions for the same degradation. We can see the outcome of using complex models, that is, models with a large number of weights: if the submitted data are close to the training data, the model performs well. However, if the submitted data differ too much from the training data, performance suffers.

- The performance of the mean approach is quite uniform with respect to data degradation. Nevertheless, symbolic approaches outperform the mean approach in almost all cases. Once again, descriptions obtained by means of a simple average are too poor, which prevents MLP models from making accurate inferences.

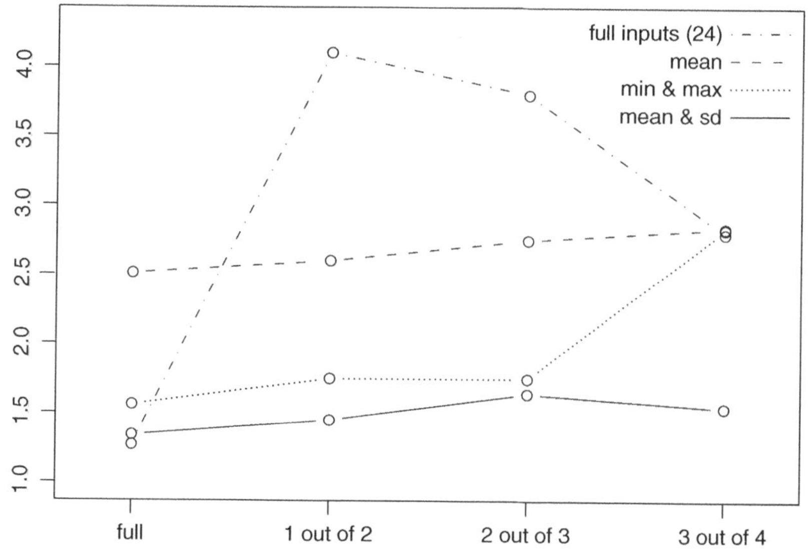

Figure 20.5 Data quality reduction: latitude.

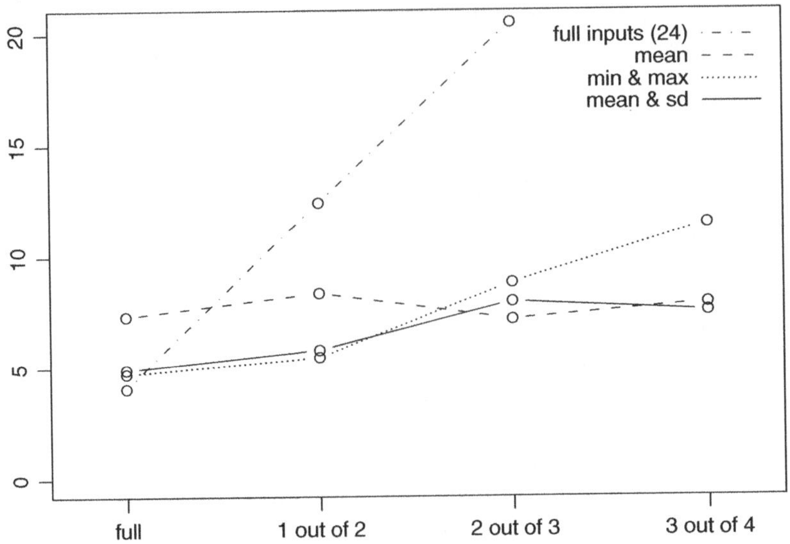

Figure 20.6 Data quality reduction: longitude.

- The symbolic approach based on (mean, sd) estimation is more robust than that based on (min, max) estimation. We can explain this result by the fact that (min, max) estimation is more sensitive to outliers (months with unusual temperature, for instance) than (mean, sd) estimation. This dependency has noticeable consequences for model inferences. We can conclude that (mean, sd) estimation should be preferred in all cases.

20.6 Conclusion

We have proposed in this chapter a simple recoding solution that allows us to use arbitrary symbolic inputs and outputs for multi-layer perceptrons. We have shown that traditional techniques, such as weight decay regularization, can be easily transposed to the symbolic framework. Moreover, the proposed approach does not necessitate specific implementation: the standard neural net toolbox can be used to process symbolic data. Experiments on semi-synthetic data have shown that neural processing of intervals gives satisfactory results. In future work, we plan to extend these experiments to categorical multi-valued variables and modal variables in order to validate all the recoding solutions. Finally, we plan also to study more precisely the adaptation of standard imputation methods (mean approach and k-nearest-neighbour approach) to the symbolic framework, especially by comparing imputation on recoded variables and imputation on symbolic variables using symbolic mean operators.

References

Beheshti, M., Berrached, A., de Korvin, A., Hu, C. and Sirisaengtaksin, O. (1998) On interval weighted three-layer neural networks. In *Proceedings of the 31st Annual Simulation Symposium*, pp. 188–194. Los Alamitos, CA: IEEE Computer Society Press.

Bishop, C. (1995) *Neural Networks for Pattern Recognition*. Oxford: Oxford University Press.

Bock, H.-H. and Diday, E. (2000) *Analysis of Symbolic Data: Exploratory Methods for Extracting Statistical Information from Complex Data*. Berlin: Springer-Verlag.

Moore, R. (1966) *Interval Analysis*. Englewood Cliffs, NJ: Prentice Hall.

Press, W.H., Teukolsky, S.A., Vetterling, W.T. and Flannery, B.P. (1992) *Numerical Recipes in C*, 2nd edn. Cambridge: Cambridge University Press.

Rossi, F. and Conan-Guez, B. (2002) Multilayer perceptron on interval data. In A.S.K. Jajuga, A. Sokolowski and H.-H. Bock (eds), *Classification, Clustering, and Data Analysis*, pp. 427–434. Berlin: Springer-Verlag.

Rossi, F. and Conan-Guez, B. (2005) Functional multi-layer perceptron: a nonlinear tool for functional data analysis. *Neural Networks*, 18(1): 45–60.

Rossi, F., Delannay, N., Conan-Guez, B. and Verleysen, M. (2005) Representation of functional data in neural networks. *Neurocomputing*, 64: 183–210.

Shiyan, T., Congbin, F., Zhaomei, Z. and Qingyun, Z. (1997) Two long-term instrumental climatic data bases of the people's republic of China. Technical Report 4699, Institute of Atmospheric Physics Chinese Academy of Sciences, Beijing, September. ftp://cdiac.ornl.gov/pub/ndp039/ (accessed May 2007).

Šíma, J. (1995) Neural expert systems. *Neural Networks*, 8(2): 261–271.

Simoff, S.J. (1996) Handling uncertainty in neural networks: An interval approach. In *IEEE International Conference on Neural Networks*, pp. 606–610. New York: IEEE.

White, H. (1990) Connectionist nonparametric regression: mutilayer feedforward networks can learn arbitrary mappings. *Neural Networks*, 3: 535–549.

Part IV

APPLICATIONS AND THE SODAS SOFTWARE

21

Application to the Finnish, Spanish and Portuguese data of the European Social Survey

Soile Mustjärvi and Seppo Laaksonen

21.1 The initial database

The European Social Survey (ESS) is designed to chart and explain the attitudes, beliefs and behaviour patterns of Europe's diverse populations (Jowell *et al.*, 2003). The survey is funded by the European Commission, the European Science Foundation, and academic funding bodies in each participating country. The data for the first round of the survey were collected in 2002 and 2003. These data cover 21 nations throughout Europe.

Among the main requirements imposed on the survey were full coverage of the target population, high response rate (70%), no substitutions, the same minimum effective sample size in the participating countries ($n_{\text{eff}} = 1500$, or 800 where population is smaller than 2 million), and a minimum net sample size of $n_{\text{net}} = 2000$. The ESS represents all persons aged 15 or over and resident within private households in each country, regardless of their nationality, citizenship, language or legal status (Häder *et al.*, 2003).

The samples were selected by strict random probability methods at each stage and the respondents were interviewed face-to-face (60 minutes average duration). There were also self-administered questionnaires (20 minutes average duration) which were sent to selected people by post. Around half of the interview questionnaire comprised 'core' items and the other half 'rotating' items. The self-administered questionnaires contained items designed specifically to help evaluate the reliability and validity of other items. The target response rate for the self-administered questionnaires was 90%.

Symbolic Data Analysis and the SODAS Software Edited by E. Diday and M. Noirhomme-Fraiture

More detailed information on the sampling and on the data from the survey is available on the website of the ESS at http://www.europeansocialsurvey.com.

21.2 Variables of interest for this study

We used six variables as background variables in this survey. These are *age, gender, country, employment status* and *residential area* (Table 21.1). We use the data from three countries – Finland, Spain and Portugal. The *age* variable has three different groups: young, middle age and old. *Employment status* has three categories: the first comprises employed and self-employed persons, the second persons not gainfully employed, and the third category contains all those who did not answer this question.

We carried out our experiments using two kinds of symbolic variables, interval-valued and modal. Modal variables were categorical (or ordinal-scaled) and divided into three or four groups. These variables are placement on left–right scale (*rightleft*), 'politicians generally care about the opinions of people like the respondents' (*pocare*), 'discusses politics' (*podisc*), 'member of political party' (*party*), 'voted in latest national election' (*voting*), 'could take an active role in a group involved with political issues' (*poact*), 'politics too complicated to understand' (*pocompl*) and degree of interest in politics (*pointr*). Table 21.2 shows how these variables are constructed.

The interval-valued symbolic variables used in this study were formed by summing the values of two or three different variables together and dividing these sums by the number of the summed values. Therefore, these variables are average-based indicators of each interest area. Since the initial scale of these variables was from 0 to 10, the scale of these variables remains the same. The new symbolic variables are trust in a European organ of government (*trust_eur*), trust in own country's organ of government (*trust_own*), satisfaction in own country's economical and democratic state (*satisfy_own*), satisfaction in current system (*satisfy_sys*), political following (*pol_ follow*), trust in other people (*trust_other*), own voice (*own_voice*) and political activity (*pol_activ*).

All these interval-valued symbolic variables were prepared by Winsorizing so that 1% of the smallest part and 1% of the largest part, respectively, were removed to these limits.

Table 21.1 Background variables for creating symbolic objects.

Country	**Gender**
F = Finland	1 = Male
E = Spain	2 = Female
P = Portugal	
Age group	**Employment status**
1 = 15-30	1 = Employed
2 = 31 - 60	2 = Not gainfully employed
3 = 61+	3 = No answer
Residential area	
1 = Urban	
2 = Rural	

Table 21.2 Modal variables.

Rightleft (placement on left–right scale) 1 = Left 2 = Middle 3 = Right	**Pocare (Politicians generally care about the opinions of people like the respondents)** 1 = Hardly 2 = Some 3 = Many 4 = No answer
Podisc (Discusses politics) 1 = Often 2 = Once a week to once a month 3 = Less often than once a month	**Party (Member of political party)** 1 = Yes 2 = No 3 = No answer
Voting (Voted in latest national election) 1 = Yes 2 = No 3 = No answer	**Poact (Could take an active role in a group involved with political issues)** 1 = No 2 = Probably 3 = Yes 4 = No answer

The purpose was to make the minimum/maximum values more robust. Secondly, we tested quartile range intervals which naturally are shorter and obviously more robust than the previous ones. Thirdly, we applied the same information to modal variables so that the initial values are categorized into four or five subgroups so that each includes an as equal as possible number of frequencies. The latter approach loses a very small amount of the information.

The initial data set also included the sampling weights that were constructed based on the sampling design only, not taking into account non-response and frame errors. These design weights vary in such countries where unequal inclusion probabilities and/or clustering are used, but for Finland all the weights are equal due to simple random sampling design. For the analysis, the weights are further transformed so that the average of all weights in each country is equal to 1. This means that when using these weights, each country will be taken into account with the same weight.

21.3 Symbolic objects

The data were distributed into two different symbolic data tables for each of the three countries. In one example the data for the three countries are combined. All the data tables include two or three different background variables within each country. The symbolic objects are constructed from these background variables and the maximum number of objects depended on the symbolic data table and on the background variables.

The background variables of the first table are *gender* and *residential area* (urban vs. rural). The maximum number of symbolic objects is $2 \times 2 \times 3 = 12$ due to the three countries. The second table also contains three different background variables which are *gender*, *employment status* and *age group* ($2 \times 3 \times 3 = 18$), leading to $3 \times 18 = 54$ symbolic objects. This full number was used for hierarchical and pyramidal clustering. Due to the fact that the oldest age group for Portugal and Finland was not well represented in the sample, we excluded this group from the analysis of divisive clustering in order to get fair results. Hence, this combined symbolic data set includes 42 objects. Most results are based on the second approach, but some descriptive analyses are given using the first approach.

21.4 Targets

Our symbolic data from the ESS were analysed with the help of various methods of the SODAS2 software, including *symbolic principal component analysis*, *generalized canonical analysis* and *symbolic Kohonen maps*, but in this chapter we only concentrate on the two clustering methods, that is, *divisive classification* and *hierarchical and pyramidal clustering*. Naturally, several graphical tools of the software were also used.

The main research problem was to compare some political opinions of the Finnish, Spanish and Portuguese citizens. We carried out this analysis using different approaches: (i) each country data table was handled separately; (ii) one combined data table based on the background variables was created and analysed; (iii) all independent country data SODAS files were pooled together and analysed as such. We consider all approaches to be interesting and to give a partially different picture to the phenomenon, but we concentrate in this chapter on the last approach which is perhaps most illustrative.

21.5 Some technicalities on the data creation

As is known, a SODAS file may be created directly using the module of the native data (module ND2SO; see Chapter 5 of this volume). In this case, we used a more traditional method as described below. The initial micro-data set was an SAS file. Using SAS, we created the appropriate background variables for constructing symbolic objects. Correspondingly, we performed the necessary operations (categories, quantiles, Winsorization) within SAS in order to later create the symbolic variables.

The SAS file was exported to Microsoft Access in which the file was converted to the format required in SODAS (for example, with unique identity code in position 1, symbolic object in position 2, sampling weight in the last position, and correct coding format). SODAS2 fairly straightforwardly creates a correct SODAS file from an Access file via module DB2SO, and this was naturally exploited in our analyses.

21.6 Symbolic data analysis strategy

Our strategy for handling the pooled ESS data is quite standard, so we start with descriptive analysis, and especially looking at the data using SODAS graphics. This step is also important for checking the data quality and analysing whether this kind of file could be useful for multivariate analysis. In some cases we rejected our file and created a new one.

Core data analysis is always related to the topic of interest. In this case, we tested, as mentioned earlier, several SODAS modules such as principal component analysis. Some interesting outcomes were found, but we excluded these results from this core analysis, because the clustering modules seemed to be of greater interest.

21.7 Some results

21.7.1 Graphics

Figures 21.1, 21.2 and 21.3 show four symbolic objects and four variables for the Finnish, Spanish and Portuguese data, represented using star visualization (see Chapter 7). The axes represent the variables, with three categories, and the bars represent the distributions. The bars have different colours according to the four different objects. The notation of these classes and objects is explained in Tables 21.1 and 21.2.

From Figure 21.1 we can see that most Finnish people voted in the last national election and most of them felt that their placement on the left–right scale was on the right or in the middle. When it came to degree of interest in politics, the commonest answer was 'quite interested'. The Finnish people also thought that politics was seldom or regularly too complicated to understand. There were no major differences between the objects.

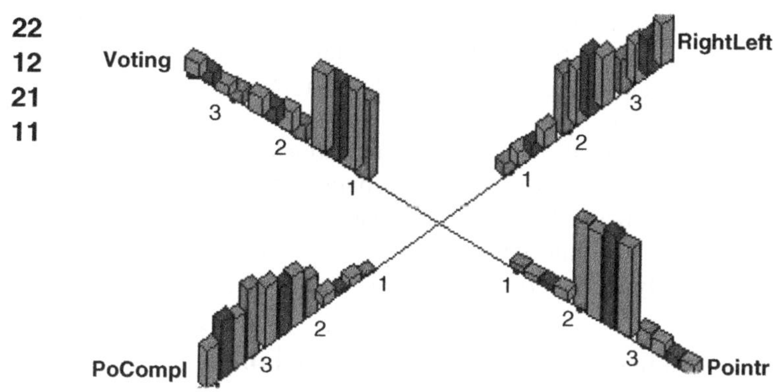

Figure 21.1 Zoom star superimposition for Finnish data when symbolic objects were formed by using *gender* (first position) and *residential area*.

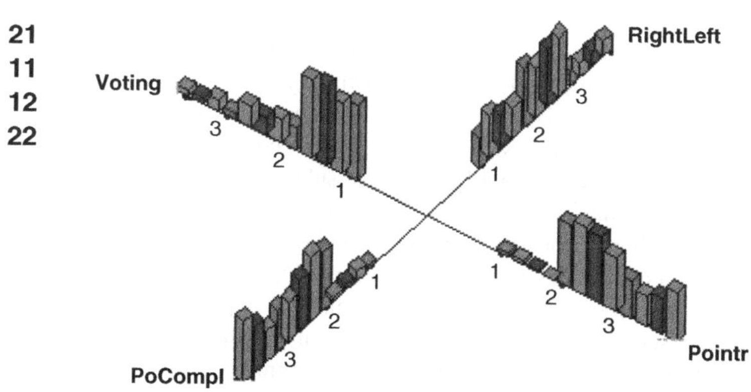

Figure 21.2 Zoom star superimposition for Spanish data when symbolic objects were formed by using *gender* (first position) and *residential area*.

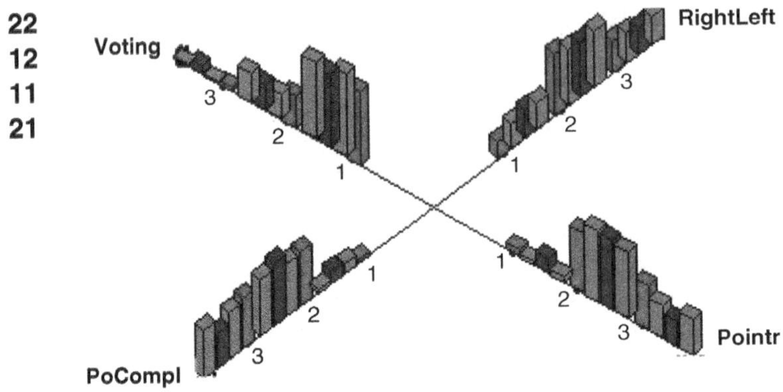

Figure 21.3 Zoom star superimposition for Portuguese data when symbolic objects were formed by using *gender* (first position) and *residential area*.

Most Spanish people voted in the last national election. They felt that their placement on the left–right scale was in the middle or on the left. Spanish people thought that politics was quite or not at all interesting and they also felt that politics was seldom or regularly too complicated to understand.

In Portugal, most people also voted in the last national election, but there was also quite a large population who did not vote. The distribution on the left–right scale is quite even between different placements, but the biggest bars are in middle of the scale. Portuguese people felt that politics was quite or not at all interesting and they also thought that politics was seldom or regularly too complicated to understand

21.7.2 Divisive classification for modal variables

The SODAS divisive classification (DIV) module was thus applied to the pooled Finnish, Spanish and Portuguese data for the ESS 'trust' and 'satisfaction' variables; see Section 21.2. This method performs an indexed hierarchy of symbolic objects based in this example on continuous and modal ordinal variables. The output of DIV, hierarchy of partitions, is a decision tree.

We tested several approaches to DIV. Three of them are presented in Table 21.3. A general observation of the table leads us to the hypothesis that the clusters are rather country-specific, but some mixed-country clusters are also found. However, there is no cluster with all three nationalities. Most mixed-country clusters are Spanish–Portuguese. This being the case, we can conclude that Finns differ substantially from the other nationalities, especially using the first and second data sets in which all Finnish groups belong to their own cluster without any other nationalities. In the second data set, Finns are divided into the two subgroups, clusters 4 and 7. The latter consists only of Finns, while cluster 4 includes Spanish non-working young women as well as several Finnish groups. We continued the clustering using the first data set, in an attempt to divide the Finns. This was successful and increased the number of clusters to eight, including the female Finnish cluster (F223,F212,F222,F213) without any other nationalities.

Some subgroups are so different that they constitute a cluster in themselves. It is interesting that *Spanish middle-aged men not working* constitute such a cluster in all applications.

Table 21.3 Clusters using DIV, and based on the three different data specifications. The variables are concerned with trust and satisfaction; see Section 21.2.

Cluster	Symbolic objects (Notation for code ABCD, A=*country* (S=Finland, E=Spain, P=Portugal), B=*gender*, C=*age group*, D=*employment status*)		
	Modal	Interval (1%, 99%)	Interval (25%, 75%)
1	S223 S113 S112 S111 S221 S123 S212 S122 S222 S211 S121 S213 all Finns	E111 E131 E211 E122 E121 E231 P213 P121	P212 P213 P323 P111 P112 P222 P121 P221 P211 P122 P113
2	E123 *Spanish middle age men not working*	E112 E133 E232 E212 E233 E132 E222	S223 S113 S112 S111 S221 S123 S212 S122 S222 S211 S121 S213 all Finns
3	E112 E111 E133 E131 E232 E212 E233 E132 E211 E122 E222 E121 E113 P111 P121	E223 E213 E221	E123 *Spanish middle age men not working*
4	**E231 E213** P212 P123 P112 P221 P211 P113	S223 S113 S112 S111 S123 S212 S122 S222 **E113**	P223
5	**E221** P213 P223 P222	E123 *Spanish middle age men not working*	E213
6	E223	P212 P123 P111 P112 P223 P222 P221 P211 P122 P113	E112 E111 E133 E131 E231 E232 E212 E233 E132 E211 E122 E222 E121 E221 E113
7	P122	S221 S211 S121 S213	E223
Explained inertia	69.0	75.6	99.3

Some other one-group clusters were not found across applications. This thus means that the results based on different databases from the same micro data vary to some extent, but not dramatically. Note that we excluded the basic interval solution from our examples, thus the data without any robustness. Moreover, the first and third results seem to be closer to each other. This is a good thing as we lose less information than in the second example. What else can we learn about these results? At least, that a user has to carefully consider how to construct the SODAS variables, so that they are not too sensitive, and so that the information loss is as minor as possible.

We can continue the interpretation of the results taking into account which background variables were behind these clusters. In this chapter we use a low profile and exclude a comprehensive subject-matter (politological) interpretation of the results. We hope that the reader will continue from our brief interpretation in the rest of the chapter, starting from the Table 21.4 that is based on the first application, being a formulation from the SODAS output.

Table 21.4 Some further analyses based on modal variables. Note that we cut this to group 4.

	Group 1: All Finns		
High satisfaction with public services in the country	Yes ↑ No ↓	**Group 2:** Spanish non-working men over 60 years	
	High satisfaction with politics and economics of the country	Yes ↑ No ↓	**Group 3:** Spanish and Portuguese men under 30 years, working or not, men, and 30–60, not working
		High trust to politicians and institutions of the country	Yes ↑ No ↓
			Group 4: Spanish and Portuguese non-working women under 30 years, and working women over 60 years

21.7.3 Hierarchical and pyramidal clustering

This application is constructed from the same database as the application for the divisive clustering already discussed. In addition, the three modal variables (*pointr, poact, pocare*) are included. We can compare these results with those from the divisive clustering, but not completely. Figure 21.4 gives our main results. In order to make some summaries from the full clustering, we have divided all symbolic objects into five groups (Ryhmä 1, Ryhmä 2, etc.), marked in the figure.

The first group includes citizens from all countries. All the Finns and Spaniards are over 30 years old but the Portuguese are from all age groups. Spanish women in this group are not working and 30–60 years old. Political opinions vary in this group, but a majority are not very interested in politics. Their left–right position is rather central. Nevertheless, most of them voted in the last election.

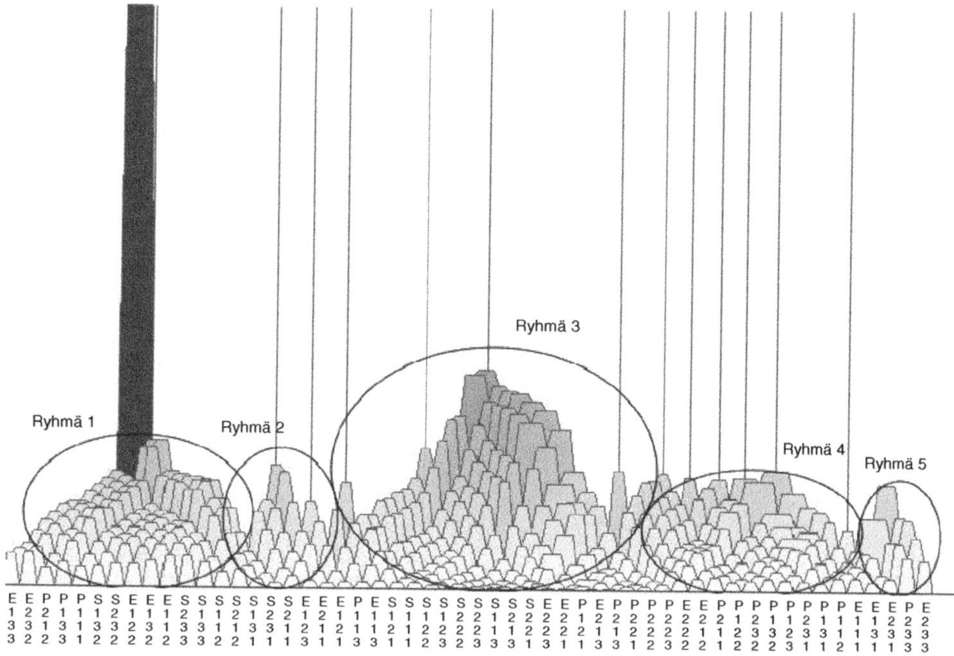

Figure 21.4 Pyramids and groups formed from the symbolic objects.

The second group is much smaller than the first. It includes only Finnish and Spanish people. All Finns are under 30 years old, and men are not working. Spanish women in this group are working and under 30 years old, but Spanish men not working and over 60 years old. This group was less politically active than the first.

There are in the third group citizens from all three countries. Politically they were not active but more active than the previous groups. The fourth group includes only Portuguese and Spaniards, and their political activity was lower than average but they were willing to discuss items of this kind. The fifth group was small and consisted also only of Portuguese (over 30 years) and Spanish (under 30 and over 60) females. Their political interest was lowest, as was their voting activity, although most of them did vote.

21.7.4 Conclusions from the results

The results differ considerably between divisive clustering, and hierarchical and pyramidal clustering, although common features are found. For example, these results also show well that Finns differ from Portuguese and Spanish substantially, although there are even the two groups with symbolic objects from each country (in contrast to the two examples with divisive clustering). From both results we also see that Finns are most satisfied and Portuguese least, Spaniards being in the middle but in the same direction as Portuguese. In some sense, divisive clustering was easier to use and interpret.

References

Bock, H.-H. and Diday, E. (2000) *Analysis of Symbolic Data: Exploratory Methods for Extracting Statistical Information from Complex Data.* Berlin: Springer-Verlag.

Diday, E. (2001) An introduction to symbolic data analysis and the SODAS software. In *The Yearbook of the Finnish Statistical Society*, pp. 36–67. Helsinki: Finnish Statistical Society.

Diday, E. (2004) From data-mining to knowledge mining: Symbolic data analysis and the SODAS software. Paper presented to the Workshop on Symbolic Data Analysis, Lisbon, 26 January. http://www.info.fundp.ac.be/asso/dissemlink.htm.

Häder, S., Gabler, S., Laaksonen, S. and Lynn, P. (2003). Sampling for the European Social Survey – Round I. http://www.europeansocialsurvey.org/

Jowell, R. and the Central Co-ordinating Team (2003) *European Social Survey 2002/2003: Technical Report.* London: Centre for Comparative Social Surveys, City University. http://www.europeansocialsurvey.org/ (accessed November 2003).

22

People's life values and trust components in Europe: symbolic data analysis for 20–22 countries

Seppo Laaksonen

22.1 Introduction and the data

The first round of the European Social Survey (ESS) was carried out during 2002 and 2003 in 22 countries: Austria (AT), Belgium (BE), Switzerland (CH), Czech Republic (CZ), Germany (DE), Denmark (DK), Spain (ES), Finland (FI), France (FR), United Kingdom (GB), Greece (GR), Hungary (HU), Ireland (IE), Israel (IL), Italy (IT), Luxembourg (LU), Netherlands (NL), Norway (NO), Poland (PL), Portugal (PT), Sweden (SE) and Slovenia (SI); for more details, see http://www.europeansocialsurvey.org and Chapter 21 in this volume. The initial microdata were mainly collected in face-to-face interviews, but self-administered questionnaires were used for some questions. This latter part included life-value factors (based on those formulated by Shalom Schwartz), among other things. Unfortunately, this was not carried out in Italy and Luxembourg, and hence we have used two types of data here: (i) 22 countries as symbolic objects with the four initial variables (converted from about 40 000 individuals), and (ii) 20 countries as symbolic objects with the eight initial variables (converted from about 36 000 individuals).

The initial variables are described in more detail in Table 22.1. Two sets of variables were thus selected. The first set consists of variables that measure people's trust in their country's government and legal system, and in general. These are available for all 22 countries. By contrast, we have Schwartz's life-value factors for only 20 of the countries. These are labelled here in our way, not in full compliance with social-psychological terminology, since this is a methodological study. The factor scores are automatically scaled so that their mean

Symbolic Data Analysis and the SODAS Software Edited by E. Diday and M. Noirhomme-Fraiture

Table 22.1 Creation of the basis of symbolic variables.

First group: 'Trust variables'	
TRUSTLEG	The average of two questions scaled from 0 to 10 relating to trust in the police and legal systems in the country of residence. A smaller value means lower trust. This was then standardized so that the mean = 0 and the standard deviation = 1.
TRUSTGOV	The average of three questions as above relating to trust in government and other administrations in the country of residence. Standardized as above.
PEOPLETRUST	One question measuring trust in other people, scale as above. Standardized as above.
PEOPLEFAIR	One question measuring people's fairness, scale as above. Standardized as above.
Second group: 'Life-value variables'	
TRADITION	This is the first factor of the 21 life-value variables in the questionnaire. It is rescaled so that a higher value means a higher belief in these values, e.g. importance of following traditions, behaving properly.
EQUALITY	The second factor as above, e.g. importance of people being treated equally and having equal opportunities, understand different people.
SUCCESS	The third factor relates to the following values: importance in life of being rich and successful, demonstrating abilities and being admired, and being recognized for achievements.
ENJOY	The fourth factor: importance of having a good time, seeking fun and things that give pleasure.

is equal to zero and the standard deviation is equal to one. We do not change this scaling but we still have a scaling problem with other variables that are initially measured within a closed interval [0, 10]. Our question is: does it matter how these are scaled? We tested this with some symbolic methods and observed that it does matter. Since we desired that each variable in a certain symbolic analysis should have the same influence in some sense, we rescaled these variables similarly to factors as far as the symbolic interval variables were concerned.

Our main emphasis is on the data for the 20 countries that give us the opportunity to use all eight variables together. Nevertheless, it is not at all clear how the symbolic variables should be created, even though they have all been scaled in the same way (this is not the only way to make scaling consistent, either). We tested several specifications, both interval and modal-based, the most important ones being the following:

(a) Averages that correspond to the strategy of classical analysis.

(b) Interval variables based on 95% confidence intervals (CIs) that do not differ essentially from specification (a) because these CIs are not very large in this case due to a well-harmonized survey and fairly large sample sizes. Hence, we do not present any of these results.

(c) Minimum vs. maximum values as intervals: this is not of interest in this case because the variation between the countries and their associated symbolic descriptions almost disappears.

(d) Intervals based on 90% vs. 10% quantiles. The variation between the countries (symbolic objects) is too small in this case, too, and consequently this analysis is not of great importance.

(e) Intervals based 75% vs. 25% quantiles (quartile ranges). This distinguishes the countries quite well and hence the majority of our experiments are based on this approach.

(f) Modal variables based on several categories created from continuous variables. We tested seven categories and thus calculated a relative frequency for each category. This strategy was not very useful, since these symbolic methods do not take into account the fact that categories are ordered. Hence, for example, the method based on this kind of modal variable clusters together countries using categories which are not connected. This, consequently, makes interpretation more difficult.

(g) Modal variables based on two categories, in which case such problems as in (f) will not occur. Our main experiments are based on the categories of below vs. above mean, but we did some tests so that the cut value was either −0.3 or +0.3, in which cases the countries far from the mean would receive more attention.

In the following sections we present a selection of results based on different SODAS modules. We start from descriptive statistics and continue towards such methods that are workable for our symbolic variables.

22.2 Descriptive methods

First, we give an illustration of the difference between a classical variable and a symbolic one in Figure 22.1. As we recall, the initial variables are normally distributed and ideally it would be good if the distribution of the symbolic ones did not differ much from this. This would mean that the information loss due to aggregation was small. As observed, the symbolic solution is much nicer than the classical average-based solution from this point of view. These two interval variables are not exceptional, for the results for all the others are quite similar.

In Figures 22.2 and 22.3 we present some examples of the use of zoom stars. Figure 22.2 is for four modal variables and Figure 22.3 for eight quartile-based interval variables.

We have chosen the four examples of the 20 countries for Figure 22.2 so that extremes in a certain sense are included. Naturally, the value may be still quite close to the cut value and thus not substantial. However, some general principles of life values are nicely visible. For Israel, all these values are more important than the average for the other countries, whereas for Sweden all are less important. For French people EQUALITY and ENJOY

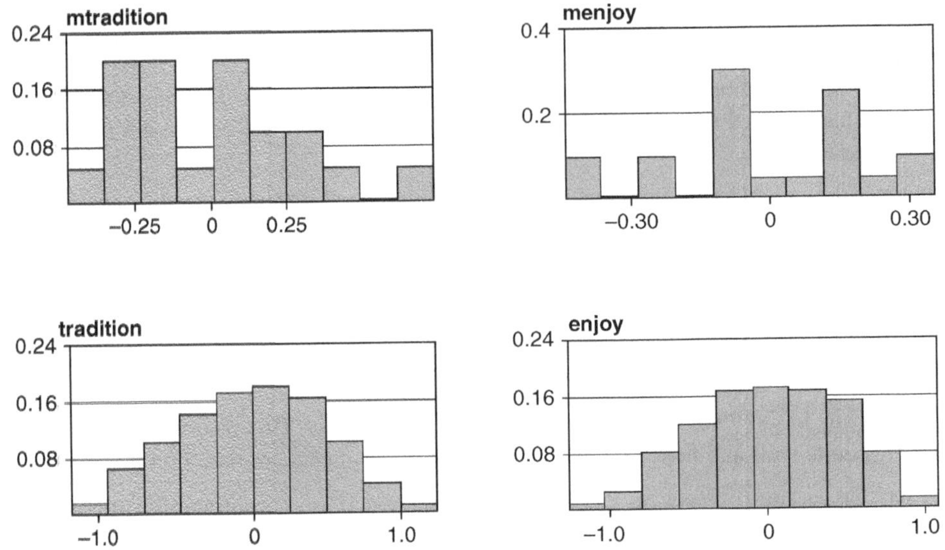

Figure 22.1 Distributions for the variables TRADITION and ENJOY. The upper figures are for averages (case (a)) and the lower ones for quartiles (case (e)). Both analyses were done by SODAS.

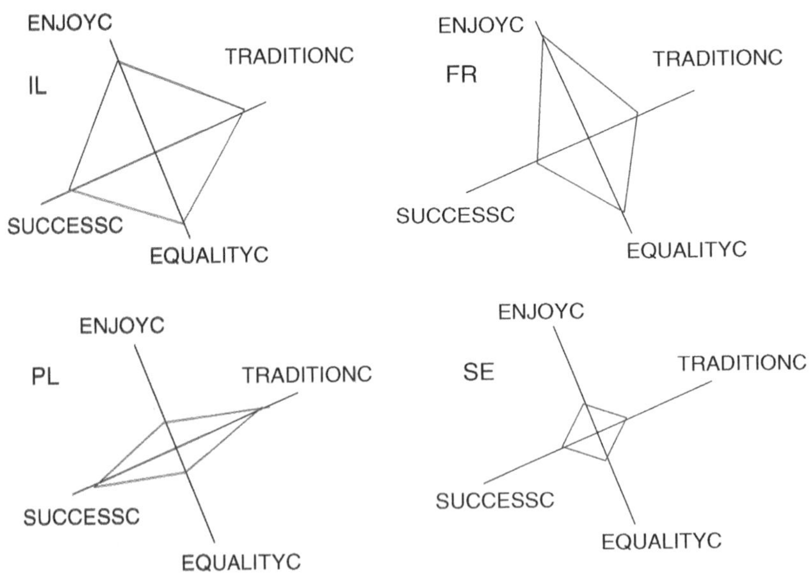

Figure 22.2 Simple graphics for modal-based factors in case (g) so that the cut value is the mean of each variable.

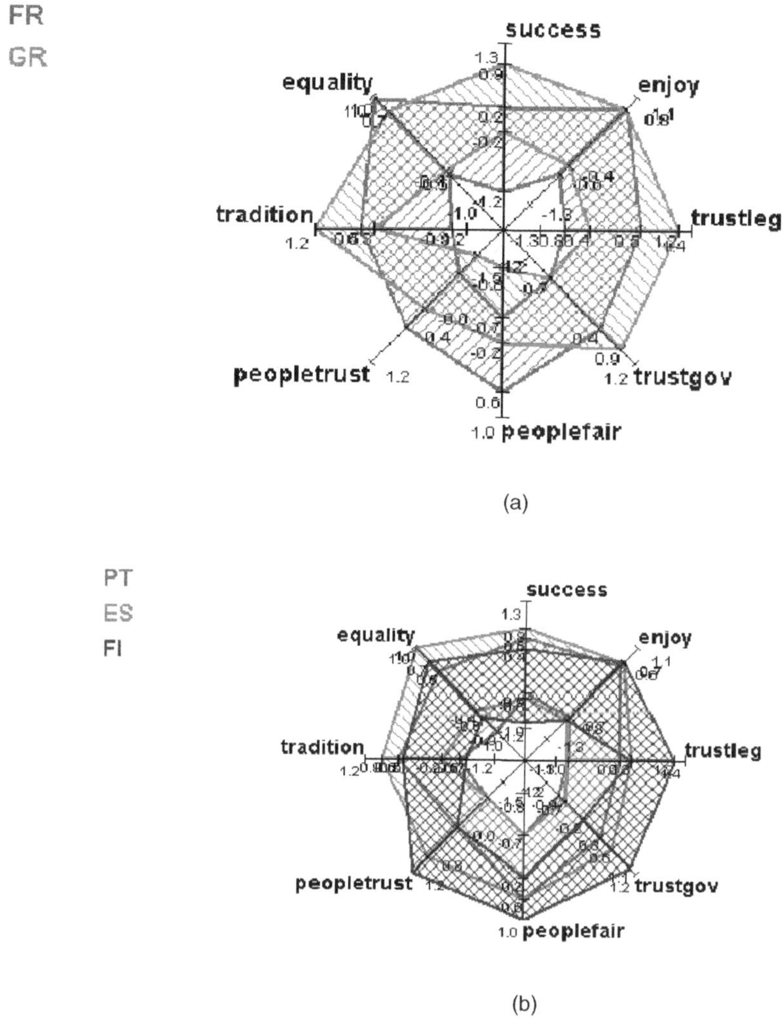

Figure 22.3 Superimposition examples for the eight interval variables including (a) France and Greece and (b) Spain, Finland and Portugal.

are relatively important but TRADITION and SUCCESS are not. Poland is a 'complete opposite' to France.

Figure 22.3(a) shows well how much Greece and France differ from each other, ENJOY being the only variable that is at the same level. The majority of Greek people trust their government and legal system much more than the French people do theirs. However, the minimum interval value is equal for TRUSTGOV, which indicates that this trust varies relatively much among people.

Figure 22.3(b) shows clearly that trust in these four aspects is much higher in Finland than in Spain and Portugal. In these four factors Portugal and Finland are fairly close to each other but in Spain traditions, equality and success are more important than in Portugal and Finland. The difference between the countries is minor for ENJOY.

These descriptive statistics are just examples but still illustrate the fact that the people in these European countries are not similar in their opinions, attitudes and life values. Next, we present various clustering and other techniques using the SODAS software. These give some explanations for these differences between the countries.

22.3 Divisive classification (DIV) vs. unsupervised classification tree (SCLASS)

Two SODAS modules, DIV (see Bock and Diday, 2000) and SCLASS, yield a tree based on interval variables. SCLASS does this automatically in the sense that it 'decides' the number of clusters of symbolic objects but the user must give a maximum number of symbolic objects for each cluster. So, if the desire is to compare results, the user has to decide how many clusters to produce. We made this decision after constructing an SCLASS tree in which we used the default value of 3. For DIV we used the Hausdorff distance and set the number of clusters equal to 11 since this number was obtained via SCLASS. In this analysis, we also wanted to see how the full number of variables worked, thus including the question of which types of variables play the most important role in clustering.

A priori, it is not clear how similar or different the trees will be, even though the number of clusters is fixed to be the same. This is due to their different approaches, although both modules have many similarities, giving an explanatory tree and thus 'monothetic' clusters, that is, described by conjunctive expressions on the values taken by the variables. Both our trees are presented in Figure 22.4 (ellipses are nodes which will be split further into other nodes and finally into terminal nodes, represented by boxes).

The trees cannot be completely similar due to the different approaches of the methods, although many common features are found. The clusters they have in common are Greece (GR), the Czech Republic plus Poland (CZ, PL) and Denmark plus Finland (DK, FI) although Norway and Sweden (NO, SE) are in the same clusters in both cases, too. Much depends on the order in which the symbolic variables are chosen for the tree construction. It is interesting that both trees start with a 'trust' variable, although the first choice is different.

The values below the 'tree' variable are cut values. The countries below this value are selected to the left and the countries above the cut value to the right side of the tree. So, for example, SCLASS first uses the variable TRUSTLEG and immediately distinguishes Denmark and Finland as countries where trust in the legal system is highest. Second, the method chooses the same variable again and finds Greece where this trust is fairly high. In the next step the SCLASS module exploits TRUSTGOV on the one hand and SUCCESS on the other, and creates new terminal nodes (clusters). The method does not need all eight variables: PEOPLEFAIR, TRADITION and EQUALITY are missing. This is due to the method trying to take advantage of the most important variables (TRUSTLEG is here the most significant variable).

The general principle of the DIV tree is similar to that of SCLASS tree but the criteria for the choice of variables are different. Hence the trees naturally differ to some extent. DIV shows more clearly the similarity of the Nordic countries. This is largely because this clustering is based on 'trust' variables, in which sense the Nordic countries are more similar than in the case of life-value variables. It should be noted that if we do not want as many

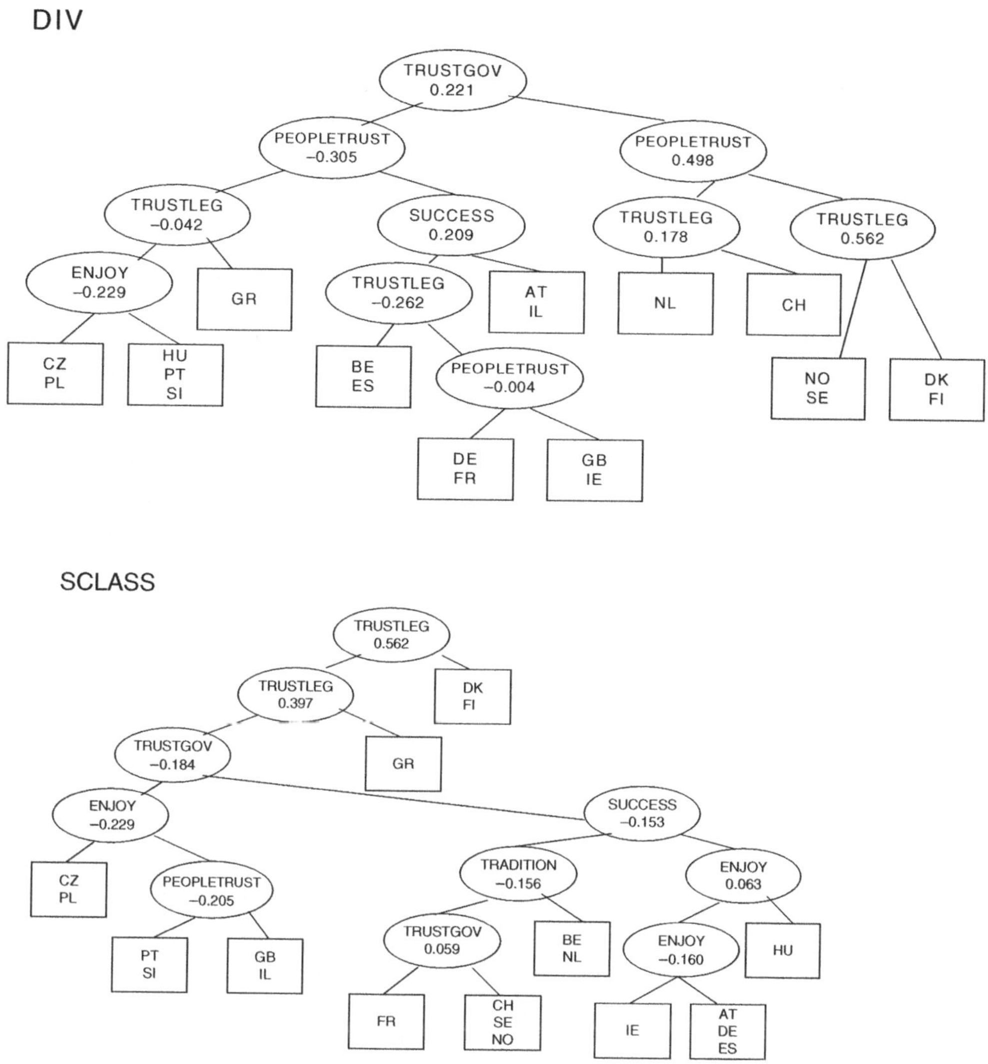

Figure 22.4 VTREE using DIV and SCLASS using the eight variables for the 20 countries. Each node gives the name of the explanatory variable and the cut value.

terminal nodes as this, we can easily go up the tree and merge the clusters within each node together (ellipse levels).

22.4 Clustering by SCLASS, DIV, SCLUST, SYKSOM and HIPYR for 'trust' variable in 22 countries

This analysis is based on the four 'trust' interval variables. We compare the four methods, although they have different approaches. SCLUST and SYKSOM produce prototypes as cluster representatives. Some of these prototypes have no exact interpretation but some are well related to the symbolic objects, that is, the 22 countries. HIPYR constructs a hierarchy or pyramid on a set of symbolic data, with the cluster structure based either on the individual-variable data set or on dissimilarity values between the elements of the data set. In the first case, the symbolic hiearchical/pyramidal clustering method described in Chapter 10 is applied. The description of each cluster is then given by the associated symbolic object, and each cluster is defined by the set of its members together with its description (as a complete SO that generalizes its members).

We first present the clusters given by each method in Table 22.2 and then illustrate the results of HIPYR in Figure 22.5. Finally, we give some examples of prototypes relating to these 'trust' variables.

This analysis illustrates how different the results can be, depending on the choice of clustering method. In fact, they discover interesting and complementary symbolic objects and prototypes by using different points of view on the data due to the different criteria they use. Naturally, there are, and should be, different strategies for approaching this question and hence, although these are two different approaches, they produce similar results in terms of general tendency. It is interesting that the clusters are very stable for some countries but not for some others. This means that they simultaneously satisfy different criteria. In particular, Greece is almost always in its own cluster, as are Denmark and Finland, although

Table 22.2 Clustering of 22 countries based on SCLASS, DIV, SCLUST, SYKSOM and HIPYR so that the number of clusters is nine, except for HIPYR in which case this cannot be done exactly in the same way.

Country	Other countries, if any, clustered together with the reference country				
	SCLASS	DIV	SCLUST	SYKSOM	HIPYR
AT	ES LU	DE FR GB IE IL	ES GB IL IT	BE CZ DE ES FR IL	IE GB
BE	DE NL	ES	DE FR	AT CZ DE ES FR IL	DE NL
CH	IE NO SE	LU SE	IE LU SE	—	LU
CZ	PT	PT	—	AT BE DE ES FR IL	SI FR PT
DE	BE NL	AT FR GB IE IL	BE FR	AT BE CZ ES FR IL	BE NL
DK	FI	FI NO	FI NO	FI NO SE	NO FI SE
ES	AT LU	BE	AT GB IL IT	AT BE CZ DE FR IL	—
FI	DK	DK NO	DK NO	DK NO SE	DK NO SE
FR	HU IT	AT DE GB IE IL	BE DE	AT BE CZ ES DE IL	PT CZ SI
GB	IL	AT DE FR IE IL	AT ES IL IT	IE	AT IE
GR	—	—	—	—	IT
HU	FR IT	PL SI	PL SI	PL PT SI	PL
IE	CH NO SE	AT DE FR GB IL	CH LU SE	GB	AT GB
IL	GB	AT DE FR GB IE	AT ES GB IT	AT BE CZ DE ES FR	—
IT	FR HU	—	AT ES GB IL	—	GR
LU	AT ES	CH SE	CH IE SE	—	CH
NL	BE DE	—	—	—	DE BE
NO	CH IE SE	DK FI	DK FI	DK FI SE	DK FI SE
PL	SI	HU SI	HU SI	HU PT SI	HU
PT	CZ	CZ	—	HU PL SI	FR CZ SI
SE	CH IE NO	CH LU	CH IE LU	DK FI NO	NO DK FI
SI	PL	HU PL	HU PL	HU PL PT	CZ FR PT

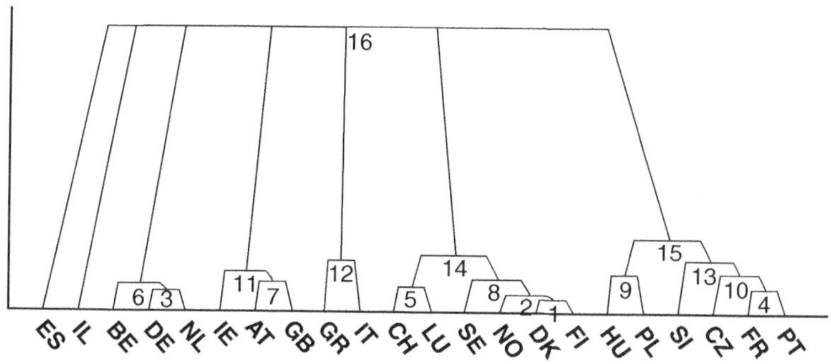

Figure 22.5 Hierarchy obtained by HIPYR based on 'trust' variables using the symbolic objects, with generality degree as criterion.

sometimes these countries have one or two companions (Norway, Sweden and Switzerland). The Netherlands also has a unique nature in three cases and in the rest of the methods it is close to Belgium and/or Germany or Sweden. The companions of some countries, for example, France and Italy, vary surprisingly. These differences could be explained quite well if the focus were on a subject-matter analysis (thus considering how people's trust, in its four senses, is related to each country) but since this chapter is a technical one we will not pursue such an evaluation.

The HIPYR graph extends the analysis illustratively. It shows well the order in which countries are clustered together. Finland and Denmark are cluster 1. This is mainly due to their high values on all the trust variables, while the interval is rather narrow. Norway is close to these countries and Sweden as well, although the trust values are lower for the latter country. Cluster 3 consists of the Netherlands and Germany. The trust values in these countries are quite similar on all four variables and slightly above the average country, while the interval is narrow. Belgium is quite close to these. On the right-hand side of the graph are first found Portugal, France, Czech Republic and Slovenia and then Poland and Hungary. In all these countries the trust variables are below the average, except for PEOPLEFAIR which is at average level. The uniqueness of Israel is due to a big interval, thus some people have a high level of trust, while others have very little. Spain is not much different from Israel, but TRUSTLEG is lower there. It is a little surprising that Greece and Italy are as close to each other as they are. In these countries, the values for TRUSTPEOPLE are very low (lowest overall for Greece), quite low for PEOPLEFAIR, but are close to the average for TRUSTLEG and TRUSTGOV. Ireland, Austria and United Kingdom are roughly average countries on all variables.

Both methods, SCLUST and SYKSOM, give the opportunity to produce prototypes. We present here an example from SYKSOM in Figure 22.6. In this graph we can see how 'trust in other people' is related to 'trust in government and other administration' by the opinions of the people in these country clusters.

The relationship between these two symbolic variables is not very close, although 'trust in people' often means 'trust in administration' as well. This holds well for the Nordic cluster 3×5 in which both are fairly high and for the eastern European cluster 1×2 in which both are rather low. However, the Greek cluster 3×3 is not in line with this: 'trust in people' is rather low in Greece but they have more trust in administration, in contrast to the general tendency in the ESS countries.

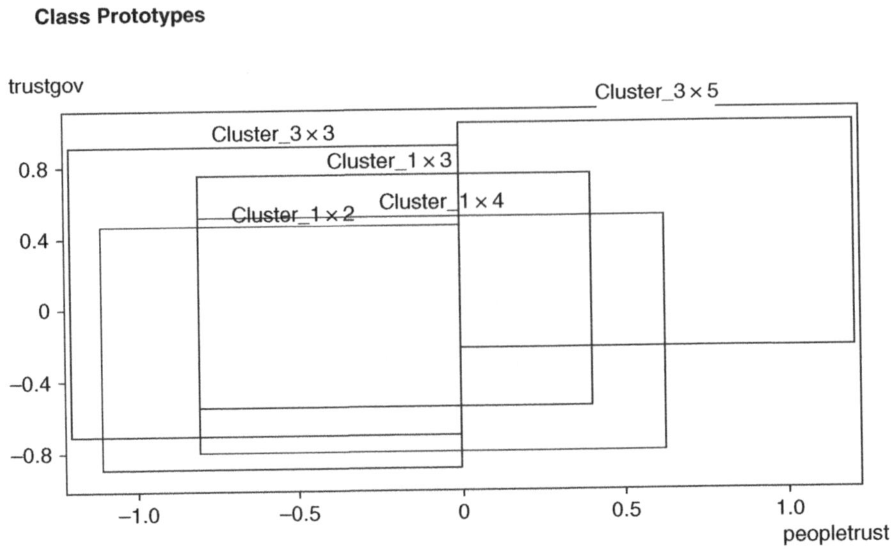

Figure 22.6 A prototype example from SYKSOM when using the same model as presented in Table 22.2. Clusters have the following interpretation: 1×2 = HU, PL, PT, SI; 1×3 = IT, 3×5 = DK, FI, NO, SE; 1×4 = AT, BE, CZ, DE, ES, FR, IL, 3×3 = GR.

22.5 DIV for life-value variables using modal and interval variables

It is useful to compare results using both categorical and interval variables from the same database. For this purpose there are not many methods available in the SODAS, but DIV is workable for both types of variables. Hence we tested it and at the same time we were able to get results from the clustering of the 20 ESS countries by life values, that is, initially based on the four main factors.

The trees in Figure 22.7 are similar in the sense that the first clustering variable is the same, TRADITION. Later, the modal tree exploits the three other variables but the interval tree does not need the variable EQUALITY. For this reason and due to the different handling of the cut value, the results differ quite substantially. The modal tree is easier to interpret although it loses more information. The number of countries in each cluster also differs.

There are three single-country clusters in the modal tree (Belgium, the Czech Republic and Austria), but only one similar case in the interval tree (Greece). This country is clustered with Israel, Hungary and Slovenia in the second graph. The first graph is preferable, however. It shows that the Nordic countries of Norway and Sweden are close to each other in people's life values, but Denmark is closer to Switzerland and France, and Finland to Austria, Germany and the United Kingdom. The majority of people in Belgium and the Netherlands are also quite similar to each other, as well as people in the Czech Republic, Ireland, Poland and Portugal. The results were not easy to see in advance but they look quite believable when taking into account the religions and the histories of the countries,

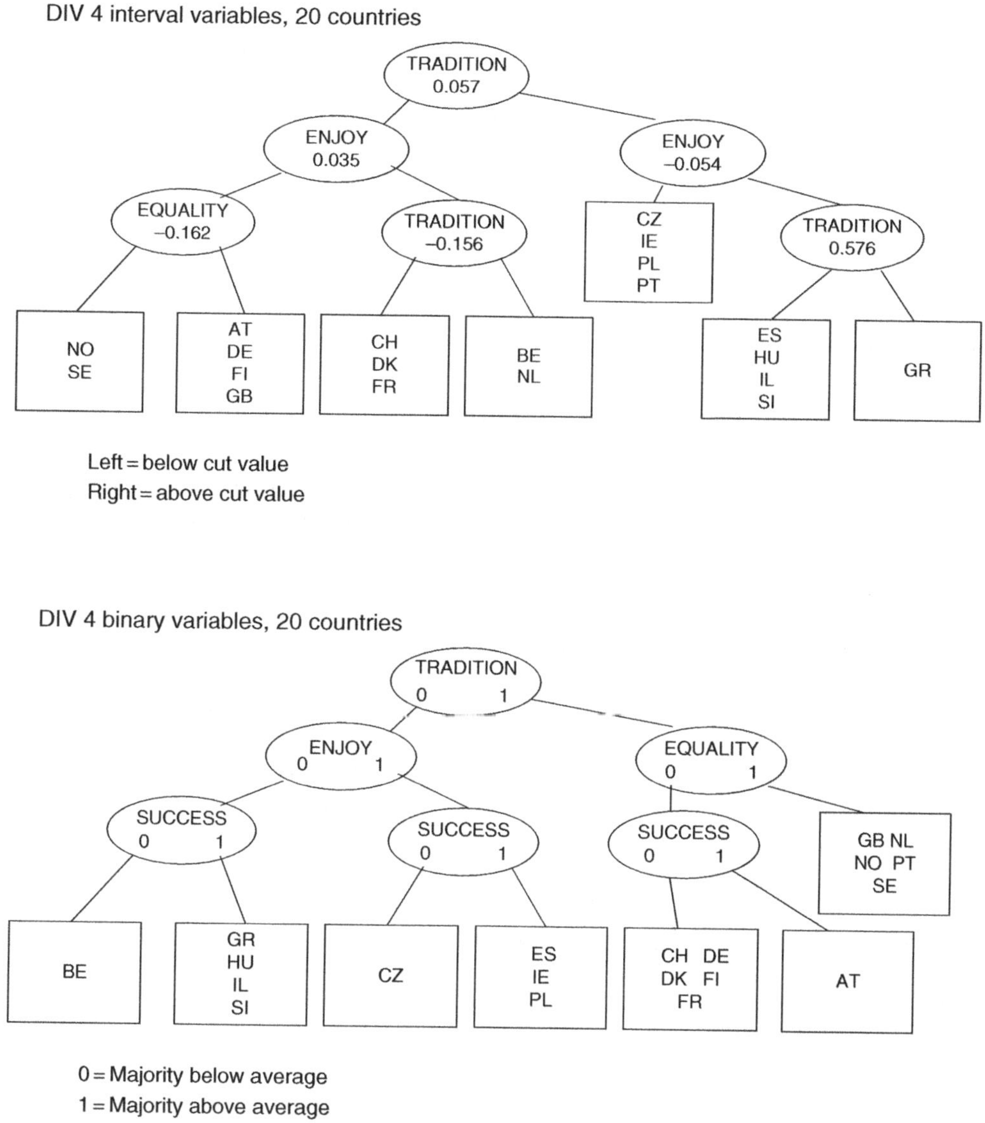

Figure 22.7 DIV tree using both interval and modal (binary) variables for life values for 20 countries.

among other things. Perhaps surprisingly, Israel is not in its own single cluster, but together with the Catholic countries of Spain, Hungary and Slovenia.

We carried out some further modal variable tests by categorizing each variable as explained in (g) in Section 22.1. When this was done so that the higher value was the cut point, a fairly big cluster was obtained (AT, BE, CH, DE, ES, GB, GR, HU, IE, IL, NL, PT, SI) due to the small variation between these countries. Consequently, the other clusters were quite small: Denmark, Sweden and Norway in their own cluster, the Czech Republic and Poland in one cluster, and France and Finland together in yet another cluster. The latter was formed as fairly large proportions of people in these countries appreciate success in their life but not so much traditions.

When more consideration is given to low appreciation in the categorization the biggest cluster consists of Germany, Finland, United Kingdom, Ireland, Netherlands, Norway, Portugal, Sweden and Slovenia. This was the result when all four variables were utilized in this order: SUCCESS (0 = less), TRADITION (1 = more), EQUALITY (0) and ENJOY (1). Denmark and Belgium were included in the same node until the third step and Switzerland and France until the second step. The rest of the countries divided into two main clusters: the Czech Republic, Spain and Hungary, on the one hand, and Greece, Poland, Austria and Israel, on the other.

22.6 Dissimilarity module DISS

This module gives the opportunity for a general description of differences or similarities between the symbolic objects. The basic output is a dissimilarity matrix between variables. We present here some results based on our eight interval variables. The dissimilarities thus show how much the whole pattern of these variables differs from one country to another. Naturally, the dissimilarity is zero in the diagonal of the matrix and in all other cases more or less positive unless all variable values are equal in the countries compared.

Our analysis thus includes both the 'trust' and 'life-value' variables and should be interpreted accordingly. Thus this does not prove anything about how good or bad the countries with a small or large dissimilarity are.

Figures 22.8 and 22.9 illustrate the two basic graphical representations of DISS. These are useful for getting a general idea of differences. For example, we see from Figure 22.8 that a specially marked country, Greece, receives quite high dissimilarity values, on average. Figure 22.9 shows this same feature since the Greek pie is fairly large, implying that this country deviates considerably from the others, on average. By contrast, the sizes of the pies for the United Kingdom and Ireland and also for France are fairly small, implying that these are quite 'average' countries.

In most cases, the user will not confine his/her analysis to the graphic representations of DISS. He/she will, of course, look at the numerical details of the matrix and, for example, can rank the symbolic objects and get some useful ideas, and maybe go to the other SODAS modules to deepen his/her study. Naturally, the matrix may be taken as input to other analyses. The SODAS software offers various facilities for this purpose, but the DISS matrix may also be used in classical analyses.

We carried out an analysis using the classical factor analysis so that all 20 countries were both statistical objects and variables, and the variable values were dissimilarities. This kind of analysis yields distance dimensions of which the first one could be interpreted as the general distance (or is, at least, the most important); see Figure 22.10.

This ranking gives quite a traditional picture of these European countries. For example, the Nordic countries constitute their own group, and next come Switzerland and Germany. Both eastern European (Slovenia, Poland, Hungary, Greece, Czech Republic) and south-western European countries (Portugal and Spain) locate on the right-hand side. France, United Kingdom, Belgium, Ireland, Austria, Netherlands as well as Israel are in the middle and do not differ much from each other.

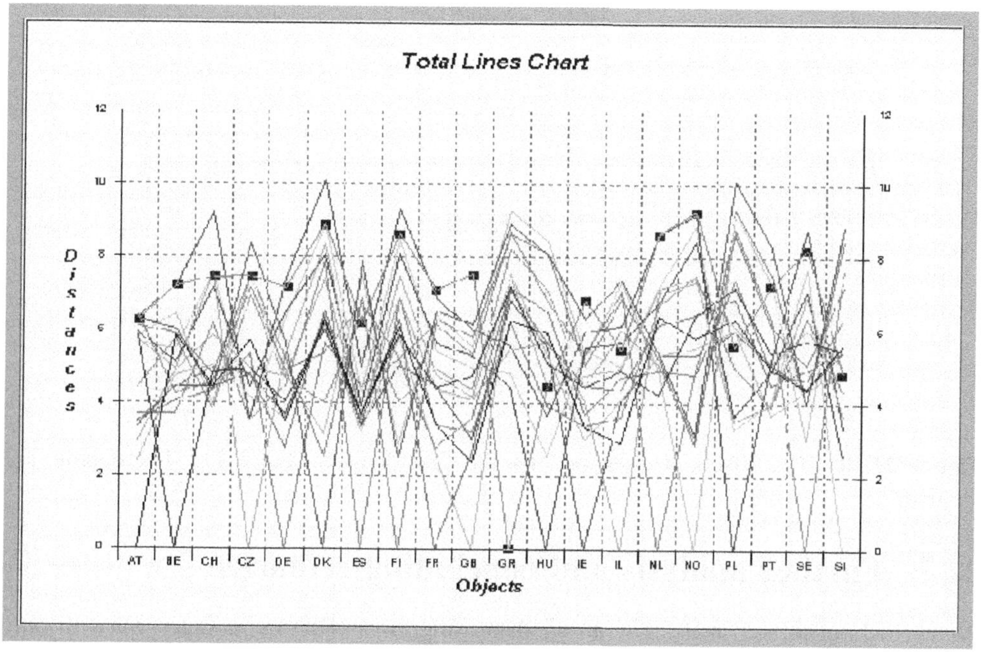

Figure 22.8 Line representation of the dissimilarity matrix. The data are from 20 countries and all eight interval variables are used. Greece (GR) is specially marked.

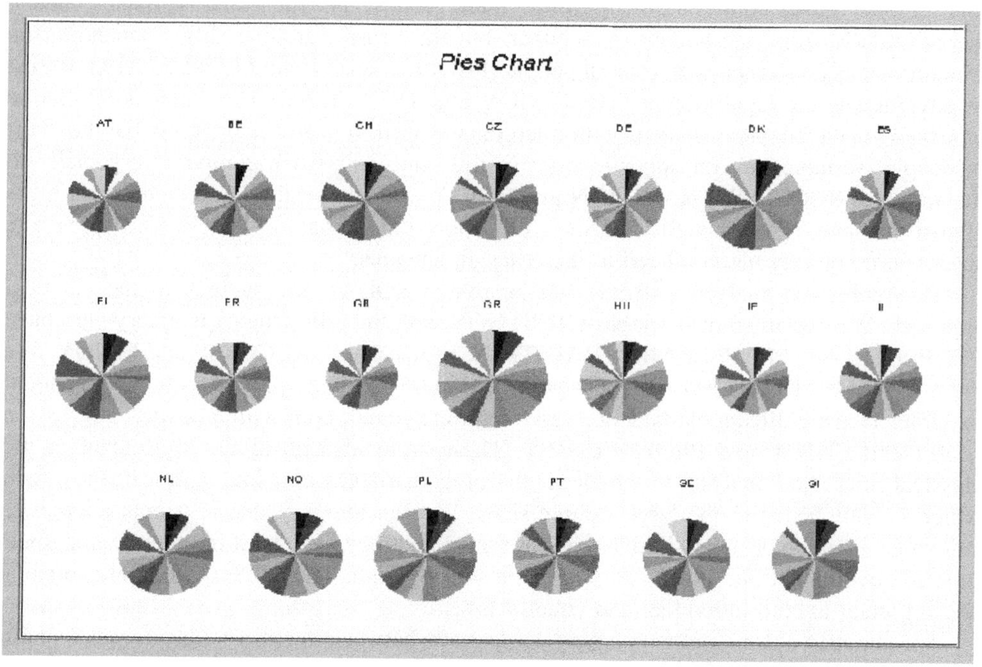

Figure 22.9 Pie representation of the dissimilarity matrix. The data are from 20 countries and all eight interval variables are used.

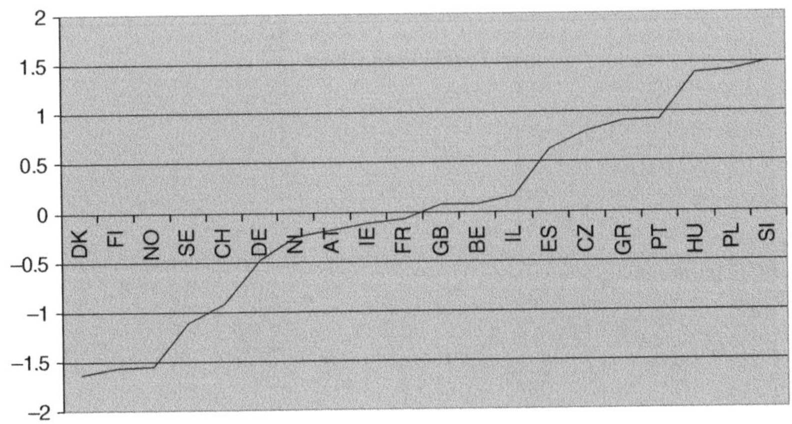

Figure 22.10 The 20 countries sorted by the first distance dimension. For details, see the text.

22.7 Further analysis and concluding remarks

We carried out further analysis with our data using other SODAS modules, but will only present partial results before drawing conclusions. Both the analyses presented are based on the data for the 20 countries, hence the larger number of variables.

The principal component analysis (SPCA; see Chapter 15) illustrates, for example, that this set of variables could be classified into three groups: the 'life-value' variables behave quite similarly from one country to another, but the 'trust' variables differ depending on whether one is concerned with trust in people (PEOPLETRUST and PEOPLEFAIR) or trust in government and legal system (TRUSTGOV and TRUSTLEG). This latter point can be observed via the classical analysis with microdata as well. It is also interesting that the 'trust in people' variables are on opposite sides of the same axis when compared especially to the variables SUCCESS, TRADITION and ENJOY. The classical microdata analysis gives almost the same result, but SUCCESS is of a more unique nature. People who appreciate success are not very much related to the other dimensions.

The regression analysis (SREG) can be run in SODAS analogously to the classical approach. We explained the variable TRUSTGOV with the following explanatory variables: PEOPLETRUST, PEOPLEFAIR, TRADITION, EQUALITY, SUCCESS and ENJOY. We did not include TRUSTLEG since it is presumed to be highly correlated with the dependent variable. Two significant explanatory variables were found, both with a positive regression coefficient: ENJOY and PEOPLETRUST. These were also positively significant in the classical microdata analysis in which all the other variables also had a significant effect, which is natural due to the large sample sizes. We can thus say that if people appreciate enjoyable, creative or other similar types of things and they have trust in other people, their trust in government and other administration will be high. Thus, these two characteristics are important at both individual and country level.

The above two examples concerning comparison between the microdata analysis and the aggregated symbolic analysis are presented here to illustrate the point that, in some cases, even if the results are not the same, which can be understood because they present micro and macro points of view, it seems evident that they cannot be totally different. This is particularly useful if the symbolic analysis is carried out because microdata are not available for confidentiality or budgetary reasons, among other things. Hence, if this is the case it is important to create symbolic data in order for this to succeed reasonably well.

On the other hand, aggregation of microdata is by no means the only reason to use symbolic data. Symbolic data with interesting symbolic units (aggregates), such as the ESS countries here, are often needed for further analysis by multivariate techniques. The multivariate analysis, of course, can be based on classical techniques but in most cases this would be too limited due to poor construction of symbolic variables. For example, the use of averages in the classical analysis leads to too simple an analysis. Naturally, the classical analysis could be tried with several simple variables (such as inclusion of medians and other quantiles, and variances) but would lead to confusion in interpretations. The symbolic approach is in such cases more capable of creating variables whose values can be intervals or frequencies but the analysis and interpretation would be simpler. Our examples in this chapter show some uses (and advantages) of the symbolic analysis. It should be noted that each symbolic analysis requires careful consideration of how the symbolic variables should be created. Our solutions for this analysis are fairly correct.

Even if the methods and SODAS software are helpful, much work remains to be done such as taking into account ordinal variables or variables expressed as a continuous distribution. Another big problem is that this 'symbolic world' is not well known, but it has made a promising start.

References

Bock, H.-H. and Diday, E. (eds) (2000) *Analysis of Symbolic Data. Exploratory Methods for Extracting Statistical Information from Complex Data.* Berlin: Springer-Verlag.

23

Symbolic analysis of the Time Use Survey in the Basque country

Marta Mas and Haritz Olaeta

23.1 Introduction

Time use surveys provide valuable information on what people do with their time, what proportion of time is spent on economically productive activities, leisure pursuits, personal care or household-related activities. Diversity in time use patterns between male and females, older and younger generations and any other different socio-economic groups is important not only for policy development and programme planning but also for product and service provision.

The amount of data collected by statistical institutes has increased enormously in recent decades for a number of different reasons. Difficulties in understanding, processing and filtering this information have emerged accordingly. Symbolic data analysis – see, for instance, Bock and Diday (2000) and references therein – allows one to work with complex structured data or symbolic objects within a coherent framework. The Institute of Statistics of the Basque country (EUSTAT) has tested the SODAS2 software for symbolic data analysis that will be used in this work.

Symbolic analysis techniques will be briefly introduced and applied to the Time Use Survey of 1998 (EUSTAT, 1999) in the Basque Country in order to process, filter and identify subpopulations with significant differences in time use behaviour.

23.2 Data description

Data refer to the Time Use Survey performed in the Basque country in 1998 (EUSTAT, 1999). The microfile consists of 5040 statistical units (Basque people over 16 years old) and 47

Symbolic Data Analysis and the SODAS Software Edited by E. Diday and M. Noirhomme-Fraiture

Table 23.1 Time use survey: microdata structure.

Socio-demographic var.					Time use var.				
Id	Day	Sex	Age	...	Cleaning	Clothing	Sleep	...	Weight

variables. These variables constitute a list of all activities carried out by individuals during the day. These activities are divided into various areas: physiological needs, work and studies, housework, family care, social life, active leisure and sports, passive leisure, and travelling. Each week is divided into working days, Fridays, and weekends. This ensures that all activities are included in the survey. Finally, data are collected in two steps, half in the spring and the rest in the autumn, in order to capture the seasonal nature of some activities.

Each statistical unit is described by socio-demographic and time use variables, the day of the week the information refers to, and the sampling weight (Table 23.1). The socio-demographic variables used in this work are categorical and can be described as follows:

- Sex: male, female.
- Age: $<$ 35 years old, 35–59 years old, $>$ 59 years old.
- Relation to labour force: inactive, employed, unemployed.

Time use variables can be considered in two different ways (Calvo, 2000):

- As quantitative – each variable represents time in minutes for each activity.
- As categorical variables with four categories – each category indicates the frequency in performing the activity: 1 = no participation, 2 = scarce participation, 3 = medium participation, 4 = high participation.

Of course, the first modality indicates no participation at all in the considered activity, which means zero values in the quantitative version. The remaining categories divide the population in groups of the same weight preserving the marginal distribution of time in each activity. That is, for example, less than 8 hours of sleep is considered scarce participation, while more than 30 minutes and less than an hour of personal care is classified as medium participation. Thus, each time variable keeps its own properties and can be compared with the others in a homogeneous way (Iraola *et al.*, 1997).

Variables which are quantitative in character are not included in the study, and we concentrate on the second approach in the following sections.

23.3 Symbolic approach

The general philosophy of symbolic analysis relies on the construction and study of a set of individuals derived from the original statistical units. These *new* second-order units, defined by group attributes (in our case, the socio-demographic variables) and described by symbolic variables (in this work, time use variables), will be treated as new statistical units. Therefore, all the classical statistical analysis techniques can be performed on them (Bock and Diday, 2000).

The maximum number of symbolic descriptions that can be constructed from the socio-demographic variables described above is the Cartesian product of their numbers of categories ($2 \times 3 \times 3 = 18$). Time use variables will describe each socio-demographic group by means of a probabilistic distribution computed taking into account the frequency of each category in the group, and the sampling weights. In the symbolic notation, these kinds of variables are known as modal variables.

The symbolic descriptions of two of these individuals are given below (with the relative frequencies of each category in parentheses):

w = "Male, 35–59 years old, employed"

d_w = (Sleep = [Scarce particip.(0.60), Medium particip.(0.30), High particip.(0.10)],

Cooking = [No particip.(0.74), Scarce particip.(0.19), Medium particip.(0.06), High particip.(0.01)],

Cleaning = [No particip.(0.92), Scarce particip.(0.04), Medium particip.(0.02), High particip.(0.02)],

Clothing = [No particip.(0.96), Scarce particip.(0.03), Medium particip.(0.01)],

Shopping = [No particip.(0.81), Scarce particip.(0.09), Medium particip.(0.05), High particip.(0.05)],

Care-children = [No particip.(0.90), Scarce particip.(0.05), Medium particip.(0.03), High particip.(0.01)],

Care-elderly = [No particip.(0.99), High particip.(0.01)],

Read-TV-Radio = [No particip.(0.11), Scarce particip.(0.40), Medium particip.(0.31), High particip.(0.17)],

Personal-care = [Scarce particip.(0.36), Medium particip.(0.48), High particip.(0.16)])

w = "Female, 35–59 years old, employed"

d_w = (Sleep = [Scarce particip.(0.71), Medium particip.(0.24), High particip.(0.06)],

Cooking = [No particip.(0.09), Scarce particip.(0.36), Medium particip.(0.41), High particip.(0.15),

Cleaning = [No particip.(0.26), Scarce particip.(0.28), Medium particip.(0.32), High particip.(0.14)],

Clothing = [No particip.(0.46), Scarce particip.(0.24), Medium particip.(0.19), High particip.(0.11)],

Shopping = [No particip.(0.35), Scarce particip.(0.24), Medium particip.(0.25), High particip.(0.17)],

Care-children = [No particip.(0.75), Scarce particip.(0.14), Medium particip.(0.09), High particip.(0.02)],

Care-elderly = [No particip.(0.97), Medium particip.(0.02), High particip.(0.01)],

Read-TV-Radio = [No particip.(0.19), Scarce particip.(0.40), Medium particip.(0.28), High particip.(0.13)],

Personal-care = [Scarce particip.(0.35), Medium particip.(0.39), High particip.(0.26)])

These two groups are, a priori, supposed to have significant differences in housework participation. A graphical representation might help to identify them easily.

23.4 Symbolic visualization

The SODAS2 software provides a user-friendly graphical interface to represent symbolic objects (Noirhomme-Fraiture and Rouard, 2000; and see also Chapter 7 of the present volume). Each symbolic description is shown as a multiple-axis star containing the distribution of symbolic variables that describe the group (time use variables in our case).

Categories with the highest frequency are joined by a line, building up a modal polygon. Differences in shape and area between polygons will reflect substantial differences between objects.

The growing presence of women in the labour market has affected their participation in some of the housework and care activities. However, as can be seen in Figures 23.1

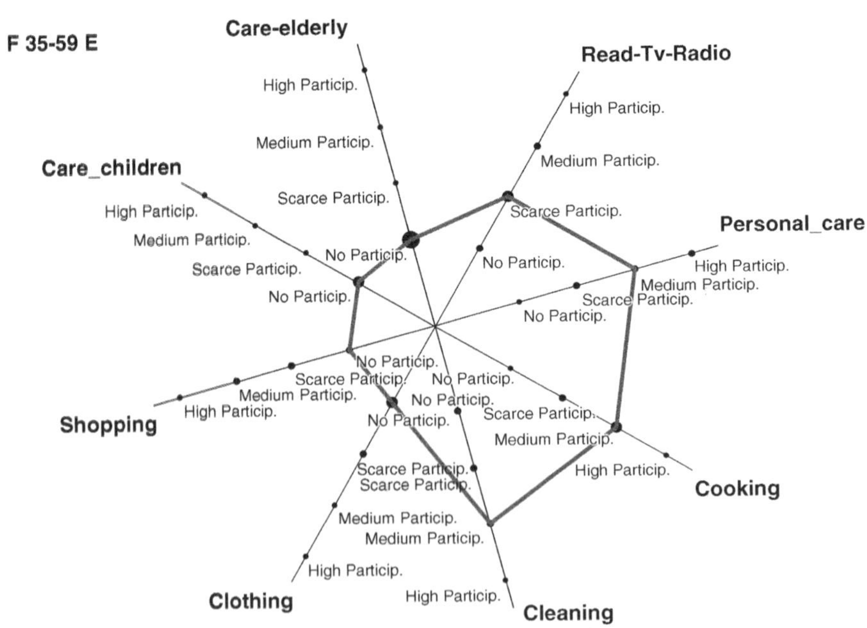

Figure 23.1 Two-dimensional zoom star representation: females, 35–59 years, employed.

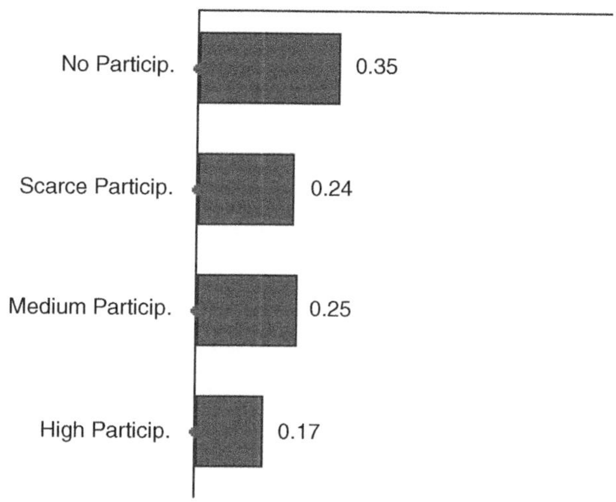

Figure 23.2 Univariate distribution for shopping: females, 35–59 years, employed.

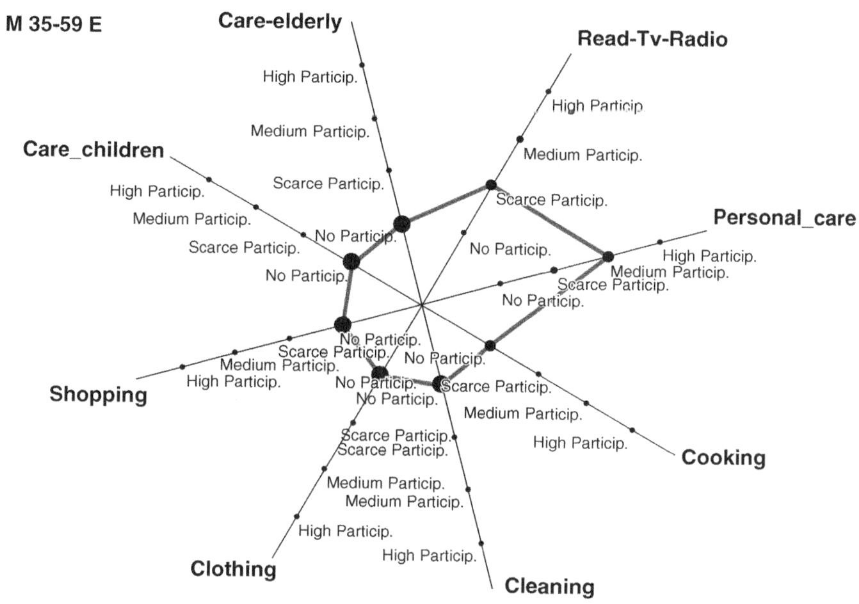

Figure 23.3 Two-dimensional zoom star representation: males, 35–59 years, employed.

and 23.3, women still spend much more time on housekeeping (cleaning, cooking, . . .) than men with the same socio-economical characteristics.

In addition, the highest concentration of men in the ‘no participation’ category, mainly in housework and care activities, contrasts with a more heterogeneous distribution of women within all the modalities in variables such as clothing or shopping (see Figures 23.2 and 23.4).

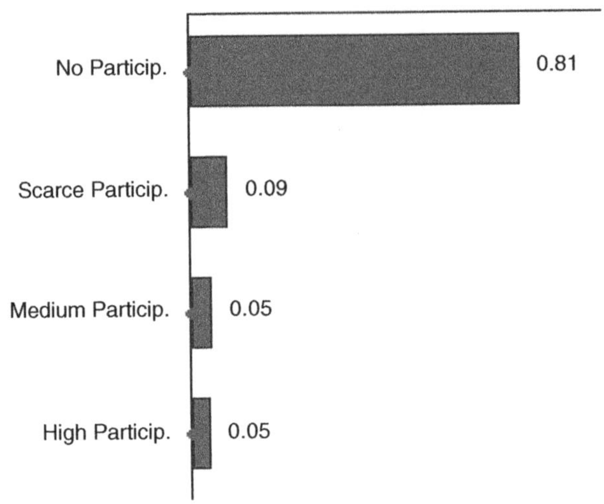

Figure 23.4 Univariate distribution for shopping: males, 35–59 years, employed.

23.5 Pyramidal clustering

A clustering analysis has been done using the HIPYR module in SODAS2 (see Chapter 11 in this book). We have focused again on housework activities. Subpopulations with significant differences in housework behaviour will be identified and classified by a pyramid. The pyramidal clustering generalizes hierarchies by forming non-disjoint classes instead of a

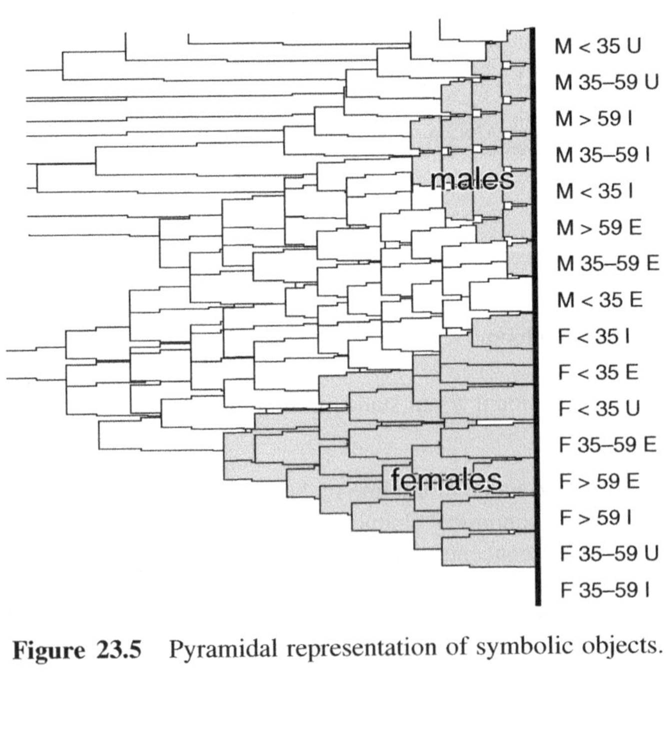

Figure 23.5 Pyramidal representation of symbolic objects.

partition. Clusters are formed by considering variation on the values taken by the variables that describe the groups. In particular, symbolic objects are joined following the 'generality degree' criterion (Brito, 1991, 2004).

Three socio-demographic variables (sex, age, and relation to labour force) have been used to construct 18 symbolic objects defined by nine time use variables (i.e. sleep, personal care, read-TV-radio, care elderly, care of children, shopping, clothing, cleaning, cooking). For this analysis only data referring to working days have been considered.

The symbolic objects described above constitute the input data for the method and, therefore, a symbolic pyramid is constructed. However, a numerical (classical) pyramid using previously computed distance or dissimilarity matrices could have been used as an input.

The graphical output (Figure 23.5) clearly shows two gender clusters and also a natural order, given by the method, which determines the 'closest' symbolic objects. Employed men between 16 and 59 years old are joined in a first step (class 1), while the first 'female cluster' (class 18) consists of women under 35 either employed or inactive. Age groups in women have more effect on housework activity behaviour than the economic situation.

Each node in the pyramid has its own symbolic description. An intermediate node that contains the closest male–female objects is described as follows:

THE CLASS EXTENSION (SYMBOLIC OBJECTS)

[Males < 35 Employed, Females < 35 Inactive]

LONG SYMBOLIC OBJECT DESCRIBING THE CLASS

[Sleep = (Scarce particip.(0.64), Medium particip.(0.30), High particip.(0.11))]

∧ [Cooking = (No particip.(0.66), Scarce particip.(0.24), Medium particip.(0.20), High particip.(0.28))]

∧ [Cleaning = (No particip.(0.90), Scarce particip.(0.22), Medium particip.(0.18), High particip.(0.18))]

∧ [Clothing = (No particip.(0.92), Scarce particip.(0.15), Medium particip.(0.17), High particip.(0.09))]

∧ [Shopping = (No particip.(0.80), Scarce particip.(0.15), Medium particip.(0.18), High particip.(0.13))]

∧ [Care_children = (No particip.(0.80), Scarce particip.(0.10), Medium particip.(0.09), High particip.(0.25))]

∧ [Care-elderly = (No particip.(1), Scarce particip.(0.002))]

∧ [Read-TV-Radio = (No particip.(0.13), Scarce particip.(0.55), Medium particip.(0.30), High particip.(0.13))]

∧ [Personal_care = (Scarce particip.(0.36), Medium particip.(0.37), High particip.(0.36))]

In general, male objects are 'closer' than female ones (see differences in pyramid volumes in Figure 23.5). This reveals a quite homogeneous behaviour of men in relation to housework activities but a changeable situation in the case of women.

Notice that frequencies in symbolic variables do not form a probabilistic distribution but an accumulative distribution. Values in the modalities reflect the 'at most' proportion of the population that presents this modality.

23.6 Conclusions

We have shown an approach that is able to deal with large amounts of data and uses software (SODAS2) that is both user-friendly and easy to interpret. A descriptive analysis and a clustering division have been performed to point out the gender differences in time use, especially in housework activities.

References

Bock. H. and Diday E. (2000) *Analysis of Symbolic Data*. Springer-Verlang, Berlin.

Brito, P. (1991) Analyse de données symboliques. Pyramides d'héritage. Doctoral thesis, Université de Paris IX Dauphine.

Brito, P. (2004) Hierarchical and pyramidal clustering. In *User Manual for* SODAS2 *Software*, public deliverable D3.4b of ASSO project (IST-2000-25161). http://www.assoproject.be/sodaslink/.

Calvo, P. (2000), Applications of symbolic objects in official statistics. Technical report, Eustat.

Iraola, J., Iztueta, A. and Pérez, Y. (1997) *Análisis de tipologías de Jornadas Laborales* Eustat, Vitoria-Gasteiz, Spain.

EUSTAT (1999) *Encuesta de presupuestos de tiempo 98*. Instituto Vasco de Estadística, Vitoria-Gasteiz, Spain.

Noirhomme-Fraiture, M. and Rouard, M. (2000) Visualising and editing symbolic objects. In H.-H. Bock and E. Diday (eds), *Analysis of Symbolic Data*. Springer-Verlag, Berlin.

24

SODAS2 software: Overview and methodology

Anne de Baenst-Vandenbroucke and Yves Lechevallier

24.1 Introduction

The SODAS2 software[1] is the outcome of the 'ASSO: Analysis System of Symbolic Official Data' project (IST-2000-25161). It is the tool dedicated to the symbolic data analysis framework described in the previous chapters of this book. It is the new and extended version of the prototype called SODAS.[2]

SODAS2 provides a wide selection of innovative statistical methods to carry out analysis on symbolic data tables and symbolic objects. It also implements data management, including metadata support. Moreover, it offers a set of tools for graphical visualization of results.

In SODAS2, a symbolic data analysis is graphically represented as a chain of icons. The top icon represents the symbolic data file or 'SODAS file'. The rest of the chain gathers the set of symbolic statistical methods applied to this file. The successive methods are executed in the order in which they appear in the chain, and subsequent methods can use the results created by the preceding one. In all cases, consistency checks are carried out if the method needs previous results. After execution, icons representing result reports and graphical visualization tools are added to the right of the method icons.

The SODAS2 software is based on a modular architecture managed by a workbench structure that links, organizes and controls the computation. It is developed mainly in C++ and runs on Microsoft Windows NT4, 2000 or XP platforms.

---

[1] The SODAS2 software may be downloaded from the ASSO project website (http://www.assoproject.be).
[2] SODAS was the result of the SODAS project belonging to the DOSIS programme (1996–99). It may be obtained from the web page of this project (http://www.ceremade.dauphine.fr/~touati/sodas-pagegarde.htm).

Symbolic Data Analysis and the SODAS Software Edited by E. Diday and M. Noirhomme-Fraiture

24.2 Features

The purpose of the SODAS2 software is to extend the standard tools of statistics and data mining to concepts modelled by symbolic objects and to generalize knowledge discovery, data mining and statistics to higher-level units described by symbolic data. Here is only a sketchy list of features offered by the software.

SODAS2 is an appropriate software environment for manipulating, joining, visualizing, comparing, analysing and explaining complex statistical data.

SODAS2 allows the construction of symbolic data from a relational database or directly through editing. It offers the possibility of retrieving new symbolic data and propagating results on the initial database. Moreover, it permits the introduction of structured statistical metadata, ensuring better quality of statistical results.

SODAS2 provides outstanding tools for the study of quality, stability and robustness of symbolic data analysis methods and offers user-friendly dissemination of statistical data by means of symbolic data. It uses, creates, propagates and designs statistical concepts by means of symbolic objects. It proposes new measures of dissimilarity between symbolic objects and new symbolic data analysis methods of supervised classification of symbolic objects.

SODAS2 includes various tools for visualizing symbolic objects and/or the results of symbolic data analysis.

24.3 Overview

24.3.1 Main windows in SODAS2

On launching SODAS2, the main graphical user interface consists of three components as shown in the screenshot of Figure 24.1. First, the main toolbar is the heart of SODAS2, containing the highest-level menu. Second, the Methods window offers the various symbolic data analysis methods in the form of drag and drop icons. For the sake of simplicity, the methods are grouped in the following way: Descriptive Statistics (2 methods), Dissimilarity and Matching (2 methods), Clustering (7 methods), Factorial (2 methods), Discrimination & Regression (8 methods). Third, the chaining window is the SODAS2 workspace where the list of operations of a symbolic data analysis is represented visually with a chaining of

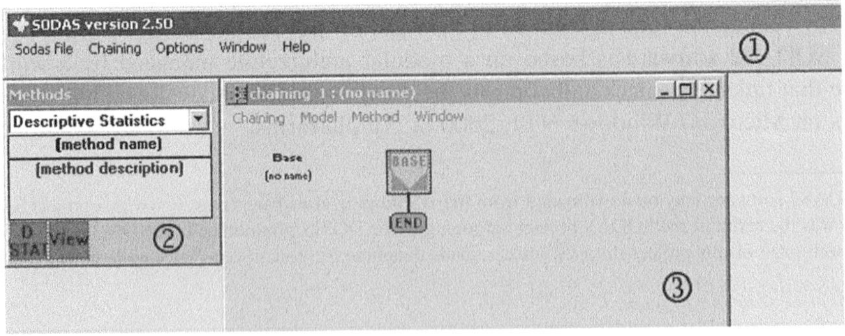

Figure 24.1 The initial SODAS2 graphical user interface.

icons representing the various elements. All the window functionalities can be reached from the window top menu but also using drag and drop facilities and the right mouse button.

24.3.2 Starting with SODAS2

24.3.2.1 Data management

The SODAS2 software works on symbolic data files or SODAS files that contain symbolic objects, variables and, possibly, metadata. The entire *data management* procedure is done through the SODAS file option in the main toolbar menu (Figure 24.2). It allows the user to create and edit a SODAS file (with SOEDIT), to import it from a native ASCII file (with ND2SO) or from a relational database (with DB2SO), as well as to export symbolic information back to the original database (with SO2DB).

24.3.2.2 Preparing an analysis

The actual *symbolic data analysis* is performed through a chaining window that opens or is created via the Chaining option in the main toolbar menu. It is possible to work with several chaining windows at the same time.

When a new chaining window is opened, it shows an empty BASE icon that represents the main SODAS file containing the SODAS database on which the analyses will be done (see Figure 24.1). The first step is to choose the SODAS file using the Chaining option from the top menu of the chaining window or by clicking on the BASE icon with the left or right mouse button.

The next step is to add the desired methods using the Method option from the chaining menu or directly by drag and drop from the Methods window. Each method icon is associated with a treatment module and the chaining of the method icons represents the desired procedure to be run on the main SODAS file.

In a chaining, visual links between method icons indicate the order of execution of the methods. This does not mean that the computation of a method requires the results of the preceding one. If this is the case, the software manages it. A method may need one or more

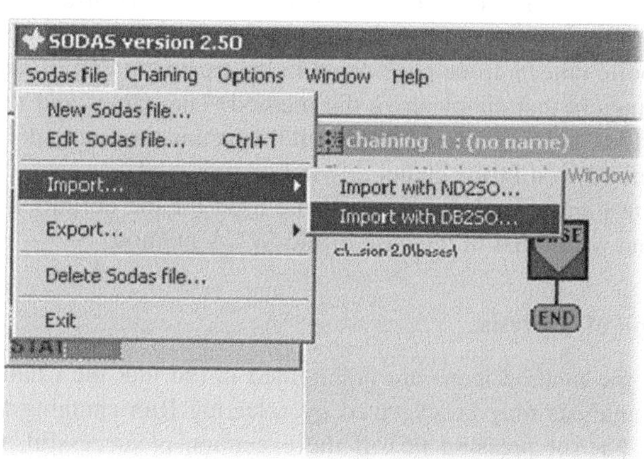

Figure 24.2 The SODAS file menu in the main toolbar.

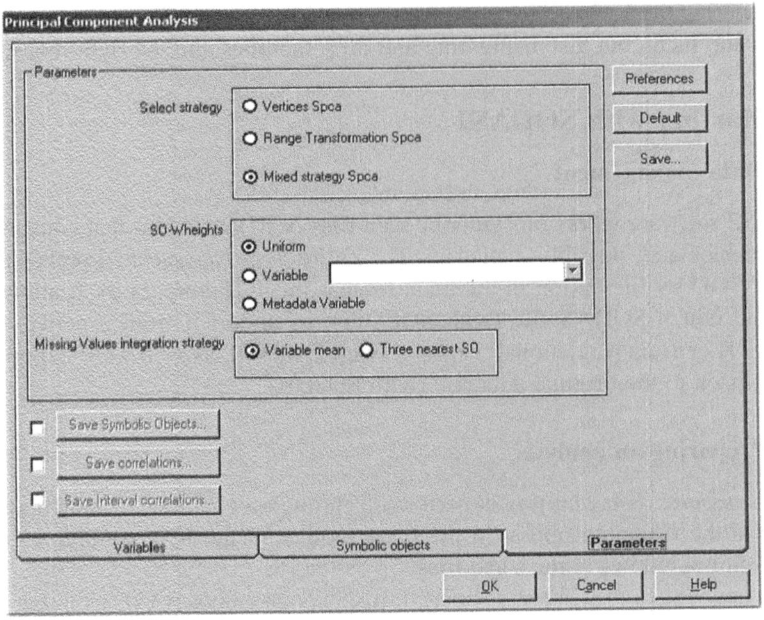

Figure 24.3 Dialogue window for defining parameters for the SPCA method.

result files from the preceding one. If files are explicit, the software asks for their names as parameters of both methods. If these files are implicit, the user has nothing to manage. In both cases, the software checks that the chaining of the two methods is allowed. A method may also need two SODAS files as input, one at the top of the chaining and another one. Moreover, a method may create a new SODAS file. The name of the additional file will also be requested from the user as a parameter. Only one SODAS file is visible in the chaining window, and it is the main one at the top of the chaining.

The manipulation on a chaining is handled by the menu bar options in the chaining window or by using the right mouse button. When a method is inserted, the method icon is in grey colour. The parameters then have to be defined. This involves selecting the variables and symbolic objects to be considered in the treatment and, if necessary, defining the additional parameters that characterize the method. The names and the location of the additional SODAS files are also defined. When all the parameters corresponding to a method are defined, the method icon in the chaining window turns red.

Figure 24.3 shows an example of the dialogue used for the definition of the variables, symbolic objects and additional in the case of the SPCA method.

24.3.2.3 Execution of analysis

It is only when all the method icons are highlighted in red that the Chaining representing the symbolic data analysis may be executed by selecting Run chaining from Chaining in the chaining menu bar (or pressing F5). If the execution is successful, the results report and the access to the relevant visualization tools are added to the chaining as icons inserted horizontally to the right of the corresponding method icon.

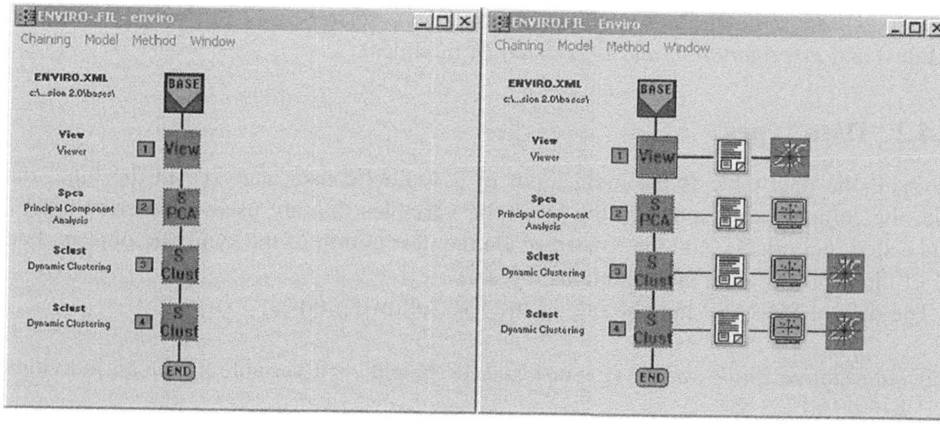

Figure 24.4 A chaining before and after a successful execution in SODAS2.

Figure 24.4 shows a chaining before (when all the parameters have been defined) and after a successful execution. All the result icons on the right of the method icons are clickable. The yellow report icons open a window containing the results report in text format. The other icons are visualization icons and give access to the corresponding visualization tools.

24.4 Methodology

24.4.1 SODAS2 structure

As already mentioned in Chapter 5, SODAS2 is based on the modular architecture illustrated in Figure 24.5. The core of the system is the *symbolic analysis* part, which consists of a set of 21 modules associated with the treatment methods allowing symbolic objects and symbolic data tables to be manipulated. The other parts encompass modules connected with data (and metadata) management (importation, exportation, editing) and graphical visualization of results.

The symbolic analysis modules are classified into five groups depending on the type of statistical method implemented in each module, namely *descriptive statistics* and visialization

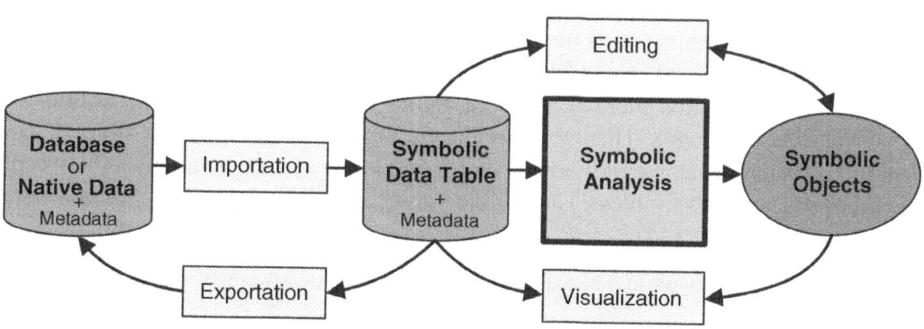

Figure 24.5 The SODAS2 architecture.

(2 modules), *dissimilarity and matching* (2 modules), *clustering* (7 modules), *factorial* (2 modules) and *discrimination and regression* (8 modules).

24.4.2 Data types

The symbolic data table is the main input of a symbolic data analysis. In the input data table, the columns correspond to the symbolic variables that are used to describe a set of units called individuals, and the rows provide the description of the symbolic objects. Each cell of the symbolic data table contains the data.

The data in SODAS2 may be one of the five following types:[3]

(a) *quantitative single-valued* (for instance, if 'height' is a variable and w an individual, $\text{height}(w) = 3.5$);

(b) *interval* (for instance, $\text{height}(w) = [3, 5]$, which means that the height of w varies in the interval [3, 5]);

(c) *categorical single-valued* (for instance, $\text{height}(w) = \text{tall}$);

(d) *categorical multi-valued* (for instance, in the quantitative case $\text{height}(w) = \{3.5, 2.1, 5\}$, which means that the height of w can be either 3.5 or 2.1 or 5);

(e) *modal* or multi-valued with weights or probabilities (for instance, a histogram or a membership function).

In addition to the five basic types of symbolic variables, SODAS2 uses other types of data, namely missing values (NUL or *), basic dependence between variables (logical or hierarchical (NA)), data structures for relationships between objects (dissimilarity matrix, matching matrix, or tree structure, and data structures for relationship between variables), correlation matrix and order between variables.[4]

Figure 24.6 shows how the symbolic variables are selected in SODAS2 in the case of the VIEW module.

24.4.3 SODAS file format

The standard used in SODAS2 for the SODAS file is XML, although it is still possible to use as input the SODAS file in the proprietary 'sds' standard defined previously.

The top of a SODAS file shows the XML declaration. It also indicates the used XML style sheet (.xsl) and schema (.xsd) files used. The information elements are nested between the start tag <assofile> and the end tag </assofile>. A first section contains information about the file (title, creation method and date, number of symbolic objects and number of symbolic variables by type). Then follows the definition of the symbolic objects named 'individus' (between the tags <individus> and </individus>), of the variables (between the tags <variables> and </variables>) and of the values for all the data table cells enumerated

---

[3] More information about data types may be found in Chapter 1. For an extensive study, see the first four chapters of Bock and Diday (2000).

[4] These other data types are described in more detail in the previous chapters as well as in the help files of the SODAS2 modules.

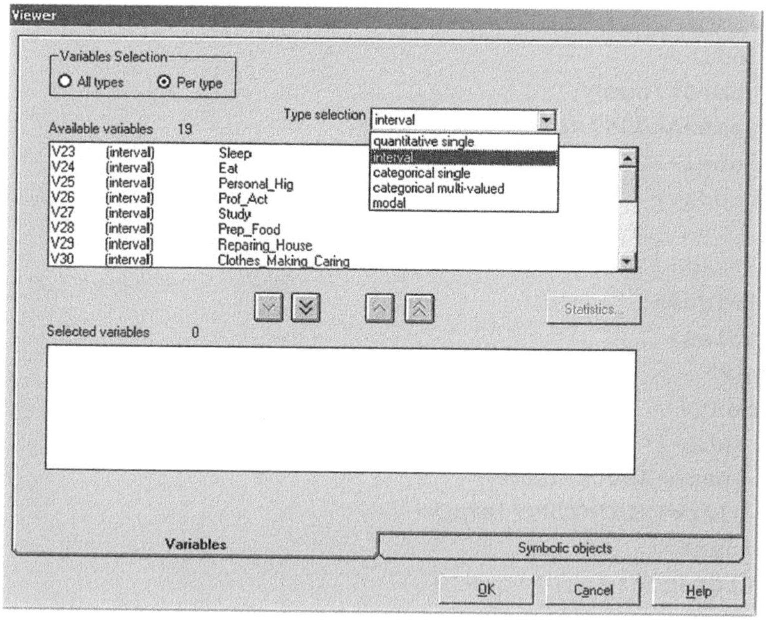

Figure 24.6 Example of the type selection for the variables in the parameters window for the VIEW module.

line by line (between the tags <indiv_mat> and </indiv_mat>). If there exist non-basic variables, they are enumerated here with the appropriate start and end tags.

The following listing excerpt, taken from the SODAS file abalone.xml provided with the SODAS2 release, illustrates this XML structure and content:

```
<?xml version="1.0" encoding="UTF-8"?> <?xml-stylesheet
type="text/xsl" href="asso2.2.xsl"?>

<assofile xmlns:xsi="http://www.w3.org/2000/10/XMLSchema-
instance" xsi:noNamespaceSchemaLocation ="asso.xsd">
   <contains FILES="YES" HEADER="YES" INDIVIDUALS="YES"
      VARIABLES="YES" RULES="NO" RECTANGLE_MATRIX="YES"
      DIST_MATRIX="NO" MATCH_MATRIX="NO"
      HIERARCHIE="NO" />
   <filed>
     <Procedure>db2so</Procedure>
     <date_creat>Wed Jun 11 09:11:39 2003</date_creat>
   </filed>
   <header title = "abalone" sub_title = "abalone"
      indiv_nb = "24" var_nb = "7" rules_nb = "0"
      nb_var_nom = "0" nb_var_cont = "0"
      nb_var_text = "0" nb_var_cont_symb = "7"
      nb_var_nom_symb = "0" nb_var_nom_mod = "0"
      nb_na = "0" nb_hierarchies = "0" nb_nu = "0" />
```

```
  <individus>
    <stindiv>
      <num>0</num>
      <name>AA00</name>
      <label>F_4-6</label>
    </stindiv>
    ...
    ...
  </individus>
  <variables>
    <stvar>
      <ident>
        <num>1</num>
        <name>AB00</name>
        <label>LENGTH</label>
      </ident>
      <inter-cont>
        <continue-desc nbna = "0" nbnu= "0" min= "0.075"
         max= "0.815"/>
      </inter-cont>
    </stvar>
    ...
    ...
  </variables>
  <indiv_mat>
   <ligmat>
     <valmat>
       <val_interv>
         <pmin>0.275</pmin>
         <pmax>0.66</pmax>
       </val_interv>
     </valmat>
     ...
   </ligmat>
   ...
  </indiv_mat>
</assofile>
```


Figure 24.7 shows how the information about the file is visualized in SODAS2. In this file, there are 24 symbolic objects and 7 variables. All the variables are of interval type.

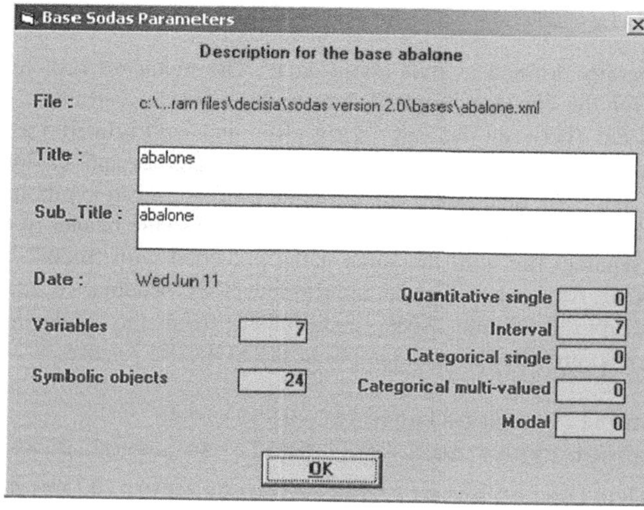

Figure 24.7 Base SODAS Parameters window for abalone.xml.

Figure 24.8 shows how the information defining the symbolic data table is visualized with the help of the SOEDIT module. The listing excerpt focuses on the first cell of the table. It concerns the symbolic object with name AA00 and label F_4-6 (first line in the table) for the variable with name AB00 and label LENGTH of type interval with no missing values and values bounded by 0.075 as minimum and 0.815 as maximum. The value of this first cell is bounded by 0.28 as minimum and 0.66 as maximum (visualized in SOEDIT as [0.28:0.66]).

	LENGTH	DIAMETER	HEIGHT	WHOLE_WEIGHT	SHUCKED_WEIGHT	VISCERA_WEIGHT	SHELL_WEIGHT
F_4-6	[0.28 : 0.66]	[0.19 : 0.47]	[0.07 : 0.18]	[0.08 : 1.37]	[0.03 : 0.64]	[0.02 : 0.29]	[0.03 : 0.34]
F_7-9	[0.31 : 0.75]	[0.22 : 0.58]	[0.01 : 1.13]	[0.15 : 2.25]	[0.06 : 1.16]	[0.03 : 0.45]	[0.05 : 0.56]
F_10-12	[0.34 : 0.78]	[0.26 : 0.63]	[0.06 : 0.23]	[0.20 : 2.66]	[0.07 : 1.49]	[0.04 : 0.53]	[0.07 : 0.73]
F_13-15	[0.39 : 0.81]	[0.30 : 0.65]	[0.10 : 0.25]	[0.26 : 2.51]	[0.11 : 1.23]	[0.05 : 0.52]	[0.09 : 0.80]
F_16-18	[0.40 : 0.75]	[0.31 : 0.60]	[0.10 : 0.24]	[0.35 : 2.20]	[0.12 : 0.84]	[0.09 : 0.48]	[0.12 : 1.00]
F_22-24	[0.45 : 0.80]	[0.38 : 0.63]	[0.14 : 0.22]	[0.64 : 2.53]	[0.16 : 0.93]	[0.11 : 0.59]	[0.24 : 0.71]
F_19-21	[0.49 : 0.73]	[0.37 : 0.58]	[0.13 : 0.21]	[0.68 : 2.12]	[0.17 : 0.81]	[0.13 : 0.45]	[0.20 : 0.85]
F_25-29	[0.55 : 0.70]	[0.47 : 0.58]	[0.18 : 0.22]	[1.21 : 1.81]	[0.32 : 0.71]	[0.20 : 0.32]	[0.47 : 0.52]
I_1-3	[0.08 : 0.24]	[0.05 : 0.17]	[0.01 : 0.06]	[0.00 : 0.07]	[0.00 : 0.03]	[0.00 : 0.01]	[0.00 : 0.02]
I_4-6	[0.13 : 0.58]	[0.09 : 0.45]	[0.00 : 0.15]	[0.01 : 0.89]	[0.00 : 0.50]	[0.00 : 0.19]	[0.00 : 0.35]
I_7-9	[0.26 : 0.67]	[0.19 : 0.50]	[0.00 : 0.19]	[0.08 : 1.30]	[0.03 : 0.60]	[0.01 : 0.32]	[0.03 : 0.39]
I_13-15	[0.32 : 0.66]	[0.25 : 0.52]	[0.08 : 0.19]	[0.16 : 1.69]	[0.06 : 0.71]	[0.03 : 0.40]	[0.05 : 0.42]
I_10-12	[0.34 : 0.73]	[0.26 : 0.55]	[0.09 : 0.22]	[0.17 : 2.05]	[0.07 : 0.77]	[0.02 : 0.44]	[0.06 : 0.65]
I_16-18	[0.44 : 0.65]	[0.33 : 0.52]	[0.13 : 0.20]	[0.44 : 1.63]	[0.16 : 0.63]	[0.07 : 0.34]	[0.13 : 0.53]

Figure 24.8 Visualization of the symbolic data table of abalone.xml in SOEDIT.

24.4.4 Metadata

Metadata are generally defined as data about data.[5] The metadata kept in SODAS2 give metainformation for the classical original data such as survey variables, statistical units, frame population, etc. (between the tags <OrigInfo> and </OrigInfo>) as well as for the symbolic data analysis procedure (between the tags <MetaMat> and </MetaMat>). These metadata refer to symbolic data tables, to symbolic objects and to symbolic variables. For the sake of simplicity, metadata other than those occurring in the header of the SODAS file are defined in a separate file with the same name preceded with 'meta'. This file is also defined using a XML format based on an appropriate XML schema.

Here are excerpts from the XML code taken from the SODAS metadata file metaabalone.xml also provided as an example in the SODAS2 release:

```
<?xml version="1.0" encoding="ISO-8859-1"?>
<?xml-stylesheet type="text/xsl" href="Metasso2.2.xsl"?>
<MetaAsso xmlns:xsi=" http://www.w3.org/2000/10/XMLSchema-
      instance"
    xsi:noNamespaceSchemaLocation ="./Metasso5.xsd">
  <OrigInfo>
    <NbOrigVar>7</NbOrigVar>
    <NbOrigMat>1</NbOrigMat>
    ...
    <ListeVarOrig>
      <OrigVar>
      ...
      </OrigVar>
      <OrigVar>
        <Num>3</Num>
        <Name>Abalone_OS </Name>
        <Label>HEIGHT</Label>
        <Source>C:\...\SODAS version 2.0\bases\Abalone</Source>
        <Computed>select * from Abalone_OS</Computed>
        <DataType>continue</DataType>
      </OrigVar>
      ...
    </ListeVarOrig>
    ...
  </OrigInfo>
 <MetaMat>
   <Author>unknown</Author>
   <MetaHistory>
     <SqlQuery>select * from Abalone_OS</SqlQuery>
     <OdbcSource>C:\...\bases\Abalone</OdbcSource>
     <History>
       <PrevFile>none</PrevFile>
```

[5] See Chapter 4 for a detailed theoretical approach to a statistical metadata model for symbolic objects.

```
      <PrevTreatment>none</PrevTreatment>
    </History>
  </MetaHistory>
  <Comment>NU</Comment>
  <NbMetaVar>7</NbMetaVar>
  <NbMetaInd>24</NbMetaInd>
  ...
  <MetaVarLis>
    <MetaVar>
    <Num>1</Num>
    <Name>AB00</Name>
    <VarOrig>LENGTH</VarOrig>
    <VarOrigLis>
      <Vo>1</Vo>
    </VarOrigLis>
    <Label>LENGTH</Label>
    ...
    <Operator>aggregated</Operator>
  </MetaVar>
  ...
  </MetaVarLis>
 </MetaMat>
</MetaAsso>
```


24.5 SODAS2 modules

SODAS2 modules are classified according to their purpose in the software: there are modules for data management; for operational symbolic data analysis, called treatment modules; and for the graphical representation of the output results, called visualization modules. Figure 24.9 gives a schematic overview of the SODAS2 software showing all the modules and their purposes, and the rest of this section gives a brief sketch. An extended technical description of the modules may be found in ASSO (2004a, 2004b).

24.5.1 Data management modules

The data management modules are those modules dedicated to the import, export and editing of symbolic data. They are opened by selecting SODAS file option from the main toolbar menu as shown in Figure 24.2. The modules are as follows:

ND2SO: From Native Data to Symbolic Objects (FUNDP). This manages the import of external data into symbolic objects. It allows native data to be imported directly from an ASCII file into a SODAS file (see Chapter 5).

DB2SO: From DataBase to Symbolic Objects (INRIA). This also manages the import of external data into symbolic objects. It allows a SODAS file to be constructed from data stored in any relational database (see Chapter 2).

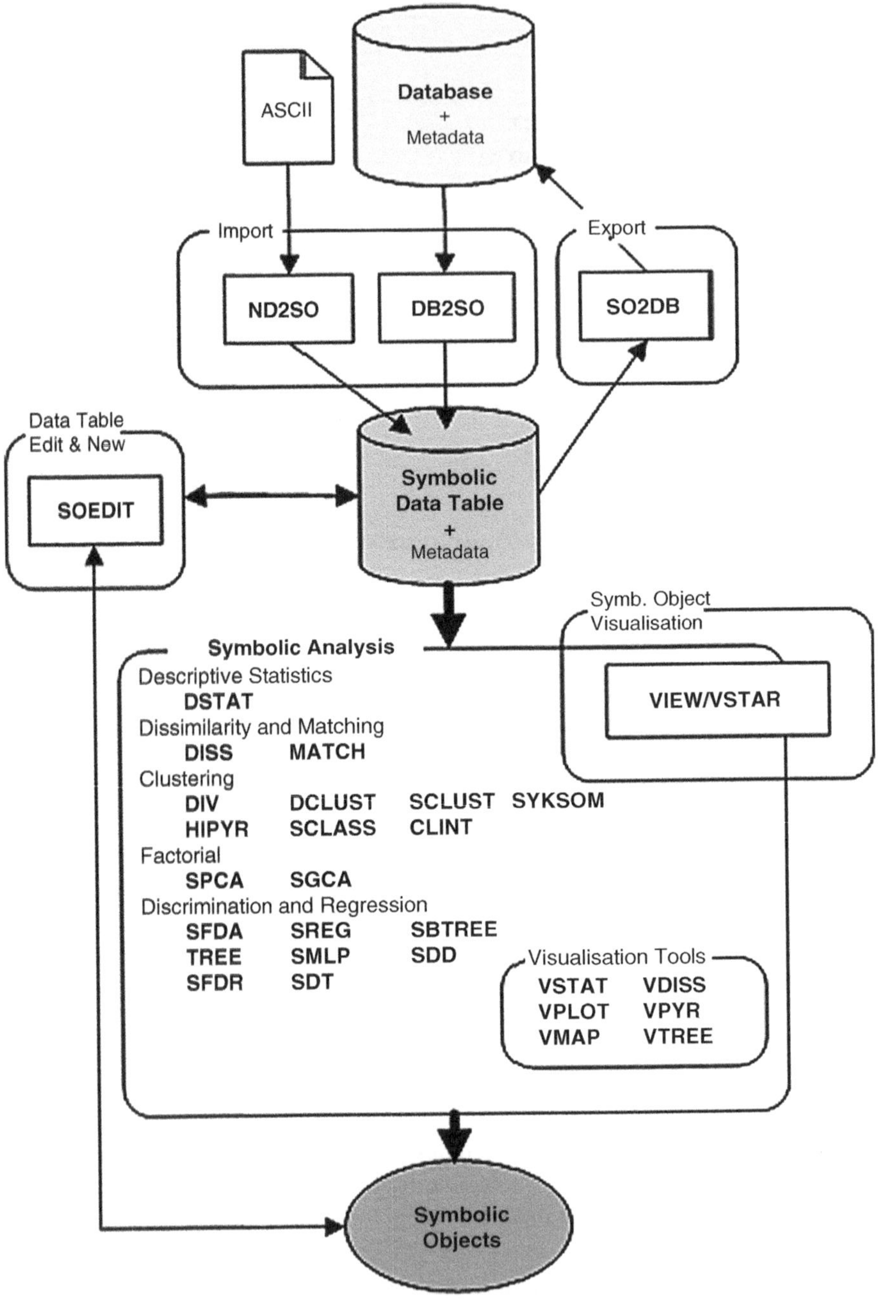

Figure 24.9 A SODAS2 software overview.

SO2DB: From Symbolic Objects to DataBase (DIB). This makes it possible to export additional data into the initial relational database if it exists (see Chapter 3).

SOEDIT: Symbolic Objects Editing (FUNDP). This allows the user to edit a SODAS file in a symbolic data table. It also offers various interactive editing facilities enabling work on the data table such as addition, selection or suppression of symbolic objects and/or variables, cells modification, symbolic object generalization and merging of data files. It also allows a SODAS file to be created interactively from scratch (see Chapter 5).

24.5.2 Treatment modules

The treatment modules are the core modules of SODAS2. They offer a set of 21 methods[6] for performing analysis on symbolic data. They are managed using the Method option in the top menu in the Chaining window or by drag and drop from the Methods window. For the user's convenience, the treatment methods have been split into the five categories mentioned in Section 24.4.1.

24.5.2.1 Descriptive statistics

The descriptive statistics modules are devoted to methods affording basic analysis tools for symbolic data analysis:

DSTAT: Descriptive Statistics (DAUPHINE). This extends to symbolic objects several elementary statistics methods (see Bertrand and Goupil, 2000).

VIEW: Symbolic Objects Viewer (FUNDP). This is a launcher allowing the visualization module VSTAR to be used as a treatment method (see Chapter 7).

24.5.2.2 Dissimilarity and matching

The aim of the dissimilarity and matching modules is to compare symbolic objects in order to quantify the existing correlations, to cluster or to discriminate between them. The results of this class of methods help to explicitly understand, measure and identify groups of symbolic objects. The modules are as follows:

DISS: Dissimilarity Measures (DIB). This compares boolean or probabilistic symbolic objects specified in a SODAS file in order to evaluate their dissimilarity (see Chapter 8).

MATCH: Matching Operators (DIB). This implements two forms of matching: the canonical matching that checks for an exact matching and the flexible matching that computes a degree of matching (see Chapter 8).

24.5.2.3 Clustering

The clustering modules are for clustering large sets of symbolic objects into a reduced number of homogeneous classes also expressed as symbolic objects, and/or visualizing them

[6] The modules DIV, TREE and SDT are added just as they were developed in the SODAS prototype. They have only been adapted to the SODAS2 standards and they do not take advantage of the graphical visualization tools resulting from the ASSO project.

geometrically in order to obtain synthetic and effective descriptions of such objects. The modules encompass three approaches that appropriately extend some classical multivariate data analysis methods to the case of symbolic objects, namely hierarchical and pyramidal clustering, partitional clustering, and divisive clustering of symbolic objects. The modules are as follows:

DIV: Divisive Classification (INRIA). This offers a hierarchical divisive clustering method adapted to symbolic objects (see Chavent, 2000).

DCLUST: Clustering Algorithm based on Distance Tables (UFPE). This interactively clusters a large set of symbolic objects into a reduced (fixed) number of homogeneous classes on the basis of a proximity (dissimilarity) table (see Chapter 11).

SCLUST: Symbolic Dynamic Clustering (INRIA). This provides several options for partitioning a set of symbolic objects into a defined number of homogeneous clusters where each class has a prototype in the form of a symbolic object. The optimality criterion used for the partition is based on the sum of the proximities between the individuals and the prototypes of the clusters (see Chapter 11).

SYKSOM: Kohonen Self-Organizing Map for Symbolic Data (RWTH). This constructs a Kohonen map that assumes a data set with a defined number of items or individuals being described by symbolic variables of interval type (see Chapter 12).

HIPYR: Hierarchical and Pyramidal Clustering (FEP). This performs a hierarchical or pyramidal clustering on a set of symbolic objects based either on a set of symbolic data type (symbolic clustering) or on a dissimilarity matrix computed by the dissimilarity module DISS (dissimilarity matrix clustering) (see Chapter 10).

SCLASS: Symbolic Unsupervised Classification Tree (FUNDPMa). This reduces the symbolic objects and/or variables in a symbolic data table by keeping the most discriminant symbolic variables and the most representative sample (see Chapter 9).

CLINT: Interpretation of Clusters (FEP). This allows the user to obtain an interpretation of classes or subsets of the data set, taking into account the roles of the symbolic variables.

24.5.2.4 Factorial

The factorial modules extend the classical factorial techniques to symbolic data analysis. They aim to visualize in a reduced dimension space symbolic objects represented in forms of hypercubes with images pointing out differences and similarities according to their descriptors. The modules are as follows:

SPCA: Symbolic Objects Principal Component Analysis (DMS). This looks for the best representation of a set of symbolic data described by symbolic variables of interval type on a factorial plane through three principal component analysis approaches extended to symbolic objects (see Chapter 15).

SGCA: Symbolic Objects Generalized Canonical Analysis (DMS). This offers an extension to symbolic objects of the generalized canonical analysis for all types of variables (see Chapter 16).

24.5.2.5 Discrimination and regression

The discrimination and regression modules cope with data that are structured according to information. They model the relationship between a target variable and other variables in order to explain the relationship and to predict the target value when it is missing. The modules are as follows:

SFDA: Symbolic Objects Factorial Discriminant Analysis (DMS). This defines the factorial subspace that best discriminates the a priori classes defined in the training set descriptions of the symbolic objects (see Chapter 18).

SREG: Symbolic Regression (DAUPHINE). This implements methods for linear regression with symbolic data (see Chapter 19).

SBTREE: Symbolic Bayesian Decision Tree (FUNDPMa). This offers a tree-growing algorithm merging approaches of density estimation and decision tree (see Chapter 17).

TREE: Decision Tree (INRIA). This offers a tree-growing algorithm applied to explicitly imprecise data (see Brito, 2000).

SMLP: Symbolic Multi-Layer Perceptron (DAUPHINE). This constructs a multi-layer perceptron neural network from a set of objects described by variables (see Chapter 20).

SDD: Symbolic Discriminant Description towards Interpretation (DAUPHINE). This proposes a procedure for marking and generalization by symbolic objects that can be used for the building of a new symbolic data table summarising an initial given one for interpretation aid in a clustering or factorial method.

SFDR: Symbolic Objects Factorial Discriminant Rule (DMS). This aims to use the geometrical rule achieved by the SFDA module to affect the new set of symbolic objects to the classes (see Chapter 18).

SDT: Strata Decision Tree (UCM Madrid). This extends binary segmentation techniques to the symbolic data analysis framework (see Polaillon, 2000).

24.5.3 Visualization modules

In order to visualize the results of the treatment modules, seven tools, the visualization modules, have been developed:

VSTAR: Zoom Star Visualization (FUNDP). This displays the symbolic data table of a SODAS file and visualizes the symbolic objects in 2D and 3D using a radial star shape, the zoom star, individually or superposed on a common window (see Chapter 7).

VSTAT: Descriptive Statistics Visualization (DAUPHINE). This gives many graphical representations of statistical indices and histograms on symbolic variables.

VDISS: Matrix Visualization (DIB). This gives a graphical visualization of the output dissimilarity matrix computed by the treatment module DISS, also showing information stored in the metadata file associated to the current running chain (see Chapter 8).

VPLOT: Biplot Visualization (DAUPHINE). This displays rectangular or circular biplot type graphs for interval variables.

VPYR: Hierarchy or Pyramid Visualization (FEP). This provides a graphical interactive display of a hierarchy or a pyramid (see Chapter 10).

VMAP: Symbolic Kohonen Map Visualization (INRIA). This displays the self-organization map and the prototypes associated with a neural network node (see Chapter 10).

VTREE: Tree Visualization (FUNDPMa) This outputs a Bayesian decision tree corresponding to a discriminant analysis rule (see Chapter 17).

References

ASSO (2004a) *Help Guide for SOSAS2 Software*. Public deliverable D3.4a of ASSO project (IST-2000-25161), April. http://www.assoproject.be/sodaslink/.

ASSO (2004b) *User Manual for SODAS2 Software*. Public deliverable D3.4b of ASSO project (IST-2000-25161), April. http://www.assoproject.be/sodaslink/.

Bertrand, P. and Goupil, F. (2000) Descriptive statistics for symbolic data. In H.-H. Bock and E. Diday (eds), *Analysis of Symbolic Data: Exploratory Methods for Extracting Statistical Information from Complex Data*, pp. 103–124. Springer-Verlag, Berlin.

Bock, H.-H. and Diday, E.(eds) (2000) *Analysis of Symbolic Data: Exploratory Methods for Extracting Statistical Information from Complex Data*, pp. 1–77. Springer-Verlag, Berlin.

Brito, P. (2000) Hierarchical and pyramidal clustering with complete symbolic objects. In H.-H. Bock and E. Diday (eds), *Analysis of Symbolic Data: Exploratory Methods for Extracting Statistical Information from Complex Data*, pp. 312–324. Springer-Verlag, Berlin.

Chavent, M. (2000) Criterion based divisive clustering for symbolic data. In H.-H. Bock and E. Diday (eds), *Analysis of Symbolic Data: Exploratory Methods for Extracting Statistical Information from Complex Data*, pp. 299–311. Springer-Verlag, Berlin.

Polaillon, G. (2000) Pyramidal classification for interval data using Galois lattice reduction. In H.-H. Bock and E. Diday (eds), *Analysis of Symbolic Data: Exploratory Methods for Extracting Statistical Information from Complex Data*, pp. 324–341. Springer-Verlag, Berlin.

Index

Note: Figures and Tables are indicated by *italic page numbers*, footnotes by suffix 'n'

Symbolic Data Analysis and the SODAS Software Edited by E. Diday and M. Noirhomme-Fraiture

Printed and bound by CPI Group (UK) Ltd, Croydon, CR0 4YY
16/09/2026
14963387-0005